1949 **Hideki Yukawa** (Japan), for prediction of the existence of mesons.

1950 **Cecil Powell** (Great Britain), for the photographic method of studying nuclear process and discoveries about mesons.

1951 **Sir John Cockcroft** (England) and **Ernest Walton** (Ireland), for work on the transmutation of atomic nuclei.

1952 **Edward Purcell** and **Felix Bloch** (both U.S.), for the discovery of nuclear magnetic resonance in solids.

1953 **Frits Zernike** (Netherlands), for the development of the phase contrast microscope.

1954 **Max Born** (Great Britain), for work in quantum mechanics; and **Walter Bothe** (Germany), for work in cosmic radiation.

1955 **Polykarp Kusch** (U.S.), for measurement of the magnetic moment of the electron; and **Willis E. Lamb, Jr.** (U.S.), for discoveries concerning the hydrogen spectrum.

1956 **William Shockley, Walter Brattain,** and **John Bardeen** (all U.S.), for development of the transistor.

1957 **Tsung-Dao Lee** and **Chen Ning Yang** (both China), for discovering violations of the principle of parity.

1958 **Pavel Čerenkov, Ilya Frank,** and **Igor Tamm** (all U.S.S.R.), for discovery and interpretation of the Čerenkov effect (emission of light waves by electrically charged particles moving faster than light in a medium).

1959 **Emilio Segrè** and **Owen Chamberlain** (both U.S.), for confirmation of the existence of the antiproton.

1960 **Donald Glaser** (U.S.), for development of the bubble chamber for the study of subatomic particles.

1961 **Robert Hofstadter** (U.S.), for determination of the size and shape of the atomic nucleus; and **Rudolf Mössbauer** (Germany), for the discovery of the Mössbauer effect of gamma ray absorption.

1962 **Lev D. Landau** (U.S.S.R.), for theories about condensed matter (superfluidity in liquid helium).

1963 **Eugene Wigner, Maria Goeppert Mayer** (both U.S.), and **J. Hans D. Jensen** (Germany), for research on the structure of the atomic nucleus.

1964 **Charles Townes** (U.S.), **Nikolai Basov** and **Alexandr Prokhorov** (both U.S.S.R.), for work in quantum electronics leading to the construction of instruments based on maser-laser principles.

1965 **Richard Feynman, Julian Schwinger** (both U.S.), and **Shinichiro Tomonaga** (Japan), for research in quantum electrodynamics.

1966 **Alfred Kastler** (France), for work on atomic energy levels.

1967 **Hans Bethe** (U.S.), for work on the energy production of stars.

1968 **Luis Alvarez** (U.S.), for the study of subatomic particles.

1969 **Murray Gell-Mann** (U.S.), for the study of subatomic particles.

1970 **Hannes Alfvén** (Sweden), for theories in plasma physics; and **Louis Néel** (France), for discoveries in antiferromagnetism and ferrimagnetism.

1971 **Dennis Gabor** (Great Britain), for the invention of the hologram.

1972 **John Bardeen, Leon Cooper,** and **John Schrieffer** (all U.S.), for the theory of superconductivity.

1973 **Ivar Giaever** (U.S.), **Leo Esaki** (Japan), and **Brian Josephson** (Great Britain), for theories and advances in the field of electronics.

1974 **Anthony Hewish** (Great Britain), for the discovery of pulsars; **Martin Ryle** (Great Britain), for radio-telescope probes of outer space.

1975 **James Rainwater** (U.S.), **Ben Mottelson** and **Aage Bohr** (both Denmark), for showing that the atomic nucleus is asymmetrical.

1976 **Burton Richter** and **Samuel Ting** (both U.S.), for discovery of the subatomic J and psi particles.

1977 **Philip Anderson, John Van Vleck** (both U.S.), and **Nevill Mott** (Great Britain), for work underlying computer memories and electronic devices.

1978 **Arno Penzias** and **Robert Wilson** (both U.S.), for work in cosmic microwave radiation; **Pyotr Kapitsa** (U.S.S.R.), for research in low temperature physics.

1979 **Steven Weinberg, Sheldon Glashow** (both U.S.), and **Abdus Salam** (Pakistan), for developing the theory that the electromagnetic force and the weak nuclear force are facets of the same phenomenon.

1980 **James Cronin** and **Val Fitch** (both U.S.), for work concerning the assymetry of subatomic particles.

1981 **Nicolaas Bloembergen, Arthur Schawlow** (both U.S.), and **Kai Siegbahn** (Sweden), for developing laser technology to study the form of complex forms of matter.

1982 **Kenneth Wilson** (U.S.), for the study of changes in matter.

1983 **Subrahmanyan Chandrasekhar** and **William Fowler** (both U.S.), for research on the processes involved in the evolution of stars.

1984 **Carlo Rubbia** (Italy) and **Simon van der Meer** (Netherlands), for work in the discovery of three subatomic particles in the development of a unified force theory.

1985 **Klaus von Klitzing** (Germany), for developing an exact way to measure electrical conductivity.

1986 **Ernst Ruska, Gerd Binnig** (both Germany), and **Heinrich Rohrer** (Switzerland), for work on microscopes.

1987 **K. Alex Muller** (Switzerland) and **J. Georg Bednorz** (Germany), for development of a ''high temperature'' superconducting material.

—3rd Ed. Wilson has additional information on
  Newton 3rd & 2nd Law p 116, 117 Ex 4.17
— 2nd Ed = 1st Ed Re Newton 3rd & 2nd Law

PHYSICS

Newton 1st Law ———————— 8f
  "   2nd " Contradiction w/3rd 4/8f
  " 3rd Law (Contradiction w/2ed) 15, 18
→ Force "Capable" ———————— 9
→ Pascal, automobile lift ——— 15a
→ Mass ————————— 11
→ Acceleration zero ————— 11

JERRY D. WILSON

Chair, Division of Science and Mathematics
Lander College, Greenwood, South Carolina

 SAUNDERS GOLDEN SUNBURST SERIES

SAUNDERS COLLEGE PUBLISHING

Philadelphia New York Chicago San Francisco Montreal Toronto
London Sydney Tokyo

PHYSICS

A PRACTICAL AND CONCEPTUAL APPROACH

SECOND EDITION

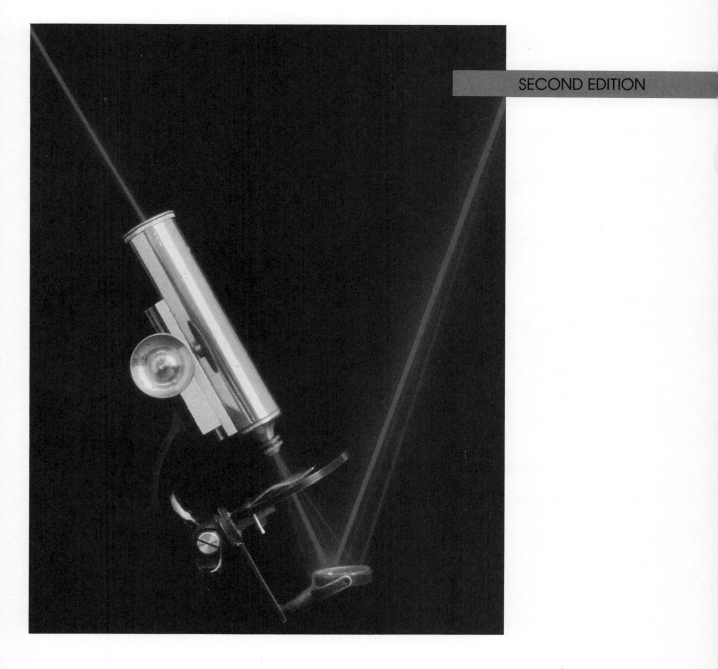

Text typeface: Garamond Light
Compositor: Progressive
Acquisitions Editor: John Vondeling
Development Editor: Ellen Newman
Project Editors: Margaret Mary Anderson, Elizabeth Knighton, Diane Zuckerman
Copy Editor: Nanette Bendyna
Art Director: Carol Bleistine
Art Assistant: Doris Bruey
Text Designer: Tracy Baldwin
Cover Designer: Lawrence R. Didona
Text Artwork: J & R Technical Services, Inc.
Layout Artist: Tracy Baldwin
Production Manager: Harry Dean

Cover Credit: Laser by Garry Gay © THE IMAGE BANK

Printed in the United States of America

PHYSICS: A PRACTICAL AND CONCEPTUAL APPROACH

ISBN 0-03-023764-5

Library of Congress Catalog Card Number: 88-043141

9012 061 98765432

To Joyce

Preface

PHYSICS: A Practical and Conceptual Approach is intended for a one-semester course in introductory physics for students majoring in the liberal arts or social sciences. Since nonscience majors are often wary of science, my purpose in writing this text is to make these students aware of and appreciative of their physical environment. A great deal of time and effort has been spent to present the fascinating subjects of physics in an interesting and understandable manner.

Although mathematics is an important tool in investigating the physical universe, students should concentrate on using their ability to read, think, and formulate questions as they progress through the course. Once they comprehend the concepts, the use of some basic mathematics will flow naturally from the established principles. The use of mathematics is kept to a minimum, although optional material with more mathematics is provided for those instructors who find it beneficial.

The physical universe is a dynamic, exciting place that becomes more meaningful to the person willing to learn the "why" as well as the "what." To help students achieve this goal, the following pedagogical features are in the second edition.

■ NEW FEATURES IN THE SECOND EDITION

- The material on the metric system, formerly Chapter 1, has been rewritten and placed in the Introduction to avoid intimidating students who have limited mathematical background. Although I have used little mathematics in the text, motivated students may enjoy this section and find that it furthers their understanding of physics. The instructor may choose whether the Introduction should be assigned reading.

- New "Extended View" sections are available for the instructor who wants a little more mathematics, but they are placed in the Appendix and may easily be omitted. The Introduction as well as Chapters 1, 2, 3, 5, 8, 11, and 17 contain topics that lend themselves to simple mathematical analysis; the "Extended View" of these chapters in the Appendix includes worked-out examples and solutions, as well as exercises with answers that are highlighted in color. These sections are cross-referenced in the text.

- Coverage has been expanded to include new developments in physics. Such topics as superconductivity, the electroweak force, Chernobyl, and the X-ray CAT scan are discussed in the text, as well as in the Special Features.

- Chapters 10, 13, 14, 15, and 25 include very simple mathematical exercises following the questions at the end of the chapter; these optional exercises can be omitted if the instructor so chooses. The answers are provided with the exercises to reinforce the student's understanding of the material.

- Answers to Selected Questions appear at the back of the book. Questions for which answers are given are marked with an asterisk in the text.

- Color has been added to highlight important concepts in the text and the illustrations.

The second edition has retained the following features:

STYLE. I have endeavored to write in a style that is clear and logical as well as somewhat informal and relaxed to assist students in their learning. New terms are carefully defined, and I have avoided using too much jargon.

ORGANIZATION. The book is divided into seven parts: I. Mechanics; II. Properties of Matter; III. Heat; IV. Sound; V. Electricity and Magnetism; VI. Light; and VII. Modern Physics. This is a well-known standard format and was chosen because one topic in physics builds on the next. For example, the concepts of motion, work, and energy (I. Mechanics) are needed to understand the concepts of temperature and heat (III. Heat), which in turn are used in explaining energy losses (joule heating) in electrical circuits (V. Electricity and Magnetism). Similarly, the basic concepts of electrical charge and magnetism are important in the study of light (VI. Light), as well as in identifying particles coming from radioactive isotopes (VII. Modern Physics).

PART OPENERS. Each opener presents a brief overview of the section that follows. Some historical matter is introduced to give students a perspective on how physics concepts developed through time.

COVERAGE. The text fully covers the realm of the physical universe. For example, in the mechanics section students will learn of the fundamentals of measurement, microgravity, and moment of inertia. In the section on modern physics the topics include lasers, masers, quasars, quarks, and black holes. Things are explained as simply as possible with little or no mathematics required.

ILLUSTRATIONS. The text is highly illustrated with many photographs and skillfully drawn pieces of artwork. Much of the artwork has been rendered in a "realistic" style that emphasizes accuracy, relative scale, and actual appearance. Students will make the connections between concept and application more easily if objects with which they are familiar really do *look* like those objects. Likewise, new information — whether visual or verbal in nature — is presented as accurately as possible. Color is used to enhance and highlight illustrations as well as text material.

SPECIAL FEATURES. These 65 Special Features, of which there is one or more in every chapter, should be of great interest to most students. They emphasize the practical aspects of the book to help to link the student's everyday experiences with physical principles. Among those included are:

The Automobile Air Bag
Microgravity
A Pioneer in Outer Space
The Greenhouse Effect
Acid Rain
Photochromic Glasses
Blood Pressure
The Microwave Oven

Galloping Gertie: The Tacoma Narrows Bridge Collapse
Noise-Exposure Limits
Ultrasonics
Superconductivity
Personal Safety and Electrical Effects
Xerography and Electrostatic Copiers
LCDs — Liquid Crystal Displays
Space-Age General Relativity
Lasers in the Supermarket
The Smoke Alarm
Case Studies: Three Mile Island and Chernobyl — The "Incident" and the "Accident"
Black Holes

QUESTION/ANSWER EXAMPLES. In these examples a question is posed and an answer is given immediately. These examples provide insight into interesting physical aspects while training students to think about physics in terms of questions as well as answers. These examples also function as models for the student's work in the Questions at the end of the chapter.

SUMMARY OF KEY TERMS. Key terms that are presented in the chapter are defined; this is an easy reference tool as well as a helpful review.

QUESTIONS. The end-of-chapter Questions are designed to stimulate student thinking by requiring that students apply what they have learned in the chapter. These are generally of the thought variety and require little or no mathematics. The Question/Answer Examples within the text will help students adapt to this kind of learning format.

■ ANCILLARY PACKAGE

Users of *Physics: A Practical and Conceptual Approach* will receive an extensive set of ancillary items that should substantially assist the instructor in the presentation of the course as well as motivate the student. These supplements include:

INSTRUCTOR'S RESOURCE MANUAL. This printed manual contains the following items:

● answers to all Questions
● a Test Bank of over 2,000 questions, composed of an average of 80 questions per chapter (multiple choice, completion, and matching)

COMPUTERIZED TEST BANK. The CTB contains the multiple choice and completion questions found in the Test Bank and is available for the IBM PC microcomputer.

CLASSROOM DEMONSTRATION DISK. This program disk contains a series of interesting computer programs that demonstrate or simulate physics phenomena. The disk is available for the Apple II microcomputer.

HOME-STUDY EXPERIMENTS IN PHYSICAL PHYSICS. This printed supplement contains simple home experiments that may be assigned by the instructor or done by interested students. The experiments provide a hands-on feeling for some of the physical principles studied in the text, and they are fun to do. This separate softcover guide is offered free to every student and is packaged with the textbook.

OVERHEAD TRANSPARENCIES. A selection of 50 figures from the textbook are reproduced on acetate sheets suitable for projection.

Acknowledgments

I would like to acknowledge all the assistance provided by various persons in the preparation of this book. My grateful thanks is given to the following professors across the country for their reviews and many constructive suggestions:

Louis H. Cadwell, Providence College
Robert Cole, University of Southern California
Jerry S. Faughn, Eastern Kentucky University
H. H. Forster, University of Southern California
David Hammer, University of California at Berkeley
Ann Hanks, American River College
David Kagan, California State University (Chico)
Robert A. Luke, Boise State University
Henry J. Mackey, University of North Texas
Robert G. Packard, Baylor University
Jerry Riley, Cuyamaca College
Ajit S. Rupaal, Western Washington University
Cecil G. Shugart, Memphis State University
Paul Varlashkin, East Carolina University
Brian Weiner, The Pennsylvania State University

I would especially like to thank James T. Shipman, Professor Emeritus, Ohio University, for reviewing the Questions and Exercises and for assisting in the preparation of the Instructor's Resource Manual.

For their help in manuscript preparation, art work, and photography, I thank Ruth Hodges, Jocelyn Sanders, David Williams, Lee Kelly, Sue Brannon, and Nicki Ciurro at Lander College. Last but not least, there was the able assistance of the staff at Saunders College Publishing, in particular, John Vondeling, Associate Publisher, and Ellen Newman, Developmental Editor. I salute and thank these people.

Jerry D. Wilson
Lander College

Contents Overview

Contents

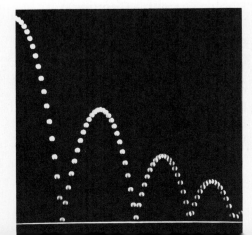

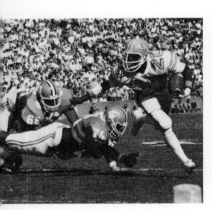

Special Features

Introduction: To the Student

I'm glad you're reading this introduction. Let me tell you why. In general, a preface, like the one on the preceding pages, is where the author tells the purpose of the book, how it's organized, what's great and wonderful about it, who helped him or her in putting the text together, and so on. An introduction for the student, on the other hand, may vary. For example, it might be a pep talk to get you all excited about reading the text. In this text, the Introduction is intended to be an informal "talk" with the student in which, as the author, but more as a teacher, I can give you some brief instructive insights about physics.

You may be a bit apprehensive about taking a physics course. This is probably your first, so a fair opening question you might have is, "What is physics?" Going to *Webster's Super New School and Office Dictionary,* we find:

physics (fiz′iks), *n.*   The science dealing with matter and energy.

This is true, but a more descriptive definition might be that physics is concerned with the description of nature—that is, the description and explanation of natural phenomena. In other words, physics is concerned with how and why things work or behave the way they do. (Have you ever wondered how something works—a microwave oven, a laser—or perhaps why an airplane flies or why the sky is blue?)

One of the first things we do toward this end is observe, with the belief that there are underlying principles or "rules" that govern things. As the Nobel Prize–winning American physicist Richard Feynman said in a lecture to undergraduates:

> *We can imagine that this complicated array of moving things which constitutes "the world" is something like a great chess game being played by the gods, and we are observers of the game. We do not know what the rules of the game are; all we are allowed to do is to* watch *the playing. Of course, if we watch long enough, we may eventually catch on to a few of the rules. The rules of the game* are what we mean by *fundamental physics. . . . If we know the rules, we consider that we "understand" the world.*

We often refer to these rules as laws, which like legal laws are sometimes repealed or changed when a greater understanding or insight is obtained. Once we understand the "rules of the game," we use them to plan strategy for our own benefit. Technology uses or applies physical laws to invent things such as motors, television, and satellites.

But let's back up a minute. How do we describe nature? Physics is a quantitative science, and to describe things we *measure* them—for example, how much of this or the size of that. Of course, we want to do this simply, so we start with the simplest or most fundamental things. Oddly enough, a vast majority of what we observe in nature can be measured or described in terms of four fundamental properties—length, mass, time, and electric charge—and their various combinations. In the text chapters there will be Questions and Answers, such as the following that addresses the idea of a *fundamental* property:

QUESTION: What is time?

ANSWER: This question gives some insight as to what is meant by a *fundamental* property or concept. In trying to define or explain what time is, you may experience some difficulty. (Have you ever tried to do so?) Sometimes the discussion of such a fundamental property leads to philosophical implications. General definitions may seem a bit vague. For example, it is sometimes stated that

> *Time is the continuous, forward flow of events.*

Marcus Aurelius, the Roman emperor and philosopher, wrote

> *Time is a strong river of passing events, and strong is its current.*

The Mad Hatter of Lewis Carroll's *Alice in Wonderland* thought he knew time:

> *"If you know Time as well as I do, you wouldn't talk about wasting it. It's him. . . . Now, if you only kept on good terms with him, he'd do almost anything you liked with the clock. For instance, suppose it were nine o'clock in the morning, just time to begin lessons; you'd only have to whisper a hint to Time, and around goes the clock in a twinkling: Half-past one, time for dinner."*

St. Augustine pondered this question too.

> *What then is time? If no one asks me, I know; if I want to explain it to a questioner, I do not know.*

A safe answer is: Time is a fundamental property or concept. This sort of masks our ignorance, and physics goes on from there to use the concept to describe and explain what we observe.

Now that we know what we measure, how do we express measurement descriptively? It's really a matter of choice. For example, a table has a certain length no matter how we describe it. One person might measure the length in feet, another in yards, and still another in meters. Certainly the length of the table doesn't change, only the choice of *units* used to describe it.

A group of popular units forms a system of units, which are adopted or made "standard" usually by governmental action. There are two major systems of units used today—the British or English system, which is predominantly used in the United States, and the metric system, which is used throughout most of the world. Some texts relegate this topic to an appendix, but not this one—it is put right up front in the Introduction for the student because it is thought to be so important. It might be said that to understand physics, one needs to understand units.

You are familiar with the units of the British system, since you have used these units probably all of your life—for example, we measure our heights in feet and inches and longer lengths in miles. However, let's take a look at some metric units. They are already

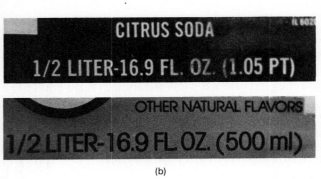

(b)

Figure I.1
Evidence of a metric change. (a) Weight-mass compari-sions on commercial items. (b) Volume comparisons on soft-drink labels. Notice that the metric unit is listed first.

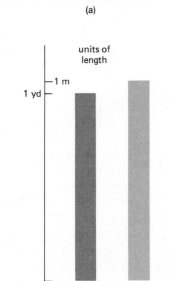

Figure I.2
The meter and the yard. A meter is slightly longer than a yard—3.37 in. longer.

part of our daily life (Fig. I–1). Have you looked at the labels on products in your supermarket lately?

LENGTH

The metric standard length is the meter. (See Special Feature I.1.) (You'll find these Special Features throughout the chapters too.) The meter (abbreviated m) is slightly longer than a yard, in fact, 3.37 inches longer (Fig. I.2).

With a length standard selected, the next job is to define sub-multiple and multiple units. For example, in our British system there are 12 inches in 1 foot, and 3 feet are 1 yard. The metric system is a decimal or "base-10" system. That is, larger and smaller units are obtained by multiplying or dividing standard units by factors of ten.

A variety of metric prefixes are used to indicate these factors. Only three prefixes, however, are usually used in everyday measure-ments:

kilo- (meaning one thousand, 1000)
centi- (meaning one hundredth, 1/100 or 0.01)
milli- (meaning one thousandth, 1/1000 or 0.001)

Applying these prefixes to length, we have the kilometer (abbre-viated km and pronounced *kil·omm'·etter*), which is 1000 meters; the centimeter (cm), which is 0.01 meter (1 meter = 100 cm); and the millimeter (mm), which is 0.001 meter (1 meter = 1000 mm) (Fig. I.3). You may have heard of another metric prefix— *mega-*. Some people talk about megabucks. If you had 1 megadollar, you'd be a millionaire.

Actually, money is a good analogy for the metric prefixes, since our money system is a decimal system. The dollar is divided into 100 cents. If a dollar is comparable to a meter, then a cent or penny is comparable to a *centi*meter. How about saying that a penny is a

The Meter

In 1790, in the midst of the French Revolution, the National Assembly of France requested that the French Academy of Sciences "deduce an invariable standard for all the measures and all the weights." The commision appointed by the Academy developed a system that was simple and scientific. The name *metre,* which we spell *meter,* was assigned to the unit of length. This name was derived from the Greek word *metron,* meaning "to measure." The length of the meter was defined as one ten-millionth of the distance along a meridian from the North Pole to the equator. A portion of a meridian running near Dunkirk in France and Barcelona in Spain was surveyed and the length of a meter determined. Based on these results, a 1-meter bar of platinum was constructed. This bar became the "Meter of the Archives," from which copies were made.

The use of metric weights and measures was legalized in the United States in 1866, and since 1893 the yard has been defined in terms of the meter. Metal bar meter lengths are used for common measurement reference standards, but these lengths are affected by temperature variations. In 1960, the meter was defined in terms of the wavelength of light. In 1983, a new definition was adopted that references the meter to a distance light travels in vacuum.

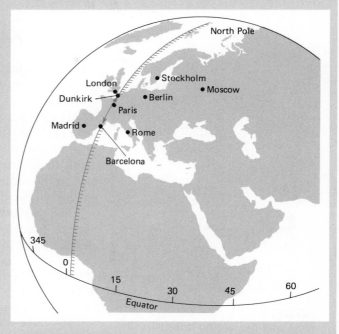

Definition of the meter. The meter was originally defined as one ten-millionth of the distance from the North Pole to the equator along a meridian that ran through France.

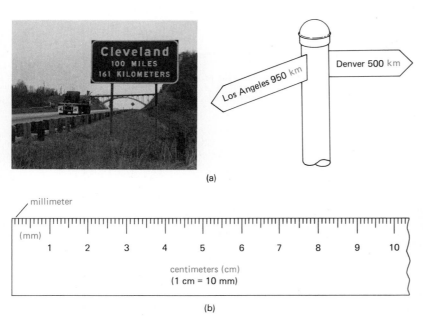

(a)

millimeter

(b)

Figure I.3
Metric prefixes. Only three metric prefixes are needed to describe most everyday measurements—*milli-, centi-,* and *kilo-.* As applied to length measurements, the units are (a) kilometers (km), and (b) centimeters (cm) and millimeters (mm).

"centidollar"? We can carry this comparison one step further. You may have heard how property taxes are assessed in mils. A mil is 1/10 of a cent, and there are 1000 mils in a dollar. Hence, a millimeter is analogous to a mil or a "millidollar."

■ MASS AND VOLUME (CAPACITY)

An object's mass is a measure of the quantity of matter it contains. In the British system, we express this in terms of weight, which is the gravitational force on a quantity of matter. Mass and weight are related, but mass is the fundamental property. The weight of an object can vary depending on the force of gravity, e.g., an object would weigh six times more on Earth than on the moon, since the gravitational attraction is six times greater on Earth. However, the mass is the same in each case. (The relationship between mass and weight will be discussed more fully in the text.)

The quantity of water in a particular metric volume was originally used to define the standard metric mass unit. A container 10 cm on a side has a volume of 10 cm × 10 cm × 10 cm = 1000 cm³ (cubic centimeter, sometimes abbreviated cc). See Figure I.4. With the container filled with water, the mass of this quantity of water was defined to be a kilogram (kg) (under particular conditions of pressure and temperature).

Since the metric prefix *kilo* means 1000, it follows that 1 kg = 1000 grams (g) and that one cubic centimeter of water has a mass of 1 gram. The gram unit is often divided into milligrams (1 g = 1000

Figure I.4
Mass units are related to length in the metric system. (a) A cube 10 cm on a side has a volume of 1000 cm³. (b) The amount of water that fills this volume has a mass of 1 kg, and hence, 1 cm³ of water has a mass of 1 g. (The volume 1000 cm³ is defined to be a liter.)

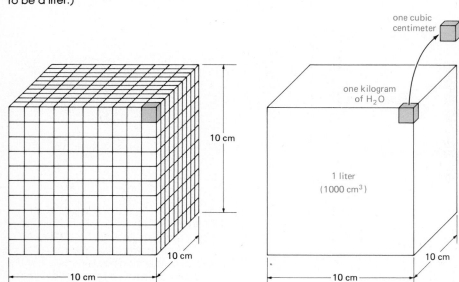

mg), which is a convenient unit for small quantities. Speaking of this, see if you can answer the following question.

QUESTION: What is the difference between a carat, a karat, and a carrot?
ANSWER: It is interesting to note that the *carat* unit used for precious stones is really a metric mass unit, with 1 carat having a mass of 200 mg. Hence, a 2-carat diamond weighs or has a mass of 400 mg. The carat unit is also "decimalized" into points, and 1 carat is equivalent to 100 points. A 50-point diamond is then one-half carat or 100 mg.

Another type of karat (with a *k*) is used to indicate the purity of gold. Pure gold is 24-karat (24-k). A piece of 12-karat gold jewelry contains 50 percent gold and 50 percent of some other metal. Pure gold is very soft, and other metals mixed or alloyed with it produce gold alloys that are harder and stronger (and cheaper).

You know what the other carrot is—a vegetable.

In case you're interested in how the metric kilogram compares with the British pound, 1 kg of mass has an equivalent weight of 2.2 pounds (lb) on the surface of the Earth. See Figure I.5.

A common metric unit of volume or capacity is the volume used in defining the kilogram. A volume of 1000 cm³ is defined to be a liter. The abbreviation for liter is either a small "ell" (l) or a capital L. The latter is generally preferred so as to avoid confusion with the numeral one. A liter is slightly larger than a U.S. quart (Fig. I.6). A common smaller unit of volume is the milliliter (mL). From the

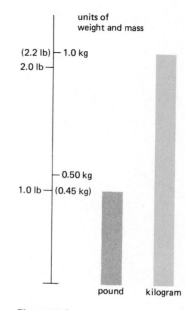

Figure I.5
The kilogram and the pound. The metric standard unit of mass is the kilogram, which has an *equivalent* weight of 2.2 lb.

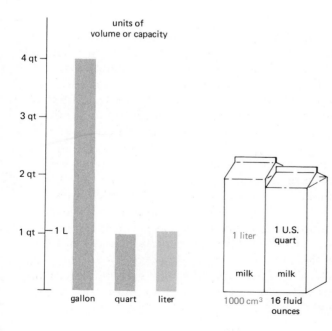

Figure I.6
The liter and the quart. The liter is slightly larger than the quart (1 L = 1.056 qt).

previous discussion, it follows that 1 L = 1000 cc = 1000 mL, and 1 cc = 1 mL.

■ TIME

The standard unit of time may give you some relief. This is the second (s) in both the British and the metric systems. Prior to 1956, the apparent daily motion of the Sun was used to define the second. A second was defined as 1/86,400 of an *apparent* solar day. (One day = 24 hours = 1440 minutes = 86,400 seconds.) Because of slight variations in the Earth's rotation, an atomic standard is now used for greater precision.

■ THE SI SYSTEM

You may have heard about the SI system. SI stands for the International System of Units (from the French, *Le Système International d'Unités*). This is the *modernized* version of the metric system established by international agreement in 1960. With increased interaction in international science and commerce, there was a need for a uniform usage of units and symbols.

The SI system uses the meter as the standard or base unit for length, the kilogram for mass, the second for time, the ampere for electric current (a measure of electric charge), and the kelvin for temperature. (The Kelvin unit is related to the Celsius degree, which is used in everyday applications.) There are a few other units in this system, which you may never hear of, so we won't belabor the point here.

In the SI system, all measurements or quantities are expressed only in the designated "base" units and their combinations. Multiple and submultiple prefix units are not used. The SI system is an international unit language in which everyone communicates with the same fixed words. However, as with the expanded capability of expression within a general language, it is convenient to use the flexibility of the general metric system in common measurements.

The metric system is the preferred system of science. So, as we go about describing nature in this text, a majority of the units will be metric. As the United States gradually adopts the metric system, you will have to become familiar with some new units in your everyday activities. This will not be as difficult as it may seem, as a governmental flier implies (Fig. I.7). Also, it is very easy to convert from one system to another. This is done by using conversion factors; the procedure is given in the Appendix. (It is omitted here because this Introduction is getting to be long enough and I know you are eager to start reading and to learn about the world around you.)

All You Will Need to Know About Metric

(For Your Everyday Life)

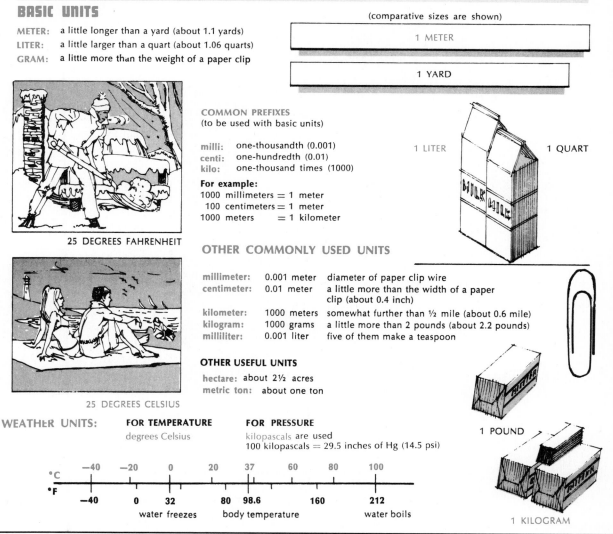

10

Metric is based on Decimal system

The metric system is simple to learn. For use in your everyday life you will need to know only ten units. You will also need to get used to a few new temperatures. Of course, there are other units which most persons will not need to learn. There are even some metric units with which you are already familiar: those for time and electricity are the same as you use now.

BASIC UNITS

METER: a little longer than a yard (about 1.1 yards)
LITER: a little larger than a quart (about 1.06 quarts)
GRAM: a little more than the weight of a paper clip

(comparative sizes are shown)

1 METER

1 YARD

25 DEGREES FAHRENHEIT

25 DEGREES CELSIUS

COMMON PREFIXES
(to be used with basic units)

milli: one-thousandth (0.001)
centi: one-hundredth (0.01)
kilo: one-thousand times (1000)

For example:
1000 millimeters = 1 meter
100 centimeters = 1 meter
1000 meters = 1 kilometer

1 LITER   1 QUART

OTHER COMMONLY USED UNITS

millimeter:	0.001 meter	diameter of paper clip wire
centimeter:	0.01 meter	a little more than the width of a paper clip (about 0.4 inch)
kilometer:	1000 meters	somewhat further than ½ mile (about 0.6 mile)
kilogram:	1000 grams	a little more than 2 pounds (about 2.2 pounds)
milliliter:	0.001 liter	five of them make a teaspoon

OTHER USEFUL UNITS

hectare: about 2½ acres
metric ton: about one ton

1 POUND

WEATHER UNITS:

FOR TEMPERATURE
degrees Celsius

FOR PRESSURE
kilopascals are used
100 kilopascals = 29.5 inches of Hg (14.5 psi)

°C	−40	−20	0	20	37	60	80	100
°F	−40	0	32	80	98.6	160		212
			water freezes		body temperature			water boils

1 KILOGRAM

Figure I.7
A flier to help people "think metric."

I have enjoyed "talking" with you, and I hope you enjoy the course and the text. Keep in mind that science in general is learned in a "building process." That is, you must crawl before you can walk, so to speak. You can't learn or explain everything at once, nor get the cart ahead of the horse. One principle builds on or adds to another in our understanding of the physics of how and why. I've tried to present these in the text in a clear and enjoyable manner so that in the end you will have a greater understanding and appreciation of your physical environment. I hope I've succeeded. Let me know what you think.

Jerry D. Wilson
Lander College
Division of Science and Mathematics
Greenwood, SC 29646

MECHANICS

Mechanics abounds in this view of St. Louis' Gateway arch.

PART I

Mechanics is the study of the motions of material bodies. Historically, it was one of the earliest exact sciences to be developed. Some mechanical principles were known to Greek scientists in the third century B.C. The tremendous growth of physics since the 1600's began with the discovery of the laws of mechanics by Galileo and Newton. Early successes were in predicting the motions of the moon, the Earth, and planets and their satellites (celestial mechanics). Now we apply the same principles to the motions of artificial satellites such as an orbiting *Space Shuttle.*

In general, the principles of mechanics can be applied to (a) the motions of celestial objects so as to accurately predict events, in some cases many years before they happen, for example, the return of Halley's comet; (b) the motions of ordinary objects on Earth, for example, an automobile or a thrown baseball; and (c) the behavior of atoms, atomic particles, and subatomic particles, with considerable success. The term *classical mechanics* is generally used to differentiate these principles from those of newer physical theories, such as relativistic mechanics and quantum mechanics. (See Part VII, Modern Physics.)

Mechanics greatly influenced the growth of later sciences such as sound and electricity. It may be said that mechanics furnishes the basic concepts of the whole physics, so quite naturally, the study of physics begins here.

A Restless World: Motion, Force, and Newton's Laws

Force, motion, and applications of Newton's laws.

Look around you. It is difficult to observe a scene in which something is not moving. For example, the hands on a clock move (or the digits change) relentlessly, and students hurry to class (sometimes a bit late). Thinking on a bigger scale, recall that you are moving through space as you read this book—the Earth is rotating on its axis and revolving about the Sun.

The study of motion has intrigued scientists since early times. The Greeks, in particular Aristotle, put forth theories on motion that were not always correct. Other scientists, such as Galileo and Isaac Newton, studied motion and helped formulate our present concepts. These were summarized into three laws of motion by Newton (see Special Feature 1.1), which will be studied in this chapter. But first, let's see how we characterize and describe motion.

■ MOTION

You may think that the description of motion would be quite simple. But suppose you were asked, "What is motion?" Could you give an answer? A typical reply is that something is in motion when it is moving. This is true, but does it really define motion?

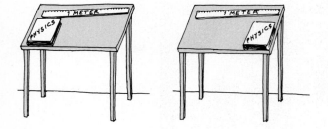

Figure 1.1
Motion is the process of a change in position. An object *in* motion is undergoing a continual change in position and a changed position is evidence that motion has occurred, as illustrated here. The book on the table has changed positions, so it was moved or was in motion. The description of motion involves length (distance) and time.

You are on the right track, however. Moving involves a change of position, and it quickly becomes clear that

Motion is a changing of position.

An object *in* motion is continually undergoing a change in position. Also, a changed position is evidence of motion. For example, if you put your book on a desk and come back later and find it in another place, you know motion has occurred (Fig. 1.1).

A distinguishing feature of the motion of objects is how *fast* a change in position occurs, or the *rate* at which this occurs. As you might suspect, we use combinations of length and time to describe motion. Length is how far one travels or the distance between two positions. Coupled with time, we get time *rates* of changes.

■ SPEED

Suppose you are in a car and it is in motion. If someone asks how fast you are moving, you might look at your *speed*ometer and say 55 miles per hour (mi/h), or about 88 kilometers per hour (km/h). This is how fast you are traveling almost instantaneously. So for practical purposes the speedometer expresses what we call instantaneous speed. **Instantaneous speed** is the speed of an object right then or at a particular instant of time.

More generally, speed is expressed by the distance traveled divided by the time required to travel the distance. Stated simply in equation form:

$$\text{Speed} = \frac{\text{distance}}{\text{time}} = \frac{\Delta d}{\Delta t}$$

where the Greek letter Δ (delta) means "change in" or "difference." For example, if you drove a distance of 100 km in 2 h, your speed would be $\Delta d/\Delta t = 100 \text{ km}/2 \text{ h} = 50 \text{ km/h}$ (50 kilometers *per* hour*). Speed then is a *rate,* in particular, the time rate of change of position, or how rapidly distance is covered [Fig. 1.2(a)].

Note that this is an **average speed** taken over an appreciable time interval. That is, on the *average* you traveled 50 km each hour. However, on a trip in your car there usually are variations in speed (stopping, starting, and so on). If, on the other hand,

---

* The term *per* essentially means "divided by."

Isaac Newton

On Christmas Day, 1642, Isaac Newton was born in Woolsthrope in Lincolnshire, England. His father had died three months earlier. When he was 14 years old, his mother became widowed for a second time and brought Isaac home from school to help run the family farm. He proved to be a lackadaisical farmer, being occupied more with mathematics than farm chores.

At the age of 18, he entered Trinity College at Cambridge and received his degree four years later in 1665. Later that year while he was preparing for advanced degree examinations, the spread of the Great Plague caused the university to close. Newton returned home, and during the next 18 months he conceived most of the ideas for his famous discoveries in science and mathematics. Chief among these were the development of calculus mathematics and studies on light and color, motion, and gravitation. As Newton later described this period,

> I was in the prime of my age for invention, and minded Mathematics and Philosophy (science) more than at any time since.

After the plague had passed, he returned to Cambridge and at the age of 26 was appointed professor of mathematics. Newton's early published works were in the field of optics. He developed a new type of telescope, a reflecting telescope, that used a mirror rather than a lens to collect light (see Chapter 22). His most notable book, *Principia Mathematica Philosophiae Naturalis* (Mathematical Principles of Natural Philosophy*), or *Principia* for short, was published in 1687. The cost of the book was borne by a contemporary, Edmund Halley, who predicted the return of a famous comet that bears his name. In *Principia* Newton set forth his theories on gravity, tides, and motion.

A bachelor, Newton lived very austerely and is reported to have been the classic absent-minded professor, being so absorbed in his work that he forgot meals and other day-to-day activities. In later life Newton moved to London and was appointed master of the mint in 1701. Queen Anne knighted him in 1705 in recognition of his numerous accomplishments. Early in 1727 he became seriously ill and died on March 20 of that year.

An insight into the character of this great scientist is given in one of his statements:

> If I have been able to see farther than some, it is because I have stood on the shoulders of giants.

One of these giants was Galileo. (See Fig. 1.7 and Special Feature 1.3.)

* Physics was once called natural philosophy.

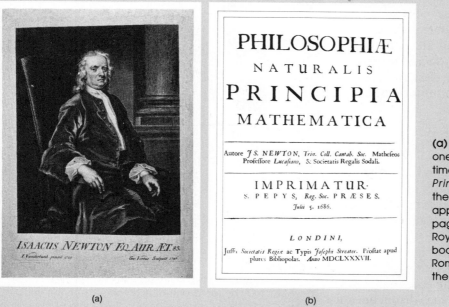

(a) (b)

(a) Sir Isaac Newton (1642–1727), one of the greatest physicists of all time. (b) The title page of his *Principia*. Notice that the name of the famous diarist Samuel Pepys appears near the middle of the page. Pepys was president of the Royal Society that sponsored the book. Can you read the date in Roman numerals at the bottom of the page?

Figure 1.2
Speed. (a) Average speed. The average speed of a trip is the distance traveled divided by the time taken to travel the distance. (b) Constant or uniform speed. When something is traveling at a constant speed, equal distances are traveled in equal times. A car's speedometer, which reads almost instantaneous speed, would have a constant reading in this case.

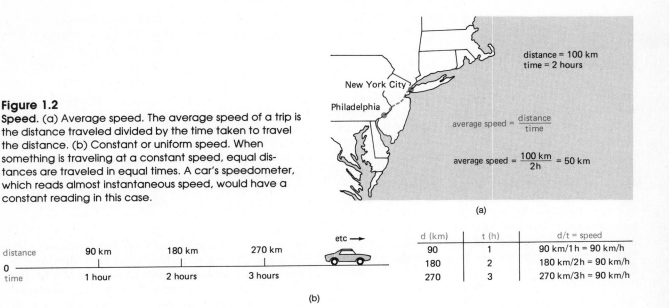

(a)

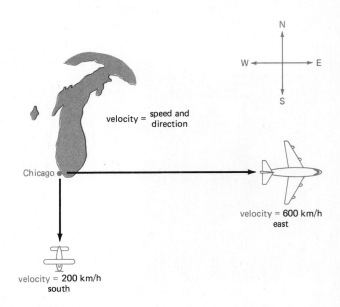

(b)

there were no variations, then you would be traveling at a *constant* or *uniform* speed. For example, if you drove at a constant speed of 90 km/h, you would travel 90 km each hour, or cover equal distances in equal intervals of time [Fig. 1.2(b)]. Your speedometer would *constantly* read 90 km/h. This is also your *instantaneous* speed—the speed at any particular instant.

> *Question:* When you are traveling at a constant speed, what is your average speed?
>
> *Answer:* When traveling at a constant speed, the average speed is the same as the constant speed. Let's illustrate this by an analogy. Suppose your class took a test and everyone scored 90 points. Then, the class average is 90, the same as the constant or uniform score.

▮ VELOCITY

We often use the terms *speed* and *velocity* interchangeably. However, there is a difference in physics. Speed tells you how fast you are moving, but not the *direction*. Velocity specifies *both* speed and the direction of the motion.

For example, if you say you are traveling 75 km/h, you are specifying speed. However, if you say you are traveling 75 km/h to the east, then you are specifying velocity. In science, such distinctions are made by referring to *scalar* and *vector* quantities.

A **scalar** quantity, like speed, has magnitude or "size" only. For example, population and temperature are scalar quantities. A **vector** quantity, like velocity, has both magnitude and direction. Graphically, we often represent a vector quantity by an arrow, where the length of the arrow is proportional to the magnitude and the arrowhead indicates the direction (Fig. 1.3).

Figure 1.3
A vector quantity has both magnitude and direction. A vector may be represented by an arrow, with the arrowhead indicating the direction and the length of the arrow being proportional to the magnitude, for example, 600 km/h east or 200 km/h south, as illustrated here.

Question: If an object travels at a constant speed, does it also have a constant velocity?

Answer: Not necessarily. If an object travels at a constant speed in a straight-line path, then it also has a constant velocity—constant speed and *constant direction*. However, an object may move with a constant speed in a curved path, for example, a car going around a circular track at a constant speed. In this case the velocity isn't constant because the direction of motion is continually changing. As a matter of fact, if the car circles the track once at a constant speed, its average velocity is zero because it has had equal and opposite velocities at each set of points on opposite sides of the track. Opposite vector velocities cancel each other.

ACCELERATION

Velocity expresses a change in position (and/or direction). Then, too, we may have a change in velocity. To express this, we talk about acceleration. For example, what is another name for the gas pedal of a

car? You got it—an accelerator. Why? Because by pushing down or letting up on the accelerator, you cause the car to *speed up or slow down*. This is a change in speed and therefore a change in velocity. Technically, acceleration is defined as the time rate of change of velocity, or

$$\text{Acceleration} = \frac{\text{change in velocity}}{\text{time to make change}} = \frac{\Delta v}{\Delta t}$$

Suppose you are traveling in your car in a straight-line path at a constant speed of 10 m/s (about 36 km/h), and you push on the accelerator so that the velocity increases 2 m/s each second— 10 m/s to 12 m/s, to 14 m/s, and so on. The acceleration, or rate of change of velocity, is then 2 m/s per second ($\Delta v/\Delta t$), or 2 m/s².* This is an example of a constant or uniform acceleration, since the change in velocity is uniform—2 m/s each second [Fig. 1.4(a)]. Here the speed changes but not the

* Acceleration is the change in velocity divided by time, $a = \Delta v/\Delta t$, or $a = v/t$ with the deltas omitted. In standard units, the unit of velocity is meters/second (m/s), so $a = v/t$ is (m/s)/s, and (m/s)/s = m/s-s = m/s², which is read as "meters per second squared."

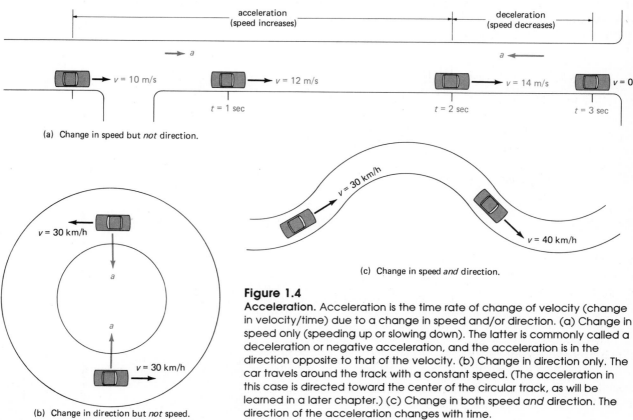

(a) Change in speed but *not* direction.

(b) Change in direction but *not* speed.

(c) Change in speed *and* direction.

Figure 1.4
Acceleration. Acceleration is the time rate of change of velocity (change in velocity/time) due to a change in speed and/or direction. (a) Change in speed only (speeding up or slowing down). The latter is commonly called a deceleration or negative acceleration, and the acceleration is in the direction opposite to that of the velocity. (b) Change in direction only. The car travels around the track with a constant speed. (The acceleration in this case is directed toward the center of the circular track, as will be learned in a later chapter.) (c) Change in both speed *and* direction. The direction of the acceleration changes with time.

direction of motion. Notice that for a constant acceleration,

Change in velocity = acceleration $\times$ time

The term *acceleration* applies to both increases *and* decreases in velocity. When a car slows down, there is a change (decrease) in its velocity. We commonly call this a *deceleration* and the acceleration is in the opposite direction to the velocity. Similar to the gas pedal of a car being called an accelerator, the brake pedal might be called a decelerator.

However, this is not the total story. As was learned from the previous Question and Answer, a change in velocity can also result from a change in the direction of motion, since velocity is a vector quantity. Hence, a change in velocity or an acceleration can result from (a) a change in speed, (b) a

change in the direction of motion, or (c) changes in both speed and direction (Fig. 1.4). Perhaps we should also call the steering wheel of a car an accelerator because it can change *direction!*

■ FORCE

Closely associated with motion is force. In describing what a force *does*, we know intuitively that a force can produce a change in motion. For example, when an object starts to move (a change in motion) we know a force is acting. Similarly, when a moving object stops (change in motion), we know a force is acting. A change in motion then is *evidence* of a force.

We are commonly familiar with contact forces such as a push or a pull. This is the case of forces

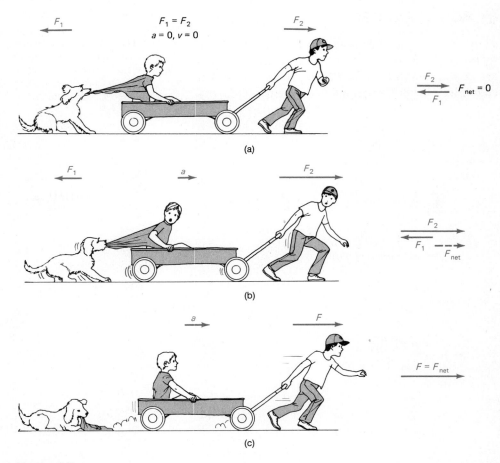

Figure 1.5
Unbalanced or net force. (a) The forces are equal and opposite, so they "cancel," and the net force is zero. (b) The net or unbalanced force is in the direction of F_2 and has a magnitude equal to the difference in the magnitudes of F_2 and F_1. (c) The applied force F is the net force. Note that the acceleration a is proportional to the net force. (Friction is neglected in all cases.)

resulting from one object being in contact with another. For example, you push or exert a (contact) force on a door to open it. There are also action-at-a-distance forces, which will be considered in more detail in later chapters. For example, you know that the gravitational pull or attraction (force) on an object causes it to fall (move) toward the Earth. Also, the gravitational forces of the moon and Sun on the Earth's oceans causes tides.

It is common in physics to describe an object's "state of motion," for example, 20 m/s *up* or 20 m/s *west,* and being at rest is also a state of motion (zero speed or velocity). Then,

> A force is something <u>capable</u> of changing an object's state of motion (including starting it from rest).

A key word here is *capable*. A force may not produce motion. You have applied many contact forces to objects and not produced motion. (Try pushing against the side of a building.)

As illustrated in Fig. 1.5(a), two forces may act against each other (in opposite directions) and effectively cancel each other with regard to producing motion. (Force is a vector quantity, so we may represent it with vector arrows.) In this case, we say that the forces are balanced or that the net force is zero. No change in motion would occur. If an object were initially at rest (no motion), it would stay at rest.

However, if one of the applied forces is greater than the other (case b), then there is a net or unbalanced force. The effect of the two forces would be the same as that of an equivalent net force *F*, which is in the direction of the larger force and has a magnitude equal to the difference of the magnitudes of the forces.

When we speak of a force changing an object's state of motion, it is understood that this may be the resultant net or unbalanced force of two or more individual forces. Usually, we are interested in the *net* motional effect produced by the net force rather than the individual forces.

Two or more forces may act in different directions at various angles to each other. In this case, there is also a net or "resultant" force. For example, in the simple case of two equal forces acting at 90° in the *x*- and *y*-directions as in Figure 1.6, the resultant force is in the direction midway between the two forces. The same effect of the two forces could be accomplished with a single force equal to the resultant force. Looking at it another way, a single force *F* can be thought of as being made up of component

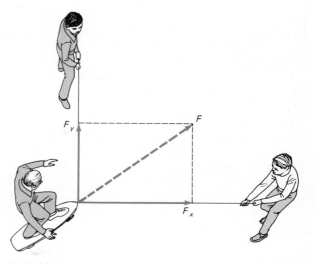

Figure 1.6
Vector addition. The effect of two (or more) forces could be accomplished by a single resultant force F *(dashed line)* equal to the vector sum of the forces (F_x and F_y).

forces F_x and F_y in those directions. A vector can be "broken up" or resolved into *x* and *y* parts or components to analyze its effects in these directions.

So, knowing that forces can produce changes in motion, the next question is, How are force and motion related? Sir Isaac Newton described that relationship in three laws. Let's look at Newton's laws of motion.

■ NEWTON'S FIRST LAW OF MOTION: THE LAW OF INERTIA

According to Aristotle's theory of motion, which had prevailed some 1500 years after his death, a body required a force to keep it in motion. That is, the normal state of a body was rest, with the exception of celestial bodies, which were naturally in motion. Aristotle observed that moving objects tended to slow down and come to rest (because of friction, we now know), so this conclusion seemed logical to him.

Galileo laid the groundwork for the first law. Rather than rely solely on intuition and general observations, Galileo tested his ideas with experiments. He observed the motions of a ball on inclined planes (Fig. 1.7). When a ball was released and allowed to roll down an incline, it would roll up

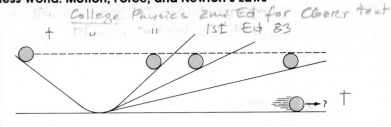

Figure 1.7
Galileo Galilei (1574–1642) and an illustration of his rolling ball experiment. Known now universally by his first name, this Italian mathematician, astronomer, and physicist made many contributions to science, including the description of motion. In one experiment, he noted that a ball would roll farther the smoother the surface. In an ideal case of no friction, the ball would continue indefinitely, since there would be no force to alter its motion.

an adjoining incline to about the same height (a little less, because of friction), being stopped by the retarding forces of gravity and friction. When the angle of incline of the adjoining plane was made less steep, the ball would roll to approximately the same height in each case, *but it rolled farther in the horizontal direction.* Then, when the ball was allowed to roll onto a horizontal plane, it rolled a considerable distance before coming to rest.

Galileo took great pains to make the surfaces of the plane and the rolling ball as smooth as possible to reduce friction, and he found that the smoother the surface the greater the horizontal distance the ball would roll. So the question arises, How far would the ball travel if friction could be removed completely and the plane made infinitely long? Galileo reasoned that in this ideal case the ball would continue to travel in straight-line, uniform motion forever, since there would be nothing (no force) to alter or change its motion.

Contrary to Aristotle's ideas, Galileo concluded that material bodies exhibited the behavior or property of maintaining a state of motion. Similarly, if a ball is at rest, it would remain so, unless something caused it to move. Galileo called this property inertia, and we say,

Inertia is the property of matter that describes its opposition to changes in motion.

That is, if an object is at rest, it seems to want to remain at rest. If an object is in motion, it seems to want to remain in motion. Here, as in the case of forces and indeed all physical quantities, the property of inertia is defined in terms of behavior or observations. The term simply describes observed effects.

Question: When a hammer head becomes loose on its handle, a person sometimes brings the hammer downward and strikes the butt of the handle sharply on a hard surface (Fig. 1.8). What does this accomplish?

Answer: When the hammer is brought downward, the handle and head are in motion. When the handle butt strikes the surface, it stops suddenly because of the large contact force of the surface on the handle. But the massive head continues in motion until it is stopped by the tapered handle. This tightens the hammer head on the handle.

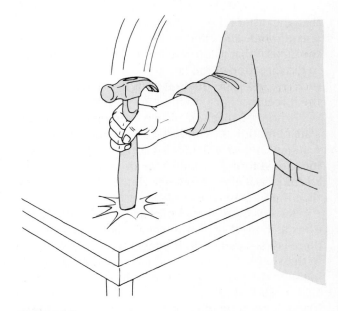

Figure 1.8
Inertia "in action." Inertia is used to tighten a loose hammer head.

Newton eventually made the idea of inertia quantitative by relating inertia to *mass*. Originally, he thought of mass as a "quantity of matter" but effectively redefined it as the measure of inertia. Today, we know that the mass of a given quantity of matter is not totally independent of conditions. (The mass of an object depends on its speed and even on its temperature, but these changes are commonly too small to measure.) Quantity of matter is still commercially acceptable as a definition of mass, but physicists usually use this definition:

Mass is a measure of inertia.

The greater the mass of an object, the greater inertia it exhibits. For example, it is easier to push and get a sports car moving than a more massive van.

Newton summarized these results in his first law of motion, which is also called the law of inertia:

*Every body preserves its state of rest, or of uniform motion in a right (straight) line unless it is compelled to change that state by forces impressed thereon.**

Or, in more modern language:

An object remains at rest or in motion with a constant velocity unless acted upon by an unbalanced force.

◼ NEWTON'S SECOND LAW OF MOTION: CAUSE AND EFFECT

Newton's first law says that a body remains at rest or in motion with a constant velocity until it is acted upon by a force. In the absence of an unbalanced force then, the acceleration of a body is zero, since there is no change in velocity. This should lead you, like Newton, to the conclusion that a force acting on a body produces an acceleration. However, Newton recognized that inertia or mass also plays a part. For a given force F, the greater the mass of a body, the less its acceleration a, or change in motion. Expressing this as a proportion,†

$$a \propto \frac{F}{m}$$

Thus, the acceleration of a body depends *both* on the net force and on the mass of the body. These relationships are illustrated in Figure 1.9.

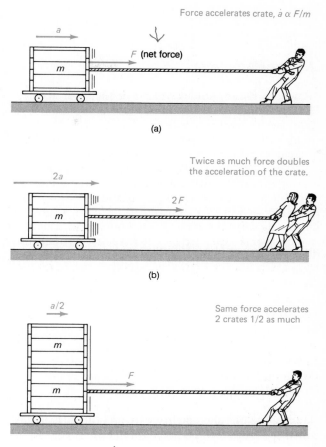

Figure 1.9
Acceleration is proportional to F/m. (a) For a given net force F, the acceleration is proportional (and equal) to F/m. (b) If the net force is doubled, the acceleration is doubled (same mass). (c) If the mass is doubled, the acceleration is one half as great (same original net force). The net force in each case is the vector sum of the applied force and the opposing frictional force.

* *Principia Mathematica Philosophiae Naturalis,* from Magie, W. F., *A Source Book in Physics,* Harvard University Press, Cambridge, Mass., 1963.

† The symbol ∝ means "proportional to." The expression $a \propto F$ means that a is *directly* proportional to F and they change in the same proportion. For example, if F is doubled, a is doubled. The expression $a \propto 1/m$ means that m is *indirectly* or *inversely* proportional to a and they change in the same inverse proportion. For example, if m is doubled, a is halved. When these are written together, we have $a \propto F/m$. A proportion gives only relative changes. An equation can be used to calculate the exact values.

An Accelerometer

A carpenter's air-bubble level can be used as an accelerometer or force meter in detecting an acceleration or a force by observing the motion of the bubble. If a force is applied toward the left to a level at rest, as shown in the figure, which way will the bubble move?

Many people say that the bubble would move to the right, but actually it would move to the left, or in the direction of the acceleration and the force. The incorrect answer arises from the fact that we are used to observing the bubble rather than the liquid. The correct answer is explained by Newton's first law. Because of inertia, the liquid resists the motion and "piles up" toward the rear of the level. This forces the bubble in

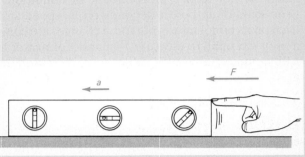

Level accelerometer. The motion of the bubble indicates the direction of the acceleration or the net force.

Newton's second law of motion is commonly expressed in equation form as

$$F = ma$$

Force = mass $\times$ acceleration

Thus, if the unbalanced force on an object (mass) is zero, its acceleration is zero, and it remains at rest or is in motion with a constant velocity, which is what Newton's first law also tells us.

The metric unit of force using SI quantities is quite appropriately called the newton (N). The second law states that a force of 1 N is the amount of force that gives a mass of 1 kg an acceleration of 1 m/s^2. (One newton is 0.225 or about ¼ lb. Recall that the pound is a unit of force.)

Newton's second law relates a force directly to the acceleration it produces—a sort of "cause-and-effect" relationship. Suppose a force of 15 N (the cause) acts on an object with a mass of 5 kg. What is the acceleration (the effect) of the object? By Newton's law, $a = F/m = 15/5 = 3$ m/s^2. The acceleration of a body is always in the direction of the applied net force. If the force is applied in the direction of a body's motion, it will increase the body's velocity. If it is applied in the direction opposite to the motion, a decrease in velocity will result (negative acceleration). When applied at an angle to the direction of a body's motion, a force will deflect or change the direction of the body's motion. This is also an acceleration or change in veloc-

ity (a change in direction certainly *and* possibly a change in magnitude as well).

Another way of looking at this is that an acceleration of a body is evidence of an applied force. Thus, we may say that a force is anything that can accelerate a body. A simple way to detect acceleration or force is given in Special Feature 1.2.

Examples of Newton's Second Law

Friction

The force of friction resists relative motion between contacting media. We usually think of friction in terms of contacting solid surfaces, but it also occurs for liquids and gases.

For solid surfaces, friction arises from the irregularities in the surfaces of objects. All solid surfaces are microscopically rough, no matter how smooth they may appear or feel. Early investigators thought friction to be primarily due to the interlocking or fitting together of surface irregularities. However, modern research shows that much of the friction between contacting surfaces of ordinary solids (metals in particular) is due to local adhesion or "sticking" between the surface irregularities rather than to their fitting together.

Generally, we represent the overall frictional effect by a friction force f. The force of friction always opposes relative motion of the surfaces in contact. We have static friction that opposes the starting of motion and kinetic or sliding friction that acts when the surfaces are in relative motion (Fig. 1.10).

Applying Newton's second law to stationary objects, we have, for static friction,

$$F - f_s = 0 \ (= ma, \text{ where } a = 0)$$

or $\qquad F = f_s$

The applied force F and the force of static friction f_s are equal and opposite, so there is no acceleration or motion. This is a case of a net force ($F - f_s$) being zero.

However, if the applied force is increased so that the object slides, then the frictional force op-

posing the motion is kinetic. The net force in the direction of the motion is

$$F - f_k = ma$$

or $\qquad a = \dfrac{F - f_k}{m} = \dfrac{F_{net}}{m}$

The acceleration a may or may not be zero, depending on the magnitudes of the applied and frictional forces. It generally takes a smaller applied force to keep an object moving than to get it moving because kinetic friction is less than static friction.

> *Question:* What would happen if the applied force on a sliding object had the same magnitude as the sliding frictional force?
>
> *Answer:* In this case, $F - f_k = F_{net} = 0 = ma$, so the net force and the acceleration are zero. This means that the object slides with a constant velocity (à la Newton's first law).

Mass and Weight

Mass and weight are sometimes confused. This comes chiefly from differences in the customary British system and the metric system. The British system is a "force" system and measures the concept of matter in terms of gravitational *force* or *weight,* whereas the metric system uses *mass.* Mass is the fundamental property. Weight is the gravitational attraction or force on a body due to a celestial object, most commonly for us, the Earth.

Newton's second law clearly distinguishes between mass and weight. Because weight is a force, it can be expressed by the equation $F = ma,$ and we commonly write the formula for weight as $w = mg.$ Note that this is a special form of Newton's second law:

$$w = mg$$
$$(F = ma)$$

Weight (w) is a force, and g is the acceleration due to gravity (given a special symbol because it is used so commonly*). This "special" acceleration has a relatively constant value near the Earth's surface of 9.8 m/s² (or 32 ft/s²).

Being a force, the metric SI unit of weight is the newton (N). The weight of a 154-lb person in the

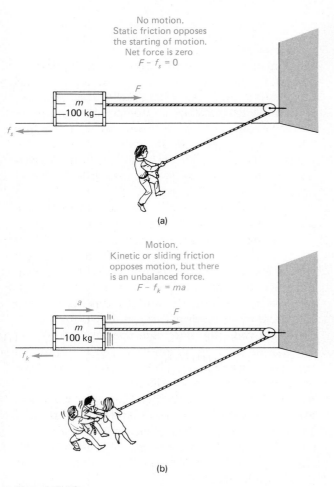

No motion.
Static friction opposes the starting of motion.
Net force is zero
$F - f_s = 0$

m
100 kg

F

f_s

(a)

Motion.
Kinetic or sliding friction opposes motion, but there is an unbalanced force.
$F - f_k = ma$

a

m
100 kg

F

f_k

(b)

Figure 1.10
Friction. (a) Static friction opposes the initiation of motion. The force of static friction is equal and opposite to the applied force on an object. (b) Kinetic or sliding friction opposes motion in progress. If the object accelerates, then there is a net force.

---

* This should not be confused with the accepted abbreviation for the gram (g).

Figure 1.11
Mass and weight. Mass is a fundamental property and is a measure of the quantity of matter a body contains. Weight is the gravitational attraction or force on a body due to a celestial object, most commonly for us, the Earth. Weight = mass × g, where g is the acceleration due to gravity.

metric system would be 686 N. However, in countries using the metric system, mass is used instead of weight force, and things are "weighed" in terms of mass, i.e., expressed in units of kilograms and grams. For example, a 154-lb person would have a mass of 70 kg, or "kilos" (Fig. 1.11). One kilogram of mass has an *equivalent* weight of 2.2 lb, or by the preceding formula, 9.8 N. [$F = ma = (1$ kg$)$ $(9.8$ m/s$^2) = 9.8$ N]

Indeed, mass is the fundamental property. Our 154-lb, or 70-kg, friend (Fig. 1.11) would still have a mass of 70 kg on the moon, where the acceleration due to gravity is one sixth of that on Earth. However, he would have only one-sixth the weight, about 26 lb, or 114 N. Why?

If you zipped up to the moon, you'd lose a lot of weight, but not mass. (This would be a great way to reduce, but you'd still have the same waistline.) On Earth, don't be confused by mass and weight. If the mass of an object increases, so does its weight. Similarly, if you lose weight, you lose mass. Keep in mind that mass and weight are different, but they only differ in magnitude by a factor of g ($w = mg$).

Just for your information and scientific trivia, there is a unit of mass in the British system. You may have never heard of it, as it is rarely used. It's called the *slug*. If you picked up a slug of mass, as Newton's second law tells you (using $g = 32$ ft/s^2), you'd be picking up 32 lb. (It is said that the slug unit was derived from the word *sluggish*. That's rather appropriate inertia-wise, isn't it?)

Free Fall

The gravitational weight force is easily demonstrated in terms of motion. When an object is dropped, it falls faster and faster (accelerates) toward the Earth. When an object falls toward the Earth with only gravity acting on it, we say it is in free fall. Air resistance and any other forces are neglected. This makes life easier and is a good approximation in some cases—for example, a heavy object falling a short distance.

All objects in free fall near the Earth have the same acceleration—the acceleration due to gravity.

It might be thought that heavy objects would fall faster than lighter objects. After all, if one object weighs twice as much as another, or the gravitational force is twice as great, why shouldn't it fall twice as fast? Aristotle thought that this was the case in his theory of motion, but Galileo showed, or at least believed, otherwise. See Special Feature 1.3.

Galileo gave no reason why objects in free fall have equal accelerations, but Newton's second law does. Recall that the acceleration of an object depends not only on the applied force, but *also* on its mass ($a = F/m$). If one object has twice the mass of another object, then it also has twice as much inertia. Hence, it needs twice as much force to accelerate it at an equal rate. This result of Newton's law is illustrated in Figure 1.12.

Thus, we say that the *acceleration* of an object in free fall is independent of its mass or weight. This means that all objects in free fall accelerate at the same rate—the acceleration due to gravity—regardless of their masses.

Galileo and the Leaning Tower of Pisa

There is a popular and well-known story that Galileo performed experiments with falling bodies by dropping objects from the Leaning Tower of Pisa. However, the authenticity of this story is questionable. The original question as addressed by Aristotle was, Why do bodies fall to the ground? According to the Aristotelian view, they were seeking their natural place at the center of the Earth, which itself was the center of the universe. How this occurred supposedly depended on the "earthiness" of a body; that is, the heavier a body, the faster it would fall in seeking its natural place. There is little doubt that Galileo questioned this view. In an early writing from about 1590, when Galileo was living in Pisa, he stated:

How ridiculous is this opinion of Aristotle is clearer than light. Who ever would believe, for example, that . . . if two stones were flung at the same moment from a *high tower,* one stone twice the size of the other, . . . that when the smaller was half-way down the larger had already reached the ground?*

And from a later passage concerning the falling of wood and lead objects,

* This and following quotations from Cooper, L., *Aristotle, Galileo, and The Tower of Pisa,* Cornell University Press, Ithaca, N.Y., 1935.

. . . but a little later the motion of the lead is so accelerated that it leaves the wood behind, and if they are let go from a *high tower,* precedes it by a long space; and I have often made this test.

Here there were probably some air resistance considerations.

But in a 1638 work we find,

Aristotle says that "an iron ball of one hundred pounds falling from a height of one hundred cubits reaches the ground before a one-pound ball has fallen a single cubit." I say that they arrive at the same time.

Free fall. Objects in free fall have the same acceleration. (a) Legend has it that Galileo demonstrated this by dropping objects of different mass or weight from the top of the Leaning Tower of Pisa. (b) A time-interval photograph of a baseball and a golf ball released simultaneously showing that they fall together, or have the same acceleration.

(a)

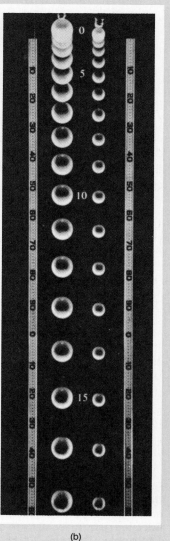

(b)

The first account of a Tower of Pisa experiment came a dozen years after Galileo's death and was written by Vincenzo Viviani, his last pupil and first biographer. Viviani relates that the falling bodies

> . . . all moved at the same speed; demonstrating this with repeated experiments from the height of the Campanile (Tower) of Pisa in the presence of the other teachers and philosophers, and the whole assembly of students . . . .

Yet there is no independent record of this from the time, and it is not mentioned in Galileo's writings. Did Galileo relate this to Viviani in his declining years, or did Viviani use his imagination in describing his teacher's experiments from a "high tower"? In debating this, many lose sight of the point. Whether or not Galileo in myth or in fact dropped objects from the Tower of Pisa, the acceleration due to gravity is independent of an object's mass.

When an object is accelerated, its velocity changes with time. For an object in free fall, its velocity changes (increases) 9.8 m/s each second. ($g = 9.8$ m/s per second, and acceleration = change in velocity per second.) This means that the distance fallen by the object increases during each time interval or second because its velocity increases each second (Fig. 1.13).

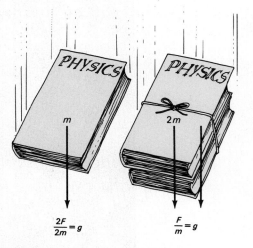

$$\frac{2F}{2m} = g \qquad \frac{F}{m} = g$$

Figure 1.12
The acceleration of objects in free fall does not depend on weight. A heavier object (mass 2 *m*) in free fall has twice the (gravitational) force acting on it as does a lighter object (mass *m*). But, having twice as much mass and inertia, the heavier object falls with the same acceleration (*g*) as the lighter object.

Nonfree Fall with Air Resistance

Technically, free fall would occur only in a vacuum because air resistance normally acts on a falling body. The resistance opposes the motion, so an object falls at a slower rate than in free fall. In some cases, the air resistance is appreciable and drastically affects the motion. For example, a feather "floats" or falls slowly because of the upward air resistance. A common lecture demonstration is illustrated in Figure 1.14. With air in the tube, the heavier coin always beats the feather to the bottom. However, if enough air is evacuated from the tube, giving a partial vacuum, the air resistance is negligible and the feather and coin fall together at the same rate.

A similar experiment with a hammer and feather was performed on the moon in 1971 by astronaut David Scott. The moon has no atmosphere, and hence there is no air resistance. When released simultaneously, the hammer and feather fell together at the same rate and hit the lunar surface at the same time. Of course, they fell quite a bit slower than they would in a vacuum chamber on Earth, since the acceleration due to gravity on the moon is one-sixth that on Earth.

The air resistance on a falling body depends on its (1) shape, (2) size (exposed area), and (3) speed.* The amount of air a body "catches" de-

---

* The air density is also a factor, but we will assume this to be relatively constant near the Earth's surface.

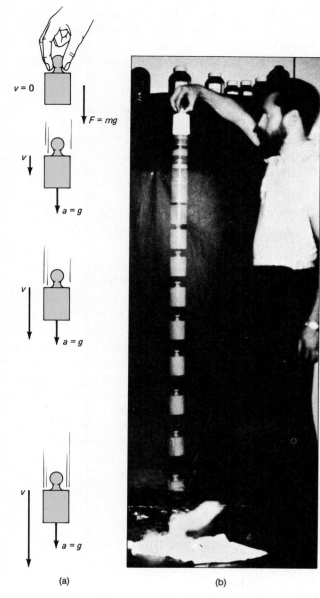

(a)   (b)

Figure 1.13
Constant acceleration, changing velocity. The distance
traveled by a falling object increases during each time
interval or second, since its velocity increases each
second. With a constant acceleration, the velocity increases uniformly with time.

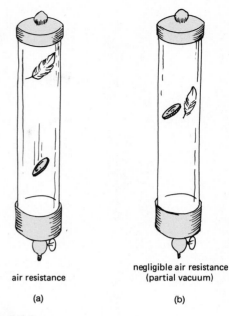

air resistance   negligible air resistance
                        (partial vacuum)

(a)   (b)

Figure 1.14
Air resistance. (a) When a feather is falling in air, its
motion is retarded more than that of a coin. (b) If most of
the air is removed from the tube, the coin and feather fall
together.

This effect can be observed by moving your
hand through water. With the palm of the hand in
the direction of motion, there is more resistance
than if you have the tips of the fingers in the direction of motion. Automobiles are now "streamlined"
in shape to reduce air resistance and improve fuel
consumption.

Air resistance is also velocity- or speed-depen-
dent. That is, the greater the speed of a falling object, the greater the air resistance. This is because
the faster an object falls, the greater the number of
air molecules it hits or collides with per second.
This gives a greater total resistive impulse force, or
air resistance. In many cases, the air resistance is
proportional to the square of the speed.

So as an object falls, it gains speed or accelerates
because of gravity, and the retarding force of air
resistance increases with speed (Fig. 1.15). This
continues until the force of air resistance equals the
weight force of the falling object. The net force is
then zero (downward weight force balanced by the
upward force of air resistance), and the object no
longer accelerates but falls with a constant speed.
We then say that the object has reached *terminal
velocity.* For a sky diver with an unopened para-

pends on its shape. For example, a sky diver with an
unopened parachute falls quite rapidly. But when
the chute opens, the shape and size of the falling
"body" change, so the air resistance is increased
and the descent is slowed, which makes skydiving a
pleasure, or at least repeatable.

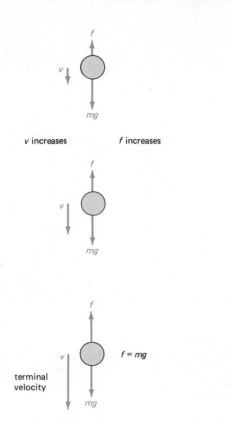

v increases   *f* increases

Figure 1.15
Air resistance is velocity-dependent. As the velocity of a falling object increases, the force of air resistance *f* increases. At a certain velocity, called the terminal velocity, the force of air resistance is equal to the weight of the object. There is then no longer a net force or acceleration, and the object falls with a constant velocity.

Figure 1.16
Air resistance in action. Sky divers assume a spread-eagle position to increase the air resistance and prolong the time of fall.

Ex 21.

* *Q. Physics 11th Ed. Young, Friedman 140, 141*

chute, the terminal velocity is about 200 km/h ($\approx$ 125 mi/h). Sky divers are aware of the shape-dependence of air resistance. In falling they use a spread-eagle position to increase the air resistance and prolong the time of fall (Fig. 1.16). When the parachute is opened, the fall is slowed by the additional resistive force to a terminal velocity of about 40 km/h ($\approx$ 25 mi/h). See Special Feature 1.4 for the effect of air resistance on falling objects of different weights.

NEWTON'S THIRD LAW OF MOTION: ACTION AND REACTION

Although we commonly talk of single forces, Newton recognized that it is impossible to have just an individual force. Rather, there is a mutual interaction, and forces always occur in pairs. An example given by Newton was that if you press on a stone with your finger, the finger is also pressed on by the stone. That is, if one object exerts a force on a second object, then the second object exerts a force on the first. This is like saying that you can't touch something without being touched.

Newton termed these forces *action* and *reaction,* and **Newton's third law** is commonly expressed as follows:

> For every action, there is an equal and opposite reaction.

Or, alternatively,

> For every force, there is an equal and opposite force.

In symbol form,

$$F_{\text{action}} = -F_{\text{reaction}}$$

where the negative sign indicates the opposite direction. Which force is the action or reaction is arbitrary and depends on how you look at the situation —it's a relative interaction.

The third law may seem contradictory to the second law. (If you have equal and opposite forces, how can there be an acceleration?) However, the second law is concerned with force(s) acting *on a particular body* and its resulting acceleration. In applying the second law, we look at only the forces acting on a given body. The force pair of the third law acts on *different bodies* (Fig. 1.17).

Terminal Velocity and Falling Objects

The effect of air resistance on falling objects is to retard their motion. The magnitude of the air resistance on an object depends on the object's shape, size, and speed. Ordinarily, heavy objects of similar shape and size dropped short distances appear to accelerate at the same rate and strike the ground at the same time. This is the essence of Galileo's alleged Leaning Tower of Pisa experiment (see Special Feature 1.3).

However, suppose the objects were dropped from a very high altitude so that speed and air resistance were appreciable factors. For example, suppose two metal balls made of different materials, but of the same size and shape, were dropped from a high-flying aircraft. Let one of the balls be much more massive (heavier) than the other. Which would strike the ground first?

Here, air resistance is a factor, and the heavier ball would hit the ground first. As the falling balls gain speed and the air resistance increases, the weight of the lighter ball would be the first to be balanced by the force of air resistance. The heavier ball would continue to accelerate downward until it reached its faster terminal velocity. So the heavier ball would be ahead of the lighter ball and would be falling faster, thereby reaching the ground first.

Air resistance and terminal velocity. Because of air resistance, a light (less massive) ball will reach its slower terminal velocity before a heavier ball of similar shape and size. As a result, the heavier ball will strike the ground first.

Figure 1.17
Newton's third law. An example of Newton's third law. For every force there is an equal and opposite force. The forces act on *different bodies.*

Let's take a look at some examples of the third law action-reaction force pair. When you are holding something quite heavy, you supply an upward force (action) *on the object*. After a short time, you may become painfully aware of the reaction force the heavy object exerts *on you* [Fig. 1.18(a), particularly if the cooler is full].

In Figure 1.18(a), there are two sets of force pairs. There is the upward action force *on the cooler handle* by the person, and the downward reaction force *on the person's hand* by the cooler. Also, there is a downward action force *on the cooler* due to gravity (its weight). Although not obvious, there is an upward reaction force *on the Earth*. Notice that there are equal and opposite forces acting *on the* held cooler, so there is no net force acting on it and it is stationary (Newton's first law).

But, suppose the cooler is dropped [Fig. 1.18(b)]. Now there is a net force on it, and it falls or accelerates downward (Newton's second law). The third law action-reaction force pair between the cooler and the Earth is still there (as it always is since we can't turn off gravity). The Earth is so massive (6×10^{24} kg), however, that its reaction motion (acceleration) is negligible.

(Note: Here's a good example of why mass is the fundamental property, as discussed in the previous section. It doesn't make much sense to talk about the weight of the Earth, does it?)

Let's look at a less massive example that involves acceleration by applying the third law to the firing of a rifle (Fig. 1.19). When the charge explodes, the bullet is accelerated down the barrel. It is acted on by a force (an action), as evidenced by its acceleration. The reaction force acts on the rifle. This force accelerates it in the opposite direction and causes the backward recoil or "kick" that is commonly experienced with a large-caliber rifle or shotgun.

By Newton's third law, the magnitudes of the action-reaction forces are equal, or $F_b = F_r$ (*b* for bullet and *r* for rifle). Then,

$$A_b = \frac{F_b}{m_b} \qquad F_b = A_b m_b$$

and

$$a_r = \frac{F_r}{M_r} \qquad F_r = a_r M_r$$

where the symbol size is meant to emphasize relative magnitudes. Hence, the acceleration of the rifle a_r is much smaller than the acceleration of the bullet

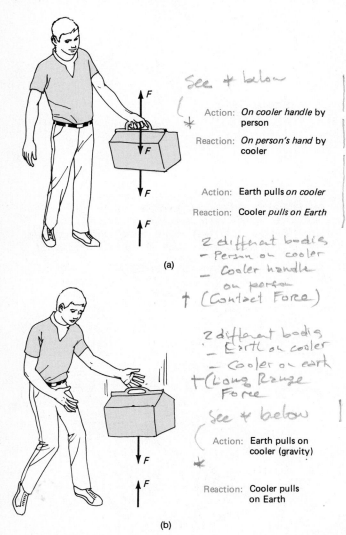

Action: *On cooler handle* by person

Reaction: *On person's hand* by cooler

Action: Earth pulls *on cooler*

Reaction: Cooler *pulls on Earth*

*[handwritten: See * below]*

[handwritten: 2 different bodies — Person on cooler — Cooler handle on person (Contact Force)]

(a)

Action: Earth pulls on cooler (gravity)

Reaction: Cooler pulls on Earth

*[handwritten: 2 different bodies — Earth on cooler — Cooler on earth (Long Range Force) See * below]*

(b)

Figure 1.18
Newton's third law force pairs. (a) There are two force pairs associated with the cooler when it is held stationary. Notice that the net force on the cooler is zero. (b) When the cooler is dropped and falling, there is still a force pair, but the net force on the cooler is not zero, since it accelerates toward the Earth.

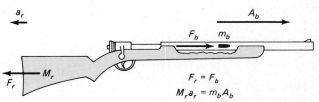

$$F_r = F_b$$
$$M_r a_r = m_b A_b$$

Figure 1.19
Newton's third law force pair for an isolated rifle and bullet. There are equal and opposite forces on the rifle and bullet. However, because the bullet has much less mass, its acceleration is much greater than that of the rifle.

*[handwritten at bottom: * Physics Fishbane et al 104]*
[handwritten: + " 10thd Young & Friedman 107]

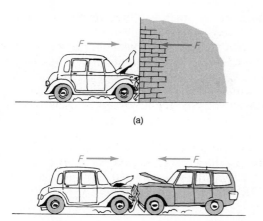

(a)

(b)

Figure 1.20
Undesirable action-reaction pairs. The reaction force of the wall on the car (a) is easily recognized when a substitution is made (b). Notice the effect is the same.

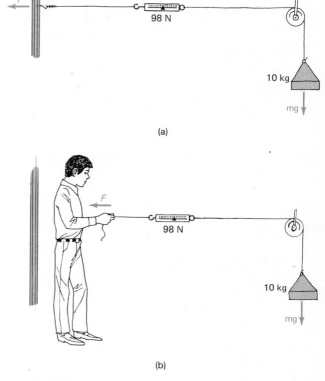

(a)

(b)

Figure 1.21
Newton's third law one more time. (a) The weight of the mass exerts a force of 98 N on the wall via the rope, and the wall "pulls" back with an equal and opposite force. (b) This can be easily shown by replacing the wall with yourself or another person.

A_b (as one would hope), owing to the difference in masses.

Sometimes a not-so-obvious reaction force can be better understood by substituting another force in its place. For example, the reaction force of a wall on a car produces the same effect as the force applied by another car (Fig. 1.20). As another example, consider a 10-kg mass suspended by a rope fastened to a wall, as shown in Figure 1.21. A spring scale is used to measure the force. This would be the weight of the mass, $w = mg = (10 \text{ kg})(9.8 \text{ m/s}^2) = 98 \text{ N}$. The force is transmitted through the rope and the scale to the wall.

To illustrate that the wall "pulls" on the rope with a reaction force of 98 N, suppose you unfastened the rope and held the mass stationary (like the wall). You would have to pull with a force of 98 N, so the wall must be doing the same.

Hence, we see by Newton's third law that forces occur in pairs acting on different things. A clue to

this is that an object, or a person, cannot exert a force on nothing. There has to be a reaction force, and there is a simple way to describe it. If you describe a given force as exerted *on* object X *by* object Y, then if you switch the two prepositions *on* and *by* and reverse the direction of the force, you have now described the reaction to that force.

SUMMARY OF KEY TERMS

Motion: a changing of position.
Speed: time rate of change of position, or how rapidly the distance is covered: Speed = change in distance/change in time.
Velocity: the rate of change of position *and* the direction of the motion. Velocity is a vector quantity.

Acceleration: time rate of change of velocity: Acceleration = change in velocity/time it takes for change.
Force: a quantity capable of producing motion or a change in motion.
Unbalanced or net force: the equivalent or resultant force F of two or more forces.

Newton's first law of motion (law of inertia): an object remains at rest or in uniform motion with a constant velocity unless acted upon by a net unbalanced force.

Inertia: the property of matter that resists changes in motion. Mass is a measure of inertia.

Newton's second law of motion: $F = ma$. Relates force to acceleration.

Friction: the force that opposes the relative motion of contacting media.

Acceleration due to gravity: the acceleration of an object falling under the influence of gravity. (Near the Earth's surface $g = 9.8 \text{ m/s}^2 = 32 \text{ ft/s}^2$.)

Weight: the gravitational force acting on a body: $w = mg$.

Free fall: an object falling with only gravity acting on it.

Nonfree fall: an object falling under the influence of gravity with another force (e.g., air resistance) acting on it.

Terminal velocity: the maximum constant velocity that a nonfreely falling object reaches when a retarding force balances the gravitational force and the object's acceleration is zero.

Newton's third law of motion: for every force (action) there is an equal and opposite force (reaction).

[For more on the description of motion, see An Extended View (Optional) in the Appendix.]

QUESTIONS

Force and Motion

1. We often refer to a moving van or a moving company. Is the adjective *moving* descriptive of what each does? Why?

2. In checkers or chess, a player's turn is called his or her move. Why is this?

3. Is it possible to be in the same place and to have moved? Explain.

•4. When a police officer using radar gives someone a speeding ticket, is the officer concerned with the driver's average speed or the approximate instantaneous speed?

5. Do speed limit signs refer to average or instantaneous speeds?

6. A car travels along a straight road with a constant speed of 75 km/h. What is its acceleration?

•7. A motorist in a moving car applies the brakes. Does the car accelerate? Explain.

8. Discuss what could be called "accelerators" and "decelerators" on a car. Could a steering wheel be a "decelerator"? How about "gearing down"?

9. A race car travels around a racetrack with a constant speed. Is the car accelerating? Explain.

10. An air-bubble level is pushed along a table surface. Where is the bubble or in which direction does it move (a) when the level is moved with a constant velocity? (b) when the applied force is removed and the level comes to a stop?

11. When observing an object, how can you tell if there is a force acting on it? Can you always "see" what is applying the force? Explain.

•12. A child sitting in a stationary car holds a helium balloon by a string. Explain what happens to the balloon when the car starts to move forward.

13. If an object at rest is acted on by a force and the object doesn't move, what can be concluded?

14. On packaged items, such as breakfast cereals, the net weight is listed. What does this mean?

Newton's First Law

•15. Is it possible to have motion without a force? Explain.

16. How did Galileo's and Aristotle's ideas of motion differ?

17. Can you actually isolate or find the inertia of a body? Explain.

18. An old parlor trick involves suddenly pulling a tablecloth from beneath a setting of plates and glasses. Rather than falling to the floor and breaking, the plates and glasses remain on the table (Fig. 1.22, which also shows a modern version that is a bit hard on the tableware. In this case, the table is mechanically pulled from under the plates and glasses.) Explain the "magic" of this trick.

•19. When a paper towel is torn from a roll on a rack, a jerking motion tears the towel better than a slow pull. Why is this? Does this method work better when the roll is large or when it is small and near the end of the roll? Explain.

20. A passenger in a car struck from behind may experience a whiplash, which results from the upper vertebrae of the spine being bent backward. What causes this?

(a)

(b)

Figure 1.22
An old trick. (a) A tablecloth is pulled from underneath a table setting. (b) Modern mechanical version—the whole table is pulled from beneath a table setting, which seems to be suspended in air (for the moment). See Question 18.

21. Common safety devices in automobiles are seat belts and shoulder straps. Explain why these devices are needed in terms of Newton's first law.

*22. A person riding in a bus cannot find a seat, so he stands in the aisle. Explain what happens to the person when (a) the bus starts up from a stop, (b) the moving bus accelerates, (c) the bus travels at a constant velocity, (d) the bus slows down, and (e) the bus turns a corner.

Newton's Second Law
(23–24 Proportionality)

*23. The amount of money ($) one receives when taking returnable bottles to a store is proportional to the number (n) of bottles returned. (a) Write a symbol relationship for this proportionality. (b) Does the relationship tell you how much money you would actually get for a certain number of bottles? (c) If you received 5 cents for each bottle returned, write the equation relationship. (d) Does this equation allow you to calculate the amount of money you would receive for a certain number of bottles, say $n = 10$?

24. Student study time is inversely proportional to extracurricular activities. Write and explain the symbol relationship for this proportionality.

*25. By Newton's second law of motion, a net force of 1 N acting on a mass of 1 kg produces an acceleration of 1 m/s². What would be the acceleration if (a) the force is doubled (same mass)? (b) if the mass is halved (same force, 1 N)? (c) if both the force and mass are doubled (2 N and 2 kg)?

26. Suppose you are standing on an icy surface (very little friction) and a heavy object and a light object of the same size and shape are also on the ice. How could you tell which object is heavier without picking them up?

*27. An astronaut in a spaceship and in a "weightless" condition wants to know which of two identical containers used to store equipment is empty. How can the astronaut tell which container is empty without opening them?

28. Newton's first law of motion may be derived from his second law of motion. Explain this statement.

TIGER

Figure 1.23
Whose side is Suzy on? See Question 37.

29. In a tug-of-war, two teams pull on the rope with equal and opposite forces. Analyze the situation (forces) in terms of (a) Newton's second law and (b) Newton's third law.

•30. A freely falling body falls with a constant acceleration. Is the distance traveled by the body constant for the same time interval, say a second? Explain.

•31. Two sky divers jump at the same time from an airplane, after having agreed to open their parachutes at a particular altitude. One diver falls with his arms and legs pulled in (fetal position) and the other in a spread-eagle position. Which diver would open his parachute first?

32. Explain what effect an updraft of air would have on the terminal velocity of a falling object.

33. Snowflakes fall more slowly than sleet. Why?

•34. Is it possible for a heavy object and a lighter object to have the same terminal velocity? Explain.

35. A sky diver in a spread-eagle position reaches her terminal velocity. She then decides to fall feetfirst. What happens?

36. Suppose you live in the future, when moon vacations become common. Someone asks you to join a parachute club he is going to form for jumping on the moon. Would you sign up?

Newton's Third Law

37. What suggestion would you give Suzy in Figure 1.23? Give a detailed suggestion and explanation as Isaac Newton might have given her.

38. Identify the action and reaction forces of Newton's third law for each of the following cases:

Figure 1.24
And away we go. . . . See Question 38(c).

a. A person pushing on a compressed spring.

b. A swimmer changing direction in starting another lap at the end of the pool.

c. A person jumping onto a bank from an untied canoe (Fig. 1.24). Why does the canoe move backward or opposite the direction the person jumps? What might happen to the person and why?

*39. When a person pushes on a wall, the wall pushes on the person à la Newton's third law. Suppose the person put a block of wood between his or her hand and the wall. Analyze the forces acting on the block of wood. Why doesn't it move?

*40. (a) A car sits stationary on a level road. Identify the action-reaction force pair(s). (b) Suppose the car starts to move down the road. What force pairs are acting now? What force causes it to accelerate?

41. What causes a rotary lawn sprinkler to rotate? What force causes a helicopter to begin to rise?

42. Two people pull with equal forces of 100 N on the ends of ropes that have the other ends tied to

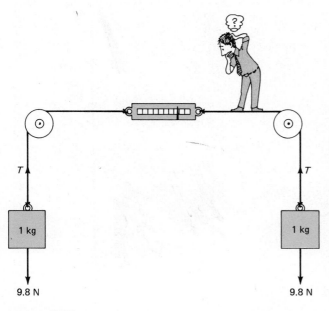

Figure 1.25
A scale in between will read. . . . See Question 43.

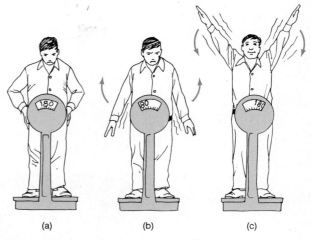

Figure 1.26
A quick way to gain and lose weight. See Question 45.

each end of a spring scale. (a) What does the scale read? (b) If one person tied his end of the rope around a post, what would the scale read?

*43. Two masses are attached to a spring scale, as shown in Figure 1.25. (a) If both masses were 1 kg, what force in newtons would the scale read? (Hint: See Fig. 1.21 and neglect the fellow on the cord and the mass and friction of the pulley and scale.) (b) We say that the force is transmitted undiminished by the rope. What would happen if this were not the case, that is, if the tension were different in different parts of the rope?

44. In approaching (falling toward) the moon, astronauts in a spaceship slow their spacecraft down by firing retrorockets in order to go into orbit and not crash into the moon. How does firing a rocket slow down the spacecraft?

*45. (a) A person places a bathroom scale (not the digital type) in the center of the floor and stands on the scale with his arms at his sides (Fig. 1.26). Keeping his arms *rigid* and quickly raising them over his head, he notices the scale reading increases as he brings his arms upward. Why? (b) Then, with his arms over his head, he brings them quickly to his side. How does the scale reading change and why? (Try this experiment yourself.)

Work and Energy

2

When work is done,
energy is transferred.

WHAT IS ENERGY?

Closely associated with force and motion are work and energy. People have a general idea about doing work, and we hear a great deal about *energy*. How many times have you read this word in a newspaper or magazine, heard it on radio or TV, or even discussed it yourself? Sometimes we speak of an "energy crisis," and we look for ways to save energy. But, what *is* energy?

If someone asked you to define energy, what would you say? If your instructor asked you to go out and bring back a bucket of energy, with what would you return?

Although energy may be difficult to explain, it is familiar to all of us. The energy needed for life processes is supplied by food. We say we are "full of energy" when ready (and willing) to do something. Fuels supply the energy needed to run factories, to bake pizzas, to provide transportation, and to do many other things.

From experience we might expect energy to be in some way related to force — something that does things or gets things done or moving. Like force, energy is a *concept*. It is defined in terms of what it does or is capable of doing. To explain this critical concept, we begin with a related concept — work. Are you ready for a little brain "work"?

WORK

Work is done when you pick up your physics book. The heavier the object you pick up, the more work you do. Holding a heavy object like a cement block is commonly considered work because it is physically tiring. However, there is no work done on the block in this case, in the scientific sense. What, then, is this seemingly unfair scientific definition of work?

Obviously there must be a force involved, but there is another ingredient — motion. The motion is expressed in terms of the *distance* traveled in the direction parallel to the force, and we define work as follows:

> The work done by a force acting on an object is defined as the product of the force and the distance through which the object moves parallel to the force.

That is,

$$\text{Work} = \text{force} \times \text{distance}$$
(moved in the direction parallel to the force)

or $\quad W = Fd$

So we see why no work is done when holding a heavy object stationary or pushing against a wall or some other immobile object [Fig. 2.1(a)]. There is no motion in any of these cases ($d = 0$).

The simplest case of work is that done by a constant force with the motion in a straight line parallel to the direction of the force [Fig. 2.1(b)]. However, when a constant force that does work is not parallel to the direction of motion, only a part or *component* of the force does work [Fig. 2.1(c)]. The component of the force parallel ($F_{\parallel}$) to the direction of motion or parallel to the direction of d does work. The vertical force component on the lawn mower in the figure does no work because the mower does not move in that direction.

How about the units of work? From the general equation ($W = Fd$), we can see that they are force times distance. In the SI system this is newton-meter (N·m) and in the British system, the pound-foot, which is customarily reversed to foot-pound (ft·lb). The SI unit of work, the N·m, is given the special name of joule (J) (pronounced "jool") in honor of the English scientist James Prescott Joule, and $1\,\text{J} = 1\,\text{N·m}$.

$$(Fd = W)$$

(SI)	newton-meter = joule (J)
(British)	pound-foot = ft-lb

Work is a *scalar* quantity and has no direction.

As will be learned shortly, work and energy are intimately related, and energy has the same units as work (Fig. 2.2).

POWER

Suppose your instructor asks you to move a stack of boxes up to the second floor. You might do this *work* in 10 minutes. Then the instructor asks another student to move the boxes up to the third floor. Not being too enthusiastic, your classmate takes 20 minutes to move the boxes.

The same amount of work is done in each instance, but there is a difference — time, or the time

Figure 2.1
Work is force times distance. (a) Work requires motion. When there is no motion (or no distance traveled, $d = 0$), no work is done. (b) The simplest case of work is that done by a constant force with the motion in a straight line parallel to the direction of the force. (c) When a force is not parallel to the direction of motion, only a part or component of the force in the direction of motion ($F_\parallel$) does work.

Figure 2.2
Joule, the SI unit of work and energy. A reminder to turn off the lights when not needed and save a joule.

rate of doing the work. To distinguish between such cases, we talk about power. **Power is the time rate of doing work**, or simply the work divided by the time it takes to do the work.

$$\text{Power} = \frac{\text{work}}{\text{time}}$$

or

$$P = \frac{W}{t}$$

Over an appreciable time interval, this is the average power, as in the case of average speed. In the example of transporting the boxes, since you did the same work in half the time, you produced or developed on the average twice as much power as your classmate.

The units of power can be seen from the defining equation. In the SI system the units are joule/second(J/s). The J/s is given a special name, the watt (W), in honor of James Watt, a Scottish engineer. In the British system, the units of power are foot-pound/second (ft-lb/s).

$$(W/t = P)$$

(SI)   joule/second = watt (W)
(British)   foot-pound/second = ft-lb/s

You may have never heard of a ft-lb/s. A more common unit used to express power in the British system is the horsepower (hp), and

1 hp = 550 ft-lb/s = 746 W

Oddly enough, the horsepower unit was originated by James Watt, after whom the SI unit is named. Watt had invented an improved steam engine. In the 18th century, horses did most of the work in the hauling of coal and pumping of water from mines in Great Britain. To characterize his new steam engine, Watt used the average rate at which a horse did work as a standard unit, which he cleverly called the horsepower. With this unit, customers for his steam engine could quickly compare power capabilities.

The horsepower unit is still the common unit used to rate motors and engines. Suppose a motor had a 2-hp output. Then this motor could do twice as much work in the same time as a motor with a 1-hp output, or could do the same amount of work in half the time (Fig. 2.3).

ENERGY

So what is the difference between work and energy? Work is not energy itself, but a transfer of energy. Thus, if a body has energy it can do work through a transfer of energy. Energy appears in many forms. We will focus on two basic forms of **mechanical energy**—potential energy and kinetic energy.

Potential Energy

One might ask, when work is done on an object, where does the work "go"? For example, you do work in compressing a spring or lifting an object (Fig. 2.4). It might be thought that the work goes

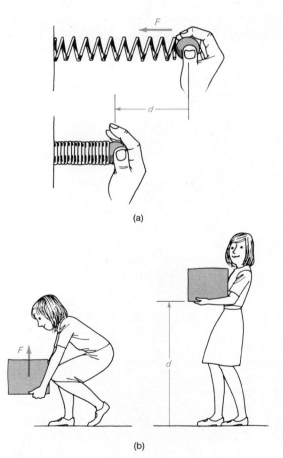

(a)

(b)

Figure 2.4
Potential energy, the energy of position. Work is done in compressing a spring (a) or in lifting an object (b), causing a change in potential energy, which depends on position.

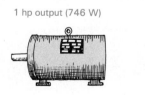

1 hp output (746 W)

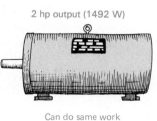

2 hp output (1492 W)

Can do 746 J of work each second (746 W).

Can do same work in half the time or twice as much work in the same time.

Figure 2.3
Power is work per time. The power ratings of motors tell how much work they can do per time.

into the spring or the object. However, work is not something a body *has,* but is something *done on* (or by) a body. After being "worked on," the spring and the object do not have work, but do possess the

(a)

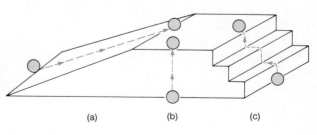

(b)

Figure 2.5
Potential energy at work. Examples of applications of (a) spring potential energy and (b) gravitational potential energy.

ability to do work, or energy. Hence, **a body has energy when it is capable of doing work.** Energy was transferred to the spring and the object in the process of doing work.

In the case of a compressed spring or a raised object, this energy is called potential energy. It was brought about by a change in position. Hence, we say

Potential energy is the energy of position.

That is, an object has more energy at one position relative to another position. Practical examples of spring potential energy and gravitational potential energy are shown in Figure 2.5. The potential "position" energy of a spring is due to a change in its length or a change from its equilibrium position (where it is neither compressed nor stretched).

Let's take a closer look at gravitational potential energy, since it is so common. The potential energy of a raised object is equal to the work done against gravity in lifting it. The upward force required is equal to (initially, slightly greater than) the weight of the object (mg). The distance through which the object is moved or raised is the height h (Fig. 2.4). The work (Fd) is then equal to mgh, and

$$\text{Gravitational potential energy} = \text{weight} \times \text{height}$$
$$PE = mgh$$

The units of energy are the same as those of work, for example, joules of energy, and energy is also a scalar quantity.

The height h is measured relative to its original position. The potential energy is "independent of path" or how the object gets to a height (Fig. 2.6). It depends only on the vertical height between the initial and final positions.

The initial position, usually taken as h = 0, is arbitrary. That is, you can put the zero reference

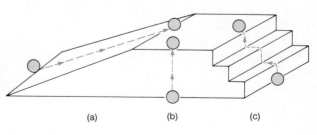

(a)   (b)   (c)

Figure 2.6
Potential energy is independent of path. All the balls on the platform have the same potential energy relative to the floor, regardless of the path used to get there.

$h = 0, PE = 0$

h

$mg(-h) = -PE$

Figure 2.7
Negative potential energy. Relative to the zero reference point, the potential energy of an object may be negative. This gives rise to the concept of a potential energy "well."

position wherever you like (Fig. 2.7). This may give rise to negative potential energy and the concept of a potential energy "well." Like a real well, energy or work is required to get higher in the energy well. On Earth, we are near the bottom of a potential energy well. Work or energy is required to put satellites into orbit at a higher energy level in the well.

Kinetic Energy

How about when a force is applied to an object and it is moved without friction on a horizontal surface? Work is done, but there is no change in potential

energy because the height doesn't change. Where does the work "go" in this case? From Chapter 1 and Newton's second law, we know that when a (net) force acts on an object it is accelerated. Without work being done, there would be no motion (or change in motion). The work must go into setting the object into motion or changing its velocity. Energy then is associated with the motion of an object, and this is called kinetic energy. We say

Kinetic energy is the energy of motion.

Hence, when an object is in motion (in any direction), it has the ability to do work. For example, a moving car may run into another one and expend its kinetic energy in doing the work of a "fender bender."

The kinetic energy of a moving object is equal to one half the product of its mass and the square of its velocity:

Kinetic energy (KE) $= \frac{1}{2}mv^2$

Notice that kinetic energy is proportional to the square of the velocity (v^2). Thus, if the velocity of an object is doubled, its kinetic energy is quadrupled or increased by a factor of four ($2^2 = 4$).

If the work done on a stationary or moving object goes only into changing its motion, then the *change* in kinetic energy is equal to the work done on the object. Similarly, because of its kinetic energy, a moving object can do work (Fd). If the object is brought to rest in doing the work, then we may write

Work $=$ kinetic energy

$Fd = \frac{1}{2}mv^2$

where F is the "stopping" force and d is the stopping distance.

This equation shows that speed or velocity is an important factor in stopping a car. For example, a car traveling 90 km/h has four times as much kinetic energy as when it is traveling 45 km/h (v^2 dependence). Then, assuming the stopping (braking) force F to be constant, it would take *four* times the distance to stop when traveling at 90 km/h than when traveling at 45 km/h (Fig. 2.8).

Say it took you 50 m to stop when going 45 km/h. It would take 200 m to stop when going 90 km/h. Add your reaction time in applying the brakes, which would increase the stopping distance, and you can see why we are required to drive 35 km/h (20 mi/h) in school zones.

Figure 2.8
Kinetic energy, the energy of motion. The change in kinetic energy depends on the square of the velocity, KE = ½mv². If the velocity is doubled, the kinetic energy increases by a factor of four. This is an important factor in the stopping or braking distance of a car.

Forms of Energy

We say that the potential energy and kinetic energy discussed thus far are forms of "mechanical" energy. This is because their uses involve the mechanical activities of work with forces and motions. In fact, if you think about it, you might come to the conclusion that *energy is the result of work being done.*

In general, all forms of energy are the result of work being done by forces or against forces. This work may be taking place right now with the "release" of energy or may have taken place sometime in the past with the "storing" of energy. On the bottom line, the forces involved are the fundamental forces of nature.

We have already seen an example of this with the gravitational force. Work is done against gravity in lifting an object, and the result is gravitational potential energy. When an object falls, the gravitational force does work, and kinetic energy results. Let's take a quick look at some other forms of energy and the forces involved. Specific forms and forces will be discussed in greater detail in later chapters.

Electrical energy is a big one. The basis of electrical energy lies in the fundamental force interactions between electrically charged atomic and subatomic particles. Electrical energy is associated with the motion of electric charges, or electric currents.

It is this form of electrical energy that operates common electrical appliances (which do work for us).

> *Question:* What do we "buy" from an electric power company?
>
> *Answer:* If you have ever paid an electric bill, you know we pay for electricity in units of kilowatt-hours (kWh). These are multiple units of power (watts) times time (seconds). Since power = work/time (P = W/t), we have work = power × time (W = Pt).
>
> Hence, the kilowatt-hour is a unit of work or energy. So we pay for or buy electrical energy (to do work) from an electric "power" company. Perhaps we should really call it an energy company.

Electrical forces also hold together the atoms and molecules that make up matter. We might think of atoms as being held together by electrical "springs," and hence there is potential energy. The elastic energy of an actual spring can be traced to the work done against the intermolecular electrical forces.

During a chemical reaction, such as the burning of fuel, electrons and atoms are rearranged (work is done), and energy is released. The result of the combination is another chemical compound in which the particles have less energy. We refer to this form of energy, the energy released, as chemical energy.

Heat is also commonly considered a form of energy.* Heat energy is associated with the random motional (kinetic) energy of the atoms and molecules in matter. When heat (energy) makes a body hotter, its molecules move faster on the average. In a solid, the molecules vibrate faster, whereas in a gas the molecules are free to move and collide with other molecules.

Light is a form of electromagnetic energy. We sometimes call this radiant energy. Something "radiates" as the result of the acceleration of electrically charged particles. Other forms of electromagnetic energy include radio and television waves. Electromagnetic energy can travel through a vacuum. This is how energy gets to us from the Sun, our

---

* More correctly, heat is "energy in motion," or the energy transferred from one body to another because of a temperature difference. See Chapter 11.

major source of energy. The Sun's energy enters into the life cycles of plants and animals and is the basis of our chemical energy in fossil fuels (coal, oil, and gas). From the oceans it also evaporates water, which through winds, clouds, and rain ends up inland behind dams with gravitational potential energy for hydroelectric energy.

Nuclear energy is the source of the Sun's and other stars' energy. This involves the fundamental nuclear forces (strong and weak). These forces do work on nuclear particles. The result is a rearrangement of nuclear particles to form different nuclei with the release of energy.

In energy-releasing nuclear processes, part of the masses of the nuclei is transformed into energy. Hence, we now consider mass to be a form of energy. Einstein showed that the amount of energy released in a mass transformation is given by his famous equation $E = mc^2$, where E is the amount of energy released, m is the part of the nuclear mass that is transformed, and c is the speed of light (a constant). This transformation can work the other way, too—energy can be transformed into mass. More about this later in Chapter 26.

Mass-energy transformation is the principle of nuclear weapons and of nuclear reactors, which supply an appreciable part of the energy used to generate electricity. Nuclear mass energy (or fossil fuel chemical energy) is transformed into heat, which is used to produce steam that turns a turbine. The mechanical energy of the turbine is transformed by a generator into electrical energy, which can again be transformed into heat, light, or the mechanical energy (by motors) that we use to do work. Indeed, energy transformations are all around (and inside) us.

Conservation of Energy

Whenever energy is transformed from one form to another, or transferred from one object to another, we find that no energy is lost in the process. This is expressed in one of the most important conservation laws—the law of **conservation of energy**.

> Energy cannot be created or destroyed. It may be transformed from one form to another, or transferred from one object to another, but the total amount of energy is constant or conserved.

Another way of saying this is that the total energy of the universe is conserved. It is always there some-

where in some form. Of course, the universe is the largest system we can think of, and its energy content might be expected to be constant. Let's apply the conservation of energy to some smaller, more practical systems.

To simplify the understanding of the conservation of energy, we often use *ideal* systems in which the energy is in only two forms—mechanical potential and kinetic energies. Then we talk about the conservation of mechanical energy.

Consider the diver in Figure 2.9. On the platform, the diver's total mechanical energy is all po-

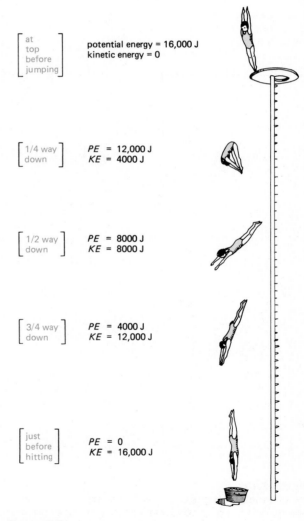

| at top before jumping | potential energy = 16,000 J
kinetic energy = 0 |

| 1/4 way down | PE = 12,000 J
KE = 4000 J |

| 1/2 way down | PE = 8000 J
KE = 8000 J |

| 3/4 way down | PE = 4000 J
KE = 12,000 J |

| just before hitting | PE = 0
KE = 16,000 J |

Figure 2.9
Conservation of energy. The total mechanical energy, or the sum of the kinetic and potential energies, is constant at any point during the dive.

tential energy. During the dive, potential energy is converted to kinetic energy, but at any point the total mechanical energy (KE + PE) is constant. Just before he hits the bucket, the diver's total mechanical energy is all kinetic energy. Of course, when he hits the bucket, it's another story. The total mechanical energy is no longer conserved, since the bucket is not part of our *ideal* system.

Another common example used to illustrate the conservation of mechanical energy is an ideal pendulum (no air friction or friction in the support). As illustrated in Figure 2.10, there is a continual conversion between kinetic and potential energies. At the maximum heights of the swing (on each side), the pendulum bob stops instaneously ($v = 0$), and

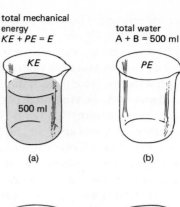

total mechanical energy
$KE + PE = E$

total water
A + B = 500 ml

KE

500 ml

(a)

PE

(b)

KE

200 ml

(a)

PE

300 ml

(b)

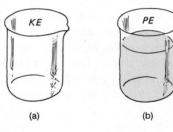

KE

(a)

PE

(b)

Figure 2.11
Water analogy of energy conservation. The total amount of water remains constant in a "conservative" beaker system, just as the total mechanical energy of a conservative system remains constant. However, if some of the water is spilled, the system is "nonconservative," just as energy is lost in a nonconservative system.

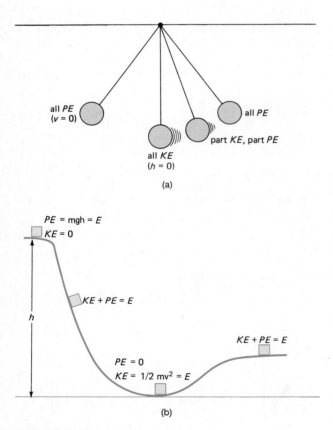

all *PE*
($v = 0$)

part *KE*, part *PE*

all *KE*
($h = 0$)

(a)

$PE = mgh = E$
$KE = 0$

$KE + PE = E$

$KE + PE = E$

h

$PE = 0$
$KE = 1/2\ mv^2 = E$

(b)

Figure 2.10
Other examples of the conservation of energy. (a) The energy of a swinging pendulum exchanges back and forth between kinetic and potential energies (friction neglected). (b) The total mechanical energy (KE + PE) of a block sliding on a frictionless surface is constant as it varies in height and speed.

the total energy is all potential. At the bottom of the swing ($h = 0$), the total energy is all kinetic. In between, the kinetic energy and potential energy always add up to equal the constant total energy, so the total mechanical energy is conserved.

The conservation of mechanical energy for a block sliding on a frictionless curved surface is also illustrated in Figure 2.10.

We call such ideal systems in which the total mechanical energy is conserved conservative systems. But what about in the real world? On Earth there is always some energy lost as a result of friction or some other cause. In this case, we have non-conservative systems.

A water analogy of a conservative system and a nonconservative system is shown in Figure 2.11. In the conservative system, the total amount of "mechanical" water in the kinetic-energy and potential-energy beakers is conserved. However, if some of the water is spilled in pouring it back and forth, the beaker system is then nonconservative. The spilled water is analogous to the pendulum energy lost as a result of friction. But even for a nonconservative system, the *total energy* is conserved. All the energy (like the water) is around somewhere in some form.

We say that the lost or wasted energy is no longer available to do *useful work*. In a conservative system work can be done and stored as energy, and the same amount of work can be gotten back. In a real, nonconservative system some energy is always lost. A demonstration of the conservation of energy (and one's belief in it) is to have a person hold a suspended bowling ball to the tip of his nose and then release it (Fig. 2.12).

The person stands perfectly still and lets the ball swing back toward him. Many are reluctant to try this. However, by the conservation of energy, the bowling ball will swing back no farther (or to no greater height) than its point of release. Since the system is slightly nonconservative, the ball will not quite reach the person's nose. Of course, he should be advised simply to release the ball and be careful not to give it a push. Why?

■ MACHINES AND EFFICIENCY

Machines

How much energy is lost is an important consideration in machine "systems" that are used to do useful work. If you ask someone what a machine is, a typical reply might be that a machine is something that does work. This answer is prompted by the experience of using machines to perform tasks, some of which require more strength than our muscles alone can supply. A person cannot lift an automobile, but with a jack (a machine) this is easily accomplished.

However, machines *do not* actually save us work. That is, we do not get something for nothing. Nature doesn't operate that way—no free lunches. Indeed, in practical applications we always get less work out of a machine than we put in.

In the mechanical sense, **a machine is any device used to change the magnitude or direction of a force.** Let's see how this makes doing work "easier." Consider a simple machine called a lever (Fig. 2.13). Neglecting any frictional losses, we have, by

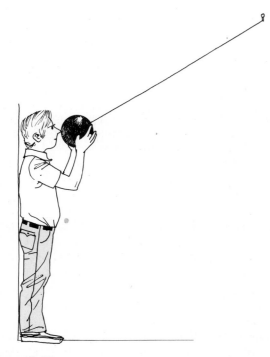

Figure 2.12
Trusting the conservation law. Will the ball hit him in the nose on the backswing?

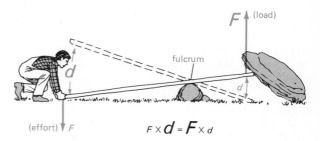

$$F \times d = F \times d$$

Figure 2.13
The lever. A simple machine that can multiply force (greater force output than input) but at the expense of distance.

Figure 2.14
Mechanical advantage. With the mechanical advantage provided by a lever, a person can lift a heavy car.

the conservation of energy in terms of work, that the work output will equal the work input, or

Work in = work out

or, since $W = Fd$,

$(Fd)_{in} = (Fd)_{out}$

But notice in the figure that by applying a small force through a large distance, a large force is exerted through a small distance. Thus, the lever increases or multiples the force, but at the expense of distance.

The force multiplication factor is called the **mechanical advantage** of a machine. As you can probably guess from Figure 2.13 or know from experience, the mechanical advantage of a simple lever depends on the location of the pivot point or fulcrum. This controls the lengths of the lever "arms" of the load (output force) and the effort (input force). By movement of the fulcrum closer and closer to the load, the mechanical advantage becomes greater and greater. This is probably the fact that caused the Greek scientist Archimedes to exclaim, "Give me a fulcrum on which to rest and I will move the Earth."

You use the lever principle when you pull a nail with a claw hammer, use a bottle opener, or jack up a

car, as shown in Figure 2.14. (Note the difference in the lengths of the lever arms.) Suppose you exerted a 200-N force (about 45 lb) to jack up a car weighing 10,000 N (more than a ton). Then your force multiplication factor or mechanical advantage would be 50.

However, the output distance (the distance the car moves) compared with the input distance (the distance you move the jack handle) would be reduced by a factor of 1/50. If you moved the jack handle through 50 cm, the car would be raised only 1 cm.

Another type of simple machine is the pulley (Fig. 2.15). A single fixed pulley doesn't give you any force multiplication (mechanical advantage of one). It simply gives a change in the direction of the force. However, if you use a system of pulleys, called a block and tackle, there is a mechanical advantage. For example, if the suspended movable pulley has three supporting ropes, to make the load rise 50 cm you would have to pull the rope down 150 cm (to shorten each support strand by 50 cm).

A moment's thought should tell you that the mechanical advantage of the system is three; that is, ideally you could lift 600 N by pulling down on the rope with a force of slightly more than 200 N.

(a)   (b)

Figure 2.15
The pulley. (a) A single fixed pulley is a "direction changer" and not a force multiplier. (b) In a system with a movable pulley, or a block and tackle, the mechanical advantage is equal to the number of strands or ropes supporting the movable pulley.

Efficiency

In the previous ideal examples, friction was neglected, but machines do have friction. By the conservation of total energy we should really write:

Work in = work out + work "lost"

where the work "lost" is due to friction. The work output is the "useful" work done by the machine.

To express how much useful work we get out for the work we put in, we talk about the **efficiency** of a machine:

$$\text{Efficiency} = \frac{\text{work out}}{\text{work (or energy) in}}$$

Literally, efficiency is "what you get out (useful work) for what you put in (work or energy)." Efficiency can also be expressed as the ratio of power out to power in. Why?

The efficiency ratio is a decimal fraction, but it is usually expressed as a percentage. For example, if 300 J is the useful work output of a machine for a 500-J input, the efficiency is $^{300}\!/_{500}$ = 0.60, or 60 per-

Table 2.1
Typical Percent Efficiencies of Some Complex Machines

Automobile	12–15
Electric motor	Up to 95
Steam engine	50–75
Steam locomotive	5–10
Turbine	Up to 40

cent. Sixty percent of the work input goes into useful work, which means that 40 percent, or 200 J, of work input is lost.

Some typical efficiencies of complex energy-work conversion machines are given in Table 2.1. Obviously, it pays to have highly efficient machines so as to save work and energy. We'll look at efficiency again in the next chapter, where energy conservation is a major consideration in terms of energy resources and dollars and cents. For now, take a look at Special Feature 2.1 to see where the energy losses are in the not-so-efficient automobile.

SPECIAL FEATURE 2.1

Energy and the Automobile*

Automobiles powered by gasoline engines are known to be very inefficient machines. Even under ideal conditions, less than 15 percent of the available energy in the fuel is used to power the vehicle. This situation is much worse in stop-and-go driving in the city.

Many mechanisms contribute to the energy losses in a typical automobile. About two thirds of the energy available from the fuel is lost in the engine. Part of this energy ends up in the atmosphere via the exhaust system, and part is used in the engine's cooling system. About 10 percent of the available energy is lost in the automobile's drive-train mechanism. This loss includes friction in the transmission, drive shaft, wheel and axle bearings, and differential. Friction in other moving parts accounts for about 6 percent of the energy loss. Approximately 4 percent of the available energy is used to operate fuel and oil pumps and such accessories as power steering, air conditioning, power brakes, and electrical components. Finally, about 14 percent of the available energy is used to propel the automobile

("useful" work). This energy is used mainly to overcome road friction and air resistance. These energy losses are summarized in Table 2.2.

Table 2.2
Energy Losses in a Typical Automobile

Mechanism	Power loss (%)
Exhaust (heat)	33
Cooling system	33
Drive train	10
Internal friction	6
Accessories	4
Propulsion of vehicle ("useful" work)	14

* From Serway, R. A., *Physics for Scientists and Engineers,* Second Edition, Saunders College Publishing, Philadelphia, 1986.

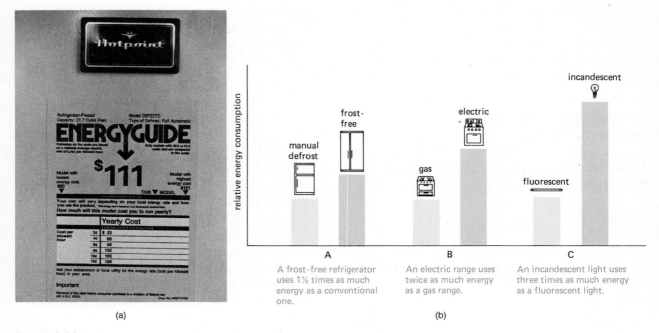

(a)   (b)

Figure 2.16
Energy comparisons. (a) Federal law requires appliances to have an Energy Guide label that compares energy costs. (b) A comparison of the energy consumption of some types of appliances and lamps.

General concern for energy efficiency and energy conservation is quite evident. (Here the conservation of energy is used in the context of reducing consumption and *costs*.) We pay a great deal for energy and energy-consuming machines and devices, so people are interested in what they get for their money. This is a "cost efficiency" rather than mechanical efficiency, but it illustrates the general idea of efficiency—what you get out for what you put in. Some examples of energy consumption or cost efficiency are shown in Figure 2.16.

As a final thought for this chapter, what would be the efficiency of an *ideal* machine, such as a motor or engine, with the work output equal to the work (or energy) input? The answer is 1.0 or 100 percent. But this is not possible, because in practice there is always some work or energy lost.

If it were possible, we could build a **perpetual motion machine**. By feeding the machine's output back into its input, it would run forever!

Even better would be a machine with greater than 100 percent efficiency. In this case energy would be created—more work out than put in. Not only could it run perpetually, but we could remove the extra energy and never have to worry about an energy shortage. Sound too good to be true? It is.

SUMMARY OF KEY TERMS

Work: the product of the force acting on an object and the parallel distance through which the object moves. $W = Fd$

Energy: the ability or capability to do work.

Joule (J): the SI unit of work and energy. $1 \text{ J} = 1 \text{ N-m}$

Power: the time rate of doing work. $P = W/t$

Watt (W): the SI unit of power.

Horsepower (hp): the British system unit of power. $1 \text{ hp} = 550 \text{ ft-lb/s} = 746 \text{ W}$

Potential energy: the energy of position. Gravitational potential energy, $PE = mgh$

Kinetic energy: the energy of motion. $KE = \frac{1}{2}mv^2$

Conservation of energy: energy cannot be created or destroyed. It can be transformed from one form to another, or transferred from one object to another, but the total amount of energy is constant or conserved.

Conservation of mechanical energy: the sum of the kinetic energy and potential energy (KE + PE), the total mechanical energy, is constant or conserved.

Conservative system: a system in which the total mechanical energy is conserved.

Machine (mechanical): any device used to change the magnitude or direction of a force.

Mechanical advantage: the force multiplication factor of a machine.

Efficiency: the ratio of work out/work (or energy) in for a machine that expresses the fraction (or percentage) of useful work done by a machine

Perpetual motion machine: an ideal machine with 100 percent efficiency (or greater) that could run forever after an initial work input.

[For more on the conservation of energy in nonconservative systems, see An Extended View (Optional) in the Appendix.]

QUESTIONS

What Is Energy?
1. Can we detect energy with our senses? That is, can we feel, smell, or taste energy?
2. Explain the concept of energy or why it is a concept.

Work
3. A weight lifter holds a set of weights over his head. Is he doing work? Has he done work? Explain.

4. A tall weight lifter and a shorter weight lifter both lift 1200 N (270 lb), as illustrated in Figure 2.17. (a) Who does more work? (b) How could they both do the same work?

*5. Is it possible to do *negative* work? (*Hint:* Although work is a scalar quantity, force and distance are vectors with directions.) What does this mean in terms of an object in motion on which the work is done?

Figure 2.17
Same weight lifted, same work? See Question 4.

6. You and another student are late to class and race to your classroom on the second floor by different routes. (a) Who does more work against gravity? (b) If you beat your classmate there, who had the greater power output?

7. How much work is required to draw a bucket of water from a well?

*8. What does work on a shuffleboard puck as it slides to rest? Why is the board dusted?

Power

9. Could motors with ¼-hp and ¾-hp outputs be used to do the same amount of work? Explain. Does either motor have an advantage over the other? If so, what?

*10. Some factory workers do piecework, i.e., they are paid according to the number of pieces or items they process or produce. Other workers are paid by the hour. Is there a power consideration involved in these two methods of remuneration? Explain.

11. An 8-hp riding tractor that is used to mow lawns cannot generally be used to plow a garden. Explain why not.

*12. A constant applied force F does work in moving an initially stationary object through a distance. If a constant force of $2F$ moves the object through the same distance, (a) are the power outputs the same? (b) If not, does the $2F$ force have twice the power output? [*Hint:* Write the powers, e.g., $P_1 = Fd/t_1$, and compare.]

Energy

13. Why are water towers very tall structures and often placed on high elevations?

14. Explain how a slingshot and an archery bow can have potential energy. What happens to the potential energy when a stone or an arrow is shot?

15. Suppose you are at the bottom of a deep hole in the ground. How much initial kinetic energy would you have to give a stone you threw for it to reach the top of the hole? (The stone's initial velocity would be its "escape" velocity.)

*16. A ball lies on the first floor of a house. A person on the first floor says the ball has zero potential energy. A person on the second floor says the ball has -20 J of potential energy. A person in the basement says it has $+20$ J of potential energy. Can they all be right? Explain.

17. A person sits in a moving car. Does the person have kinetic energy? Explain.

18. A moving object has work done on it by a force parallel to its motion. How does this affect its kinetic energy if the force is (a) in the direction of the motion and (b) opposite the direction of motion?

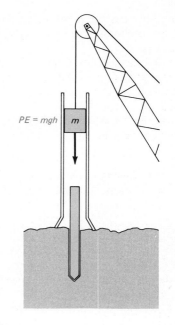

Figure 2.18
A pile driver uses energy and does work. See Question 22.

19. Explain the work and energy considerations in lifting an object and letting it (free) fall. What happens to the energy when the object hits the ground?

20. Why is it easier to drive a large nail with a carpenter's hammer than with a small tack hammer?

*21. A person on a trampoline can go higher with each bounce. Explain the energy considerations of a trampoline and how this is possible. Is there a maximum height the person can go? Explain.

22. (a) A pile driver is used to drive a pile into the ground (Fig. 2.18). Explain the principle of the pile driver in terms of energy and work. (b) If the "stopping" force of the pile were the same with each strike, how would the distance the pile is driven into the ground be affected with each strike? Explain.

23. How much work is required to bring a moving object to rest?

*24. An automobile traveling with a velocity v accelerates until its velocity is tripled. How does this affect its kinetic energy?

25. A state patrol officer investigating the head-on collision of two automobiles of the same make and model notes that the length of the skid marks caused by braking one car before collision is four times that of the skid marks of the other. A witness testifies that both drivers locked brakes at the same time before collision. What can be inferred from this evidence?

Forms of Energy

26. What happens to the energy supplied to electrical devices?
•27. What happens when energy is released in a chemical reaction?
28. Sound is considered to be a form of energy. Explain.
29. Name four or five "forms" of energy.

30. The law of conservation of mass says that matter cannot be destroyed. Is this true in light of nuclear reactions?
•31. Discuss the conversion of the forms of energy for the situations shown in Figure 2.19. What is the ultimate source of energy for the environmentally related example?

(a)

(b)

(c)

(d)

Figure 2.19
Different forms of energy. See Question 31.

Figure 2.20
Loop-the-loop. See Question 39.

Conservation of Energy

32. When a moving car comes to a stop on a level road, what happens to its kinetic energy?

33. A car traveling with a constant velocity on a level road coasts up a hill and comes to a stop. What happens to its kinetic energy?

34. A baseball player slides into home plate. What happens to his kinetic energy?

35. A ball is dropped from a height where it has 50 J of potential energy. How much kinetic energy does the ball have just before it strikes the ground?

•36. A pendulum is taken to the moon, and its bob is released from the same height above the ground as it was on Earth . Would there be any difference(s)? [Consider work, speed, and maximum heights of swing.]

37. On a cold day you may rub your hands together to warm them. How does this work?

38. A basketball player running down the floor stumbles and gets a "floor burn" on his knee. Could the floor burn be worse in one case than another?

•39. Many amusement-park roller coasters now have loops (Fig. 2.20). (a) Does the roller coaster have potential energy at the top of the loop? (b) Is the initial starting height of the roller coaster critical? Explain.

Machines and Efficiency

40. Explain how the mechanical advantage of a lever is determined by the relative lengths of the lever arms. What would you guess the mechanical advantage to be if the lever arms were equal?

•41. Neglecting friction, the mechanical advantage of a lever is given by ratio of the lengths of the input lever arm (L_{in}) and the output lever arm (L_{out}). Is this ratio L_{in}/L_{out} or L_{out}/L_{in}? Justify your answer.

42. In an acrobatic act, two people jump onto a seesaw to propel another acrobat, as shown in Figure 2.21. (a) Why do two people jump on the seesaw? (b) Are the weights of the acrobats a consideration? Explain. (c) Suppose the board was moved so it was shorter on the single acrobat's side of the support. What effect would this have?

43. Sketch a block and tackle that has a mechanical advantage of five.

•44. An inclined plane, such as a loading ramp, is a simple machine. Explain how an inclined plane has a mechanical advantage. Does the angle of incline affect the mechanical advantage?

45. Explain how efficiency is related to the conservation of energy.

46. In Figure 2.3, the power outputs of the motors were given. (a) Does this tell you anything about the efficiencies of the motors? (b) Suppose the motors in Figure 2.3 were very poor and each had an efficiency of 50 percent. What would be the power inputs of the motors?

*47. Which is more efficient, machine A that has a work output of 500 J for a work input of 700 J, or machine B that does 900 J of useful work with an input of 1400 J?

48. To make a single cup of coffee, some people fill a kettle almost completely before heating the water. Discuss the energy efficiency of this practice.

One acrobat is propelled into the air

Two people jump onto seesaw from platform

Figure 2.21
Up and away! See Question 42.

Momentum

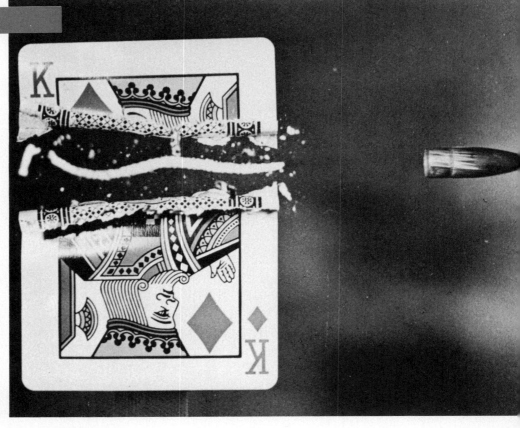

A small object with a
great deal of momentum.

3

IMPULSE AND MOMENTUM

Having considered force and work, and energy and motion, let's take a look at force and motion from another point of view. According to Newton's second law, $F = ma$, an object or mass accelerates when a net force acts on it. Such a force is often applied through a contact collision of objects, such as hitting a ball with a bat or "shooting" a pool ball with a cue stick (Fig. 3.1). The resulting motion or change in motion of an object depends not only on the applied force (and the mass of the object), but also on the *time* of contact or application of the force. We are all aware of this from pushing a stalled car. You have to push for some time to get it rolling well.

Assuming the object starts from rest, the quantities are related by

$$F \Delta t = mv$$

The left side of this equation is called **impulse**,

$$\text{Impulse} = F \Delta t$$

and the right side is called **momentum**,

$$\text{Momentum} = mv$$

The applied force produces an acceleration or a *change* in velocity, so in general we should write,

$$F \Delta t = m \Delta v$$

Hence, **impulse is equal to the change in momentum.***

The force involved in an impulse is not usually a constant force but varies with time. For example, when a bat hits a ball, the force on the ball increases rapidly from an initial zero value as the ball is deformed (Fig. 3.2). The force then decreases as the ball recoils, and the ball returns to its original shape on leaving the bat. (This is an example of elasticity. All solid materials are elastic to some degree, even steel. Recall that we make springs from coiled steel.)

Since the impulse is equal to the change in momentum, we don't need to know the force to find the impulse. By determining the initial and final velocities (and multiplying each by the mass), the change in momentum, and hence the impulse, can

---

* Δv is the change or difference in velocity, $\Delta v = v - v_o$, and $m \Delta v = m(v - v_o) = mv - mv_o$ is the change in momentum. With $v_o = 0$, then $F \Delta t = mv$.

be determined. Momentum is a vector pointing in the same direction as the velocity.

Newton called momentum a "quantity of motion." It takes into account not only the inertia of an object but also its motion. We sometimes say that a

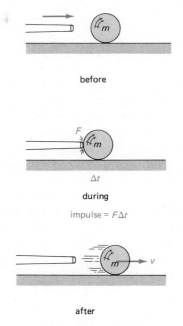

before

during

impulse = $F \Delta t$

after

momentum = mv

Figure 3.1
An example of impulse. The change in motion or momentum of an object depends not only on the applied force, but also on the contact time (Δt) of the applied force.

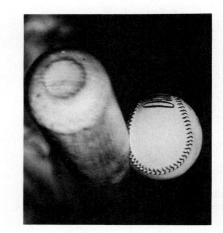

Figure 3.2
Contact! Objects usually deform in contact collisions.

football player running down the field has a lot of momentum because he is difficult to stop. This is usually attributed to the football player's large mass, although his velocity also contributes to the momentum. However, a bullet (small mass) with a large velocity can have an appreciable momentum (Fig. 3.3) and is also difficult to stop. A stationary object has zero momentum since it has zero velocity.

In most instances, a change in momentum involves a change in velocity, since the mass is constant. That is, an object either slows down or speeds up and/or its direction of motion changes. This is easily observed or measured and gives us a "handle" on the impulse, or the product of force and time. Let's see how this works in some common applications and manipulations of the impulse quantities—force and time.

Suppose someone throws you a hard ball with a velocity v and you catch it with your arms rigidly extended. As we all know, the ball "stings" your hands. The change in momentum is just mv, since when the ball stops $v = 0$ and its momentum is zero. (Change in momentum $= mv - 0 = mv$.) Then,

$$F\Delta t = mv$$

where the small Δt indicates that you stopped the

ball in a short time. This makes the impulse force large, which stings your hands (Fig. 3.4).

After a few stinging catches, you get smart and learn to move your hands backward while catching the ball; that is, you learn to manipulate the impulse (Fig. 3.4). Assuming the ball is thrown with the

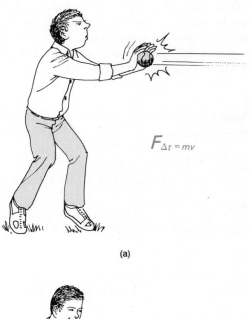

$$F_{\Delta t = mv}$$

(a)

$$F\Delta t = mv$$

(b)

Figure 3.4
Impulse in action. In stopping (catching) a ball, the change in momentum is mv, which is equal to the impulse $F \Delta t$. (a) If the contact time is small, F is large and the ball "stings" the hands. (b) If the contact time is increased by moving the hands along with the ball, the force is smaller and there is little or no sting.

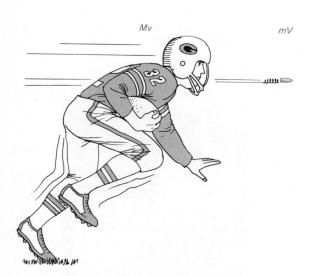

$$Mv \qquad mV$$

Figure 3.3
Momentum is mass times velocity. It is common to think of massive, moving objects to have a lot of momentum, e.g., a football player (Mv). However, a small (mass) object moving at a high velocity, such as a bullet, can also have appreciable momentum (mV).

The Automobile Air Bag

Air bags are scheduled to be installed in new automobiles as a safety feature; however, this has been delayed for some time for economic and manufacturing reasons. When they *are* installed, air bags in cars will prevent many injuries in accidents, particularly for people in the front seat who now do not "buckle up" for safety with seat belts and shoulder straps. The principle involves the automatic inflation of an air bag on the hard impact of an automobile so as to prevent the driver (and passenger) from hitting the steering wheel, the dashboard, and/or the windshield in accordance with Newton's first law (figure below).

In terms of impulse, the air bag increases the stopping contact time, thereby reducing the impact force and preventing injury.

Automobile air bag. The air bag increases the collision contact time in stopping a person, thereby reducing the impulse force and the possible injury.

same velocity, the change in momentum is the same for each catch. However, when you move your hands in the direction of the motion of the ball, the contact time is increased, which lessens the impulse force and the sting. That is, in symbol form,

$$F \Delta t = mv$$

Another example of manipulating the contact time to control the impulse force is in jumping from a high place onto a solid surface or floor. If you were to land stiff-legged, you would stop suddenly (small Δt), and the large impulse force might hurt your knees, legs, or spine. You quickly learn to bend your knees when landing so as to increase the contact time and reduce the force.

Impulse plays an important role in automobile safety. For example, cars were once built without padded dashboards, but now padded "dashes" are the rule. If you were not wearing your seat belt or shoulder strap and the car came to a sudden stop, you would continue in motion according to Newton's first law. (The frictional force on the seat of your pants would not be enough to stop you.) In this case, you would probably hit and be stopped by the dashboard. The padding of the dash would increase your contact time and reduce the impact force, possibly preventing injury. See Special Feature 3.1 for another application of the same principle.

Impulse and momentum also are involved in sports. For example, in baseball or golf when it is desired to hit the ball a long distance, the player "follows through" with the swing so as to increase the contact time. With a fixed force or maximum swing, the ball receives a larger impulse and a larger change in momentum or a greater velocity.

Other examples of contact forces occur in boxing and karate. A KO punch in boxing is characterized by a long follow-through. The force is distributed over the relatively large area of the boxing glove. Rather than doing physical damage, the idea is to accelerate the head backward with a large acceleration so that one's "senses" are left behind, so to speak. A karate blow, on the other hand (or foot), focuses the force to a small area so as to do damage. In the popular karate demonstration of breaking wooden boards or concrete blocks, impulse plays a factor. There is a relatively short contact time that gives a large force to cause the breaking. However, there are several other important considerations for this demonstration, which will be discussed in more detail in a later chapter (Chapter 8). You may be able to guess one already from the physics you've learned thus far in terms of Newton's third law. Would a karate expert do this demonstration if the reaction force broke his hand?

■ CONSERVATION OF MOMENTUM

Recall that Newton's first law tells us that a body remains at rest or in motion with a constant velocity unless it is acted on by an unbalanced force. Then, if the velocity is constant (even zero), the momentum of a body is also constant (if its mass is also constant). Hence we have the condition necessary for the **conservation of momentum** of a body:

> In the absence of an unbalanced force, the momentum of a body is conserved.

By conserved, we mean unchanged or constant with time. That is, each time we measured the momentum we would obtain the same value. This is similar to the conservation of energy discussed in Chapter 2. Because nature is so difficult to analyze, scientists look for conserved quantities and treasure them. When something is conserved or constant, it makes the job and life a bit easier for scientists.

The beauty of the conservation of momentum is that it also applies to more than one body, in fact, to any number of bodies. We call a collection of objects a system. In the case of more than one body, we speak of the total momentum of the system. Since velocity is a vector, so is momentum (commonly represented by the symbol p, i.e., $p = mv$). As such, the total momentum (P) is the vector sum (the net or unbalanced momentum) of the momentum vectors of the individual bodies. Like the individual

forces that produce net or unbalanced forces (which we saw in Chapter 1), the individual momentum vectors may add and/or cancel, depending on their directions.

So we may extend the conservation of momentum to a system of bodies.

> In the absence of an unbalanced, *external* force, the total momentum of a system is conserved.

Notice that we distinguish between *external* or outside forces that act on the system and *internal* forces that act within a system.

An external force or impulse is required to change the momentum of a body or a system. Internal forces, for example, the internal molecular forces of a billiard ball, have no effect on the ball's momentum. Or, suppose you are a passenger in a car moving with a constant velocity and you push on the floorboard with your foot (as some passengers do to apply the "brakes"). There is no change in the car's velocity or momentum. This is because these forces are internal ones. By Newton's third law, they are equal and opposite, and the net force or impulse within the system is zero, so there is no change in momentum.

Hence, the bodies of the system may move around and bump into each other, but nothing that happens within a system can change the total mo-

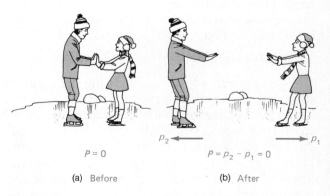

$p_2 \longleftarrow$ $\longrightarrow p_1$

$P = 0$ $P = p_2 - p_1 = 0$

(a) Before   (b) After

Figure 3.5
Conservation of total momentum. (a) The total momentum P of the skaters' system is initially zero. (b) With no external forces ("pushing off" involves internal forces), the total momentum is conserved and remains zero.

mentum of the system. The total momentum of the system is conserved — in the absence of an unbalanced, *external* force.

Let's take a look at some examples of this "before and after" conservation effect. Consider two skaters standing on a frictionless, icy surface (Fig. 3.5). Initially, the total momentum of the system (the two skaters) is zero, since they are not moving. Now, suppose they "push off" each other (internal forces). They move apart, but the total momentum is still zero — the momentum vectors cancel. Because the skaters have different masses, their speeds are different. However, the product of *mv* is the same for each — but in opposite directions, which makes the net momentum equal to zero. (How would their speeds compare if they were both the same mass?)

> *Question:* In reality, there was some friction between the skaters' skates and the ice in the previous example. Is momentum conserved?
>
> *Answer:* No. Friction would be an external force, so the momentum would not be conserved. Each skater would gradually slow down and lose momentum.

In Chapter 1 we used Newton's third law to explain the recoil or "kick" of a rifle (see Fig. 1.19). We can also look at this in terms of momentum. Let's think big this time and consider a cannon (Fig. 3.6). When the powder charge explodes (an internal force), the cannon ball accelerates down the barrel. The total instantaneous momentum of the system is still zero. This is easier to see after the cannon ball leaves the barrel. We call the velocity with which a projectile leaves the barrel of a rifle or cannon its muzzle velocity. The speed of the cannon ball is much greater than that of the cannon, but the total momentum is still zero. Why?

A similar situation occurs in jet propulsion. When fuel is combusted in a rocket or jet engine, the exhaust gas acts on the rocket and the rocket acts on the exhaust gas (Newton's third law action-reaction). As a result, exhaust gas is accelerated out the back of the rocket, and the rocket or aircraft is accelerated in the opposite direction by the reactive force.

Another way to look at this is in terms of momentum (Fig. 3.7). Assuming the rocket blasted off

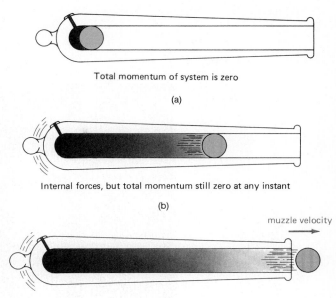

Total momentum of system is zero

(a)

Internal forces, but total momentum still zero at any instant

(b)

muzzle velocity

(c)

Figure 3.6
Conservation of momentum. The total momentum of the system is zero in each idealized case. (The cannon's "recoil") force is part of a Newton's third law force pair. Why?)

from rest, the total momentum of the molecules of the exhaust gas (assuming they don't strike anything) *at any instant* is equal and opposite to the momentum of the rocket. This is a more difficult case to analyze completely because one body, the rocket, is continually losing mass; and the other body, the expelled gas, is continually gaining mass. Your instructor may wish to explain this case more fully. A simple example is given in Special Feature 3.3 at the end of the chapter.

Another application of momentum is given in Special Feature 3.2.

> *Question:* When a rocket blasts off, what causes it to lift off the launching pad? Is it the exhaust gases "pushing against" the pad?
>
> *Answer:* The launching pad is just there as a launching place. It is actually the reactive force of the gases pushing on the rocket that causes the rocket to accelerate upward. If this were not the case, there would be no space travel, since there is nothing to "push against" in space.

(a)   (b)   (c)

Figure 3.7
Jet propulsion and conservation of momentum. If originally zero, the total momentum is
zero at any instant (neglecting gravity and the exhaust gases striking anything). The
magnitude of the momentum of the exhaust gas molecules $(\Sigma m)V$ is equal to the magnitude
of the momentum of the rocket, Mv. (The symbol Σ means "sum of" and Σm is the sum of
the masses of the exhaust molecules or the total mass of the exhaust gases.)

■ COLLISIONS

Things are always bumping into each other or
colliding—sometimes purposefully, like billiard
balls, and sometimes not so purposefully, like automobiles. We ordinarily think of objects coming into
contact during a collision with contact forces involved. However, this is not always the case. Action-
at-a-distance forces can be involved. For example, a
meteor may "collide" with the Earth (gravity acting) and miss it completely.

Because there is always an interaction between
colliding objects, momentum is involved. Therefore, in a broad sense, a **collision** is any interaction
in which momentum and energy are exchanged or

transferred. Since collision forces are internal
forces of the system, we can analyze collisions in
terms of the conservation of momentum, and if
there are no external forces,

Total momentum before collision
= total momentum after collision

Let's take a look at some simple, head-on, contact
collisions to see how momentum helps us understand what happens. Usually, we know or can measure the masses and velocities of the colliding objects before collision, so the conservation of
momentum allows us to predict the motions of the
objects after collision. We distinguish between
elastic and inelastic collisions.

Reverse Thrust

After touching down when landing a jet plane, the pilot "revs" up the engines in order to apply braking action. This is called applying reverse thrust, and one method is to use clamshell doors that deflect the exhaust gases from the jet engines in the forward direction (see figure below). With the doors in the in-flight open position (a), the exhaust gases go out the back of the engine, and the plane receives momentum in the forward direction, or a forward thrust. With the clamshell doors closed (b), the exhaust gases are deflected toward the forward direction, and there is a (vector) component of momentum in that direction. By the conservation of momentum, the engine and the airplane receive a momentum in the reverse direction, which gives an impulse or reverse thrust in that direction, thus helping to slow and stop the plane. (The airplane brakes would never do it alone.)

The deflector doors may also be exterior to the engine and form part of the engine casing. Smaller commercial jet aircraft use reverse thrust to "back away" from loading docks. Larger planes are pushed away with tow motors. (Why?) The next time you are at a large airport, watch to see if you can see the deflector doors when a plane backs out.

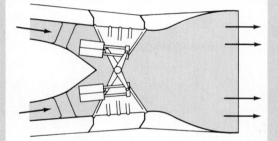

clamshell doors in forward thrust position

(a)

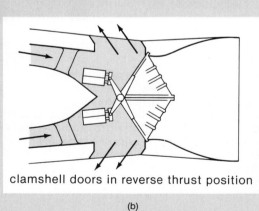

clamshell doors in reverse thrust position

(b)

(c)

Thrust. (a) With the exhaust gases going out the back of the engine, there is a forward thrust (momentum) on the engine. (b) When the clamshell doors are closed and the gases are deflected forward, there is a reverse thrust that slows a jet plane (a similar effect as if the engines could be turned around). (c) Actual jet engine clamshell doors.

Elastic Collisions

In elastic collisions, as in all collisions in which there are no external forces, the momentum is conserved. The colliding objects have a certain total kinetic energy before collision, but while the objects are in contact, some of this energy may go into deforming the objects. However, if the objects are perfectly *elastic*, they rebound without deformation or have their exact original shapes. That is, all of the elastic potential energy of the deformation is returned to the kinetic energies of the objects as they

Figure 3.8
Collisions and conservation of momentum. For collisions of balls of equal mass, there is simply an exchange or complete transfer of momentum—each ball after collision has the same momentum as the other ball before collision. The vector arrows represent the momenta of the balls.

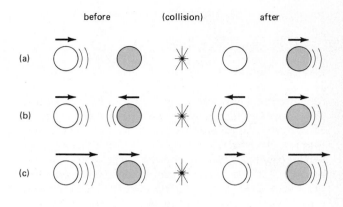

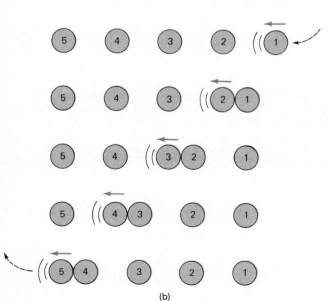

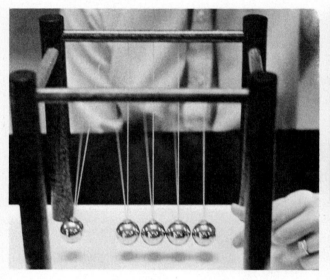

(a)

Figure 3.9
A case of momentum transfer in elastic collisions. Momentum is transferred down the line of balls of equal mass; when one ball swings in, one swings out at the opposite end of the line (a). When two balls swing in, two swing out; and so on. The line drawing (b) illustrates the collision transfer process for the case of a single ball swinging in. When two or more balls swing in, the first colliding ball stops, with momentum transferred down the line. This ball is then hit by the second ball, and so on.

separate, such that the total kinetic energy after collision is the same as the total kinetic energy before—i.e., conservation of kinetic energy. This means that no energy is lost as heat or sound. Hence,

> In an elastic collision, both momentum and kinetic energy are conserved.

In general, elastic collisions are ideal cases, but many hard objects, such as billiard balls, steel balls, bowling balls, and marbles, have nearly elastic collisions. So elastic collisions are useful approximations in many instances. In actuality, we find that only atoms and subatomic particles can have truly elastic collisions.

Consider the elastic collisions of balls of equal mass, as shown in Figure 3.8. During a collision, only internal forces are involved, so the momentum is conserved. There is an exchange or transfer of momentum during the collision, but the total momentum of the system before collision equals the total momentum after collision (as does the kinetic energy).

This is the principle of the novelty item shown in Figure 3.9. When one ball swings in, one swings out; when two balls swing in, two balls swing out; and so on. You can now understand why this happens.

During the collisions, momentum is conserved in the horizontal direction. Since the balls are of equal mass, there is, ideally, a complete transfer of momentum. When one ball swings in, it stops as it transfers all of its momentum to the first ball in the row, and so on down the line. Because the balls in the row are in contact or almost in contact, we do not "see" the collisions, but they occur just as though the balls were separated and the motions before and after could be observed [see Fig. 3.9(a)]. In the absence of a commercial apparatus, you can study collisions with the simple arrangement shown in Figure 3.10.

So, momentum is passed down the row by the collisions, and the final ball flies out with the same momentum of the ball that swung in (assuming per-

fectly elastic collisions). How about two balls in? Here there is a double shot of momentum passed down the row. The first incoming ball collides, transfers its momentum, and stops. It is then instantaneously hit by the second incoming ball, and another "shot" of momentum is passed down the line. At the other end two balls swing out, the second instantaneously after the first.

This seems simple enough, but after thinking about it for a while, you might ask the folowing question.

Question: If two balls of equal mass m swing in and collide with a velocity v, why doesn't one ball swing out at the other end with twice the velocity? Or, why does the same number of balls always swing out as swing in?

Answer: Certainly two balls in and one ball out with twice the velocity does not violate the conservation of momentum, since the total momentum before equals the total momentum after $(2m)v = m(2v)$.

But, remember that there is another condition for an elastic collision—the kinetic energy must be conserved. Let's assume that one ball did swing out with a velocity of 2v. Checking to see if the kinetic energy before equals the kinetic energy after for this case,

$$\overset{?}{KE} = KE$$
$$\text{(before)} \quad \text{(after)}$$

$$\tfrac{1}{2}(2m)v^2 \overset{?}{=} \tfrac{1}{2}m(2v)^2$$

and
$$mv^2 \neq 2mv^2$$

So, kinetic energy would not be conserved if this happened. Since the collisions are elastic and kinetic energy is conserved, it can't happen this way. Moreover, look what the last inequality tells you. There would be more energy afterward than before, so energy would have to be created, which is a no-no.

Figure 3.10
A simple apparatus for studying collisions. All you need is a grooved ruler and some marbles.

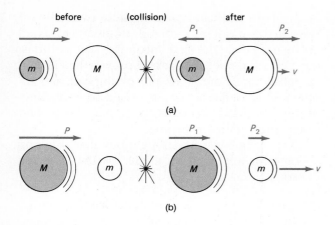

Figure 3.11
Collisions of balls of unequal masses. (a) When a small ball collides with a much more massive one, the small ball rebounds and the massive ball moves off more slowly (smaller v) in the opposite direction with a larger momentum due to its mass. The arrows over the balls represent momenta. (b) A massive ball that collides with a ball of small mass transfers a small amount of momentum, but the small ball has a relatively large velocity and so speeds ahead of the larger, more massive ball.

Returning now to the elastic collision of only two balls, suppose the masses of the balls aren't equal. The momentum is still conserved (as is the kinetic energy). Two extreme cases are illustrated in Figure 3.11. In the first case (a), the incoming ball is much less massive than the stationary second ball. After collision, the incoming ball rebounds in the opposite direction, while the big (massive) ball moves slowly to the right. It has a large

momentum, chiefly because of its mass, which conserves the total momentum, but it has a small velocity.

In the second case (b), the incoming ball is much more massive than the stationary second ball. After collision in this case, the incoming ball continues in the same direction after having transferred a small amount of momentum to the little ball. However, because of its small mass, the little ball has a relatively large velocity and so speeds ahead of the larger, more massive ball.

Not all collisions are head-on collisions. There are glancing collisions where the objects go off at angles to the direction of the initial motion (Fig. 3.12). Here, we look at the conservation of momentum in different directions through the vector components.

Recall from Chapter 1 how a vector can be resolved or "broken up" into x- and y-components. The symbol p is used for momentum, and at any time $p = mv$. Notice that before the collision all of the momentum is in the x-direction. We say that the momentum in the y-direction is zero, since there is no motion in that direction.

After collision, the momentum must be the same as before in both the x- and y-directions. Looking at the components of the momentum vectors after collision, we see that the y-components are equal and opposite. These component vectors cancel each other, and the momentum in the y-direction is still zero. The two components of momentum in the x-direction add, and their sum is equal to the momentum before collision.

Figure 3.12
A glancing collision. The total momentum is conserved before and after collision in the component (x and y) directions. Note that the momentum in the y-direction is zero before and after collision.

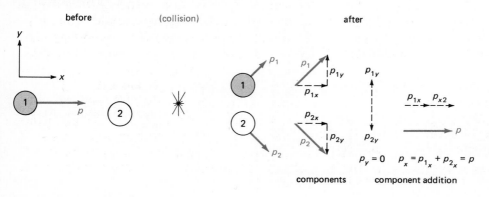

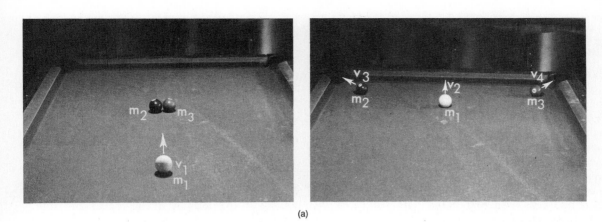

(a)

(b)

Figure 3.13
Three-body collision. The total momentum is conserved and so is the same before and after collision. This means that momentum is also conserved in the component (x and y) directions. Note that there is only momentum in the y-direction before collision as well as after collision (the x-components cancel).

Collisions can involve more than two objects. A collision of three pool balls is shown in Figure 3.13. As the component vector drawings show, the total momentum is conserved. Also, it was a pretty good shot.

Inelastic Collisions

The collisions we observe in everyday life are generally inelastic collisions. In an inelastic collision, there is usually some permanent deformation of the object(s) or heat is generated (energy is lost). This means that the kinetic energy is not conserved. Even so, momentum is conserved, provided there is no net external force. The internal forces of the collision process may be quite complicated, but the total momentum before and after collision remains unchanged even though kinetic energy is lost.

When colliding objects stick together after collision, we say the collision is "completely inelastic." Even then, the total momentum is conserved. Only an amount of kinetic energy consistent with the conservation of total momentum can be lost in any case.

SPECIAL FEATURE 3.3

The Rocket*

A rocket or spaceship consists of two primary parts: (1) the payload—the rocket hull, astronauts, instruments, and so on that we wish to propel somewhere—and (2) the fuel and anything else that gets ejected in propelling the rocket.

To simplify the situation, let's consider a rather primitive type of spacecraft of mass 1001 kg far out in space, where no forces are acting on the craft (Fig. 1). Let m_2 be the mass of the rocketship *at any time* and m_1 be the mass of a 1-kg "fuel pellet."

Suppose an astronaut throws one of the fuel pellets out the back of the rocket with a velocity of 20 m/s. We take the position where the pellet was thrown out as our zero position (see figure). Then, by the conservation of momentum,

$$m_1 v_1 = -m_2 v_2$$

or $(1 \text{ kg})(20 \text{ m/s}) = -(1000 \text{ kg})v_2$

and $v_2 = -0.02 \text{ m/s}$

where the minus sign indicates that the change in the velocity of the rocket is in the opposite direction to that of the fuel pellet, or to the right in the figure.

Hence, the rocket has changed its velocity by a small amount relative to the position in space where the fuel pellet was thrown out, and the rocket has 1 kg less mass.

* Adapted from Highsmith, P. E., and A. S. Howard, *Adventures in Physics*, W. B. Saunders Co., Philadelphia, 1972.

The astronaut then throws out another 1-kg chunk of fuel with the same velocity, and

$$m_1 v_1 = -m_2 v_2$$

or $(1 \text{ kg})(20 \text{ m/s}) = -(999 \text{ kg})v_2$

and $v_2 = -0.02002 \text{ m/s}$

The change in the velocity of the rocket this time is slightly greater because the rocket lost mass in the first throw. Relative to the original zero position, the change in the rocket's velocity is now $-0.02 + (-0.02002) = -0.04002 \text{ m/s}$.

If the astronaut continued to throw out fuel pellets, the rocket's velocity would change slightly each

Figure 1.
Rocket jet propulsion by the particle. An "impulse" rocket propelled by the ejection of "fuel pellets." See text for description.

SUMMARY OF KEY TERMS

Impulse: the product of the force and contact time. Impulse = $F\Delta t$.

Momentum: the product of a body's mass and velocity. Momentum $p = mv$.

Conservation of momentum: the momentum of a body or system remains constant or unchanged. Condition: there is no net unbalanced external force acting on the body or system.

System: a collection of masses or objects.

Total momentum: the vector sum of the individual momenta of all of the objects in a system.

External force: an outside force that acts on a body or system.

Internal force: a force that acts within a body or a

system. By Newton's third law, internal forces occur in equal and opposite pairs.

Collision: any interaction in which momentum and energy are exchanged or transferred.

Elastic collision: a collision in which the colliding objects are not permanently deformed and/or heat is not generated (no energy lost). Both momentum and kinetic energy are conserved in an elastic collision.

Inelastic collision: a collision in which there is usually permanent deformation of one or more of the colliding objects and/or heat is generated (energy lost). Momentum is conserved in an inelastic collision, but kinetic energy is not.

[For more on collisions, see An Extended View (Optional) in the Appendix.]

time and it would continue to lose mass, which would give the rocket a slightly greater velocity change each time. For example, by the time the rocket had a mass of 401 kg, the next-thrown fuel pellet would give the rocket a change in velocity of

$$m_1 v_1 = -m_2 v_2$$

$$(1 \text{ kg})(20 \text{ m/s}) = -(400 \text{ kg})v_2$$

and $$v_2 = -0.05 \text{ m/s}$$

The change in velocity for this throw is two and a half times greater than for the first fuel pellet.

If the astronaut continued to throw out fuel pellets, the rocket would continue to lose mass, and the change in its velocity would increase each time. If the fuel pellets were thrown out often enough, the change in velocity would appear smooth, which the astronauts would interpret as an increasing acceleration.

For a real rocket during a "burn," the exhaust gas molecules are ejected out the rear of the rocket at speeds of several thousand meters per second. The average velocity of the exhaust gases and the amount of material ejected are relatively constant. Therefore, the thrust and the force on the rocket are constant. The increasing acceleration is due to the rocket's decreasing mass.

Of course, the masses of the exhaust gas molecules are much less than that of the 1-kg fuel pellets of our simplified rocketship, but there are a lot of them. To show the effect of high exhaust velocities, suppose the

Figure 2.
In-flight mass reduction. In a multistage rocket, the burned-out stages are jettisoned for an in-flight reduction of mass. The photo shows the interstage section that connects the first and second stages of a *Saturn V* rocket falling away after the separation of the two stages.

astronaut had thrown out the first fuel pellet with a speed of 20,000 m/s. Then,

$$m_1 v_1 = -m_2 v_2$$

$$(1 \text{ kg})(20,000 \text{ m/s}) = -(1000 \text{ kg})v_2$$

and $$v_2 = -200 \text{ m/s}$$

A large velocity change indeed!

To further reduce the in-flight mass of a rocketship, multistage rockets are used. Burned-out stages are jettisoned for an in-flight reduction in mass (Fig. 2).

QUESTIONS

Impulse and Momentum

1. Pole vaulters and gymnasts use padded mats for landings. What is the purpose of these mats?
2. Most new automobiles are equipped with bumpers that collapse under a large impact. What is the purpose of such bumpers?
3. Guardrails along roadsides collapse (bend and crumple) when a car runs into them. Wouldn't it be better to install stronger guardrails so they wouldn't have to be replaced so often?
4. A golfer using a 9-iron to chip onto a green or a wedge to get out of a sand trap often uses a short "chopping" swing. Why is this?
*5. In the science fiction series *Star Trek,* when the warp drive goes out on the Starship *Enterprise* the slower impulse engines are used for propul-

sion. Describe how an impulse engine might work.
6. Prizefighting was once done with bare knuckles. Why are padded boxing gloves now used? (Some people believe that there should be more padding in the gloves or that boxing should be banned.)
7. A boxer quickly learns to move his head backward when he sees he is going to receive a jab to the head. What does this head motion accomplish?
8. Also in boxing, sharp jabs are used as "punishing" blows, whereas a boxer usually follows through with a KO punch. Explain the difference.
9. On September 8, 1974, the daredevil motorcycle rider Evel Knievel tried to jump the Snake River Canyon on a jet-powered motorcycle (Fig. 3.14).

The nose cone of the motorcycle rocket was purposely constructed so that it would crumple on impact. What was the purpose of this? (In case you don't remember, the rocket malfunctioned and Evel crashed to the bottom of the canyon — with only minor injuries.)

*10. When a balloon is blown up and released, it flies around in a zigzag fashion. What causes this?

11. While you are driving along, another car pulls out in front of you so you can't avoid hitting it. Would there be any advantage to trying to sideswipe the car rather than hit it head-on? Explain.

Figure 3.14
Daredevil rocketry.
Momentum and impulse. See Question 9.

Figure 3.15
Going places with a hand rocket. An astronaut on a space walk using a hand-held, self-maneuvering unit, or hand rocket. See Question 17.

Conservation of Momentum

12. When a rifle or gun is fired, there is generally more recoil or "kick" the greater the caliber (more massive the bullet) of the gun. Why is this? Would you want to fire a rifle that was only a few times heavier than the bullet?

•13. A bazooka, or rocket launcher, used as a weapon against tanks is essentially a tube open at both ends that a soldier holds on the shoulder when firing a heavy shell. There is very little recoil on the person firing the rocket launcher. Why is this? Also, it is very important that another person loading the launcher not stand behind it when it is fired. Why is this?

•14. The army uses large-caliber "recoilless" rifles. Explain how a firearm might be made recoilless. (Actually, the rifles are not completely recoilless, but the recoil is greatly reduced.)

15. Explain why it is difficult for a fire fighter to hold a high-pressure hose when trying to put out a fire. What must the fire fighter do in holding the hose stationary?

16. A person standing on the Earth jumps vertically upward. According to the conservation of momentum, what happens to the Earth? (You can move the Earth!) Can we observe this effect on the Earth? Explain.

17. Analyze in terms of momentum the maneuverability of an astronaut using a hand rocket in space (see Fig. 3.15).

•18. An astronaut approaching the moon must fire retrorockets to slow the craft down so it will not crash (fall) into the moon. Which way are the exhaust gases of the retrorockets directed? Explain how firing these rockets slows down the spacecraft.

•19. A rubber ball is thrown horizontally against a wall, and the ball rebounds in the opposite direction with essentially the same speed as it had just before hitting the wall. Is momentum conserved? Explain. [*Hint:* Remember momentum is a vector.]

20. Suppose an astronaut on a space walk is stranded a short distance from the spaceship. How could the astronaut get back to the ship?

•21. A person in a sailboat that is becalmed on a lake has a large battery-operated fan on board. Getting a bright idea, the person aims the fan at the sail to blow the boat to shore (Fig. 3.16). However, he finds that the boat barely moves. Getting an even brighter idea (and applying physics principles he learned in college), he takes the sail down and turns the fan around. The boat then moves forward. Explain the motion (or lack of it) in each case.

Figure 3.16
And away we go! But not so good when the sail is full. See Question 21.

Boat goes nowhere

Boat moves forward

*22. A cannon ball is fired horizontally from a cannon with a constant muzzle velocity. Is the horizontal momentum conserved thereafter? Is the vertical momentum conserved? (*Hint:* Think in terms of components of momentum.) Is the total momentum conserved?

23. In testing explosive charges in space, an astronaut leaves a bomb there that is later detonated by remote control. The bomb explodes into many fragments that fly off in various directions. What is the total momentum of the bomb fragments?

Collisions

*24. In a collision of balls of equal masses, the momentum is conserved. We might also say that the velocity is conserved. Why is this? If the balls were not of equal mass, would the velocity be conserved? Would the momentum be conserved?

25. A ball of mass m and velocity v collides with a stationary ball of unknown mass. If, after the collision, the incoming ball is stationary and the other ball moves off, what is the mass of the originally stationary ball?

26. What occurs in the collision of (a) a light ball and a heavy stationary ball and (b) a heavy ball and a light stationary ball?

27. In a suspended ball set with five balls (cf. Fig. 3.9), if three balls are allowed to swing in toward two stationary balls, explain why three balls swing out. How about four in and four out? Five in and five out? (This last one should be easy.)

*28. (a) Is it appropriate to say that momentum and mechanical energy are conserved in elastic collisions? (b) Kinetic energy is not conserved or lost in an inelastic collision. Objects stick together in *completely* inelastic collisions. Is it possible to lose all the kinetic energy in a completely inelastic collision?

29. Two cars of the same model and year (same mass) traveling at the same speed approach an intersection. One is traveling north and the other east. At the intersection, one runs a stop sign, and they collide in a completely inelastic collision. In what direction do the cars move after collision? (*Hint:* Use a vector diagram.)

30. Two railroad cars of equal mass roll toward each other on a flat or horizontal track with the same speed. What happens after collision?

Projectile, Circular, and Planetary Motions

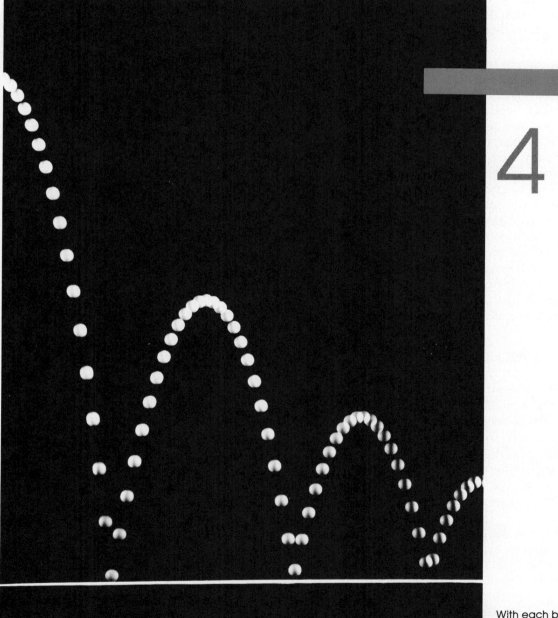

With each bounce, the ball becomes a projectile.

■ PROJECTILES

Now that you have the basics of force and motion under your belt, we can consider some special, but common, cases of objects in motion. One such case is projectile motion.

When you throw a ball or shoot a rifle, you put the ball or bullet in motion — projectile motion. A projectile is just an object that has been thrown or projected by some means. We do it all the time.

In describing projectile motion, we usually consider only the motion of the projectile after it has been thrown or launched. That is, a projected object has an initial velocity in some direction. For example, the initial velocity of a fired bullet would be its muzzle velocity on leaving the barrel. Often we are not concerned with the force that caused the initial motion, only what happens afterward.

Other forces may act on a projectile after it is launched. In particular, here on Earth there is gravity, and normally, projected objects fall to the ground. (Space launches and Earth satellites are topics of the next chapter.)

In a sense, a falling object dropped from rest is a projectile, and an object may be projected downward with an initial nonzero velocity. However, such one-directional motions were considered in Chapter 1. To see how general projectile motion is described, we will first consider objects that move vertically both up *and* down (in one dimension) and then expand our analysis to two dimensions.

■ VERTICAL PROJECTIONS

We often throw or toss things directly upward. This is a vertical projection. There is an initial velocity v_o upward, but the acceleration (due to gravity) is downward, as always. This means that the acceleration due to gravity for the upward motion is a deceleration, or tends to slow down a vertically projected object. Vectorially, the velocity and acceleration vectors are in opposite directions on the way up (Fig. 4.1).

As we all know, the deceleration slows the object to an (instantaneous) stop at its maximum height. Hence, at its maximum height a vertically projected object stops instantaneously ($v = 0$), as it must to change directions. However, the acceleration is not zero here. Gravity is still acting on the object as always, and $a = g$.

Thereafter, it is as though someone dropped the object from its maximum height (initial velocity downward is zero). At the instant a vertically projected object reaches its maximum height, it essentially becomes a dropped object in free fall.

An interesting fact for a vertical projection without air resistance is that the object returns to its starting point with the same speed as it had initially (directions and hence velocities are different — equal in magnitude but opposite in direction). Hence, in the absence of air resistance, if you shot a rifle bullet or something else vertically upward, it would return with the same speed, and you might shoot yourself.

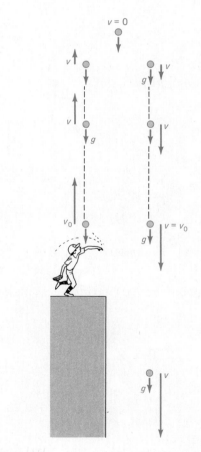

Figure 4.1
Vertical projection. Neglecting air resistance, the acceleration of a vertically projected object is constant (g). The velocity varies — decreasing on the way up and increasing on the way down (note vector arrows). At the maximum height, the velocity is zero and the ball returns to its starting position with the same initial speed. (Downward motion offset for clarity.)

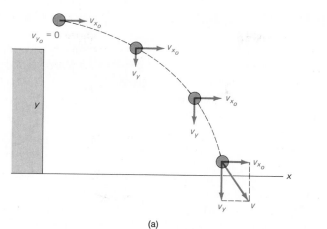

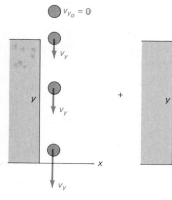

(b)

Figure 4.2

Horizontal projection. (a) The velocity in the horizontal direction is constant, since there is no acceleration in that direction, and the downward velocity component increases with time. (b) The projectile motion is a combination of an object falling from rest and moving horizontally at the same time. (c) A photograph of dropped and projected balls. Notice that the downward motions of the balls are the same in each case.

▮ HORIZONTAL PROJECTIONS

Another common type of projection is one in a horizontal direction. Such projections include a ball rolling off a level table, a horizontally thrown ball, or a bullet fired from a rifle held horizontally. For this type of projection, there is an initial velocity in the horizontal, or x-direction, but none in the vertical, or y-direction, $v_{y_o} = 0$ (Fig. 4.2). However, there is an acceleration in the downward direction due to our old friend gravity.

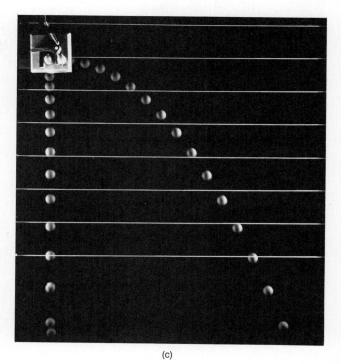

(c)

A horizontal projectile follows an arc-like path until it finally hits the ground. To analyze the motion, it is convenient to consider the individual components of motion in the two directions — much the same as we did with components of momentum in the last chapter. The actual motion is a combination or the resultant of the individual component motions.

Since there is no acceleration or force in the x-direction after it is projected, the projectile moves in this direction with a constant speed v_{x_o} (Fig. 4.2). As the object moves horizontally, it also falls (accelerates) downward owing to the force of gravity at a rate of 9.8 m/s² (acceleration due to gravity). In the downward direction, the motion is the same as that of a dropped object.

So the combination or resultant of these component motions is the actual arc motion of the projectile. To project an object a greater distance, we need only to give it a greater horizontal velocity and/or

start at a greater distance above the ground. The latter gives the object a longer time to fall, and it travels farther horizontally during this longer time. The horizontal distance a projectile travels is called its **range**.

■ PROJECTIONS AT AN ANGLE

The most general type of projection is at some angle to the horizontal (Fig. 4.3). Here the initial velocity v_o has components in both the horizontal (v_{x_o}) and the upward vertical directions (v_{y_o}). You can probably guess what happens in the component directions.

In the x-direction with no acceleration, the object moves along with a constant speed. Viewed from above, it would appear that the projectile travels in a straight line.

(a)

Figure 4.3
Projection at an angle. (a) The motion is a combination of vertical and horizontal motions. (b) The object describes a parabolic path.

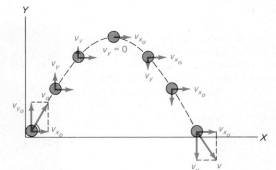

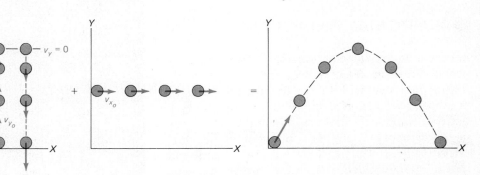

(b)

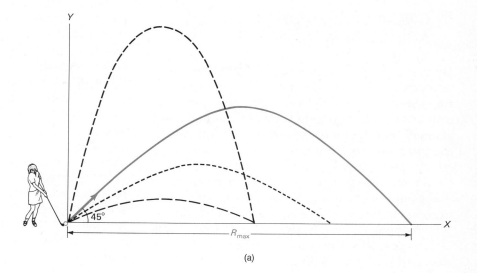

Figure 4.4
The projection angle for maximum range is 45°. This is important in several sports, such as (a) golf and (b) shot-put.

In the y-direction, the motion is the same as that of a vertically projected object. If a person moved along under the projectile at the same horizontal speed, that's what the person would see. Notice that at its maximum height the vertical velocity of the projectile is zero.

The combination of the component motions causes the projectile to describe a curved path called a parabola. We say that the path of a projectile is a parabolic arc. Notice that the second half of the motion is the same as that of a horizontal projection.

We are often interested in getting the maximum range (horizontal distance) for a projectile. A moment's thought should tell you that this depends on the initial launch speed and the angle of projection (and g, over which we don't have much control). Suppose the launch speed were fixed. Which angle would give you the maximum range? As you might guess, it is 45° (Fig. 4.4). This is a consideration in several sports—for example, shot-put, baseball, javelin throwing, and golf—when maximum ranges are desired. (Notice in the figure that projections with complementary angles—i.e., add up to 90°, e.g., 30° and 60°—have equal ranges.)

In actual practice, other considerations such as air resistance and spin come into play. The projection angle for maximum range is less than 45° with air resistance. A reverse spin on a driven golf ball provides lift, and a typical launch angle is considerably less than 45°. Also, when the launch height is above the ground or the landing height, the angle of projection for the maximum range is less than 45°.

CIRCULAR MOTION AND CENTRIPETAL FORCE

Another common type of motion is circular motion, for example, a car traveling around a circular track or yourself on the rotating Earth — you have a circular path in space. In analyzing circular motion, we will generally consider uniform circular motion. An object traveling in uniform circular motion travels in a circular path with a constant or uniform speed.

As was learned, an acceleration produces a change in velocity due to a change in speed and/*or* a change in direction. For an object in uniform circular motion, its speed is constant, but its velocity is *not,* because the direction of motion is continually changing (Fig. 4.5). Where is the acceleration and thus the force that produces this change in velocity?

If we add the velocity vectors to find the direction of the change in velocity Δv (Fig. 4.5), it is found to be instantaneously directed *toward* the center of the circle. This "center-seeking" acceleration is called **centripetal acceleration** (a_c). Since an acceleration is evidence of a force, we have that

A centripetal force is required for circular motion.

It can be shown that the magnitude of the centripetal acceleration is given by $a_c = v^2/r$ and then $F = ma_c = mv^2/r$. Hence, the force depends on the speed and radius of the circular orbit.

The centripetal force that supplies the centripetal acceleration for circular motion comes from various sources, depending on the situation. For an automobile on a circular track or rounding a corner, the centripetal force is provided by the friction force between the tires and the road. For you on the rotating Earth, the force is your weight. For a ball on a string being swung in a circle [Fig. 4.6(a)], the center-seeking force is supplied by the person via the string. A centripetal force then is just any force that produces circular motion. The force is directed toward the center of the circular path or at right angles (90°) to the instantaneous velocity. Notice that the object is accelerated toward the center of the circle but never gets any closer to the center.

It is easy to show that a center-seeking or centripetal force is required for circular motion. Suppose you were swinging a ball on a string as in Figure 4.6 and the string broke or you let go of it (centripetal force and acceleration on the ball become zero). What happens? The ball would no longer follow a circular path but would fly off in the direction of the instantaneous velocity. This direction is tangential to the circle (at a right angle to the radius of the circle at that point), and we refer to this velocity as the **tangential velocity.**

Question: A car going around a circular curve at high speed hits a patch of ice on the road. What happens?

Answer: When the car hits the icy surface, the friction supplying the centripetal force is reduced. With $F = ma_c = mv^2/r$, the centripetal force required for the car to negotiate the curve depends on its speed and the "radius of curvature" of the circular curve. At a high speed, the required centripetal force is large.

If the frictional force supplying this is reduced below the required value on the icy surface, then the car "slides outward," be-

Figure 4.5
Centripetal "center-seeking" acceleration. The speed of an object in circular motion may be constant, but because of changes in direction, the velocity is not and there is an acceleration. Uniform circular motion requires a centripetal force or acceleration (a_c).

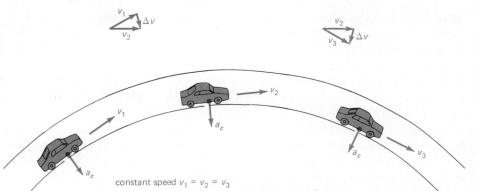

constant speed $v_1 = v_2 = v_3$
but vectors $\mathbf{v}_1 \neq \mathbf{v}_2 \neq \mathbf{v}_3$ because of changes in direction

cause the reduced frictional force is not enough to change its direction so as to go around the curve. This is like swinging a ball horizontally on a string and letting the string slip through your fingers, rather than releasing it completely.

Also, notice that the centripetal force depends on the square of the speed ($F = mv^2/r$). Hence, if the speed of an object in uniform circular motion is doubled (from v to $2v$), then the centripetal force required to keep the object traveling in the same circular path increases by a factor of four. Now you know why the speed limit is usually reduced going around a curve.

Centripetal force, or the lack of it, is used in various practical applications. For example, in a washing machine's spin cycle, water is separated from the clothes [Fig. 4.7(a)]. The tub of the washer rotates rapidly, but the force exerted on the water by the clothes is not great enough to make the water travel in a circle with the clothes. The water flies off, leaving the clothes less wet. This is similar to the lack of centripetal force in the case of a car traveling too fast on a icy curved roadway, as discussed in the preceding Question and Answer.

A less desirable case of lack of centripetal force is when the rear wheel of an automobile spins in mud [Fig. 4.7(b)]. Have you ever been pushing a stuck car and gotten a bit muddy? Stay away from behind the spinning wheel(s). The adhesion of the mud to the wheel (the centripetal force) is not great enough to hold the mud on the tire, so off it comes tangentially to the tire's circular motion.

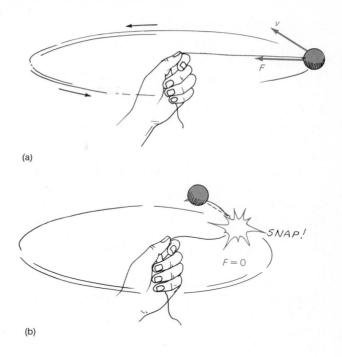

(a)

(b)

Figure 4.6
An example of centripetal force. (a) The centripetal force for the circular motion of a ball on a string is supplied by the person "pulling inward." The velocity of the ball at any point is called the tangential velocity. (b) If the string breaks (or the string is released), the centripetal force is removed or goes to zero and the ball flies off tangentially (at a right angle to the radius of the circular path).

Figure 4.7
Not enough centripetal force. (a) In a washing machine, water is separated from the clothes by a spinning action. The (centripetal) force exerted on the water by the clothes is not enough to make the water travel in a circle with the clothes, so the water flies off. (b) Because of a lack of centripetal force, one can get a bit muddy.

(a)

(b)

CENTRIFUGAL FORCE

You have probably heard the term *centrifugal force* or *centrifugal acceleration.* When in a fast-moving car rounding a sharp curve or on a rotating ride in an amusement park, we "feel" a force "pushing" us outwardly or away from the center of curvature. This is the so-called centrifugal or "center-fleeing" force.

Some people avoid using the term *centrifugal force* because they say it is a false or pseudo-force that doesn't exist according to Newton's laws. You may argue that it is "real" enough for you. The problem is a matter of definition and distinction between frames of reference, or how one looks at the situation.

To illustrate this, consider a car rounding a corner after having traveled along a straight roadway with a constant velocity (Fig. 4.8). An outside, stationary observer (such as youself) would describe the situation as follows: The car and its occupants were moving along with a constant velocity. The driver turned the wheels and the frictional force between the tires and the road supplies the necessary centripetal force for the car to round the corner.

The passenger, who is not wearing his seat belt, continues to move in a relatively straight line in accordance with Newton's first law. (The friction on the seat of his pants is not great enough to supply the necessary centripetal force.) However, the car turns "in front" of the passenger, and the door pushes against him. This supplies the centripetal force needed for him to accelerate inwardly and around the corner with the car. Notice that in this description the passenger has no outward force acting on him.

Now, imagine yourself to be the passenger in the car. Your immediate frame of reference is the car. Then the situation may be described like this: When moving along the straightaway with a constant velocity, you are at rest with respect to your reference frame (the car). As the car rounds the corner, you notice yourself moving with respect to the car—outwardly toward the door. From your knowledge of Newton's law, you think that this requires a force to be acting on you, which could conveniently be called a centrifugal or center-fleeing force.

Finally, as a result of this "force" you push up against the door and it pushes back, stopping your outward motion. You are now in equilibrium, with your outward centrifugal force being balanced by the inward (centripetal) door force, and you go around the curve with the car—your frame of reference.

Notice that the centripetal and centrifugal forces are *not* the action and reaction force pair of Newton's third law, since they act on the same body (you in the previous example). Centrifugal force then is something that is "invented" to make $F = ma$ work in an accelerating reference frame (the car changing direction or velocity).

It is preferable to describe situations from the point of view of an outside observer, for whom *centrifugal force* is a meaningless term. However, it is sometimes helpful to use an accelerating reference frame. We build machines called *centrifuges* to separate materials of different sizes and densities by spinning action. The spinning drum in a washing machine used to separate water from clothes is a centrifuge. From the spinning reference frame point of view, it would be said that "the centrifugal force throws the water outward" as the drum rotates with wet clothes. [The description with centripetal force has already been given. See Fig. 4.7(a)].

Other applications of centrifuges are in separating blood cells from plasma and cream from milk in dairy separators (Fig. 4.9). Ultracentrifuges with

Figure 4.8
Centrifugal force is a pseudo-force. A person in a car rounding a curve experiences what is believed to be an outward force, but to a stationary observer, the passenger in the car is moving in accordance with Newton's first law of motion, and there is no "center-fleeing" force.

(a)   (b)

Figure 4.9
Centrifuge applications. (a) A general-purpose refrigerated centrifuge capable of separating blood plasma and cells. (b) A dairy centrifuge used to separate cream and milk.

speeds on the order of 500,000 rpm (revolutions per minute) are capable of concentrating viruses in solutions.

In a centrifuge, the heavier materials generally migrate to the outer end of a spinning horizontal container, such as a capillary tube or a test tube. (The tube may be laid horizontally when possible or the centrifuge container may be piovoted so that a tube is horizontal when the centrifuge spins rapidly.) For example, when whole blood samples are centrifuged, the red cells are in the bottom of the tube with the white cells on top. This is because as the cells move through the viscous (resistive) liquid plasma, the red cells reach their terminal velocity later, similar to heavier falling objects with air resistance (Chapter 1). The red cells are first to "reach the ground" in their reference frame, so to speak.

In the rotating frame of the centrifuge, the greater the rotational speed, the greater the outward centrifugal force. With very large rotational speeds, the pseudo force can give rise to a pseudo acceleration many times greater than the standard acceleration due to gravity. [This effect may be used to sup-

ply "artificial gravity" for rotating space colonies as will be discussed in the next section.] For a nonrotating observer, this effect is due to an inward centripetal force, since a centrifugal force does not exist in his or her frame of reference. This observer might call the machine a "centripuge."

Let's look at another example of circular motion you may have experienced. There is an amusement park ride called the "Rotor" or some other name, where riders are spun in a large drum-like room [Fig. 4.10(a)]. When the Rotor reaches a certain speed, the floor drops out. But the riders do not fall! To an outside, nonrotating observer, a rider standing away from the wall cannot maintain the circular motion (lack of centripetal force) and flies off tangentially up against and exerts a force on the wall, and the reaction force supplies the centripetal force needed for circular motion [Fig. 4.10(b)].

But what keeps the people from falling? The centripetal force wouldn't do this because it is directed inward. The only force to balance the downward gravitational weight force is the upward frictional force between a rider and the wall [Fig.

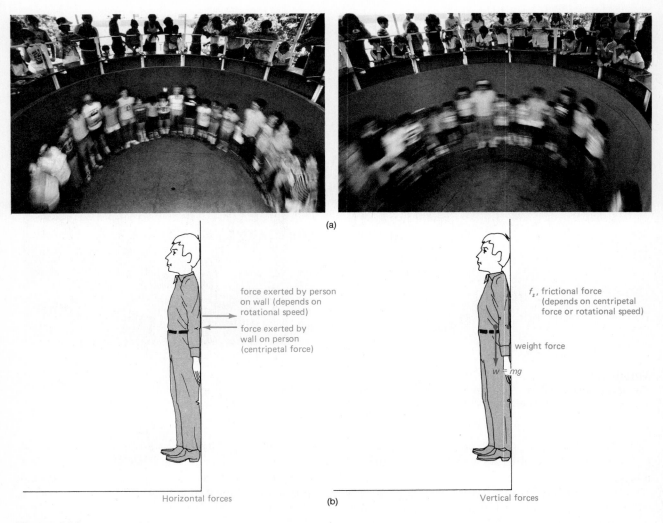

(a)

force exerted by person on wall (depends on rotational speed)

force exerted by wall on person (centripetal force)

f_s, frictional force (depends on centripetal force or rotational speed)

weight force

$w = mg$

Horizontal forces

(b)

Vertical forces

Figure 4.10
An amusement park Rotor ride. (a) At a certain critical speed, the floor is dropped out, but the riders are "stuck" to the wall and do not fall. (b) The forces involved. The reaction force of the wall on a person supplies the necessary centripetal force, and the upward friction force equals the downward weight force.

4.10(b)]. This frictional force is proportional to the reaction force the wall exerts on a person.

The reaction force increases with speed (recall that $F = ma_c = mv^2/r$), so the reaction force and the friction force build up as the Rotor speed increases. At a certain critical speed, the friction force equals the gravitational force in magnitude, and above this speed the floor may be dropped without causing the person to fall. If the wall were smooth, a person would fall even in a fast-spinning Rotor.

The frictional force also depends on surface properties. This would be an important consideration in selecting a material for the Rotor wall—

you'd want a rough wall so the critical speed wouldn't have to be excessively large.*

* It is interesting to note that analysis shows the critical speed is independent of the masses of the riders. (Your instructor may wish to show you this. It is similar to the acceleration due to gravity g being independent of the mass of a falling object.) So, at this minimum critical speed, the weights of all of the riders are balanced by equal and opposite frictional forces, regardless of their masses. A small child would have a small balancing frictional force, and a large adult would have a large balancing frictional force. This is because the frictional force *is* dependent on mass, as well as the rotational speed (similar to the weight force being mass dependent, $w = mg$).

Simulated Gravity

A potential use for centripetal force or centrifugal force, depending on how you look at it, is to provide *simulated* gravity in space. (We'll take a look at the real thing in the next chapter.) Have you ever wondered how the crews of spaceships in science-fiction movies are able to walk about normally when on missions in free space? (So has the author.) We Earthlings live in a gravitational environment, and many of our normal activities and functions depend on gravity. Think of how things would be without it.

Our awareness of gravity comes primarily from falling objects and weight. The reaction force to your weight force acts on the Earth when you are falling or not falling (see Fig. 1.18). To avoid a "weightless" feeling, you must *experience* the reaction to the force of your feet (or bottom) pushing against the floor (or chair). In an ideal gravity-free environment, your feet would not push against the floor, and there would be no reaction force. Of course, nowhere is gravity-free (see Chapter 5). In free fall, however, one seems weightless.

To simulate gravity, we might have a rotating spacecraft (Fig. 4.11). When it is rotated at the proper speed so that the centripetal reaction force of the spacecraft on a person was equal to that experienced under normal conditions of gravity, the effect would be the same as standing on Earth. Objects would also "fall" toward the person's feet (owing to a centrifugal force?). Don't be concerned about standing on the "wall." This would be the floor,

Figure 4.11
Rotating space colony. The reaction force on the people's feet would provide simulated gravity. Rotating the structure at the right speed would give a "gravity" environment similar to that on Earth.

Johannes Kepler

Tycho Brahe was fortunate. In 1576 the king of Denmark financed a fine observatory for him on the island of Hven northeast of Copenhagen. There for the next 20 years Brahe made extensive observations of the stars and planets. These were done without the benefit of a telescope, which had not yet been invented. Even so, Brahe's unaided-eye observations were more accurate than previous ones because of better instrumentation. As a result, he is considered to be one of the greatest practical astronomers of modern times. On the basis of observations that supported a mistaken idea of the time, Brahe tried to convert Galileo, a contemporary, from his belief in the Copernican heliocentric theory, but to no avail.

A new king withdrew Brahe's royal support, and Brahe left Denmark, taking his instruments and records with him. In 1599 he arrived in Prague, where the emperor provided him support in the capacity of the court mathematician. Brahe invited and was joined by a young German mathematician and astronomer, Johannes Kepler, in 1600.

Kepler, who had just celebrated his 29th birthday, had been a professor of mathematics at the University of Tubingen, where he had become acquainted with the heliocentric concept of Copernicus. His interest in planetary motion was in part associated with astrological work and predictions, for which he was responsible at the university. The positions of stars and planets supposedly exerted an influence on such predictions.

Because of religious unrest between Catholics and Protestants, Kepler and his wife had to leave the university in 1600, despite his wife's wealthy Protestant influences. Friends suggested that Tycho Brahe might help and he did. (Because of his publications Kepler had corresponded with Brahe and Galileo.) And so, Kepler went to assist Brahe in Prague.

Brahe died the next year, and Kepler was appointed his successor as court mathematician. Kepler inherited Brahe's records of the positions of the planets that he had made over many years. These were to prove a storehouse of information and to provide the basis for Kepler's laws of planetary motion for which he is famous.

Applied initially only to Mars, the first two of Kepler's three laws were announced in 1609. (In the same year, Galileo built his first telescope.) This was done in a book in which he also theorized that some kind of force on the planets emanated from the Sun. He presumed the orbital motions of the planets to be in some way associated with the rotation of the Sun. Although incorrect, it was one of the first attempts to associate force with the motions of the planets.

Kepler's attention then turned to the use of the newly invented telescope. Having had previous interest in optics, he published a work in 1611 on telescope design that later came into wide use. With intervening theological writings and a treatise on comets, in 1619, ten years after the publication of his first two laws, Kepler published his third law of planetary motion in the work *De Harmonica Mundi* (Harmony of the World). In this work, Kepler proposed a mathematical concept of "harmony" in the solar system.

His work on documenting the positions of the planets was to continue for another ten years. Kepler died in 1630, having made major contributions to astronomy, among them the dethroning of the Earth as the center of the universe and solar system.

Johannes Kepler.

since downward is defined by the direction of gravity or the direction in which things fall. (Think about two people standing on Earth, one at the North Pole and one at the South Pole. "Downward" to them would be in opposite directions.)

Space colonies of the future may have a circular or wheel design so as to provide simulated gravity by this means. Of course, such space stations would have to be quite large. Otherwise, they would have to spin so rapidly to simulate normal gravity conditions that the inhabitants would feel dizzy because of the effects on the inner ear.

■ PLANETARY MOTION

In Aristotle's theory of motion, the Earth was considered the center of the universe (geocentric theory). The apparent motions about the Earth of celestial objects such as the Sun, other stars, planets, and the moon were considered to be *natural* motions. They just "did their thing" without benefit or need of forces. However, Newton and others recognized that this could not be the case if the laws of motion were valid.

Newton's study of celestial motions led him to develop his theory of the gravitational force, which will be considered in the next chapter. The force of gravity supplies the necessary centripetal force for planetary motion. But the general description of planetary motion was developed before this. In a book published in 1543, Nicolaus Copernicus (1473–1543), a Polish astronomer, proposed a heliocentric or Sun-centered solar system. In the new system, the Earth became one of the planets revolving around the Sun. Copernicus also proposed the daily rotation of the Earth from west to east so that the daily circling of celestial bodies around us became simply apparent motions. Some scientists of the day opposed the heliocentric theory because it was not supported by convincing proof. But this was soon in coming.

■ KEPLER'S LAWS

During the 16th and 17th centuries much attention was focused on the motions of celestial objects, in particular the planets. Tycho Brahe (1546–1601), a Danish astronomer, spent most of his life studying the planets and stars and is considered to be one of the greatest practical astronomers since the Greeks. His measurements of the positions of planets and stars, all made with the unaided eye (the telescope had not yet been invented), proved to be more accurate than any previously made.

Brahe's data were analyzed by his colleague Johannes Kepler (1571–1630), a German mathematician and astronomer, who had joined Brahe during the last year of Brahe's life. See Special Feature 4.1. After Brahe's death, his wealth of observations were at Kepler's disposal, and they proved invaluable in the formulation of the empirical "laws" we know today as **Kepler's laws of planetary motion.***

Kepler found that Brahe's data fit three simple rules or laws that describe planetary motion in a heliocentric or "Sun-centered" theory.

1. First law, or **the law of elliptical orbits**. All planets move in elliptical orbits about the Sun, with the Sun at one focal point of the ellipse.

An ellipse has the appearance of an oval or a flattened circle. One can be drawn using string and thumbtacks, as shown in Figure 4.12. The tacks are at the positions of the foci (singular: focus) of the ellipse. As the focal points are moved closer together, the ellipse approaches a circle, which is a special case of an ellipse.

Although the orbits of the planets are ellipses, they are almost circular, with the exception of those of Mercury and Pluto.

2. Second law, or **the law of equal areas**. Kepler also found that the planets do not have uniform orbital speeds. They move faster as they orbit closer to the Sun than when they are farther away from the Sun. The Earth has its greatest orbital speed during the winter, when it is closest to the Sun (about January 4).† This was stated in terms of equal orbital areas: An imaginary line

---

* *Empirical* means derived from experiment or experience rather than from theory. An empirical law gives no explanation of why things occur in a particular manner, only that they do. It is interesting to note that the three laws of planetary motion, which Kepler spent a great deal of his life formulating by trial and error, can be derived on a couple of sheets of paper using advanced mathematics.

† We are closer to the Sun during our winter, but because of the inclination of the Earth's axis, the Northern Hemisphere is tilted away from the Sun at this time. The Southern Hemisphere, tilted toward the Sun, receives the more direct rays, so it is summer there.

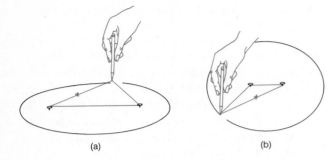

(a)   (b)

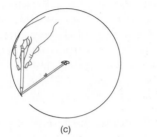

(c)

Figure 4.12
How to draw an ellipse. (a) A loop of string over two thumbtacks allows an ellipse to be easily drawn. The locations of the tacks are the focal points of the ellipse. (b) As the focal points are brought closer together, the ellipse approaches a circle, and when they are at the same point (c), they are at the center of a circle.

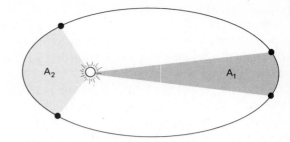

Figure 4.13
Equal areas. A planet sweeps out equal areas A_1 and A_2 in equal times; hence, it must move faster or have a greater orbital speed when it is nearer the Sun.

from a planet to the Sun sweeps out equal areas in equal periods of time (Fig. 4.13).

3. Third law, or **the law of periods** (also known as the harmonic law). It was also found that there is a common relationship between the average distances of the planets from the Sun and their orbital periods. (A period is the time it takes for a planet to make one revolution about the Sun.) This relationship is that the ratio of the square of the period of a planet (T^2) and the cube of the average distance (R^3) is a constant and the same for all planets ($T^2/R^3 =$ a constant).

SUMMARY OF KEY TERMS

Projectile: a body that has been thrown or projected by some means.

Vertical projection: an object thrown or protected vertically upward.

Horizontal projection: an object thrown or projected horizontally to the Earth's surface.

Range: the horizontal distance traveled by a projectile.

Projection at an angle: an object projected at an angle to the horizontal.

Uniform circular motion: the motion of an object traveling in a circle with a constant speed.

Centripetal force (acceleration): a center-seeking force (acceleration) necessary for an object to travel in circular motion.

Tangential velocity: the instantaneous velocity of an object in circular motion, which is at right angles to the radius of the circular path.

Centrifugal force: the fictitious center-fleeing force in a rotating or accelerating system that is "invented" to describe the motion in this system in terms of Newton's laws.

Kepler's laws: empirically derived laws that describe planetary motion.

1. Law of elliptical orbits: All planets move in elliptical orbits about the Sun, with the Sun at one of the focal points of the ellipse.
2. Law of equal areas: An imaginary line from a planet to the Sun sweeps out equal areas in equal times; i.e., planets do not have uniform orbital speeds and have greatest speed when closest to the Sun.
3. Law of periods: The period T and the average distance from the Sun R of a planet have the relationship $T^2/R^3 =$ a constant, where the constant is the same for all planets.

QUESTIONS

Projectiles

1. If a marble and a golf ball were projected vertically with the same initial velocity, (a) which would have the greater maximum height? (Neglect air resistance.) (b) Which would return to the starting point first?

•2. While sitting in a stationary car, you flip a coin upward and it falls back into your hand. Suppose the car were traveling on a level road with a speed of 90 km/h and you flipped the coin in the same manner. What would happen? What would an observer standing by the road say the path of the coin was?

3. What is the angle of projection for a horizontal projection?

4. An object is dropped from an airplane, as illustrated in Figure 4.14. Explain why the object hits the ground directly below the plane. What must be true about air resistance for this to be the case?

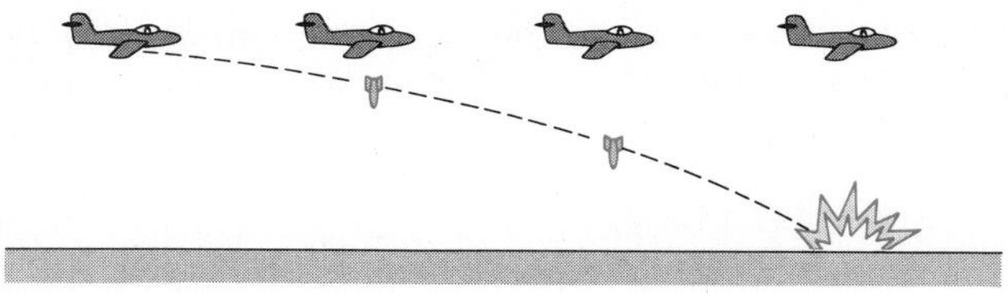

Figure 4.14
Always over the bomb? See Question 4.

5. A coyote is trying to catch a roadrunner again, this time wearing Acme jet-powered skates (Fig. 4.15). If the roadrunner makes a sudden turn at the cliff, but the coyote doesn't, sketch the path of the coyote's descent to the canyon floor below when (a) his skates fall off at the edge of the cliff and (b) his skates are still on and in operation when he is in "flight."

•6. Discuss the effect (make a sketch) of air resistance on (a) a vertical projection, (b) a horizontal projection, and (c) a projection at an angle.

7. Several years ago, a pilot of a jet plane fired the plane's cannon at an imaginary target, then put the plane into an evasive dive. Rather embarrassingly, the pilot shot himself down. How did this happen?

8. Is it possible for a vertical projection and a projection at an angle to have the same maximum height? Explain.

9. In throwing a baseball to home plate from the outfield, a fielder does not usually throw the ball exactly horizontally. Why?

10. Suppose you and another student are playing catch with a ball. He throws the ball at an angle of 35°, and you catch it at the same height it was thrown while you are standing still. Could you throw the ball back with the same initial speed but at another angle so he could catch it as you did? If so, at what angle, and which throw would have the greater maximum height?

11. A projection with an angle near 45° is important to a shot-put thrower. Give some examples of other situations in which this is a critical angle.

Figure 4.15
Watch out! See Question 5.

Coyoté Stupidus   Chicken Delightus

Circular Motion and Centripetal Force

*12. Show that $a_c = v^2/r$ has the units of acceleration.

*13. In a washing machine during a spin cycle, the clothes stay against the drum in circular motion. However, in a dryer the clothes are tumbled and do not go in complete circular motion. Why is this?

14. A race car travels around a circular racetrack with a constant speed v (Fig. 4.16). If the driver increases the speed to $3v$, how many times greater is the centripetal force on the car? What supplies the centripetal force?

*15. Curved roadways and racetrack curves are often "banked" so that cars can travel around the curves at higher speeds than on level curves without sliding outward. (The bank slope is inward toward the center of curvature.) How does banking increase the centripetal force? (*Hint:* Consider the components of the reaction force acting on the car.)

16. Do Newton's laws of motion hold on Earth? Explain.

17. On a Rotor ride, suppose the Rotor slowed down below its critical speed before the floor came back up. What would happen to the people on the ride?

18. On the rotating Earth, a person at the equator weighs slightly less than a person of equal mass at the North Pole. Explain why.

19. In a dairy centrifuge used to separate the cream from the milk, which is on the bottom or farther from the spin axis?

*20. The *Coriolis force* is a fictitious or pseudo-force that we use here on Earth. It "deflects" moving objects to the right (viewed in the direction of motion) in the Northern Hemisphere and to the left in the Southern Hemisphere. As a result, air circulation is counterclockwise around low-pressure areas ("lows") and clockwise around high-pressure areas ("highs") in the Northern Hemisphere, as viewed from above. The next time you see a satellite weather picture on TV, notice the rotation around a low, which is outlined by clouds. (A hurricane is a low.)

Suppose you are at the North Pole and fire a high-speed, long-range projectile toward the equator along a meridian. Prove that the projectile would hit the Earth to the west or to the right of meridian, and explain why. What is required

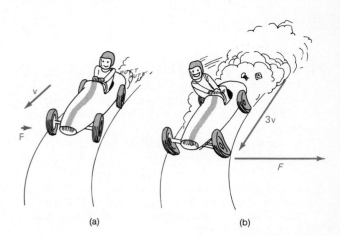

Figure 4.16
Enough centripetal force? See Question 14.

to cause this "deflection" according to Newton's laws? Which way would the projectile be deflected if you were at the South Pole and shot it toward the equator?

21. The inhabitants of a rotating space colony have become overweight, and the person in charge wants to put everyone on a diet. As an alternative, one clever fellow suggests a surefire way that everyone can lose "weight" without going on a diet. How might this be done?

Kepler's Laws

22. In drawing an ellipse with thumbtacks and a loop of string (cf. Fig. 4.12), what would you get for the two extreme cases of the tacks being as close together and as far apart as possible?

23. During which seasons does the Earth have its slowest and fastest orbital speeds?

*24. If the Earth's orbit about the Sun were exactly circular, how would its orbital speed vary?

25. Do planets nearer to or farther from the Sun have greater periods? Explain in terms of Kepler's third law.

*26. Evaluate the constant of Kepler's third law using Earth data. [*Hint:* A convenient unit for R is the astronomical unit (AU), which is defined as the average distance from the Earth to the Sun, i.e., 1 AU. What would be a convenient unit for T, days or years?]

Gravitation
and Earth Satellites

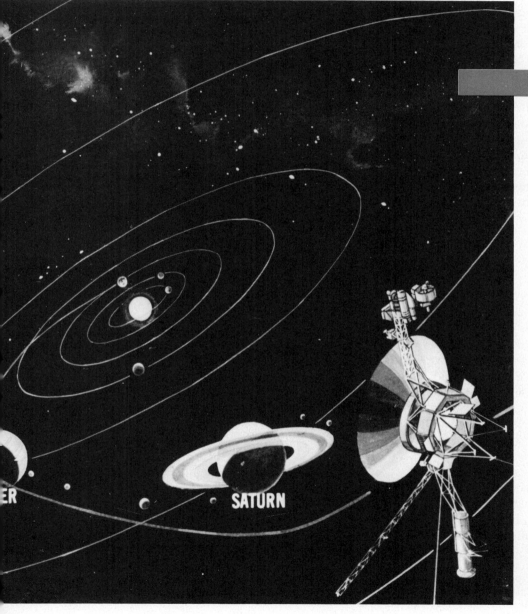

SATURN

5

The gravitational attractions of Jupiter and Saturn are used in sending satellite probes to the outer reaches of the solar system.

■ GRAVITATION

Gravity, or gravitation, is one of the fundamental forces of nature. This force acting between all masses holds the universe together, so to speak, on a large scale. There are three other fundamental forces — the *electromagnetic* force and the *strong* and *weak* nuclear forces. The electromagnetic force is dominant on the atomic level and is responsible for holding the atoms together that make up molecules. It also binds the electrons of an atom to the nucleus. Strong and weak nuclear forces act only *within* the nucleus.*

More will be said about these forces in later chapters. Here we will be concerned with gravity, intrinsically the weakest of the fundamental forces. But even though it is the weakest force, it causes objects to fall, determines the motions of celestial bodies, and gives rise to the ocean tides.

Isaac Newton, who gave us the laws of motion, also formulated an expression for the fundamental interaction that takes place between all objects or masses. This law of gravitation was the result of Newton's study of the motions of planets and other celestial objects.

■ NEWTON'S LAW OF GRAVITATION

In his study of planetary motion, Newton started a bit closer to home and focused his attention on the motion of the moon in its orbit about the Earth. Obviously, the motions of the moon and the known planets could not be "natural," as Aristotle had proposed, or without a force of some kind. If no force were exerted on them, then their motions would be straight lines by Newton's own first law.

As you know, and Newton knew, circular motion requires a centripetal force. But what supplies this centripetal force? Newton pondered this problem while he was at home on his mother's farm because of the outbreak of the Great Plague in Cambridge and the closing of the university (see Special Feature 1.1). Allegedly, his insight was sparked by observing an apple falling to the ground. If gravity

---

* Recent research suggests that the electromagnetic force, the weak nuclear force, and possibly the strong nuclear force may all be the same type of interaction or a unified (fundamental) force. Scientists continually look for ways to unify their theories, thereby simplifying the description of nature.

Figure 5.1
Newton and the apple. Legend has it that Newton's insight into gravity was helped by observing a falling apple. Did this force extend to the moon and provide the centripetal force needed to keep it in orbit?

attracts an apple toward the Earth, perhaps it also attracts the moon, and the moon is "falling" or centripetally accelerating toward the Earth under the influence of gravity (Fig. 5.1)

Actually, in its nearly circular orbit the moon is continually "falling" toward the Earth, since it is accelerated toward the center of its orbit. But it never gets any closer because it is also moving tangentially, which by itself would cause the moon to move farther from the Earth (Fig. 5.2). The combination of these component motions is a circular or elliptical orbit. The planets are also "falling" toward the Sun in their orbits.

Using his knowledge of centripetal force and Kepler's third law, along with amazing insight, Newton formulated an expression for the attractive gravitational force that acts between the Earth and the moon, the Sun and the planets, and indeed, between any two point masses m_1 and m_2. This is known as **Newton's law of gravitation**

$$F \propto \frac{m_1 m_2}{r^2}$$

where r is the distance between the two point masses.

Gravity is always an attractive force. Thus, Newton's law of gravitation states that every particle in the universe is attracted to every other particle and that this force of attraction is directly proportional to the product of the masses ($m_1 m_2$). Notice that the gravitational force is inversely proportional to the distance between the particles, in fact, the *square* of the distance ($1/r^2$). We call this type of relationship an *inverse-square law*. So, the farther apart the masses are, the smaller the force of gravity, and this falls off (decreases) as the square of the distance.

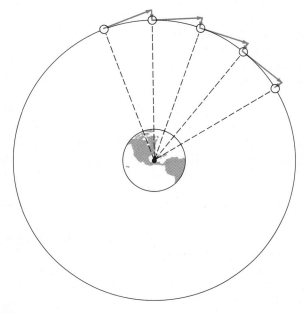

Figure 5.2
Circular motion has tangential and radial components.
Circular motion is a combination of a tangential component and one toward the center of the circle. Because of the latter, the moon is continually "falling" toward the Earth in its nearly circular orbit.

For example, if a mass were twice as far from another, the gravitational force would be only one quarter as great because of the inverse-square relationship.

$$\frac{1}{(2)^2} = \frac{1}{4}$$

For nearly spherical objects like the Earth, we measure the distance from the center of the Earth just as if all of its mass were concentrated there as a particle (Fig. 5.3). This is only true for a homogeneous sphere, but it is a good approximation for our purposes.

At the surface of the Earth, the force of gravity on an object is equal to its weight, and if dropped, the object would fall to the Earth with an acceleration due to gravity g (9.8 m/s²). The gravitational force of the Earth also acts on the moon and provides the centripetal acceleration needed for the moon's nearly circular motion about the Earth.

■ UNIVERSAL GRAVITATIONAL CONSTANT

To do direct calculations with the law of gravitation, one must express it in equation form. This is commonly written as

$$F = Gm_1m_2/r^2$$

where G is a constant, called "big" G to distinguish it from "little" g (the acceleration due to gravity).

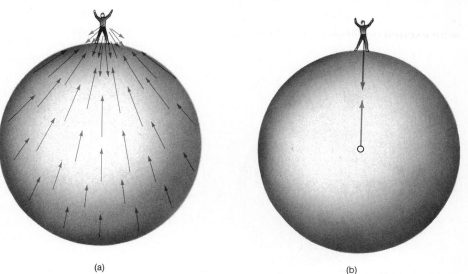

Figure 5.3
Gravity. (a) Mutual gravitational attraction exists between any two particles. (b) For a uniform spherical object, the effect is as though all the mass of the sphere were concentrated at its center.

(a)   (b)

Newton could not use his law to calculate the gravitational force between two masses because he didn't know the value of G. Nor was he able to determine its value experimentally with the equipment of his day.

It was not until 1798 (some 71 years after Newton's death) that the gravitational constant was experimentally determined by Henry Cavendish, an English physicist. He used a sensitive balance that could measure the force between two masses. The value of G is 6.67×10^{-11} N-m²/kg².

Since it is believed that Newton's law of gravitation applies to all masses everywhere, G is called the ***universal* gravitational constant**. Newton's law of gravitation is sometimes referred to as the universal law of gravitation.

> *Question:* Why do you not experience an attraction (gravitational) toward a textbook or another person?
>
> *Answer:* The gravitational force is too small for you to detect. Consider a couple of $m = 100$-kg (220-lb) football players standing a distance of $r = 1$ m apart. The order of magnitude of the force between them can be calculated by Newton's law. The result is
>
> $$F = \frac{Gm_1 m_2}{r^2} \approx 10^{-7} \text{ N (or 0.0000001 N)}$$
>
> A force on the order of 10^{-7} N is very small (much less than the weight of a flea). It would be even smaller between you and a textbook ($m \approx 1$ kg). Because G is so small, you must interact with a very large mass, for example, the Earth, at a relatively close distance to detect the gravitational force.

OCEAN TIDES

If you have visited an ocean beach, you probably noticed the periodic rise and fall of the water, or tides—two high tides and two low tides per day. Tides have long been associated with the moon, and it is easy to see how the moon would gravitationally attract the ocean water toward it to give a tidal bulge on the side of the Earth nearest the moon. However, as the Earth rotates beneath it, this would seem to give only one high tide and one low tide (the water depression on the opposite side of the Earth) per day. So what causes the other bulge for the second daily high tide?

Newton showed that the two high tides per day are caused by the differences in the gravitational pulls on the opposite sides of the Earth (Fig. 5.4). Keep in mind that the gravitational pull of the moon acts on the Earth itself, as well as on the water on both sides of the Earth, with the attraction getting weaker the greater the distance from the moon (inverse-square law).

The water nearest the moon has the greatest attraction, and this pull forms one tidal bulge. The Earth itself is attracted toward the moon with more force than the water on the side opposite the moon. As a result, the Earth is effectively pulled away from the water on the distant side, which gives a second and opposite tidal bulge. So, as the Earth rotates, two high tides "travel" around the Earth daily. The intervening water depressions are the two low tides.

Notice in Figure 5.4 that the two daily high tides at a particular location need not be equally "high." The tidal bulges can be above and below the equator because of the inclination of the moon's orbit.

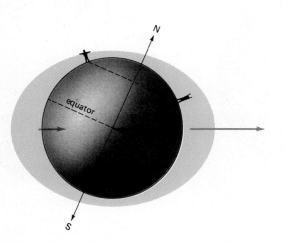

Figure 5.4
Tides. There are two tidal bulges, or two high tides, per day because of the difference in the gravitational attraction of the moon at different distances.

Question: The gravitational pull on the Earth by the massive Sun is about 200 times greater than that of the moon. Why doesn't this produce mammoth tides?

Answer: Although the Sun's gravitational attraction is greater than the moon's, the *difference* between the Sun's pull on opposite sides of the Earth is appreciably less. This is because the Earth's diameter is a relatively small fraction of the distance from the Sun as compared to that of the moon. The Sun's gravitational pull on each side of the Earth is relatively large, but the difference is relatively small.

To illustrate this idea, consider the fractional difference between two large numbers like 20,000 and 20,002, which is relatively small compared with the fractional difference between two small numbers like 20 and 22.

The gravitational pull of the Sun does have a "helping" effect at certain times to produce higher-than-usual and lower-than-usual tides. When the Sun and moon are on the same or opposite sides of the Earth, the added pull of the Sun on the bulges makes them higher than usual. This causes higher high tides and lower low tides, which we call *spring tides.* These occur at the times of the full and new moons and have nothing to do with the season of spring.

Similarly, when the Sun and moon are at right angles (90°) relative to the Earth, or at the times of the first-quarter and third-quarter moons, the Sun's pull tends to attract water to the tidal depressions and decreases the bulges. These lower high tides and higher low tides are called *neap tides.*

Actual tides involve complicating factors other than those presented in the preceding simple description, which assumed the Earth to be completely surrounded by water. The presence of land masses that stop the flow of water, "tidal" friction between the oceans and the ocean floors (which slows down the Earth's rotation about 1/1000 of a second per century), and the variable depths of the oceans are some of these.

Generally, the tidal range (the difference in height between high and low tides) in an ideal open ocean would be a meter or two. However, as this tidal bulge comes "ashore" at certain places, the tidal range can be quite different. A classic example occurs in the funnel-like Bay of Fundy between New Brunswick and Nova Scotia, Canada. The highest tides on Earth occur at the head of the Bay of Fundy, where under favorable circumstances the tidal range can be as much as 16 m (Fig. 5.5).

Figure 5.5
A high tide. The photos show a 16-m (52-ft) tidal range at Hantsport, Nova Scotia, at the head of the Bay of Fundy. The objects between the posts in the high-tide photo *(right)* are floating pieces of ice — not the top of the car. (The car was removed.)

It is interesting to note that the tidal forces of the moon and the Sun also produce *Earth tides* as well as water tides. The Earth is not totally rigid; its surface rises and falls in the same manner as water tides. However, these tides generally go unnoticed. It is believed that Earth tides may be a factor in initiating the relief of stress along cracks of faults in the Earth, which results in earthquakes.

■ DISCOVERY OF PLANETS — NEPTUNE, PLUTO, AND PLANET X (?)

Newton's law of gravitation also assisted in the discovery of the planets Neptune and Pluto. Not only does the Sun's gravity act on the planets, but there are also mutual interactions among the planets themselves. These lead to slight departures from smooth elliptical orbits.

Uranus, the seventh planet from the Sun, was discovered in 1781 by telescopic observation. Its motion was studied, and after allowing for perturbation effects of the known planets, scientists determined that something was affecting Uranus' orbit. In the 1840's Urbain Leverrier in France and John C. Adams in England, using Newton's law, independently calculated where an unknown eighth planet should be located to account for the perturbation.

Both Adams and Leverrier had little success in convincing the astronomers in their countries to look for the new planet. However, Leverrier wrote to a colleague, Johann Galle, at the Berlin Observatory. Galle trained a telescope on the position predicted by Leverrier on the very night of the day he received the letter and discovered Neptune!

Even after the discovery of Neptune, the perturbation of Uranus was not fully explained. Perhaps there was another planet. Percival Lowell, an American astronomer, calculated the position of such a planet. He searched for it until his death in 1916. With the development of a more advanced telescope, Pluto was discovered by Clyde Tombaugh at the Lowell Observatory in Arizona in 1930.

But Pluto's mass is not sufficient to account completely for Uranus's perturbation. Is there a "Planet X" beyond Pluto? Some astronomers believe that this might be the case. With calculations using Newton's law of gravitation, the search goes on.

As long as we are looking at history, we might mention Edmund Halley, a contemporary of Newton's. Halley began a friendship with Newton that

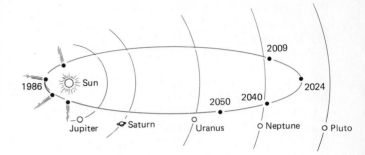

Figure 5.6
Halley's comet. With an orbital period of about 76 years, in this century Halley's comet passed near the Sun in 1910 and 1986. It will return in 2062. Notice that a comet's tail is away from the Sun, not behind the comet or in the direction opposite its motion.

resulted in the publication, at Halley's expense, of Newton's *Principia* in 1686. The *Principia* contained Newton's laws of motion and the law of gravitation.

Halley is perhaps most famous for the calculation of the orbit of a comet he observed in 1682, which has been named in his honor — Halley's comet. (Kepler wrote about the comet in 1607.) Halley's correct prediction of its return in 1758 was one of the first applications of Newton's laws.

Halley's comet is in a large elliptical orbit about the Sun, with a period of about 76 years (Fig. 5.6). The period of the comet varies somewhat because of gravitational effects of Jupiter and Saturn on its orbit as it passes by these large, massive planets. In this century, Halley's comet was visibly seen (near the Sun) in 1910 and 1986. You may have been fortunate enough to see this perhaps once-in-a-lifetime occurrence in 1986.*

A comet (from the Latin, *cometes,* meaning "long-haired") is generally characterized by a tail that extends *away* from the Sun. As a comet passes near the Sun, evaporated particles driven away from the Sun reflect light, giving rise to a long "tail." Notice that a comet's tail is not behind it or in a direction opposite to its motion (Fig. 5.6).

---

* Some people seem to have a unique relationship with Halley's comet. Mark Twain was one of them. He told his biographer, "I came in with Halley's comet in 1835. It is coming again next year, and I expect to go out with it. The Almighty has said, no doubt: 'Now here are these two unaccountable freaks; they came in together, they must go out together.'" Mark Twain died on April 10, 1910, one day after Halley's comet had circled the Sun and begun its long journey toward the outer reaches of the solar system.

A CLOSER LOOK AT g

We have talked about the acceleration due to gravity quite a bit. Let's take a closer look at g with the help of Newton's law of gravitation.

It has been stated several times that the acceleration due to gravity is constant near the Earth's surface. Why is this, when we now know that the force due to gravity, and hence the acceleration due to gravity, varies with distance? Also, it was learned that the acceleration of a freely falling body is independent of its mass. Newton's law explains.

The gravitational force on an object on Earth is its weight force, $F = mg$. But this must also be the same as Newton's law of gravitation for the special case of two interacting masses, m the mass of an object and M_e the mass of the Earth. Hence, we may equate the two equations

$$F = mg = \frac{GmM_e}{R_e^2}$$

where the separation distance is the radius of the Earth, R_e. Then, canceling the m's,

$$g = \frac{GM_e}{R_e^2}$$

So, g is independent of the mass m of the object (m does not appear in the equation), and it is also constant since G, M_e, and R_e are constants. (The value of g does vary slightly at different locations due to different mass distributions within the Earth, e.g., rock formations, and different radial distances, since the Earth has an equatorial "bulge" and all locations are not the same distance from the center.)

Question: To determine your mass, you weigh yourself and divide your weight by the acceleration due to gravity, $w = mg$ and $m = w/g$. The mass of the Earth is listed in many references. How is the mass of the Earth determined?

Answer: Determining the mass of the Earth might appear to be a sizable task. One thing for sure, we do not weigh it. Instead, the preceding expression for g is used. Rearranging the equation, we have

$$M_e = \frac{gR_e^2}{G}$$

The values of all of the quantities on the right side of the equation are known from experi-

mental measurements. Plugging in these values, the mass of the Earth can be calculated to be about 6.0×10^{24} kg.

We consider g to be constant near the Earth's surface. However, it does vary with altitude, $g = GM_e / r^2$, where r is the distance from the center of the Earth. But this variation is not great percentagewise over the heights of our normal activities.

At an altitude of 1.6 km (1 mi), the acceleration due to gravity and hence the gravitational force or weight (mg) is only 0.05 percent less than g at the Earth's surface. At an altitude of 160 km (100 mi), g, and therefore weight, is still 95 percent of its value on Earth. No matter how high we go, gravity still acts, and we still have weight (Fig. 5.7).

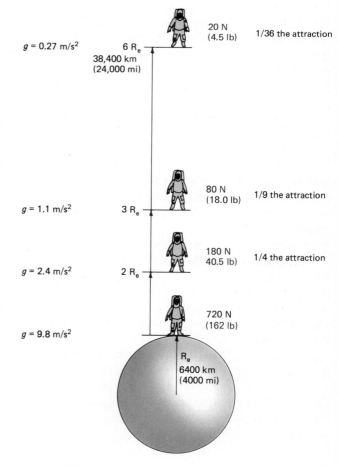

Figure 5.7
Weight and gravity. The force of gravity and g decreases as $1/r^2$. This results in small values, but an object experiences a force of gravity or weight at any height.

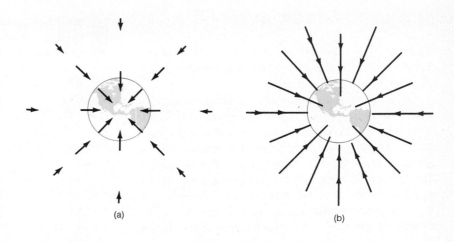

Figure 5.8
Gravitational field. (a) Around a mass, the gravitational force per mass ($g = F/m$) can be represented by vector arrows. (b) Joining the vector arrows forms lines of force.

(a)

(b)

GRAVITATIONAL FIELD

Another way of looking at the gravitational interaction is to consider only the effect rather than the cause. This is done in terms of a gravitational field. Essentially, we map the gravitational force in space due to some body like the Earth, then forget the body itself and consider only the field. That is, we think of an object interacting with the gravitational field rather than with the Earth or some other body responsible for it.

A gravitational field can be represented pictorially by vector arrows (Fig. 5.8). We compute the **gravitational force per unit mass** at points in space. Note that this is really g, since $F = mg$ or $g = F/m$ (force/mass). As shown by the lengths of the vector arrows in the figure, the magnitude of g decreases with increasing distances from the Earth.

We commonly join the arrows to form "lines of force" [Fig. 5.8(b)]. The closer together the lines of force are, the stronger the gravitational field or force in that region. Thus, a body in a field would experience an acceleration in the direction indicated by the arrowheads on the lines of force.

Einstein extended this idea in his model of gravity in his general theory of relativity, which was quite different from Newton's description of gravity. He perceived a gravitational field as a warping of four-dimensional space and time. Large bodies like the Earth put dents in this space.

To help visualize this effect, consider a heavy object on a loosely stretched rubber sheet (Fig. 5.9). The object puts a dent or depression in the sheet. If a marble were rolled on the sheet, several things could happen. It would roll in a straight line until it came sufficiently close to the "space warp" in the sheet. Depending on the direction of the marble's motion and speed, it may curve around the outer side of the dent and continue on, or it may follow a closed path and orbit the dent in an elliptical or circular path. In a similar manner, the curvature of space determines the gravitational effects.

You may have heard about the incredibly strong gravitational field in the vicinity of a black hole in space — so strong that even light that comes sufficiently close cannot escape its gravitational attraction. Black holes are believed to be burned-out stars that have gravitationally collapsed until they have incredibly high densities. The gravitational attraction near the black hole is so enormous that nothing can escape.

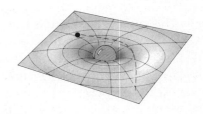

Figure 5.9
An analogy of gravity as a warping of space and time. A large mass warps space-time in a manner analogous to a heavy object denting a stretched rubber sheet. An object rolling on the sheet is deflected as it would be in a "space warp" in a gravitational field.

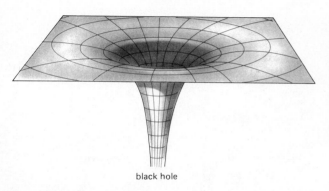

black hole

Figure 5.10
A black hole warp. A black hole would be represented in the rubber-sheet analogy of a space-time warp as an object that sinks an enormous distance into the sheet.

In terms of our space-time warp analogy of the rubber sheet, it would be as though the heavy object sank an enormous distance into the sheet (Fig. 5.10). If a marble rolled into the dent, it would be gone. Similarly, anything that comes too close to a black hole disappears from the observable universe — even light. It has been suggested, particularly in popular literature, that matter going into a black hole may come out in a parallel universe through a "white hole." (More on black holes in Chapter 27 on astrophysics.)

Question: If black holes are invisible or can't be seen, how do we know they exist?

Answer: If we were lucky, we might be able to observe the bending of light that passed near a black hole, which would indicate its presence. But we have never been so lucky. However, their presence can be detected by their gravitational influence on other stars. Most stars are binary stars — two stars orbiting about each other. One star may have gravitationally shrunk into a black hole, while the other is still luminous. The luminous star would then appear to be orbiting about an empty point in space.

But this is not enough. The black hole candidate may simply be a faint star that cannot be seen next to a bright companion. Near an object of high density (like a black hole), matter falling toward it or into it is accelerated to a high speed. Charged particles accelerated toward a black hole can emit X-rays, which are energetic enough to escape.

Thus, we must look for X-ray sources associated with binary stars with invisible companions. Some candidates have been discovered. The first was in the constellation Cygnus, called Cygnus X-1. Of course, we cannot be certain that Cygnus X-1 is a black hole, but many astronomers think it is.

APPARENT WEIGHTLESSNESS

The terms *weightlessness,* or *zero gravity,* and *g's of force* have become common as a result of the space program. Let's investigate the meaning of these terms.

Weightlessness, or the condition of zero gravity, generally refers to the situation in which astronauts "float" in space, apparently because they have no weight (Fig. 5.11). This would be the case if $g = 0$. But a more correct term for this condition is **apparent weightlessness**, because gravity does act on astronauts in space, and therefore they do have weight. (Recall the definition of weight.)

In the case of an astronaut in a spacecraft in circular orbit about the Earth, we know that gravity provides the necessary centripetal force for the circular motion. The astronaut therefore has weight. The "weightless" feeling arises because the upward reaction force normally provided by the floor or a chair is missing. This is because the floor of the spacecraft is "falling" toward the Earth just as fast as the astronaut. In the astronaut's frame of reference, he "floats" relative to the spacecraft.

To help understand this, consider a person standing on a scale in an elevator (Fig. 5.12). When stationary, the scale reads the person's weight as indicated by the reaction force R. When the elevator accelerates downward, the reaction force is less, and the scale reads less weight. Finally, if the elevator cable broke and the elevator (and the person) were in free fall, the reaction force or the scale reading would be zero. The scale is falling as fast as the person. One might call this a "weightless" condition because the scale reads zero — but is it really?

A relatively new term used to describe the apparent weightless conditions experienced by astronauts orbiting the Earth is *microgravity*. See Special Feature 5.1.

Figure 5.11
Apparent weightlessness. In orbiting *Skylab*, astronauts "float" in space, apparently weightless. However, gravity supplies the necessary centripetal force to keep them in orbit with *Skylab*, so they do have weight.

Figure 5.12
Weight in an elevator. (a) In a stationary elevator, the reaction force of the scales indicates the person's weight. (b) When the elevator accelerates downward, the reaction force is less, and a person appears to weigh less. (c) In an elevator in free fall, the scales would indicate zero. Is the person weightless?

(a) stationary

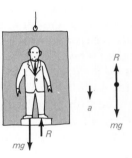

(b) descending

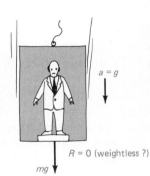

(c) descending with $a = g$

Microgravity*

Because the trajectory of a spacecraft, such as the Space Shuttle, compensates for the force of Earth's gravity, the craft approaches a state of free fall. In a state of free fall, all objects in a craft are "weightless." However, gases venting from the Shuttle and the minute drag exerted by a tenuous atmosphere at orbital altitudes create nearly imperceptible forces, which are collectively called microgravity.

Once you adjust to it, microgravity can be intriguing. You can perform slow-motion somersaults and handspins. You can float with the greatest of ease. You can push off from one side of your craft's interior and drift to the other side. You can lift or move normally heavy objects, which are nearly "weightless" in microgravity. You need never worry about dropping anything. Whatever slips out of your hand will float.

Living in space calls for special design technology. Here are some problems and solutions:

A beverage or water in a normally open container would leave the container, if shaken, as free-floating droplets. The droplets are not merely annoying; they could inconvenience people and pose a hazard to equipment. A beverage is served in a flexible dispensing unit with a plastic straw, which is clamped when the drink is not being sipped. Sponge baths rather than showers or regular baths are available. Because water adheres to the skin in microgravity, a little of it can go a long way. Water and other wastes are directed by air flow to a drain or are flushed into a sealed container.

Perspiration can be annoying in space. In the absence of proper air circulation, perspiration can accumulate layer by layer on the skin. It doesn't "roll" off.

Shaving could cause problems if whiskers ended up floating around. The whiskers could damage delicate equipment or irritate eyes or lungs. The solution: shaving cream and a safety razor. The whiskers adhere to the cream until wiped off with a disposable towel. Also available is an electric razor with a built-in vacuum pump.

Microgravity calls for special considerations in the handling of solid foods. Crumbly foods are provided only in bite sizes to prevent the crumbs from floating around the cabin. The food trays are equipped with magnets, clamps, and double-adhesive tape to hold metal, plastic, and other utensils.

Furniture is bolted in place. Tether lines, belts, and handholds enable people to move around and keep

Acrobatics in microgravity.

themselves and other objects where they are supposed to be. If you try tightening a screw or turning a nut without a restraint, you rotate in the opposite direction.

A microgravity environment causes a variety of bodily changes. For example, people in space seem to have smaller eyes because their faces become puffy. Many of the effects of microgravity are apparently due to shifts in body fluids from the lower to upper parts of the body. Some people in space suffer temporarily from a condition resembling motion sickness on Earth. Prolonged periods in space may also result in changes in the rate of formation of blood cells and bone tissue. The good point about most of these effects of microgravity is that they seem to level off eventually in space and to reverse themselves after return to Earth.

* Adapted from NASA Educational Briefs for the Classroom, an Educational Publication of the National Aeronautics and Space Administration (U.S. Government Printing Office, pub. no. 1982–361–570/3255) dealing with apparent weightlessness conditions in space, which NASA calls microgravity.

g's of Force

The constant value of *g* near the Earth's surface (9.8 m/s²) is sometimes used as a standard unit of acceleration. For example, an acceleration of 3 *g*'s would be an acceleration three times that of *g*. We also speak of ''*g*'s of force.'' On the surface of the Earth, you experience a gravitational force of 1 *g*, which is your weight. A force of 2 *g*'s would simply be some force equal to twice your weight.

During the takeoff of a commercial jet aircraft, you experience an average horizontal force of about ⅕ *g*. This means that as the plane accelerates down the runway, the seat back exerts a force of about one fifth your weight against you to overcome your inertia. The reaction force (Newton's third law) is the force you exert against the seat, which gives you the feeling that you are being pushed back into the seat. On takeoff at an angle of 30°, this increases to about ⁷⁄₁₀ *g* with a component of normal gravity helping to push you into the seat.

g's of force are also well known to pilots flying in an arc or loop. Suppose a plane is put into a circular vertical loop with just enough speed to go around the loop (Fig. 5.13). At the top of the loop, the pilot has a downward force of 1 *g* or his weight, which supplies the needed centripetal force. However, at that instant, the pilot would feel ''weightless'' because there is no reaction force as ordinarily exerted by the seat. [similar to the case in Fig. 5.12(c)]. Both he and the aircraft are ''falling'' toward the center of the loop (as an astronaut in orbit ''falls'' toward the Earth).

This condition is (incorrectly) called a state of ''zero gravity,'' as discussed previously. When astronauts are trained, the apparent weightlessness condition in space is simulated by flying a plane in a parabolic arc, which gives a greater time duration of the ''weightless'' condition. The plane is nosed down and dives to pick up speed. Pulling out of the dive and following a parabolic trajectory, 30 to 40 seconds of weightless conditions can be achieved. Essentially, the passengers are traveling inside a projectile with the same velocity as the projectile, so they can ''float'' around the cabin as when in orbit. You may have experienced a similar effect when riding in a car that goes over the crest of a hill at a high speed. It feels as though the car is ''taking off'' and you have a strange feeling of reduced gravity. Another good example of where this occurs is on a roller coaster ride.

At the bottom of the loop the situation is different. Here the pilot's weight is opposite to the direction of the centripetal force (supplied by the reaction of air against the air foil surfaces of the plane). The upward force exerted on the pilot by the seat must support the pilot's weight *and* supply the necessary centripetal force for circular motion (which is still 1 *g*), so a reaction force of 2 *g*'s is experienced.

You can get an idea of the *g*'s of force in a similar situation by swinging a pail of water in a vertical circle. Here, you are supplying the centripetal force. When you swing the pail at the minimum speed needed for the water to go in a circle, the pail would seem ''weightless'' to you at the top of the swing, but at the bottom it would be a bit different. Why doesn't the water fall out of the pail at the top of the loop? What would happen if you were to swing the pail more slowly than the minimum speed?

In the preceding example, the centripetal force acting on the pilot and the plane is not excessive because of the minimum speed. However, this may be critical in the case in which an airplane diving at a high speed makes a ''tight'' circular turn to pull out

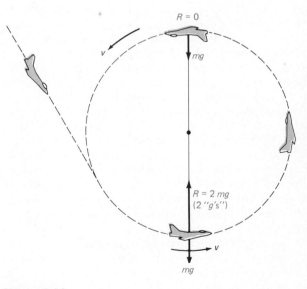

Figure 5.13
Loop-the-loop. In an airplane flying in a vertical circular loop at minimum speed, at the top of the loop the reaction force on a pilot is zero, so a ''zero g'' or ''weightless'' condition is experienced. (The weight *mg* provides the necessary centripetal force.) At the bottom of the loop, the reaction force is 2 *mg* and the pilot experiences a force of 2 ''g's.''

of the dive (large v, small r, and $a = v^2/r$). The excessive g's of force may cause structural damage to the plane, to the point that the wings may actually shear off. In addition, the pilot may "black out." During the dive, the pilot's blood is accelerated downward, and in the turn it tends to continue in that direction (Newton's first law), draining from the pilot's head. Unconsciousness and, in extreme cases, death can result.

■ EARTH SATELLITES

In our space age, many artificial satellites orbit the Earth—the moon is the only natural satellite. Most of us have watched (usually on TV) rocket "blast-offs" that send satellite payloads toward their orbital positions. Manned satellites are now common.

Putting a satellite into orbit about the Earth involves gravitational interaction. Gravity must be overcome to get the satellite to the planned orbital altitude, and the satellite must be given sufficient tangential velocity so it will not fall back to Earth. Once this is done, the gravitational force provides the needed centripetal force to maintain the satellite in orbit.

Circular Orbits

It is interesting to note that more than 300 years ago Sir Isaac Newton considered the possibility of putting an object into orbit about the Earth. He theorized about firing a cannon from the top of a mountain. If the cannonball had sufficient tangential velocity, it would not fall back to Earth, only toward it (Fig. 5.14).

Newton had the right idea, but he also realized that a cannonball could never be put into orbit like this. A cannon could never do the job of supplying enough tangential velocity. How fast do you think an object would have to be projected horizontally for it to go into a circular orbit just above the Earth's surface?

Near the Earth's surface, the object's weight force (mg) would supply the centripetal force ($F = mv^2/r$), and, $mg = mv^2/R_e$, where the radius of the orbit is the radius of the Earth, R_e, and air resistance is neglected. Using known values of g and R_e and solving for v, we find (as Newton did) that the required tangential speed is about 8 km/s (5 mi/s or 18,000 mi/h).

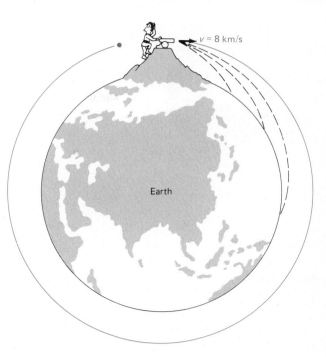

Figure 5.14
Earth satellite. "The greater the velocity . . . with which (an object) is projected, the farther it goes before it falls to the Earth. We may therefore suppose the velocity to be so increased, that it would describe an arc of 1, 2, 5, 10, 100, 1000 miles before it arrived at the Earth, till at last, exceeding the limits of the Earth it should pass into space without touching it." (From Newton's *System of the World*.)

If the cannonball were fired with a smaller velocity, it would fall back to Earth in an elliptical path that is part of a Kepler ellipse with one of the focal points at the Earth's center (see Chapter 4).

Similarly, if a satellite is given a tangential velocity of about 8 km/s, it will go into circular orbit about the Earth. A rocket with a satellite payload might be sent upward, then turned horizontally so there is a tangential speed of 8 km/s at burn-out or when the rocket engines are shut off. Actually, the required tangential speed decreases with altitude.

Elliptical Orbits

What would happen if a satellite projectile were given a horizontal speed greater than required for a circular orbit? In this case, the projectile would curve around the Earth and start to move away from it. But gravity slows it down and attracts it back

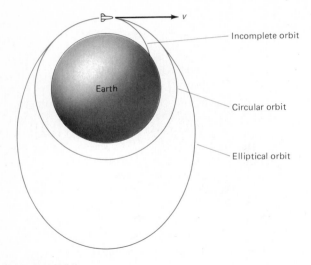

Figure 5.15
Earth orbits. Without sufficient tangential speed, a satellite would "orbit" into the Earth. A particular tangential speed at a given altitude is required for a circular orbit. Given a tangential speed greater than this, but less than the escape speed, an elliptical orbit would result.

toward Earth. The satellite then would gain speed and come back to its starting point in an elliptical orbit (Fig. 5.15).

Orbit Decay

Although satellites orbit for some time, they eventually return toward the Earth. Even at altitudes of several hundred miles, there is *some* atmosphere that supplies frictional "drag." This causes the satellite orbit to "decay," and it eventually spirals toward the Earth if the perturbation of its orbit is not corrected (by on-board rockets). A descending satellite may either burn up in the atmosphere (like a meteor) or fall to Earth (like a meteorite), depending on its size.

We have had several notable cases of the latter. In 1978 a Soviet *Cosmos* satellite with a nuclear reactor on board fell to Earth, spreading radioactive material over parts of Canada. The U.S. satellite *Skylab* came down in 1979. When astronauts returned from it in 1974, they left it in an orbit that was supposed to be stable until the 1980's.

However, unexpected solar activity caused the Earth's atmosphere to heat up slightly and expand.

This put more atmosphere into *Skylab's* orbital altitude and increased the frictional drag. Orbital maneuvering was attempted so as to reduce the drag, but to no avail. *Skylab* came tumbling down. Then, early in 1983, another *Cosmos* satellite came down. This time the nuclear reactor portion fell into the Atlantic Ocean.

Escape Velocity

The next logical question is, How much velocity must a projectile be given so it does not go into orbit but escapes the Earth completely? We know that ordinary vertically upward projections are decelerated by gravity to a stop, then are accelerated downward so they return to Earth.

But as the distance from the Earth increases, the force of gravity becomes less. Although a projected object would never escape the Earth's gravitational field, there is a critical launching speed whereby the object would never quite be slowed to a stop. This turns out to be about 11 km/s and is called the **escape velocity**.

Hence, if a projectile had an initial speed of 11 km/s or greater, it would escape from the Earth. Of course, in practical applications a spacecraft or space probe eventually comes under the gravitational influence of another body, such as the moon or another planet, and would not return to Earth. (The process is reversed in getting astronauts back from the moon.)

Using the gravitational forces of other planets is the principle of interplanetary flybys (Fig. 5.16). These are launched at a time when the outer planets are lined up in such a way that the large gravitational field of Jupiter can be used to deflect the probe in a flyby path toward other planets. The probe may also be slowed and deflected so that it goes into orbit about a large planet.

A probe may eventually go into orbit about the Sun or leave the solar system and travel into interstellar space. For example, as a probe approaches and whips around Jupiter on its way to a Pluto flyby, it gains speed. If its speed is then equal to or greater than the escape velocity from the Sun and from Jupiter at this distance, the probe will trip out into outer space — and this has happened. See Special Feature 5.2.

A Pioneer in Outer Space

On June, 13, 1983, Pioneer 10 crossed the orbit of Neptune — over 4.5 billion kilometers from the Sun — into interstellar space, becoming the first spacecraft to leave the solar system.* This unique event in human history climaxed an 11-year series of impressive achievements.

Launched on March 3, 1973, and designed for a 21-month lifetime, the flyby of Jupiter in December 1973 was Pioneer 10's primary mission. It was the first spacecraft to fly beyond Mars and to cross the asteroid belt, to fly by Jupiter, to chart that planet's intense radiation belt, to measure the mass and density of its four planet-sized moons, to find that it is a liquid planet, to take close-up pictures of its atmosphere and Great Red Spot, and the first to cross the orbits of Uranus and Pluto before reaching Neptune's. (Several other unmanned probes have since joined Pioneer 10 in outer space.)

Scientists predict that Pioneer will travel among the stars almost indefinitely because interstellar space is so empty. Some 10,500 years from now it will pass a star, and in 32,000 years it will have its nearest encounter with one — 3.2 light-years distant. Pioneer may outlast the solar system itself.

* Although Pluto is usually the outermost known planet, its highly elliptical orbit actually goes inside Neptune's orbit. Currently, Pluto is inside Neptune's orbit, which now defines the boundary of the known solar system or "outer" space.

Multistage Rockets

The escape velocity of an object from the Earth is the *initial* velocity it must be given, for example, the muzzle velocity of a cannon. However, satellites are not launched by cannons. Rockets are used. The burning rocket engines continually accelerate the rocket itself and the payload. Nor does the rocket take the satellite vertically upward to its orbital altitude and then give it the required tangential velocity. After blast-off, the rocket veers away from the vertical (Fig. 5.17). This gives a component of the rocket force in the horizontal direction and an increasing velocity.

Figure 5.16
A gravitational "slingshot" for interplanetary flybys. The large gravitational field of Jupiter can be used to deflect probes and flybys of the outer planets when they are properly aligned.

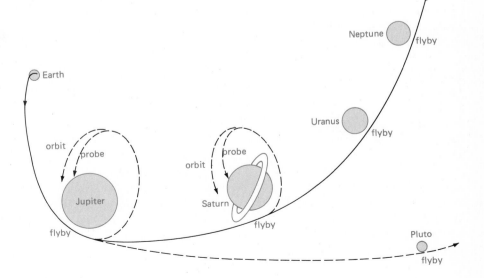

So that the necessary speeds can be achieved efficiently, multistage rockets are used. This permits an in-flight reduction in the mass of the rocket-ship by jettisoning the burned-out stages. The next stage then has less mass to accelerate. (See Special Feature 3.3.)

Figure 5.17
Multistage rockets. Multistage rockets are used to put satellites into orbit. Jettisoning the burned-out stages gives an in-flight mass reduction, and the next stage has less total mass to accelerate.

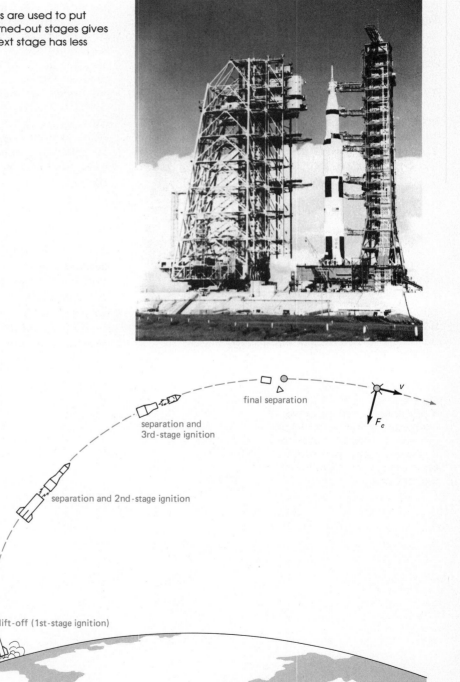

SUMMARY OF KEY TERMS

Gravitation (gravity): the attractive interaction or force between two masses.

Newton's law of gravitation: an expression for the magnitude of the force of gravity between two masses separated by a distance r, i.e., $F = Gm_1 m_2 / r^2$.

Universal gravitational constant, G: the constant in Newton's law of gravitation, which is believed to be constant throughout the universe.

Ocean tides: the periodic rising and lowering of the ocean's level owing to the gravitational attraction of the moon (and Sun) and the Earth's rotation.

Spring tides: higher high tides and lower low tides caused by the added pull of the Sun at the times of new and full moons.

Neap tides: lower high tides and higher low tides caused by the canceling effect of the Sun's pull at the times of first-quarter and third-quarter moons.

Gravitational field: a mapping of space that represents the gravitational force per mass owing to some body or bodies.

Weightlessness: the *apparent* weightlessness experienced in an accelerated or ''falling'' system in the absence of a reaction force.

g's of force: a unit of force using a body's weight on Earth as a standard; for example, 2 g's of force acting on a body is a force with a magnitude equal to twice the weight of the body.

Escape velocity: the initial or launching velocity an object would have to be given to escape from a large body like the Earth.

[For more on inverse-square laws or relationships, see An Extended View (Optional) in the Appendix.]

QUESTIONS

Newton's Law of Gravitation

1. There are four fundamental forces. Why are they ''fundamental''? (*Hint:* Compare with fundamental properties in Student Preface.)

*2. If the moon is ''falling'' toward the Earth, why doesn't it reach it?

3. If the force due to gravity on you due to the Earth is mg, what is the force due to gravity on the Earth due to you? Explain.

4. According to Newton's law of gravitation, how is the force of gravity affected when the distance between two masses is doubled?

5. G is a *universal* constant. Cite another universal constant. (*Hint:* Consider the ratio of the circumference and the diameter of a circle.)

6. If you drive down a street past a large skyscraper, does the force of gravity due to the skyscraper accelerate your car on approach and decelerate it after you go past?

*7. The Sun is much more massive than the Earth (about a million times more). Why doesn't the gravitational attraction of the Sun pull us off the Earth?

8. If the law of gravitation were somehow repealed, what would happen to the solar system?

*9. Suppose a hole is drilled through the center of the Earth to the other side. If you dropped a stone down the hole, what would happen? Remember, an object is attracted (gravitational force) toward the center of the Earth.

Tides

10. Explain why there are two high and low tides daily.

*11. The high tide on the side of the Earth nearest the moon occurs sometime later than when the moon is highest in the sky. What causes this time lag?

*12. Explain why spring and neap tides are related to particular phases of the moon.

13. Would you expect to weigh more at full moon or at new moon? Why is this effect not detected?

14. What is the tidal range and how does it vary?

15. Why are the twice-daily Earth tides unnoticed?

Planets

16. It was once speculated that the Earth had a sister planet at the same distance from but on the opposite side of the Sun, so it could never be seen. How could it be proved that there is no such planet, both directly and indirectly?

*17. If pertubation calculations indicate the possibility of a Planet X beyond Pluto, why hasn't it been discovered?

18. A grand alignment of planets occurs periodically in which the planets are in a general line with the Sun as viewed from above. When this occurred in the 1970's, it was popularized that Earth tides would cause devastating effects on Earth by the so-called ''Jupiter effect'' (which

Figure 5.18
Moon walk. See Question 21.

didn't happen). Explain why this might be a remote possibility.

A Closer Look at g

*19. What is the acceleration due to gravity at an altitude of $3R_e$ (three Earth radii) above the Earth's surface?

20. Why is the acceleration due to gravity on the moon one sixth of that on Earth?

21. An astronaut on a "moon walk" easily picks up a lot of equipment that would be too heavy for him on Earth (backpacks of 300 lb, see Fig. 5.18). Explain how this is possible.

22. The accelerations due to gravity are different on the surfaces of different planets. Explain why.

23. Is it correct to speak of the weight of a planet?

*24. The planet Saturn is 94 times more massive than Earth, but the acceleration due to gravity on the surface of Saturn is about the same as g on Earth. What does this tell you in general about the size of Saturn? Can you estimate the radius of Saturn compared with that of Earth?

Gravitational Field

25. Sketch the gravitational field around two particles of equal mass that are separated by a short distance.

26. Sketch the gravitational field for the Earth-moon system in terms of (a) a Newtonian field and (b) an Einsteinian field.

*27. The gravitational field of a black hole and the original star would be identical in a certain region. Identify this region.

28. A black hole is no more massive than the original star. Why is the gravity of a black hole so great near the "hole" (within original star radius)?

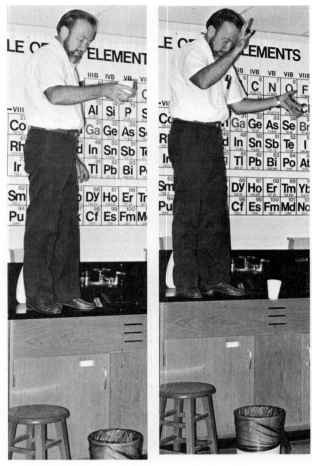

(a)   (b)

Figure 5.19
Water drains from the hole—sometimes. (a) Water can be seen draining from a hole in the bottom of a Styrofoam cup. (b) However, when the cup is dropped, no water comes out of the hole. (There is still water in the cup.) See Question 31.

Apparent Weightlessness

29. How far from Earth would you have to be in order to be truly weightless?

30. Suppose you were standing on a chair and jumped off. Are you weightless until you hit the ground? Explain.

*31. Water drains from a hole in the bottom of a cup because of its weight [Fig. 5.19(a)]. However, when the cup is dropped, no water is observed coming from the hole. Is the water weightless during the fall?

32. If you were standing on a scale in an elevator and the elevator accelerated upward, how would the scale read and why?

33. Describe the sensations you feel in an elevator when it starts up and suddenly accelerates (a) upward and (b) downward. Compare these to how an astronaut feels (a) on blast-off and (b) in a spacecraft in orbit about the Earth.

34. The plans for a new super *Skylab* suggest that a handball or racquetball court be included so the astronauts can get exercise. Describe such a game with the *Skylab* in orbit about the Earth.

35. When a pail of water is swung in a vertical circle, why doesn't the water fall out of the pail at the top of the loop above a certain minimum speed? What would happen below this speed?

Earth Satellites

*36. When a satellite is put into circular orbit about the Earth after it reaches an altitude of several hundred kilometers, would a minimum tangential velocity of 8 km/s be required? Explain.

37. Would a tangential velocity of 8 km/s be required to put a spacecraft in a circular orbit near the surface of the moon? Would air resistance be a consideration?

38. All planets do not have the same escape velocity. Why not?

*39. It is believed that the moon had an atmosphere when the Earth-moon system was formed. Why did the moon lose its atmosphere?

Rotational Motion

6

A rotation about an axis.

In our description of motions thus far, we have for the most part neglected or ignored one important type, namely rotational motion.

Special cases of rotational motion—circular motion and orbiting satellites—were considered in Chapters 4 and 5. But there are a lot more, for example, the spinning (rotating) Earth and many other common things that rotate: wheels, helicopter rotor blades, and doorknobs. This chapter will require a little rotational thinking, so get your head and mind spinning.

■ PARTICLES, CENTER OF MASS, AND CENTER OF GRAVITY

In previous chapters, physical "things" were commonly referred to as objects, bodies, or masses. In general, when we studied the motion of objects no mention was made of possible rotational motion. However, it is difficult to throw a baseball or football without rotation or spin.

Sometimes rotational considerations are ignored by talking about the motion of "particles." A particle is the physical counterpart of a "point" in mathematics. It has no physical dimensions and can be accurately localized in space. We can also say a particle has mass. But since a particle has no physical dimensions, we don't have to worry about it spinning. There is no such thing as a particle (as thus defined) in nature.

Even so, we use the concept of particles to build up extended solids or rigid bodies that can rotate. A **rigid body** is defined as a system of particles in which the particles are fixed or constant distances apart. Thus, we don't have to worry about internal particle motions in a rigid body. A quantity of water would obviously not satisfy this criterion. But if the water were frozen, the ice would be more like a rigid body.*

Another important particle concept of a rigid body is the **center of mass**. This can be thought of as the average location of all the mass particles that make up the body. That is, it is the point at which all the mass of a body can be considered to be concen-

---

* This definition of a rigid body is an idealized one because the atoms and molecules of a solid undergo thermal vibrations and because of elasticity. Even so, most solids macroscopically approximate the idealized case.

trated. For example, a uniform, symmetric object such as a sphere can be thought of as having all of its mass concentrated at its center. However, an unsymmetric object such as a hammer has more mass toward one end. The center of mass of a hammer is therefore toward the head of the hammer.

Since weight and mass are related by a constant for objects near the surface of the Earth ($w = mg$), the center of mass is also the center of gravity for such objects. The **center of gravity** is simply the average location of the weight distribution of a body. That is, it is the point at which all the weight of a body can be considered to be concentrated.

You have located the center of gravity (and center of mass) of an object when you have balanced it on your finger—for example, when you balance a stick (Fig. 6.1). If the mass of the stick is uniformly distributed, the center of gravity is at the center of the stick. You balance or support the entire weight of the stick by pushing upward through the center of gravity. It's as though all the weight of the stick were concentrated at its center of gravity.

Suppose you were blindfolded and the stick was replaced by a small object of equal mass or weight, say a cube. As far as weight goes, you would not be able to tell the difference. For a nonuniform body such as a hammer, the center of gravity is nearer the heavier end.

The center of gravity of a freely suspended body lies directly below the point of suspension (when it is not suspended at its center of gravity). When two points of suspension are used (Fig. 6.2), the intersection of the two vertical lines of suspension locates the position of the center of gravity. This method is useful in finding the center of gravity of a flat, irregularly shaped object.

Question: Does the center of gravity always lie within a body?

Answer: No. In some cases the center of gravity is outside the body itself. For example, consider a uniform donut (Fig. 6.3). Here, the center of gravity is at the center of the donut hole. You could prove this by using two points of suspension or by balancing the donut at a couple of places.

Can you think of other bodies in which the center of gravity lies outside the body? How about a horseshoe?

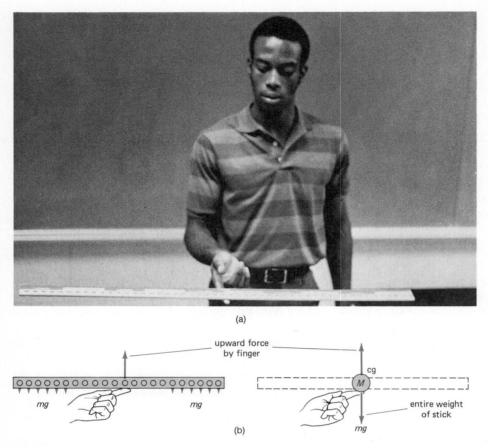

(a)

(b)

upward force by finger

cg

entire weight of stick

mg mg mg

Figure 6.1
Center of gravity. A uniform stick's center of gravity is at its center, where it can be balanced. It is as though the entire weight (mass) of the stick is at the center point.

(a)   (b)

Figure 6.2
Location of center of gravity by suspension. (a) The intersection of two (or more) vertical lines locates the position of the center of gravity of a flat, irregularly shaped object. The object could be balanced on the student's finger at the point of intersection. (b) Location of the ''center of state'' (South Carolina). Of course, states aren't flat objects, but it makes a nice demonstration.

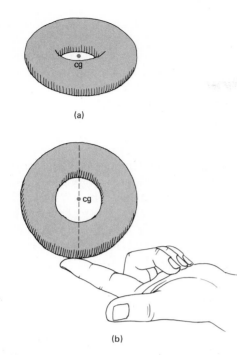

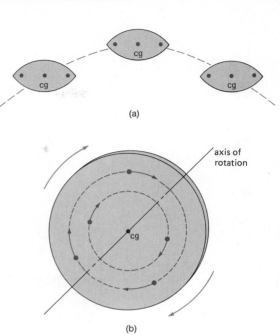

Figure 6.3
Example of center of gravity not located within the body of an object. (a) The center of gravity of a uniform donut is at a point in the space at the center of the donut (hole). (b) This can be shown by suspension or support methods.

Figure 6.4
Translational and rotational motions. (a) In pure translational motion, all the particles of an object have the same instantaneous velocity, so there is no rotation. (b) In pure rotational motion, the particles of a body move in circles about a line called the axis of rotation. The orbital speed of the particles on each circular path is different.

■ TRANSLATIONAL AND ROTATIONAL MOTIONS

In general, we have previously considered objects in translational motion and ignored spinning rotations. In **pure translational motion**, every particle of a body has the same instantaneous velocity. As illustrated in Fig. 6.4(a) for a thrown football that is not spinning or turning end over end, all the particles, including the center of gravity, have the same instantaneous velocity.

But there is also the possibility of rotational motion. In **pure rotational motion**, the particles of a body move in circles about a line called the axis of rotation. The instantaneous velocity (and speed) of the particles on each circular path is different. This is illustrated in Fig. 6.4(b) for a wheel or disk. The axis of rotation is through the center of mass, but it need not be.

The axis of rotation may be either inside or outside of the body. Also, a body may rotate about more than one axis at a time. For example, the Earth rotates daily about an axis through its center. It also rotates yearly about an axis through the Sun. We commonly say that a body "rotates" when the axis of rotation is inside or passes through the body. When the axis of rotation is outside the body, we commonly say the body "revolves," as the Earth revolves about (an axis through) the Sun.

Hence, the general motion of a rigid body is a combination of translational and rotational motions. Examples of this general motion are shown in Figure 6.5 for a tumbling football and a dropped hammer. Notice that the center of gravity of the football follows a parabolic projectile path and that the center of gravity of the hammer falls in a straight line as though they were "particles" representing the total weight of the objects.

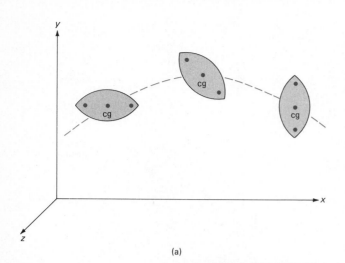

(a)

Figure 6.5
Combined translational and rotational motions. The objects rotate about an axis through the center of gravity (cg) while the center of gravity translates. Notice that the centers of gravity of the football and the hammer follow paths as though they were "particles."

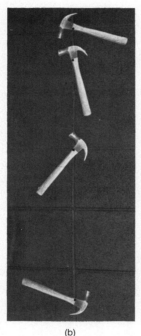

(b)

Another example of this combination of motions is a rolling object (Fig. 6.6). If you put a pure translational motion and a pure rotational motion together without slipping or spinning, you have rolling motion. Look at the velocity vector addition in the figure. A rolling object has an instantaneous axis of rotation at its point or line of contact (for example, a sphere or cylinder, respectively). The particle at this point of contact is or those along a line of contact are instantaneously at rest. Above this, the particles have increasingly greater speeds, such that the object "falls over itself" and rolls. Notice also that the center of gravity of the rolling object (assumed to be uniform) moves with a velocity v, as if it were a particle in uniform motion according to Newton's first law (Chapter 1).

Description of Rotational Motion

The description of rotational motion is analogous to that of translational motion. We have angular distance, angular speed and velocity, and angular acceleration. There are also rotational analogs of force, energy, and momentum, as we will see.

The particle nature of rigid bodies is helpful in analyzing rotational motion. For example, you might think of yourself as a rotating "particle" because you are on a rotating Earth. As part of a rotating body, you and other "particles" on and in the Earth move in circles about the axis of rotation.

Consider the particles of a rotating disk, as shown in Figure 6.7. All of the particles move in circles around the axis of rotation and rotate through

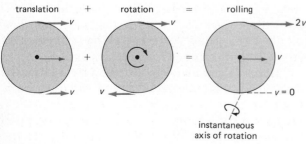

Figure 6.6
Rolling. Rolling is a combination of translational and rotational motions. The instantaneous axis of rotation of a rolling object is through its point or line of contact.

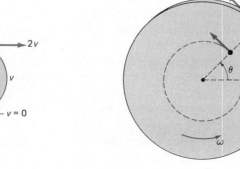

Figure 6.7
Describing rotational motion. A disk in circular motion traveling through an angle θ in a time t has an angular speed, $\omega = \theta/t$. The tangential speed of a particle is $v = r\omega$.

(a)   (b)

Figure 6.8
Rotational inertia depends on mass distribution. (a)
When the weights are close together or near the axis of
rotation, the bar is easy to twist or rotate. (b) When the
weights are farther apart or away from the axis of rotation,
getting the bar to rotate is more difficult because there is
greater rotational inertia.

an angle θ (Greek theta) in a certain time. The angle
may be measured in degrees (360° in one rotation
or a complete circular path). Scientists often use
another unit for angular measure called the radian
(rad).* There are 2π radians in one rotation or
circle, and 2π rad = 360°.

Since the particles move through an angle θ in a
time t, they have an **angular speed** ω, and $\omega = \theta/t$
(analogous to regular speed, $v = d/t$). The units of
angular speed are radians per second (rad/s). Other
common units are revolutions per second (rev/s or
rps) and revolutions per minute (rev/min or rpm).
For example, there are several types of phonograph
records, one of which is called 33⅓. This is the
record's angular speed on the turntable — 33⅓ rpm.

Notice that all of the particles in a rotating rigid
body have the same *angular* speed. They all travel
through one circle or 2π radians in the same time.
(Think about the objects on a rotating merry-go-
round.) However, the particles farther from the axis
of rotation travel in greater circular paths in the
same time. So the farther the particles are from the

* When a particle moves through an arc length s that is
equal to the radius r of its circular path, it moves an angular
distance θ of one radian (57.3°).

axis of rotation, the greater their tangential speeds.
The tangential speed of a particle is related to its
angular speed by the relationship $v = r\omega$, where r is
the radial distance from the axis of rotation.

There is also **angular velocity**, which is a vector
and hence has direction. This is rather involved, so
for simplicity we will say the velocity is in one circular direction or the other (similar to positive and
negative directions in linear motion).

Since the angular velocity can change, we can
have an **angular acceleration** α, and $\alpha = \Delta\omega/\Delta t$
(analogous to linear acceleration $a = \Delta v/\Delta t$). If the
angular acceleration is opposite to the direction of
the angular velocity, then it is a ''negative'' acceleration or a deceleration, and the rotating object is
slowed down. Friction commonly causes negative
angular accelerations that cause rotating objects to
slow down and stop. Tidal friction causes the Earth's
rotational speed to decrease somewhat. This
lengthens the rotational period or day by about
1/1000 s per century.

Rotational Inertia

Just as the property of inertia resists changes in
translational motion, there is also rotational inertia
that resists changes in rotational motion. Rotational
inertia, like inertia in the translational case, depends on the mass of the body. But rotational inertia
also depends on the mass distribution about the axis
of rotation.

Consider a barbell on which the weights can be
moved to different positions (Fig. 6.8). If the
weights are close to the axis of rotation, the bar is
relatively easy to rotate. However, if the weights are
moved farther apart, the bar is more difficult to rotate. The barbell has the same mass in both cases,
but the rotational inertia is different. When the mass
is distributed farther from the axis of rotation, there
is greater rotational inertia. We say that there is a
greater moment of inertia, I. **Moment of inertia** is
the rotational analog of mass and a measure of rotational inertia.

The mass distribution dependence of rotational
inertia or moment of inertia is the principle of the
flywheel. Flywheels generally have most of their
mass concentrated near the rim to give a greater
moment of inertia (Fig. 6.9). Once a flywheel is
rotating, there is a greater tendency for it to continue rotating because there is a greater rotational
inertia. Automobiles have flywheels to promote an

(a)   (b)

Figure 6.9
Flywheels. Flywheels have most of their mass concentrated near the rim to give a greater moment of inertia.

even power output between the firings of the cylinders. The rotational inertia of flywheels will also figure in the proposed electric cars of the future (Fig. 6.9).

> *Question:* Why is it easier to balance yourself with your arms outstretched when walking on the rail of a railroad track, as in Fig. 6.10, or on a narrow wall or fence?
>
> *Answer:* In extending your arms, you increase your rotational inertia by distributing more mass farther from the axis of rotation (along the rail). When you start to "rotate," the increased inertia resists the change and gives you more time to regain your balance.
>
> It is wiser to do this on a railroad rail. If you rotated too far around the top of a narrow wall or fence, gravity would take over and you'd become a falling object. As we know, the vertical motion (acceleration) of your center of gravity is independent of mass or inertia in this case—until you hit the ground.

■ TORQUE

Now that we know how to describe rotational motion, the next obvious question is, What produces it? A (net) force produces translational motion and the rotational counterpart is called a torque (pronounced tork). A torque might be descriptively re-

ferred to as a "twist" or twisting action.

As you might guess, a torque involves a force, but there is something else. The distance from the axis of rotation is always an important factor in rotational motion. In the case of a torque, this distance from the axis of rotation to the "line of action" of an applied force is called the lever arm (Fig. 6.11). A **torque** is defined as the product of the lever arm and the applied force that tends to produce a rotation about an axis.

Torque = lever arm × force

Figure 6.10
Outstretched arms make walking on the rail easier. The girl has a greater moment of inertia. See Question and Answer.

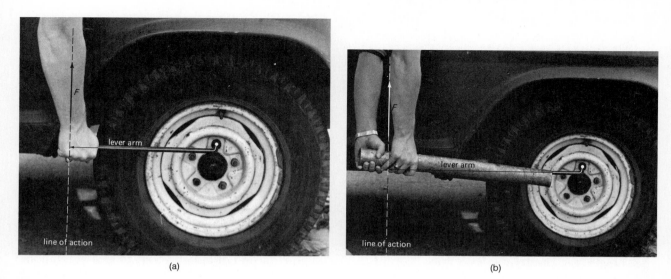

Figure 6.11
Torque = **lever arm × force.** The lever arm is the perpendicular distance from the axis of rotation to the line of action of the force. With the same force, the torque in (a) is less than that in (b) because the lever arm is shorter. To loosen a really stubborn lug, the lever arm is extended with a piece of pipe in (b) so as to get a greater torque.

Notice that the lever arm is the perpendicular distance from the axis of rotation to a line along which the force acts. The forces in the figures may be the same in each case, but the torques would be different because the lever arms are different.

Every mechanic knows that it is easier to loosen or to get a tight bolt or nut to rotate by applying more torque. This can be done by extending the lever arm of the wrench by putting a section of pipe on the wrench handle. Also, notice that if a force acts through the axis of rotation, the torque is zero. Why?

Just as the production of, or a change in, translational motion requires an *unbalanced* force, the production of, or a change in, rotational motion requires an *unbalanced* or net torque. In the situation shown in Figure 6.12, the net torque is zero, because the torques are equal and opposite. That is, one torque tends to produce a clockwise rotation, and the other a counterclockwise rotation. Since the torques have equal magnitudes, the net effect is zero, or no net torque.

So, we have a version of Newton's first law for rotational motion:

A rigid body remains at rest or in motion with a constant angular velocity unless acted on by an unbalanced (net) torque.

There is also a rotational form of Newton's second law ($F = ma$). Using the symbol notation for rotational quantities, we have

$$\tau = I\alpha$$

or

Torque = moment of inertia
$$\times \text{ angular acceleration}$$

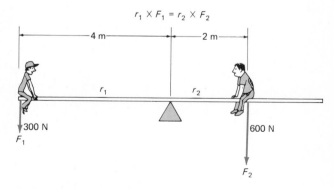

Figure 6.12
Zero net torque. The net torque is zero because the torque tending to produce a clockwise rotation is equal to or balanced by the torque tending to produce a counterclockwise rotation.

Notice the almost one-to-one correspondence between translational and rotational quantities. Of course, they are different, but the analogies make rotational effects easier to understand. We will consider energy and momentum shortly. Here, too, there are similarities and many interesting rotational effects.

Question: Why is it easier to open a door when pushing on it farther away from the hinges?

Answer: We have all experienced this, particularly when mistakenly pushing on the hinge side of a glass door. The axis of rotation of the door is along the hinges, and the closer to the hinges, the shorter the lever arm.

Suppose the same pushing force was used in both cases (Fig. 6.13). Near the hinges, the torque would be $\tau = r \times F$, whereas the torque with a greater lever arm (farther from the hinges) would be greater, $\tau = R \times F$. With the moments of inertia of the doors the same, the door with the larger torque would swing open faster (greater angular acceleration), as can be seen from the equation $\tau = I\alpha$.

If the doors swing open at the same rate in each case, then the torques would be the same, but a smaller force would be required with a greater lever arm. Hence, the larger the lever arm, the "easier" it is to open the door. Check this out on a door for yourself and become familiar with torques and lever arms.

STABILITY AND EQUILIBRIUM

Closely associated with torque and center of gravity is the stability of a rigid body. By **stability**, we mean whether an object will fall over or topple when it is slightly moved or tipped about an axis of rotation.

Stability depends on the location of the center of gravity of an object, and we describe an object as being in stable or unstable equilibrium (Fig. 6.14). If an object in **stable equilibrium** is slightly displaced, its center of gravity is still vertically above and inside its base of support. Because of the weight acting through the center of gravity, a restoring torque tends to bring the body back to its equilibrium position. Notice that when an object in stable

Figure 6.13
The lever arm makes a difference. See Question and Answer.

$\tau = R \times F$

(a)

$\tau = r \times F$

(b)

Mechanics

The Foucault pendulum at the Smithsonian Institution in Washington, D.C. Such a pendulum was used by Jean Foucault, a French physicist, to verify the Earth's rotation experimentally in 1851. As the pendulum swings, its plane of oscillation appears to rotate, since the bob swings through all points of the compass, as evidenced by the successive knocking over of the red indicators arranged in a circle. In reality, the pendulum's swing is fixed in space; it is really the Earth (and hence the building and the floor) rotating beneath the swinging pendulum that causes successive indicators to be toppled. (Smithsonian Institute photo No. 75-3965)

Pioneer leaves the solar system. The five outermost planets appear in relationship to the Sun and to each other. On June 13, 1983, Pioneer was farther from the Sun than any of its planets—a distance of over 4.5 billion kilometers (2.8 billion miles)—and, in effect, left the solar system.

Putting things in perspective.

A Space Shuttle blast-off involves the application of many physical principles in the areas of mechanics (motion, force, and momentum), thermodynamics (heat transfer and heat engines), and electricity and magnetism in the thousands of electric circuits and components.

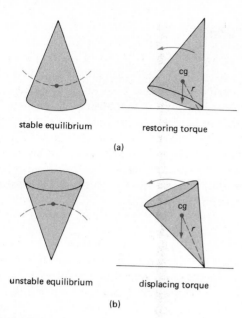

stable equilibrium   restoring torque

(a)

unstable equilibrium   displacing torque

(b)

Figure 6.14
Equilibrium. (a) If a body in stable equilibrium is slightly displaced, a restoring torque tends to bring it back to its equilibrium position. Notice that the center of gravity acts like a particle in a bowl. (b) If a body is in unstable equilibrium, a slight displacement will cause it to topple due to a displacing torque.

equilibrium is slightly displaced, the center of gravity is raised and its displacement path is somewhat like that of a particle in a bowl.

Hence, we say that **an object is in stable equilibrium and will not topple if, when slightly displaced, its center of gravity lies vertically above and inside its base of support.** For the inverted cone-shaped object in Fig. 6.14(a) to fall over, the center of gravity would have to be raised to the edge of the "bowl," i.e., so the cone would be on its circular edge. A slight further displacement, and over it would go.

If we could just balance the cone on a circular edge or on its point [Fig. 6.14(b)], then it would be in **unstable equilibrium.** In this condition, its center of gravity is still vertically above its base of support, but this is a point or edge. A slight displacement would cause it to topple, owing to a displacing torque that rotates the object about an axis through its base. In this case, the path of displacement of the center of gravity is like that of a particle sitting on top of a bowl or inverted dome.

To further understand stable and unstable equilibrium, consider the examples shown in Figure

6.15. For an L-shaped body, the center of gravity lies outside the body. When sitting on one of its long sides, the body is stable. But when sitting on an end side, it will topple over (rotate).

Hence, we see that objects with wide bases and lower centers of gravity are more stable and less likely to be tipped over. These conditions are evident in the design of high-speed racing cars, which have wide wheel bases and are very close to the ground.

An example of human stability, and perhaps instability, is also shown in the figure. When you are standing, your base of support is the area under you as bounded by the edges of your feet. Do you think the fellow in the figure could touch his toes? Probably not. His center of gravity is a bit forward, owing to an ample diet. If he bends over and his center of

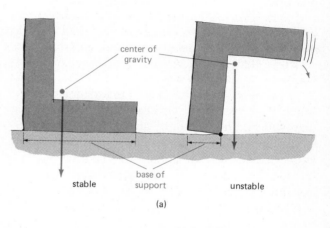

center of gravity

stable   base of support   unstable

(a)

(b)

Figure 6.15
Stability. (a) A body is stable when its center of gravity lies vertically above and inside its base of support. (b) Can he touch his toes?

gravity falls outside his base of support, over he'll go. Maybe he would then touch his toes, but not in the desired manner.

When watching someone touch his toes, notice that his hind parts are extended backward somewhat so that the center of gravity is adjusted to be over the base of support. Ask him to try touching his toes while standing with his heels against a wall and see what happens.

When you stand erect, your center of gravity lies within your body. The center of gravity of an average woman is lower than that of an average man of the same height. This is because women generally have larger pelvises and men have broader shoulders (different mass distributions). As a result, women have greater stability.

For example, women can often touch their toes more easily than men. This center of gravity property is used in a common classroom demonstration and party game. First you stand with your feet together two foot lengths (length of one's feet) from a wall with your head against the wall, and then you pick up a chair (Fig. 6.16). The challenge is to exert yourself and stand up straight.

In general, women can do this and men cannot. When bent over with the head against the wall, a man will find his center of gravity to be more forward than a woman's. When the chair is picked up, it is enough to shift the combined center of gravity outside the feet base of support for most men, but not quite or only slightly for women. Hence, by exerting backward (shifting their muscles), women

can stand up and men cannot. Try this with someone of the opposite sex and prove it to yourself.

As shown in Fig. 6.16(b), a male student not wishing to seem less agile than his female friend can use scientific principles to show that he can "stand up" to the challenge. The instructions are usually to "put your head against the wall, pick up the chair, and stand up." So, this student picked up the chair and swung it behind his back. This shifts the center of gravity of the combination backward, and he too can stand up. Next time the rules will probably be more specific.

Question: Suppose you and your classmates get together and stack your textbooks as shown in Figure 6.17, with each book displaced 4 cm from the one beneath it. How many books could be stacked before they topple over?

Answer: The stack will fall over when the center of gravity of the stack is not above the base of support (the table). The textbooks are fairly uniform, so their centers of gravity are at or near their centers. (How could you determine this for sure?) Hence, when the second text is placed on the first one, their centers of gravity are 4 cm apart. The center of gravity of the combination of both books is midway between, or displaced 2 cm to the right.

Figure 6.16
A test of stability. (a) In general, men cannot pick up a chair, as shown, and stand up, but women can. For a man, when leaning against the wall and holding the chair, the center of gravity falls outside his base of support (feet). (b) However, by swinging the chair behind his back and shifting the center of gravity, this male student was also able to stand up.

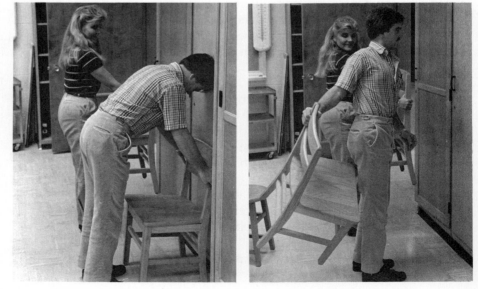

(a)   (b)

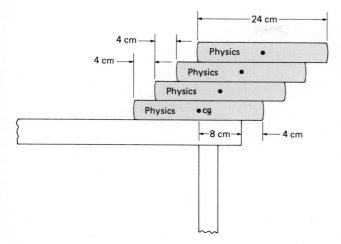

Figure 6.17
How many books can be stacked before they topple?
See Question and Answer.

The third book will displace the combined center of gravity another 2 cm to the right, and so on. The edge of the table is 8 cm from the center of the bottom book, so when four books are stacked on the table, the center of gravity is still above the edge of the base of support (the table), having been displaced 6 cm to the right.

But the fifth book will put the stack in unstable equilibrium. The stack may not fall if the fifth book is positioned *very* carefully, but it is doubtful if this could be done in practice. Just placing the fifth book on the stack would probably cause it to fall. Try it and see.

ROTATIONAL ENERGY

When we talked about kinetic energy in Chapter 2, we ignored rotational energy. We were generally focusing on the translational motion of the center of mass "particle." But a rotating object has rotational kinetic energy. Consider, for example, the pure rotational motion of a disk rotating about a fixed axis through its center (of mass). There is no translational motion, but there is motion, and hence kinetic energy—the energy of motion.

The **rotational kinetic energy** is given by $\frac{1}{2}I\omega^2$, where I is the moment of inertia of the body and ω its angular velocity. (Notice the similarity to the expression for translational kinetic energy, $\frac{1}{2}mv^2$.)

A body may have both translational and rotational kinetic energies. For example, when a ball or cylinder rolls down a hill or incline, the center of mass "particle" moves translationally and the other particles of the rigid body are in rotation.

The conservation of energy also applies. The potential energy of the body at the top of the incline (mgh) goes into both translational and rotational kinetic energies. For example, as illustrated in the figure in Special Feature 6.1, if a pair of cylinders start from rest and one rolls while the other slides freely, then the sliding cylinder moves down the incline faster than a rolling cylinder because the potential energy goes only into translational kinetic energy. For the rolling cylinder, part of the potential energy goes into the rotation of the body. A solid cylinder rolls faster than a hollow one. See Special Feature 6.1.

ANGULAR MOMENTUM

Translational or linear momentum *(mv)* involves objects moving in straight lines. In the case of circular motion, there is rotational or angular momentum. For a particle of mas *m* moving in a circle, the **angular momentum** is defined as follows:

Angular momentum = *mvr*

where *r* is the radius of the circular path and *v* is the instantaneous tangential velocity. For a rigid body, you could probably guess by analogy that the angular momentum is *Iω* (compare with *mv*).

Like the conservation of linear momentum (in the absence of an unbalanced force), the **conservation of angular momentum** is helpful in explaining many things. The conservation of rotational or angular momentum is stated as follows:

In the absence of an unbalanced torque, the total angular momentum of a system is conserved.

There are many interesting and important examples of the use of conservation of angular momentum. For example, the changes in the orbital speed of the Earth can be explained by the conservation of angular momentum. These changes were explained by Kepler in terms of his law of equal areas (Chapter 4). However, the necessity for different speeds in different parts of the orbit is more easily seen using the conservation of angular momentum. Recall that the planets follow elliptical

The Great Cylinder Race

The race is between two cylinders, one solid and one hollow (see figure). They are released at the same time from rest at the top of the incline, and the race is on. Which is your favorite? Which rolling cylinder will reach the bottom of the incline first?

Actually, the race is "fixed" (in the scientific sense). The solid cylinder will roll faster. In fact, any solid cylinder *always* beats any hollow one to the bottom. It makes no difference whether they have the same mass or outer diameter.

You may be quick to point out that a hollow cylinder has its mass concentrated farther from its axis (of symmetry) and therefore has greater rotational inertia. This is a good point, but it is possible for a solid cylinder to have a greater moment of inertia than a hollow cylinder by selecting appropriate masses and radii. It would seem that such a solid cylinder would have more rotational inertia and lose the race, but it doesn't. A solid cylinder may have a greater moment of inertia

than a hollow one; however, the moment of inertia *per unit mass* of a solid cylinder is smaller.

There is no quandary if the conservation of energy is analyzed. It turns out that for a solid cylinder, one third of the potential energy always goes into rotational motion and two thirds into translational motion. More than one third of the potential energy goes into the rotational motion of the hollow cylinder. (Your instructor may wish to show you this with some simple mathematics.)

You can experiment with some solid and hollow cylinders yourself and prove that solid cylinders will always beat the hollow ones to the bottom of an incline. Food cans make good solid cylinders, and the cans with the food removed and the ends cut out are hollow cylinders, as are napkin rings, and so on. Keep in mind that this is a rigid-body experiment. A can of juice or soda isn't a rigid body. Why?

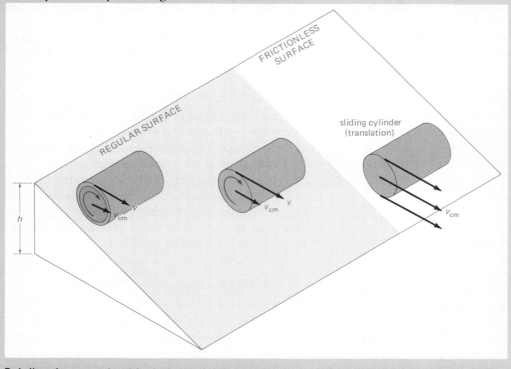

Rotational energy. An object at rest at the top of the incline has potential energy (mgh). When the object rolls down the incline, the potential energy goes into both translational and rotational kinetic energies, $KE = \frac{1}{2}mv^2 + \frac{1}{2}I\omega^2$. A solid cylinder will always beat a hollow cylinder when both are released together from rest. If an object slides down a frictionless surface, the potential energy goes only into translational kinetic energy.

orbits (Kepler's first law). Referring to Figure 6.18, there is no net torque on the Earth. The gravitational force acts through the axis of rotation (revolution) through the Sun, and so the lever arm is zero. Hence, the angular momentum is conserved.

Then, the product of mvr is constant, and if r becomes smaller, v becomes larger and vice versa. Consequently, the Earth's orbital speed is greater when it is nearer the Sun.

Notice that if the $I\omega$ expression for angular momentum is constant, then I and ω can change and the product will still be constant. For example, the products $3 \times 4 = 2 \times 6 = 1 \times 12 = 12$ are all constant (equal to 12). Varying I and ω is the basis of several applications of the conservation of angular momentum, as discussed in Special Feature 6.2.

Angular momentum, like its linear counterpart, is a vector. Hence, if angular momentum is conserved, the magnitude *and* direction do not change. That is, the direction of the angular momentum vector remains fixed in space.

This is the principle used in spirally passing a football and in giving a rifle bullet a spiral rotation (by the rifling in the rifle barrel) to prevent them from tumbling in flight. In the absence of a torque, the angular momentum vector is constant in direction, which gives greater accuracy in throwing a football or shooting a rifle bullet. This is also the

principle of the gyrocompass used to navigate ships and airplanes. If you've got a known, constant direction, you've got it made.

Another very practical application of conservation of angular momentum is in helicopters (Fig. 6.19). Suppose a helicopter had only one set of rotor blades. When the helicopter is sitting on the ground with the engine off, its angular momentum is zero. The helicopter engine supplies the (internal) torque to start and maintain the rotor blades in motion. If the helicopter lifts off the ground, the helicopter body would rotate in the opposite direction to that of the rotor blades to conserve angular momentum (equal angular momentum vectors in opposite directions), which is a bit undesirable.

As a result, large helicopters are equipped with two sets of rotor blades that rotate in opposite directions. On smaller helicopters with a single set of overhead rotor blades, an "antitorque" tail rotor provides a torque to prevent the rotation of the helicopter body.

(a)

(b)

Figure 6.19
Conservation of angular momentum one more time.
(a) Large helicopters have two rotors that rotate in opposite directions to balance the angular momentum. With a single rotor, the helicopter body would rotate to conserve angular momentum. (b) On small helicopters, an "antitorque" tail rotor provides a torque to prevent the rotation of the helicopter body.

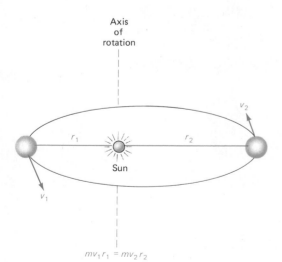

Figure 6.18
Conservation of angular momentum. The Earth revolves about an axis of rotation passing through the Sun. Since there are no external torques, the angular momentum is conserved and the Earth travels faster when closer to the Sun, which was the result of one of Kepler's laws.

Conservation of Angular Momentum in Action

When angular momentum is conserved, $I\omega$ is constant. I and ω can change, but the product is still constant. In particular, the angular speed of an object can be changed by changing I by rearranging the mass distribution of the object. If this is done within or *internal* to the system, the angular momentum is still conserved (no *external* torques). This is what a diver does in a "tuck" (smaller I) to rotate faster during a dive (see figure). To reduce the rotation, the diver may "pike" (greater I), then "layout" (even greater I) to break the water cleanly. Gymnasts use similar techniques for twisting somersaults, for example, in dismounts from the horizontal bar and parallel rings where landing on the feet is essential.

(a)   (b)

Figure 1.
The conservation of angular momentum in action. (a) A diver spins by "tucking" and decreasing the moment of inertia. By the conservation of angular momentum, this causes the angular speed to increase. (b) An ice skater uses the same principle in doing toe spins.

Cats use a similar technique in always landing on their feet when dropped from a sufficient height. While falling, a cat changes its moment of inertia by instinctively reorienting its legs and tail. By the proper twisting reorientation it is possible for the head to rotate one way and the feet the other, so that the feet are downward or underneath when the cat lands.

The conservation of angular momentum is also used by ice skaters in doing spins (see Fig. 1) and by ballet dancers in doing pirouettes. Sweeping motions of the arms and legs start the body in rotational motion. When pivoting on the tip of the skate blade or on the tip of the toe, the skater or dancer is virtually free of external torques. The arms are then "tucked in" or raised vertically overhead to reduce the moments of inertia of their bodies, and the skater and dancer spin rapidly to conserve angular momentum.

Should you lack these talents, you can still get in the act (of conservation of angular momentum). All you need is a rotating stool or turntable and some weights (see Fig. 2). This is a popular lecture demonstration. With the feet off the floor, you'll need someone (an external torque) to get you started. Try to get yourself rotating when isolated on the stool, as shown in the figure (or by standing on a turntable). You'll learn to appreciate Newton's rotational first law.

Don't start rotating too fast at first. When you bring in the weights toward your body (reducing I), you'll spin quite fast. You don't want to get dizzy, go into unstable equilibrium, and fall off the stool.

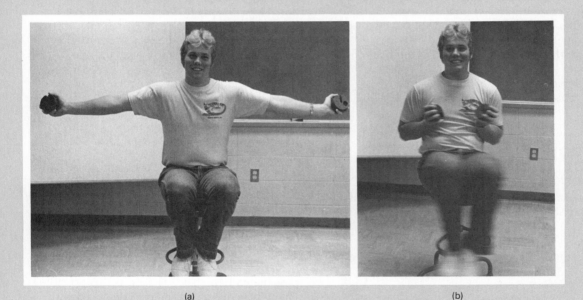

(a)   (b)

Figure 2.
A demonstration of the conservation of angular momentum. (a) When the person's arms are extended, there is a larger moment of inertia and the rotation (angular speed) of the stool is slow. (b) When the arms and weights are brought inward, the moment of inertia decreases and the angular speed increases.

SUMMARY OF KEY TERMS

Particle: a physical point concept that has no physical dimensions but may have mass and can be accurately localized in space.

Rigid body: a system of particles in which the particles are fixed, constant distances apart.

Center of mass: the average location of all the mass particles of a body, or the point at which all the mass of a body can be considered to be concentrated.

Center of gravity: the average location of the weight distribution of a body, or the point at which all the weight of a body can be considered to be concentrated. In a uniform gravitational field the location of the centers of gravity and mass is the same.

Pure translational motion: motion in which every particle of a body has the same instantaneous velocity, and the path of each particle is the same as that of the body as a whole.

Pure rotational motion: motion in which every particle of a body moves in a circle about an axis of rotation.

Angular speed: the time rate of change of angular position, $\omega = \theta/t$.

Angular acceleration: the time rate of change of angular velocity, $\alpha = \Delta\omega/\Delta t$.

Rotational inertia: the property of matter that resists changes in rotational motion.

Moment of inertia: a measure of the rotational inertia of a body.

Lever arm: the perpendicular distance from the axis of rotation to a line along which a force acts.

Torque: the product of the lever arm and force. Torques are capable of producing rotational motion and changes therein.

Newton's first law for rotational motion: a rigid body remains at rest or in motion with a constant angular velocity unless acted on by an unbalanced torque.

Newton's second law for rotational motion: torque = moment of inertia × angular acceleration, $\tau = I\alpha$.

Stable equilibrium: the condition of a body when its center of gravity is vertically above an edge or point such that a slight displacement will cause the body to topple.

Rotational kinetic energy: energy of rotational motion, $KE = \frac{1}{2}I\omega^2$.

Angular momentum: (particle in circular motion) mvr, (rigid body) $I\omega$.

Conservation of angular momentum: in the absence of an unbalanced torque, the total angular momentum of a system is conserved.

QUESTIONS

Center of Mass and Center of Gravity

*1. Is it possible for a rigid body to have a center of mass and not a center of gravity? Explain.

Figure 6.20
Up and over, but does his *cg* go over? See Question 4.

2. Where is the center of gravity of a boomerang located? How could you determine the location of a boomerang's center of gravity?

3. Is it possible for a person's center of gravity to fall outside his body? Explain.

4. When a pole vaulter goes over the bar (Fig. 6.20), his or her center of gravity may actually pass beneath the bar. (a) How is this possible? (b) Why is this an advantage compared with having the center of gravity go over the bar? (*Hint:* Think in terms of energy.)

5. In finding the center of gravity of an object by suspending it from different points, do you detect any advantage in suspending it from three or four points? Explain.

*6. Where is the center of gravity of the mobile shown in Figure 6.21?

Translational and Rotational Motions

*7. Suppose someone in your class said that it was possible for a body to have pure translational motion and pure rotational motion at the same time. Would you agree with them? Why?

Figure 6.21
The center of gravity is here somewhere. See Question 6.

8. When you pitch a softball or baseball, does it ever have pure translational motion?
9. Discuss the motions of a wheel of an automobile.
10. Describe the motion(s) of a Frisbee in flight (Fig. 6.22). Also discuss its energy and angular momentum.

Description of Rotational Motion

*11. How many radians are there in (a) 360°, (b) 180°, and (c) 90°? (*Hint:* Express in terms of π.)

Figure 6.22
What kind of motion have we here? See Question 10.

*12. What are the angular speeds of (a) the second hand, (b) the minute hand, and (c) the hour hand of a clock?
13. Consider the relative motion of a phonograph needle in the grooves of a record. Does the needle have a greater angular speed near the outer portion of the record or near the center of the record? How about tangential speed?
14. Two children ride on a merry-go-round, one on a seat near the outer edge and the other on a horse nearer the center. Which has the greater (a) angular speed and (b) tangential speed?
*15. Two songs of equal length are put on a phonograph album, one at the beginning or outside of the record and the other near the center of the record. Will the band widths of the selections be the same? Explain.
*16. A woman was given a speeding ticket because a patrol officer's radar showed her car to be exceeding a 55 mi/h speed limit. At the hearing, three witnesses in the car at the time testified that the speedometer read exactly 55 mi/h. Yet the judge imposed a fine when he learned that new, oversized tires had recently been installed on the car. Was this a fair ruling? Explain. Would the new tires have any effect on the odometer (mileage meter) reading?
17. A linear acceleration can change the magnitude and/or direction of a linear velocity. Can an angular acceleration have the same effects on the rotational velocity of a body? If so, describe what the effect might be.

Rotational Inertia

18. A type of potter's wheel consists of a heavy "kickwheel" connected by a shaft to a smaller and lighter working wheel on which the clay is spun (Fig. 6.23). The kickwheel is rotated by foot, leaving the potter's hands free to work the clay. The working wheel rotates evenly, even though the energy from the kicking is delivered periodically. Explain why.
*19. Compare the rotational inertia of a baseball bat about axes through (a) the length of the bat, (b) the handle end, and (c) the opposite (larger) end.
*20. When a front-engine automobile is stopped suddenly, why does the front end go down and the back end come up? How would a rear-engine car behave?

Torque

21. Is it possible to have forces applied to a body such that the net force is zero but the net torque is not? (*Hint:* Consider a rod that can be rotated about a fixed axis at its center and equal forces applied perpendicularly to the ends of the rod.)

Figure 6.23
Foot power. See Question 18.

*22. When pedaling a bicycle, when do you apply the maximum and minimum torques? (Assume the downward force of your foot is constant.) What is the value of the minimum torque?

*23. A stubborn screw won't come loose when you try to remove it with a screwdriver. If you had two other screwdrivers available, one with a metal shaft of the same diameter and a bigger handle and one with the same diameter but a longer metal shaft and the same size handle, which one would you use to give the screw another try?

24. Why is the steering wheel of a tractor trailer rig generally bigger than that of an automobile? Wouldn't a smaller steering wheel give the driver more room? Would power steering of the type on an automobile that allows the steering wheel to be rotated quickly with one finger be a good thing to put on trucks? Explain.

25. The center of gravity of a rod or stick can be located by balancing, as in Figure 6.1. Explain this in terms of torque. (*Hint:* Note that a balanced stick does not rotate about the finger axis.)

26. Explain the principle of a double-pan balance in terms of torque. Such balances are used to weigh gold and precious stones.

*27. A rod with end weights is balanced, as shown in Figure 6.24. If the big ball weighs 4 N and the little ball weighs 1 N, how far is the center of gravity of the system from the little ball? (Assume the rod is 1 m long, and neglect its mass.)

28. Where would you say the centers of mass of the following systems would be? (a) the Earth and moon, (b) the Sun and the Earth, (c) the solar system. (Check with your instructor to see how close you are.)

*29. When an object, such as a ball or cylinder, rolls down an inclined plane from rest, what produces the rotational motion? (*Hint:* Discuss in terms of center of gravity.)

30. Would a rotational form of Newton's third law be valid? (For every torque is there an equal and opposite torque.)

Stability and Equilibrium

31. Why do a lower center of gravity and wider base of support give an object greater stable equilibrium?

*32. We have all balanced a vertical pole or stick on our fingers and quickly learned that a longer pole is easier to balance than a shorter one (for example, a meter stick and a pencil). Why is this so when a shorter pole has a lower center of gravity? (*Hint:* Think in terms of rotational inertia.) Also, explain how you keep the long pole balanced.

33. A tightrope walker often carries a long, droopy pole. As was learned in the chapter, the pole increases the rotational inertia. Does the droopiness of the pole make a difference?

34. A common child's toy with a rounded, weighted bottom always rights itself when knocked over. Explain why this toy can't be knocked down to stay.

*35. Given a bunch of bricks 21 cm long, how many bricks could be stacked on the bottom brick

Figure 6.24
Where do we tie the string? See Question 27.

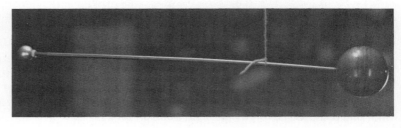

before the stack falls if each stacked brick is displaced 2 cm from the edge of the one beneath it?

36. Is a tall person with big feet more stable than a shorter person with smaller feet? (Don't forget height and center of gravity.)

37. (a) What is the most stable position you can assume? (b) Why is it easier to do sit-ups when your arms are extended in front of you than when your hands are behind your head?

38. Does putting "big wheels" on a pickup truck make it more stable (Fig. 6.25)? Explain. (What is the purpose of the bar behind the cab in the photo?)

•39. How does standing with your legs apart, for example, on a moving bus, increase your stability? Is the stability increased in all directions? (Better hang on.)

40. Does it make any difference how the trailer of a tractor-trailer rig is loaded? (Consider both sideways and vertical-load distributions for a truck that will go around some banked curves on a trip.)

41. The Leaning Tower of Pisa leans because of settling under one side of its base [see figure in Special Feature 1.3]. (a) If this continues, when will the tower fall? (b) Corrective actions have been taken.* What else might be done to keep the tower from toppling?

•42. How many positions of stable and unstable equilibrium are there for a cube? (Consider that resting on each side or edge is a different position.)

43. Explain the balancing act shown in Figure 6.26. (There's no one holding the ends of the fork and spoon.)

Angular Momentum

44. Considering the particles making up a rotating disk, do they all have the same angular momentum? Explain.

45. Explain how a high diver can do several spins while falling toward the water.

•46. Suppose you are standing on a large turntable near the outer edge and you start to walk around the outer circumference of the table. How would you move relative to someone watching

---

* The Leaning Tower (56.2 m high) was begun in 1174 and completed in the 14th century. The settling began in the early stages of construction, and the third, fifth, and top (eighth) stories were straightened somewhat to compensate for the lean. Its foundation is only as wide as the circumference of the tower. The foundation has been injected with cement, but the tower is still threatened with collapse. The top of the tower leans about 5 m from the vertical.

Figure 6.25
Big wheels and stability. See Question 38.

you who is not on the turntable? What would occur if you walked in a spiral path toward the center of the turntable? (Consider the turntable bearings to be frictionless.)

•47. Unlike helicopters, single-engine aircraft have only one propeller (or set of rotor blades). What keeps the body of the airplane from rotating? (*Hint:* Think in terms of torques from air foils.)

48. It is difficult to balance yourself when sitting on a stationary bicycle, yet it is quite easy to do so when riding. The base area (of the tires) is the same in both cases, so why does a stationary bicycle tend to fall over and a moving bicycle does not? (*Hint:* When you lean to one side while riding a bicycle, what does the bicycle do?)

Figure 6.26
A balancing act. See Question 43.

THE PROPERTIES OF MATTER

II

The 450-mile California Aqueduct is an engineering achievement.

PART II

The structure and properties of matter have been major topics of science from the earliest times. We may simplistically define matter as a fundamental property or anything that occupies space. However, the questions of what is the basic constituent of matter and why different substances behave differently still remain.

The early Greeks addressed the problem of the makeup of matter in terms of atoms. According to Democritus, who lived about 450 B.C., "The only existing things are the atoms and empty space; all else is mere opinion." As scientific focus expanded to include the atomic microcosm as well as the celestial macrocosm, theories and new discoveries began to fall into place. Today, the general properties of matter are described and understood in terms of atomic theory. This involves not the indivisible atom of the Greeks, but an atom with structure and particle makeup.

The ultimate constituent of matter is still a question of modern physics. However, atomic theory provides us with an insight into the general makeup of substances. That is, how the various atoms of matter are combined generally determines the properties of the solids, liquids, and gases we experience in everyday life.

Atoms, Molecules, and Matter

7

Processed grains of table salt are round and smooth to ensure free pouring from the container. (Scanning electron micrograph.)

ATOMIC THEORY

We have discussed the concept of matter, which is generally defined as anything that occupies space. You, the kitchen sink, your physics book, the stars —everything we see is matter, and so are some things we don't see, like air. But what is matter made of?

Suppose you cut a piece of matter in half, then cut one of the halves in half, and so on into smaller and smaller pieces. Would you ever reach a "tiniest" piece of matter, or is matter a continuous thing? This was a major question of science down through the centuries.

The notion that matter consists of discrete particles is an old one. The idea was expressed by Democritus, a Greek philosopher, around 450 B.C. According to Democritus, there was a limit to the subdivision of matter. He called the ultimate particle an **atom**, after the Greek word *atomos,* meaning "indivisible."

The atoms of Democritus were in perpetual motion and could not be created or destroyed. He implied that all of our experience is the result of the rearrangements of atoms. All these ideas have a ring of truth according to our modern concepts, as we shall see. (Democritus also thought that atoms could not fuse or combine, which wasn't such a good idea.)

However, Democritus's ideas were rejected by Plato and Aristotle, who believed matter to be continuous. Because of their influence, the atomic theory wasn't accepted at the time and lay dormant for about 20 centuries. In the 1600's the concept of atoms was suggested again by the Italian physicist Gassendi (Fig. 7.1). This time there was some influential support for the idea. Sir Isaac Newton wrote:

> It seems probable to me that God, in the Beginning, formed Matter in solid, massy, hard, impenetrable, movable Particles, of such Sizes and Figures, and with such other Properties, and in such Proportions to Space, as most conduced to the End for which He formed them. . . .

Prior to the 1800's the concept of the atomic nature of matter was basically speculative. Then in 1808 John Dalton, an English chemist and school teacher, developed explanations of several laws of chemistry using atomic theory. Some of his ideas had to be discarded, but the essentials of atoms have

Figure 7.1
Big names in atomic theory.

withstood the test of time and are with us today. In his atomic theory, Dalton referred to the chemical combination of two or more atoms as a **molecule**.

More direct evidence for the atomic theory was accidentally discovered in 1827 by the Scottish botanist Robert Brown. While observing pollen grains suspended in water under his microscope, he noticed that the tiny grains zigzagged about in an agitated manner (Fig. 7.2). Dust particles were also observed to do a similar dance.

This phenomenon is called **Brownian motion**, in honor of the observant Scot, and is readily explained by atomic "particle" theory. The pollen grains are knocked about by a barrage of randomly moving water molecules (collisions and momentum transfer as in Chapter 3). Oddly enough, it was

Figure 7.2
Brownian motion. Pollen grains suspended in water being knocked about by a barrage of randomly moving water molecules give evidence of the atomic "particle" theory.

(a)

Albert Einstein who first explained Brownian motion in 1905, the same year he published another paper on his theory of special relativity (Chapter 23). Perhaps the great man relaxed by thinking of slower-moving objects such as pollen grains.

Brownian motion and other evidence gathered in the last century have established the atomic theory of matter. Today, we even have pictures of atoms (Fig. 7.3). These are not ordinary pictures taken using visible light. The dimensions of the light waves are too big to be affected by individual atoms. To be "visible" or detected, a particle must be large enough to influence the wave or disturbance. To "see" atoms, one must use the high-energy electron beam in an electron microscope (Chapter 24). Here, the electron beam disturbances or waves are more than a thousand times smaller than those of visible light, and atoms can be detected.

So, considering the atomic theory, typical questions might be: How big are atoms and molecules? How many are there in a quantity of matter? Do atoms have structure, or are they "billiard ball" particles? Let's take a look at the answers to these questions.

Sizes and Numbers of Atoms and Molecules

Atoms are extremely small. From quantitative studies of Brownian motion, Einstein estimated atoms to be on the order of 0.00000001 cm (1×10^{-8} cm) in diameter. This was a good "ballpark" figure. More modern measurement methods have shown nitrogen to be the smallest atom, with a diameter of about 0.0000000106 cm (1.06×10^{-8} cm). The largest atom is cesium, a real monster in size with a diameter of about 0.000000054 cm (5.4×10^{-8} cm).*

The size of a molecule of course depends on the sizes of its constituent atoms and how they are put together. For example, a water molecule (H_2O) has a length of about 3×10^{-8} cm. This is a curved length. The hydrogen atoms hook on the oxygen atom at an angle, so the molecule has a bent crescent-moon shape (see Fig. 7.7).

* To avoid these small decimals, scientists often use a smaller units of length, the nanometer, 1 nm = 10^{-9} m = 10^{-7} cm, and angstrom (Å). 1 Å = 10^{-10} m = 10^{-8} cm. Hence, the nitrogen atom has a diameter of 0.106 nm or 1.06 Å and the cesium atom a diameter of 0.54 nm or 5.4 Å.

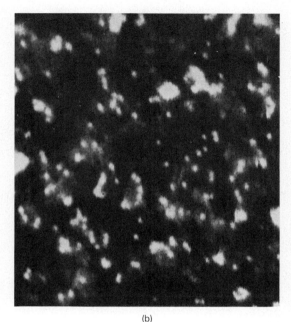

(b)

Figure 7.3
Atoms. (a) An electron microscopy image of a complex organic molecule. Four mercury atoms *(dark spots)* are clearly visible. (b) Single uranium atoms shown as small bright spots magnified more than five million times by an electron microscope. The large white spots are clusters of atoms.

There are also huge macromolecules found in plastics and living organisms. Here, the molecules are long chains of atoms (Fig. 7.4). For example, if one of the DNA molecules in the human cell were stretched out, it would be several centimeters long.

How many atoms are there in the world? So many that the number would be difficult to comprehend. Let's think on a smaller scale. Every time you breathe, you take in about a half liter of air. In this quantity of air there are about 10,000 billion billion gas molecules — 10^{22} molecules and about twice that many atoms, since the gas molecules of the air are mostly diatomic (two-atom) molecules.

On a hot day, you may easily drink a liter of water. (Recall that this is a little more than a quart.) How many molecules are you consuming? About

Figure 7.4
DNA model showing a two-turn helix. The DNA molecule is a macromolecule with a spiraling helix form. These molecules in human cells would be several centimeters long if stretched out.

Table 7.1
Relative Abundance of Atoms

Order of Atomic Abundance (number (% by weight) of atoms)	In Human Body	On Earth's Surface*
	(% by weight)	
1	Hydrogen (10)	Hydrogen (5.4)
2	Oxygen (60)	Oxygen (66)
3	Carbon (18)	Silicon (14)
4	Nitrogen (3.0)	Aluminum (4.1)
5	Calcium (1.8)	Sodium (1.9)
6	Phosphorus (1.1)	Calcium (1.8)
7	Sulfur (0.22)	Magnesium (1.1)
8	Sodium (0.13)	Iron (2.5)
9	Potassium (0.20)	Potassium (1.3)
10	Chlorine (0.15)	Chlorine (0.96)

* All of the oceans plus an equal mass of crust. (Turiel, I., *Physics, the Environment and Man,* Prentice-Hall, Englewood Cliffs, N.J., 1975.)

3.3×10^{25}.† Being a sharp student, you might be quick to ask, How do you know? Who counted the molecules? No one, of course. The numbers are computed from mass considerations and chemistry principles.

Table 7.1 shows the order of the abundance of atoms found in the human body and on the Earth's surface. Take a look and see what you're made of. Also, notice how similar the atomic abundances between you and your environment are. There are some differences. Silicon and aluminum are tightly bound to oxygen in the environment and are not available for chemical reactions in the body. An aspect of this atomic interchange is discussed in Special Feature 7.1.

Atomic Structure

Atoms are not the "indivisible" particles of Democritus and Dalton. Atoms have structure and are made up of subatomic particles—electrons, protons, and neutrons. A common, oversimplified model of the atom resulting from an early modern theory (see Chapter 24) is often referred to as the solar-system model (Fig. 7.5).

Similar to the planets orbiting the Sun, the electrically negative (−) electrons orbit about the much more massive, electrically positive (+) charged protons and uncharged neutrons, which are grouped very closely in the center nucleus of the atom. The electrons orbit the nucleus in three dimensions, with the electrical force between an electron and the nuclear protons supplying the necessary centripetal force to keep each electron in orbit.

Unlike planets and Earth satellites, the electrons can only be in certain orbits at particular distances from the nucleus and cannot orbit at intermediate distances. These specific orbits, sometimes called shells, can contain only a certain number of electrons—the innermost shell, two electrons; the next shell, eight electrons; the next, 18 electrons; and so on.

Most of the volume of the atom is empty space—again, much like the solar system. The electrons contribute very little to the mass of the atom. A proton or neutron is almost 2000 times more massive than an electron, so most of the atomic mass is in the nucleus (over 99.9 percent).

† Written out so as to look impressive, this is 33,000, 000,000,000,000,000,000,000 molecules.

Recycled Atoms

Atoms are ageless. They just get recycled around, usually in molecules, but they never change (except in relatively infrequent nuclear processes). Suppose you were a carbon atom in a plant that lived millions of years ago. Then you got buried and layers of dirt piled up on top of you over the years in geological processes. Its weight drove away some of your water molecule friends and neighbors, and you found yourself in a bed of coal.

After lying around for awhile, someone in the 20th century dug you up and threw your lump into a fire. The next thing you know you find yourself rubbing shoulders and combining with a couple of oxygen atoms in the atmosphere (becoming CO_2).

Then, some plant doing its photosynthesis thing takes you in, and you wind up in a sugar molecule that the plant uses for growth. But a cow comes along and eats the plant. After trips through two stomachs, you eventually end up as a bovine carbon atom, but not for long.

The next thing you know, you're sitting on a plate as a part of a USDA Choice cut of meat and are about to be consumed by a strange-looking creature with a knife and fork in hand.

After a hard day or two of chemical processes, you proudly become an integral part of the highest order of animal life on Earth. Who knows, if you're lucky you may make it to the top and become part of a brain cell.

Where does our carbon atom character go from there? You can write the next scenario. There are many possibilities.

But how is one kind of atom distinguished from another kind, for example, a carbon atom from a nitrogen atom? Or, put another way, what makes an atom that of a particular species? For instance, what makes an oxygen atom an oxygen atom?

The species or kind of atom is determined by the atomic number or proton number, which is simply the number of protons in the nucleus of the atom. For example, an oxygen atom is oxygen because it has eight nuclear protons. All atoms with eight protons are oxygen atoms. Similarly, a carbon atom can be distinguished from a nitrogen atom because the carbon atom has six protons and the nitrogen atom has seven protons, by definition.

You might think that you could use the number of electrons to determine the atomic species, since

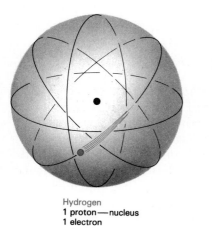

Hydrogen
1 proton—nucleus
1 electron

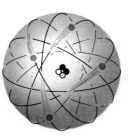

Helium
2 protons } nucleus
2 neutrons
2 electrons

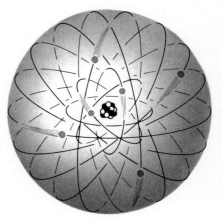

Carbon
6 protons } nucleus
6 neutrons
6 electrons
(in two shells)

Figure 7.5
Atomic structure. The solar-system model of the atom views the electrons in orbits about a central nucleus, analogous to the planets orbiting the Sun.

there is the same number of electrons as protons in a neutral atom. This is true, but an atom may be ionized by removing or adding electrons. An **ion** is an atom or molecule that has a net electrical charge. But, even if one or more electrons are added or removed from an atom, it is still the same atomic species, for example, a nitrogen ion (N^+) with seven protons and six electrons. However, if the proton (atomic) number of an atom somehow changed by a nuclear process, the atomic species would change.*

Then there are the neutrons. One species of atom may have different numbers of neutrons. These are called isotopes; they have the same number of protons but different neutron numbers. For example, natural uranium is 99.3 percent uranium-238 (92 protons and 146 neutrons) and 0.07 percent uranium-235 (92 protons and 143 neutrons). These atoms are (nuclear) isotopes of uranium.

■ ELEMENTS AND THE PERIODIC TABLE

Elements are a special class of substances. Chemically, an element cannot be decomposed into a simpler substance. From the previous discussion, the definition of an element is easy. An **element** is a substance in which all of the atoms have the same number of protons.

The ancient Greeks thought that all matter was composed of four "elements" — earth, air, fire, and water. Just four elements would have made science a lot easier but a lot less exciting. Of course, this notion was inadequate and incorrect. Even in the very early civilizations, nine of what we now call elements were isolated: gold, silver, copper, iron, lead, tin, mercury, sulfur, and carbon.

Until about the 17th century, chemistry developed rather slowly. Prior to that period, it was chiefly a by-product of alchemy. A main goal of the alchemists was to change common metals into gold. There also was a sideline of trying to develop a substance that would restore youth. They didn't have much luck with either goal, but their chemical experiments added five new elements to the known list: arsenic, antimony, bismuth, phosphorus, and zinc.

Today, the elements listed number around 106. Table 7.2 gives an alphabetical listing of the elements along with their chemical symbols, atomic

(proton) numbers, and atomic masses. There is a dispute between American and Soviet laboratories as to who discovered elements 104, 105, and 106 (not listed). The suggested names for elements 104 and 105 are Rutherfordium and Hahnium, respectively. No name has yet been suggested for element 106. The atomic mass is a relative measure based on the common carbon atom (six protons and six neutrons), which is given a value of 12.0000.*

You will notice that carbon (and other elements) do not have whole-number atomic masses. This is because of isotopes. For example, natural carbon has three isotopes, but it is mainly composed of the common isotope with six neutrons. The other two isotopes, with seven and eight neutrons, contribute to the atomic mass so that the average is not a whole number.

Notice the convenient symbol notation for the elements. In general, the first letter of an element's name is used, along with the next or another letter when there is more than one element beginning with the same letter, for example, C, Ce, and Cs for carbon, cerium, and cesium, respectively. But we have Cl and Cr for chlorine and chromium, and there is no element with the symbol Ch. There is really no fixed method for assigning symbols. The official symbols are adopted by an international chemistry organization.

> *Question:* If the first letter of an element's name generally appears in its symbol, why are K, Na, and Pb used for potassium, sodium, and lead, respectively?
>
> *Answer:* Because the symbols were taken from their original Latin names, *kalium, natrium,* and *plumbum,* respectively.

By 1869 some 63 elements had been discovered. Chemists tried to categorize or classify the growing list of elements, for example, as metals and nonmetals. However, this classification was much too broad, and a better method was desired. It had been noticed in the early 1800's that the elements could be listed in such a way that similar chemical properties recurred periodically throughout the list. In 1869 the Russian chemist Mendeleev (pronounced "Men-duh-lay-eff") formulated a table of

---

* Such nuclear transformations are possible and will be discussed in Chapter 26 on nuclear physics.

* A small unit for mass is thus defined: $\frac{1}{12}$ of a common carbon atom is 1 atomic mass unit (amu). A common carbon atom has a mass of 12 amu, and 1 amu = 1.66×10^{-27} kg.

Table 7.2
The Chemical Elements

	Symbol	Atomic No.	Atomic Mass*		Symbol	Atomic No.	Atomic Mass*
Actinium	Ac	89	227.0278	Mercury	Hg	80	200.59
Aluminum	Al	13	26.98154	Molybdenum	Mo	42	95.94
Americium	Am	95	[243]†	Neodymium	Nd	60	144.24
Antimony	Sb	51	121.75	Neon	Ne	10	20.179
Argon	Ar	18	39.948	Neptunium	Np	93	237.0482
Arsenic	As	33	74.9216	Nickel	Ni	28	58.70
Astatine	At	85	[210]	Niobium	Nb	41	92.9064
Barium	Ba	56	137.33	Nitrogen	N	7	14.0067
Berkelium	Bk	97	[247]	Nobelium	No	102	[259]
Beryllium	Be	4	9.01218	Osmium	Os	76	190.2
Bismuth	Bi	83	208.9804	Oxygen	O	8	15.9994
Boron	B	5	10.81	Palladium	Pd	46	106.4
Bromine	Br	35	79.904	Phosphorus	P	15	30.97376
Cadmium	Cd	48	112.41	Platinum	Pt	78	195.09
Calcium	Ca	20	40.08	Plutonium	Pu	94	[244]
Californium	Cf	98	[251]	Polonium	Po	84	[209]
Carbon	C	6	12.011	Potassium	K	19	39.0983
Cerium	Ce	58	140.12	Praseodymium	Pr	59	140.9077
Cesium	Cs	55	132.9054	Promethium	Pm	61	[145]
Chlorine	Cl	17	35.453	Protactinium	Pa	91	231.0359
Chromium	Cr	24	51.996	Radium	Ra	88	226.0254
Cobalt	Co	27	58.9332	Radon	Rn	86	[222]
Copper	Cu	29	63.546	Rhenium	Re	75	186.207
Curium	Cm	96	[247]	Rhodium	Rh	45	102.9055
Dysprosium	Dy	66	162.50	Rubidium	Rb	37	85.4678
Einsteinium	Es	99	[252]	Ruthenium	Ru	44	101.07
Erbium	Er	68	167.26	Samarium	Sm	62	150.4
Europium	Eu	63	151.96	Scandium	Sc	21	44.9559
Fermium	Fm	100	[257]	Selenium	Se	34	78.96
Fluorine	F	9	18.998403	Silicon	Si	14	28.0855
Francium	Fr	87	[223]	Silver	Ag	47	107.868
Gadolinium	Gd	64	157.25	Sodium	Na	11	22.98977
Gallium	Ga	31	69.72	Strontium	Sr	38	87.62
Germanium	Ge	32	72.59	Sulfur	S	16	32.06
Gold	Au	79	196.9665	Tantalum	Ta	73	180.9479
Hafnium	Hf	72	178.49	Technetium	Tc	43	[98]
Helium	He	2	4.00260	Tellurium	Te	52	127.60
Holmium	Ho	67	164.9304	Terbium	Tb	65	158.9254
Hydrogen	H	1	1.0079	Thallium	Tl	81	204.37
Indium	In	49	114.82	Thorium	Th	90	232.0381
Iodine	I	53	126.9045	Thulium	Tm	69	168.9342
Iridium	Ir	77	192.22	Tin	Sn	50	118.69
Iron	Fe	26	55.847	Titanium	Ti	22	47.90
Krypton	Kr	36	83.80	Tungsten	W	74	183.85
Lanthanum	La	57	138.9055	Uranium	U	92	238.029
Lawrencium	Lr	103	[260]	Vanadium	V	23	50.9415
Lead	Pb	82	207.2	Xenon	Xe	54	131.30
Lithium	Li	3	6.941	Ytterbium	Yb	70	173.04
Lutetium	Lu	71	174.967	Yttrium	Y	39	88.9059
Magnesium	Mg	12	24.305	Zinc	Zn	30	65.38
Manganese	Mn	25	54.9380	Zirconium	Zr	40	91.22
Mendelevium	Md	101	[258]				

* Atomic masses given here are 1977 IUPAC values based on carbon-12.

† A value given in brackets denotes the mass number of the longest-lived or best-known isotope.

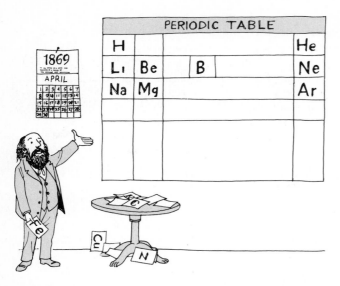

Figure 7.6
The periodic table. Mendeleev formed the periodic table by arranging the elements on the basis of the periodic recurrence of chemical properties. The elements in a vertical column have similar properties.

elements based on this periodic property (Fig. 7.6). His periodic table of the elements in its modern version is still used today and can be seen on the walls of every science building, as well as in Table 7.3.

Mendeleev arranged the known elements in rows, which are called **periods**, in order of increasing atomic masses. When he came to an element with chemical properties similar to those of one of the previous elements, he went back and put this element below the similar, lighter one. As he did this, the columns formed **groups** or families of elements with similar properties. The table was later rearranged in order of increasing atomic or proton number (the numbers at the top of the boxes in Table 7.3) because of some inconsistencies. Notice the atomic masses of cobalt and nickel (atomic numbers 27 and 28).

When Mendeleev made his table with 63 elements, there were vacant spaces in it, which he reasoned to be due to missing or undiscovered ele-

Table 7.3
Periodic Table of the Elements*

IA																VIIA	0
1 **H** 1.0079	IIA											IIIA	IVA	VA	VIA	1 **H** 1.0079	2 **He** 4.00260
3 **Li** 6.941	4 **Be** 9.01218											5 **B** 10.81	6 **C** 12.011	7 **N** 14.0067	8 **O** 15.9994	9 **F** 18.998403	10 **Ne** 20.179
11 **Na** 22.98977	12 **Mg** 24.305	IIIB	IVB	VB	VIB	VIIB	←——VIII——→			IB	IIB	13 **Al** 26.98154	14 **Si** 28.0855	15 **P** 30.97376	16 **S** 32.06	17 **Cl** 35.453	18 **Ar** 39.948
19 **K** 39.0983	20 **Ca** 40.08	21 **Sc** 44.9559	22 **Ti** 47.90	23 **V** 50.9415	24 **Cr** 51.996	25 **Mn** 54.9380	26 **Fe** 55.847	27 **Co** 58.9332	28 **Ni** 58.70	29 **Cu** 63.546	30 **Zn** 65.38	31 **Ga** 69.72	32 **Ge** 72.59	33 **As** 74.9216	34 **Se** 78.96	35 **Br** 79.904	36 **Kr** 83.80
37 **Rb** 85.4678	38 **Sr** 87.62	39 **Y** 88.9059	40 **Zr** 91.22	41 **Nb** 92.9064	42 **Mo** 95.94	43 **Tc** (98)	44 **Ru** 101.07	45 **Rh** 102.9055	46 **Pd** 106.4	47 **Ag** 107.868	48 **Cd** 112.41	49 **In** 114.82	50 **Sn** 118.69	51 **Sb** 121.75	52 **Te** 127.60	53 **I** 126.9045	54 **Xe** 131.30
55 **Cs** 132.9054	56 **Ba** 137.33	57 ★**La** 138.9055	72 **Hf** 178.49	73 **Ta** 180.9479	74 **W** 183.85	75 **Re** 186.207	76 **Os** 190.2	77 **Ir** 192.22	78 **Pt** 195.09	79 **Au** 196.9665	80 **Hg** 200.59	81 **Tl** 204.37	82 **Pb** 207.2	83 **Bi** 208.9804	84 **Po** (209)	85 **At** (210)	86 **Rn** (222)
87 **Fr** (223)	88 **Ra** 226.0254	89 †**Ac** 227.0278	104 **Unq** (261)	105 **Unp** (262)	106 **Unh** (263)	107 **Uns**		109									

★ Lathanide Series														
58 **Ce** 140.12	59 **Pr** 140.9077	60 **Nd** 144.24	61 **Pm** (145)	62 **Sm** 150.4	63 **Eu** 151.96	64 **Gd** 157.25	65 **Tb** 158.9254	66 **Dy** 162.50	67 **Ho** 164.9304	68 **Er** 167.26	69 **Tm** 168.9342	70 **Yb** 173.04	71 **Lu** 174.967

† Actinide Series														
90 **Th** 232.0381	91 **Pa** 231.0359	92 **U** 238.029	93 **Np** 237.0482	94 **Pu** (244)	95 **Am** (243)	96 **Cm** (247)	97 **Bk** (247)	98 **Cf** (251)	99 **Es** (252)	100 **Fm** (257)	101 **Md** (258)	102 **No** (259)	103 **Lr** (260)

* Atomic masses shown here are 1977 IUPAC values.

ments. Since the missing elements had properties similar to those of the other elements in a group, Mendeleev was able to predict their masses and properties. Three of the missing elements were discovered within 20 years after Mendeleev formulated his table.

There is a great deal of interesting and useful information in the periodic table. For example, all the elements above 92 are artificial; they do not occur naturally. These elements are made by nuclear processes in particle accelerators and nuclear reactors. They are radioactive, and most have very short "lives" before they disintegrate or "decay" into something else (see Chapter 25).

The elements with atomic numbers greater than 92 are called "transuranic" elements. Notice the names of the next two elements above uranium — remind you of the planets of the solar system? You'll also find that the names of the rest of these elements have a very modern ring. Check Table 7.2 and see if you can associate a person or place with their names.

So, is it safe to say that we have 92 naturally occurring elements? Well, many textbooks say that there are only 90 such elements. This is because technetium (43) and promethium (61) are not currently found in nature. They can be prepared artificially but are radioactive and decay. Some scientists believe that these elements did occur naturally along with the other 90 when the Earth was formed, but are long since gone. Incidentally, the name *technetium* comes from the Greek word meaning "artificial," and it was the first unknown element to be created by artificial means.

Only two elements, oxygen and silicon, make up 75 percent of the Earth's crust by weight. You pick up these elements in compound form when you scoop up a handful of sand (SiO_2). Many of the common elements, such as mercury and sulfur, are relatively scarce.

■ MOLECULES AND COMPOUNDS

Atoms combine to form molecules, or on a larger scale, elements combine to form compounds. A **molecule** is a group of two or more atoms held together by forces called chemical bonds (Fig. 7.7). A substance made up of more than one element is called a **compound**. Some elements occur naturally

water (H_2O)   ammonia (NH_3)   methane (CH_4)

Figure 7.7
Molecules. Molecules are groups of two or more atoms held together by forces called chemical bonds.

as molecules, for example, H_2, O_2, and N_2 (diatomic molecules of hydrogen, oxygen, and nitrogen).

The chemical bonding is associated with the electrons of the atoms of a molecule. In fact, all ordinary chemical reactions involve only the outermost electrons in atoms. The main reason that the elements in a vertical group in the periodic table have similar chemical properties is that they have the same number of electrons in their outermost shells. Molecules are formed by atoms that combine so as to have filled shells (or subshells, not discussed here).

Two general types of chemical bonds are ionic bonds and covalent bonds.

In **ionic bonds**, electrons are transferred between atoms (Fig. 7.8). The resulting ions are then held together by attractive electrical forces. In general, we cannot isolate an ionic "atom" such as NaCl because the ions are in an array and each ion is associated with several others (see Chapter 8). The ionic chemical bond is quite strong. This is evidenced by the relatively high melting-point temperatures of ionic compounds.

It is interesting to note how a simple electron transfer makes such a big difference in properties. For example, sodium (Na) is a soft, gray metal that is so reactive that it is stored under oil. Otherwise, it would react with the oxygen and moisture in the air. A piece of sodium placed on your hand would burn you quite badly. Chlorine (Cl), on the other hand, is a greenish yellow gas and is equally injurious. It was

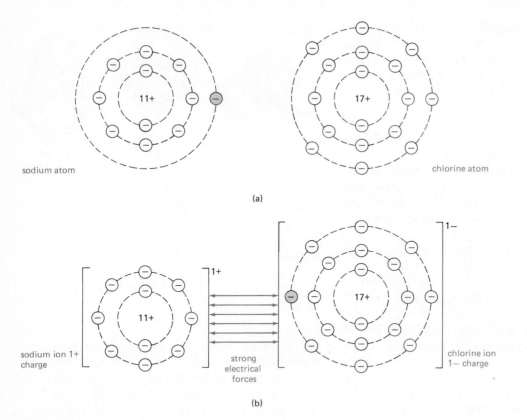

(a)

(b)

Figure 7.8
Ionic bonding. An ionic bond is formed when an electron is transferred between atoms, forming oppositely charged ions.

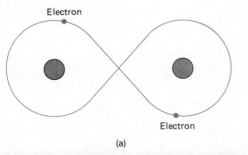

(a)

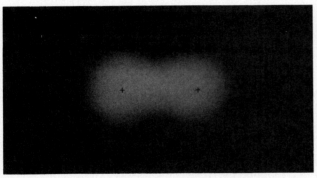

(b)

used as a poison gas in World War I. We have all smelled a hint of chlorine from bleaching solutions used in the washing of clothes or from city drinking water in which it is used to kill bacteria.

Yet, we can cause sodium and chlorine to react together chemically, and with tiny electron transfers between the atoms we get NaCl or common table salt, which most of us eat every day.

In **covalent bonds**, the electrons of atoms are mutually shared rather than permanently transferred (Fig. 7.9). Covalent bonds are generally not as strong as ionic bonds.

Figure 7.9
Covalent bonding. (a) In covalent bonds, the electrons of atoms are mutually shared rather than permanently transferred. (b) Equal sharing gives rise to a symmetric charge distribution.

The sharing of the electrons between atoms may or may not be equal. That is, the electrons may spend more time on the average with one atom than the other. As a result, such molecules have unsymmetric charge distributions, and we refer to them as having (covalent) **polar bonds** or being polar mole-cules (Fig. 7.10). The water molecule is polar, with regions of electric charge. The electrons spend more time with the oxygen atom, making this region of the molecule negatively charged and the hydrogen atom regions positively charged. This can be demonstrated as shown in Figure 7.10.

Since compounds are composed of elements, they can be decomposed chemically into their component elements. For example, ordinary sugar (sucrose) has the chemical formula $C_{12}H_{22}O_{11}$. By decomposition, the hydrogen and oxygen can be removed, and carbon is left (Fig. 7.11). Many elements are obtained from naturally occurring compounds, for example, metal ores.

Special Feature 7.2 discusses some compounds that are creating a serious problem.

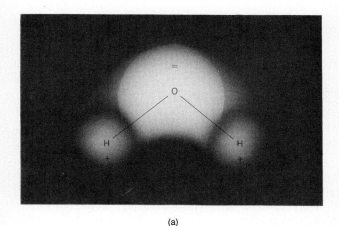

(a)

(b)

Figure 7.10
Polar molecules. (a) The water molecule is a polar molecule. Because of unequal sharing of electrons between atoms, there are regions of charge. (b) This can be demonstrated using electrical forces. The rod is electrically charged and attracts the polar water molecules (cf. Chapter 16).

Figure 7.11
Chemical decomposition. Sugar is composed of carbon, hydrogen, and oxygen. When the hydrogen and oxygen are chemically removed, carbon is left.

Acid Rain

Rain is normally slightly acidic as a result of carbon dioxide (CO_2) in the air. Water vapor and carbon dioxide combine to form carbonic acid (H_2CO_3), a mild acid we all drink in the form of cabonated beverages (carbonated water). However, the term *acid rain* refers to a more serious and abnormal source.

Coal generally contains sulfur compounds. When coal is burned, the sulfur combines with oxygen, forming sulfur oxides (SO_x). Vented to the atmosphere, the sulfur oxides SO_2 (sulfur dioxide) and SO_3 (sulfur trioxide) combine with water vapor in the air to form H_2SO_3 and H_2SO_4, sulfurous acid and sulfuric acid, respectively. Also, nitrogen oxides (NO_x) formed from the normal nitrogen and oxygen in the air in the presence of high-temperature combustion can form nitric acid. The exhaust gases from automobiles and trucks are a major source of nitrogen oxides.

Precipitation from contaminated clouds is then acidic — "acid rain" (and also acid snow, sleet, fog, and hail). Sulfur and nitrogen emissions from the industrialized midwest and northeast United States are believed to be responsible for acid rain in New England and eastern Canada. The acid levels in lakes and ponds in these regions are increasing.

Acid levels are measured on a pH scale, on which pure water has a pH of 7.0. The lower the pH, the more acidic a solution. Rainfall with a pH of 1.4 has been

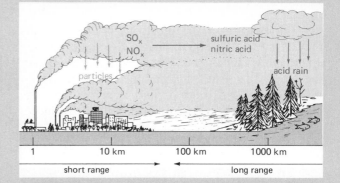

Acid rain. Acid rain arises from emissions of sulfur oxides (SO_x) and nitrogen oxides (NO_x) combining with water vapor in the air to form acids.

recorded in the United States. This is more acidic than lemon juice (pH 2.2). Sufficient concentrations of acids kill fish and aquatic plants, and lakes and ponds "die."

In addition to acid rain, there are acid snows. Over the course of a winter, acid precipitations build up in snowpacks. During the spring thaw and runoff, the sudden release of these acids gives an "acid shock" to streams and lakes.

Nor is acid rain confined to the eastern United States. Acid fogs with pHs as low as 1.7 have been measured along the west coast near Los Angeles. The culprit here is believed to be nitrogen oxides coming primarily from auto emissions.

Figure 7.12
Phases of matter. Examples of the three phases (states) of matter, solid, liquid, and gas, of a common substance.

PHASES OF MATTER

Everyone learned in grade school that at ordinary temperatures matter exists in three (physical) states or phases: solid, liquid, and gas.* We may not be familiar with the phases of some substances as they exist at various temperatures and pressures, for ex-

* Although matter is sometimes referred to as being in a particular *state*, the term *phase* is preferred because a substance can be in a particular phase with different properties that are called states, e.g., a gas with different temperature states.

ample, liquid nitrogen, which boils at a temperature of $-196°C$.

However, we are all familiar with the three phases of water (Fig. 7.12). Note that the "steam" seen coming from the teakettle is condensed water droplets (liquid phase) and not steam (invisible gas near the spout).

In general, we describe the phases of matter as follows. **Solids** have definite shapes and definite volumes. Their molecules are held together by comparatively strong forces. Many elements occur naturally as solids (Fig. 7.13).

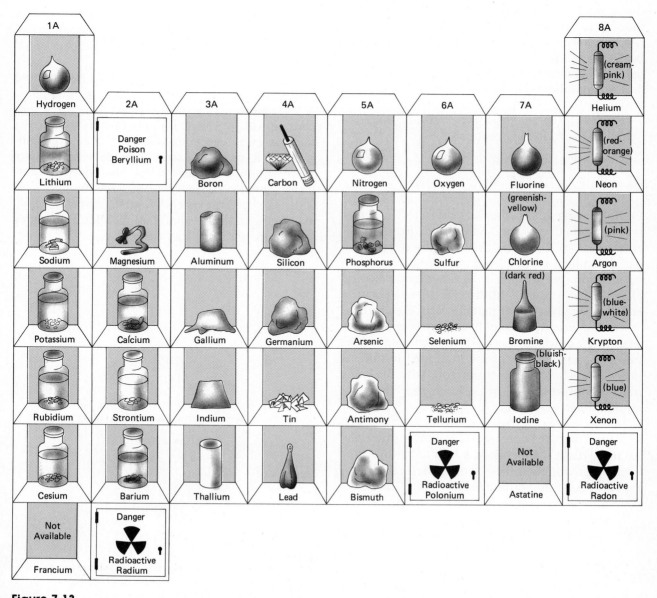

Figure 7.13
Solid, liquid, or gas? Many elements occur naturally as solids, but some occur as liquids and gases.

Liquids have definite volumes and their molecules have significant attractive forces between each other. However, a liquid has no definite shape. It assumes the shape of its container.

Gases spread out and fill the entire volumes of their containers. Therefore, a gas has no definite shape or volume. The gas molecules are relatively far apart, so the force between them is very small (except during molecular collisions).

If enough energy is added to a molecular substance, the molecules will separate into atoms, and the atoms themselves will come apart, giving a "gas" of free electrons and ions or nuclei. We call this a **plasma**, and plasma is considered to be a fourth phase or state of matter.* The plasma phase is not as common as the other three phases of matter on Earth, but it exists in various applications and natural phenomena. For example, the ionized gas in fluorescent lamps and the matter of the Sun is in a plasma state.

More will be said about the various phases of matter in the following chapters.

■ ANTIPARTICLES AND ANTIMATTER

Before leaving this chapter on matter, let's mention briefly something you may have heard of—antimatter. In 1932 a particle was discovered that had the same mass as an electron and behaved like an electron, except that it had a *positive* electric charge. This "positive" electron is called a positron and is said to be the antiparticle of an electron.

All subatomic particles have been found to have antiparticles, which are observed in cosmic rays from outer space and/or are produced in nuclear processes. There is the antiproton, with the same mass as a proton but a negative charge. There are also antineutrons.

However, the antiparticles do not get along well with their particle counterparts. When a particle and its antiparticle collide, they vanish or "annihilate" each other, and their mass is converted *completely* into energy or into other particles (Fig. 7.14).

This is another example of mass-energy conversion and conservation according to Einstein's $E = mc^2$ relationship, with a possible 100 percent con-

* This plasma is pronounced and spelled the same as the liquid part of blood (blood plasma), but it is entirely different.

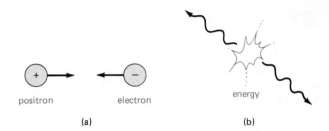

Figure 7.14
Annihilation. A particle and its antiparticle coming together are "annihilated," and their mass is converted to energy (carried away by gamma "particles").

version. In the uranium fission reaction used in nuclear reactors for energy production for electrical generation, there is only about a 0.1 percent mass-to-energy conversion.

Since antiparticles exist, it could be possible that somewhere in the universe there are quantities of antimatter, whose antiatoms are made up of antiparticles (Fig. 7.15). This might be the substance of stars or even entire galaxies (antistars and antigalaxies, to be more exact). Science-fiction writers have written about antiworlds composed of antimatter, but no one really knows. One thing for sure is that we couldn't visit our antineighbors, nor they us. There would be annihilation when the matter and antimatter came into contact.

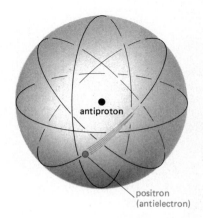

Figure 7.15
Antiatoms. Antiatoms would make up antimatter. As illustrated here, a hydrogen antiatom would have a nuclear antiproton and an orbiting positron.

SUMMARY OF KEY TERMS

Atom: the smallest unit of an element that can exist alone or in combination with other atoms.

Element: a substance in which all the atoms have the same number of protons.

Brownian motion: The erratic motion of small particles (e.g., pollen grains, dust particles) in suspension because of collisions with the molecules of the suspension medium.

Solar-system model: the atomic model that pictures the protons and neutrons to be in a central core or nucleus with the electrons in orbits (or shells) about the nucleus.

Atomic (proton) number: the number of protons in a nucleus or atom. The atomic or proton number determines the species of an atom.

Ion: an atom or molecule with a net electrical charge due to the transfer (loss or gain) of one or more electrons.

Isotopes: atoms or nuclei of the same species (same number of nuclear protons) having different numbers of neutrons.

Periodic table of the elements: an arrangement of elements based on atomic number and recurring chemical properties that are periodically repeated.

Period: a horizontal row in the periodic table.

Group: a vertical column in the periodic table. The elements of a group or "family" have similar chemical properties.

Molecule: a group of two or more atoms held together by forces called chemical bonds.

Compound: a pure substance made up of one or more elements.

Ionic bond: a bond formed by a transfer of electrons between atoms.

Covalent bond: a bond formed by a sharing of electrons between atoms.

Polar bond: a covalent bond in which there is an unequal sharing of electrons between atoms such that there is an unsymmetric charge distribution or molecular regions of net charge.

Phases of matter: the forms of matter in terms of being a solid, liquid, gas, or plasma.

Plasma: an ionized gas of free electrons and ions, which is considered to be a fourth state of matter.

Antiparticle: a particle in which the electrical (and magnetic) properties are reversed from those of ordinary particles.

Annihilation: the complete conversion of mass into energy that occurs in the collision of a particle and antiparticle.

Antimatter: matter made up of antiparticles.

QUESTIONS

Atomic Theory and Structure

1. Why doesn't a small piece of wood floating in water exhibit Brownian motion?

2. The Austrian physicist Ernst Mach (1838–1916) was one of the last holdouts against the atomic theory. He thought that it was meaningless to believe in atoms since they could not be sensed directly. If Mach were alive today, how would you convince him of the existence of atoms?

*3. Compare the populations of (a) the United States (about 250 million) and (b) the world (about five billion) to the number of molecules in a half liter of air.

*4. On the average, a person breathes in about 6 L of air per minute. (a) What is the molecular intake per minute? (b) Are the same molecules expelled? Explain.

5. Compare the magnitude of the national debt (over $2 trillion) with the number of molecules in a milliliter of water.

6. Are the common atomic model and the solar system similar or different with respect to (a) mass distribution, (b) empty space, (c) the nature of centripetal force, (d) three-dimensional orbital shells, (e) the distance of orbiting "particles" from the center of the system?

*7. Does an atom always have as many electrons as protons? Explain.

Elements and the Periodic Table

8. Sketch the electron shell arrangements for the atoms of Be and Mg, and explain why these elements would have similar chemical properties.

9. The elements sodium (Na) and fluorine (F) form the ionic compound NaF (sodium fluoride). Sketch the electron shell arrangements for the atoms of these elements, and explain why an ionic bond is readily formed.

10. What elements do the following chemical symbols represent? (a) Ag, (b) Au, (c) Hg, (d) Fe

*11. Which one of the elements in each of the following sets has the greatest number of protons in its atomic nucleus? (a) Sulfur, copper, and calcium, (b) thorium, krypton, and stron-

tium, (c) einsteinium, mendelevium, and nobelium.

12. Nickel and platinum are metals. What would you predict the lesser-known element palladium to be, and why?

•13. Suppose a U-238 atom lost two protons, two neutrons, and two electrons. What would it be then?

Molecules and Compounds

14. *Webster's Dictionary* defines a molecule as "the smallest quantity of an element or compound which can exist separately and still retain the properties and character of the element or compound." One of the characteristics of sugar is its sweetness, but one molecule or even several molecules of sugar cannot be detected by taste. Is there something wrong with Webster's definition? Explain.

•15. Suppose common sugar or sucrose were decomposed into carbon and water. How many carbon atoms and water molecules would be obtained from a dozen sugar molecules?

16. A quantity of table salt (NaCl) is dissolved in a volume of water. Is this a chemical change, or is the solution a mixture? Explain. (*Hint:* A mixture is a combination of substances that can be separated by physical means, rather than

chemical means, as is required for the separation [decomposition] of a compound.)

Phases of Matter

•17. What determines the phase of a substance?

•18. Can all three phases of matter go directly from one to the other? Explain. What is the name for each change of phase? (Does direction make a difference?)

19. Explain on the atomic or molecular level why adding or taking away heat causes a change of phase.

20. Which phases of matter have (a) definite shape and (b) definite volume?

Antiparticles and Antimatter

•21. If positrons are created by cosmic rays and nuclear processes, why are they not commonly found in nature?

22. Sketch the solar-system model for hydrogen and carbon antiatoms.

23. Could the problem of handling the antimatter for an antimatter engine be solved by having the antimatter created directly by the engine itself? (*Hint:* Consider the efficiency of such an engine.)

24. If a spaceship came to Earth from an antimatter universe, what would happen if it landed? Would it even be able to land? Explain.

Solids

8

A stalactite (from the top) + a stalagmite (from the bottom) = a column (when the dripstones meet).

We say that solids have definite shapes and volumes. It is tempting to think of a solid as an ideal rigid body in which all the particles are fixed distances apart, but the particles in a solid actually move or vibrate around their equilibrium positions.

From atomic theory we know that the particles (atoms, ions, or molecules) are held together by attractive electrical forces. It is something like the particles being held together by springs (Fig. 8.1).

The degree of vibration depends on the internal energy of the material. For example, when heat is added, the particles vibrate more. If enough energy is added, the particles may no longer be held in place by the interparticle forces. Macroscopically, the solid would no longer retain its definite shape and would be observed to melt. Hence, the melting-point temperature of a solid depends on the strength of the interparticle forces.

Solids can be described as either crystalline or amorphous. Crystalline solids have a regular arrangement of particles; amorphous solids have a completely random particle arrangement.

■ CRYSTALLINE SOLIDS

Most substances exist as solids in some characteristic crystalline form. For example, the "spring" solid in Figure 8.1 has a simple cubic crystalline lattice. Each particle is at the corner of a cube. The orderly

Figure 8.2
An X-ray diffraction pattern of ice. The pattern is characteristic of the crystalline structure of ice.

arrangement of the particles in a solid defines a pattern or lattice structure. You might think of the fixed-chair seating arrangement in a full classroom as being "crystalline," since there is an "orderly" array of student "particles."

The crystalline nature of some solids is evident from their external appearance. The crystalline structure of other solids may not be so evident. Scientists use X-rays as a tool to study the crystalline structure of solids. When a crystalline solid is illuminated with X-rays, distinct patterns appear on photographs (Fig. 8.2). These are called X-ray diffraction patterns, and each pattern is characteristic of a particular crystalline lattice. (Diffraction is discussed in more detail in Chapter 20.)

Because of the orderly particle arrangement, some crystalline solids can be split or cleaved along certain planes in the crystal lattice. For example, a solid with a simple cubic lattice can be cleaved along the molecular planes of the crystal (Fig. 8.3). Sodium chloride (NaCl, table salt) has a simple cubic lattice. If you tried to cleave the crystal along a plane diagonal to the cubic structure, things

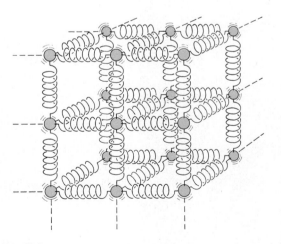

Figure 8.1
Spring model of a solid. The particles of a solid are held together by attractive electrical forces and vibrate as though they were interconnecting springs between the particles.

Figure 8.3
Cleavage. Some crystalline solids can be split or cleaved along certain planes in the crystalline lattice, as shown here for NaCl, which has a cubic lattice.

wouldn't work so well. The crystal would shatter or fracture, and you'd end up with a lot of broken pieces.

In various crystalline substances, the particles have different lattice structures, which give rise to different properties. For example, diamond is one form of carbon. The other common form of carbon is graphite. Both diamond and graphite consist only of carbon atoms but in different crystalline structures (Fig. 8.4).

Diamond is very hard and has complicated cleavage planes. Diamond cutters must study the structure before cutting or cleaving a diamond. Not cutting along a cleavage plane results in fracture, which isn't too desirable. The fractured pieces are no longer good for gemstones, but could be used for industrial grinding or etching. (Diamond is one of the hardest substances known.)

In graphite, the carbon atoms are arranged in lattice layers. The atoms within any one layer are strongly bonded to each other. However, the bonding *between* layers is much weaker, and the layers can readily slide past each other. As a result, graphite is soft and slippery and can be used as a lubricant. A common application is in the lubrication of latches on car doors. The "lead" in a lead pencil is actually graphite that has been mixed with other materials to make it more firm. When you write with a pencil, you are spreading graphite out on the paper.

Pure, finely divided graphite powder is called carbon black. It is used as the pigment in India ink and makes up about 30 percent of the weight of a typical auto tire.

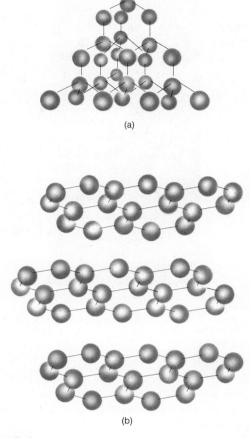

Figure 8.4
Carbon. Diamond and graphite have different lattice structures, but both are carbon. (a) When in a complicated diamond structure, the carbon atoms form a hard solid. (b) As graphite, the carbon atoms are arranged in layers that easily slide over each other, so graphite makes a good lubricant.

■ AMORPHOUS SOLIDS

Amorphous solids have a random particle arrangement. Instead of picturing the orderly "crystalline" seating arrangement in a classroom, you might think of a crowd sitting on a grassy slope at an outdoor rock concert as an "amorphous" array.

Truly amorphous solids are rare. Many solids that were once thought to be amorphous have been found to have a partially crystalline structure. However, materials like glass and paraffin may be considered amorphous (Fig. 8.5). These materials have the properties of solids but lack the sharply defined melting points of crystalline solids. In many re-

Figure 8.5
Amorphous solid. Glass is an amorphous substance with a random particle makeup.

(a)

(b)

Figure 8.6
LCD. LCD (*liquid crystal display*) applications.

spects, they resemble liquids and flow very, very slowly at normal temperatures. As a result, the bottoms of old window panes (as in cathedrals) are somewhat thicker than at the tops.

Common applications of glass include mirrors and lenses in eyeglasses that many of us wear (Chapter 21). A special eyeglass application is discussed in Special Feature 8.1.

■ LIQUID CRYSTALS

In general, when a crystalline substance melts, the resulting liquid no longer has an orderly particle arrangement. However, in certain cases, there is a carryover of the crystalline properties of the solid state to the liquid state, where the liquid shows some degree of molecular order. This is an in-between state known as the liquid crystal state.

A liquid crystal behaves like a liquid, inasmuch as it can flow and assume the shape of its container. However, the optical properties of the liquid crystal depend on the order of its molecules. For example, certain liquid crystals with orderly molecular arrangements are transparent. But this crystalline structure is very weak and can be scrambled or disordered by applied forces (e.g., electrical or mag-

netic). The disordered liquid crystal then scatters incident light.

These properties are used in LCDs (*l*iquid *c*rystal *d*isplays), which are now quite commonly used in calculators and digital watch readouts (Fig. 8.6) You can also see LCDs on gas pumps. More will be said about how these displays work in Chapter 20.

■ DENSITY

The masses of particles and the spacing between them determine the density of a material. Density is a measure of the compactness of the particles in a material (either solid, liquid, or gas). That is, density expresses the amount of mass per unit volume:

Density = mass/volume

SPECIAL FEATURE 8.1

Photochromic Glasses

Small amounts of various substances are often added to glass to produce different-colored glasses such as those used in sunglasses. For example, Cr_2O_3 (chromium oxide) gives green glass, and MnO_2 (manganese dioxide) gives a violet glass. This shading or coloring is permanent. However, photochromic glasses that darken and lighten automatically are now commonly available.

Photochromic glass darkens on exposure to ultraviolet light (cf. Chapter 19) and becomes lighter again when removed from exposure. Eyeglasses made of photochromic glass darken in sunlight, which contains ultraviolet light, and so act as sunglasses. They become lighter again when the wearer goes indoors.

This special glass contains finely dispersed crystals of silver chloride (AgCl), which is an ionic compound. When exposed to ultraviolet light, some of the silver ions (Ag^+) are converted to atoms of metallic silver (Ag), which is dark and almost black. This process is similar to that used in the exposure of a photographic film that contains silver bromide in the photographic emulsion.

However, the silver crystals formed in photochromic glass are many times smaller than those in a photographic emulsion, and in glass the chlorine atoms are not removed by side reactions and do not migrate away from the reaction zone, so the reaction is reversible. When removed from ultraviolet light, the silver atoms and chlorine atoms in photochromic glass recombine to form silver chloride again, and the glass becomes lighter in color.

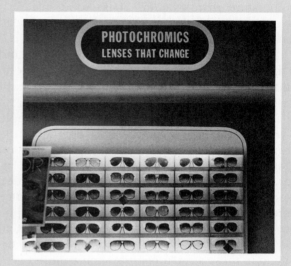

Darkening lenses. Photochromic eyeglasses darken automatically in bright sunlight.

Density may be expressed in gram/cubic centimeter (g/cm^3) or kilogram/cubic meter (kg/m^3). The densities of some substances are listed in Table 8.1.*

The densest solid on Earth is the metal osmium. Another metal, iridium, runs a close second. These solids are so dense because their atoms are closely packed. Atoms of gold, lead, and uranium are all more massive than those of osmium and iridium (see periodic table, Chapter 7). However, in the crystalline structures of these substances, the atoms

* Since mass and weight are related ($w = mg$), density can also be expressed in terms of weight, or weight density = weight/volume. Weight densities are not uncommon in the customary British system. For example, the weight density of fresh water is 62.4 lb/ft³ (or 9812 N/m³). In the SI system, mass density is commonly used. The SI density units are kg/m³, but the smaller units of g/cm³ are sometimes used for convenience.

Table 8.1
Densities of Some Substances

Material	Density	
	g/cm³	kg/m³
Air	0.00129	1.29
Aluminum	2.7	2700
Brass (70% Cu)	8.5	8500
Copper	8.9	8900
Glass (general)	2.6	2600
Gold	19.3	19,300
Ice	0.92	920
Iridium	22.42	22,420
Iron	7.9	7900
Lead	11.5	11,500
Mercury	13.6	13,600
Osmium	22.5	22,500
Water	1.00	1000 (62.4 lb/ft³)
sea water	1.03	1030 (64.3 lb/ft³)
Wood (oak, general)	0.70	700
Zinc	4.3	4300

are not as close together as in osmium and iridium, and their densities are smaller — less mass in a given volume.

TYPES OF SOLIDS

Diamond is much less dense than osmium because its atoms are lighter and are less compact in its lattice. Yet, diamond is harder and melts at a higher temperature than osmium does. Solid materials have a variety of properties. Steel is strong, lead bends easily, quartz is brittle, and talc is soft.

All of these properties are in some way related to the nature of the particles (atoms, ions, or molecules) of a solid and the forces between the particles.

In ionic solids — NaCl (table salt), for example — oppositely charged ions are held together by the strong electrical forces of ionic bonds (electron transfer). In an ionic solid with a simple cubic lattice, such as NaCl, each ion (Na^+ and Cl^-) is bonded to six other ions (Fig. 8.7). There are ionic lattice structures other than simple cubic, for example, face-centered cubic and body-centered cubic. (Can you imagine how the various ions would be arranged in these structures from their names?)

Because of the strong ionic bonding, ionic solids have relatively high melting points (typically

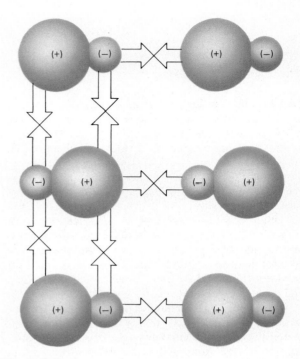

Figure 8.8
Molecular solids. Molecular solids are generally held together by intermolecular forces between polar molecules.

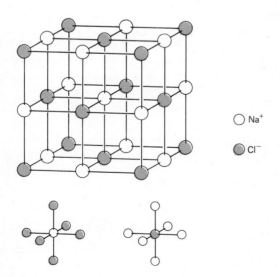

○ Na^+

◉ Cl^-

Figure 8.7
NaCl lattice. Each ion (Na^+ or Cl^-) is bonded to six other ions.

600 to 2000°C). The ionic bond must be broken to melt the solid by separating oppositely charged ions from each other. Only at high temperatures do the ions acquire enough kinetic energy for this to happen.

Molecular solids are made up of molecules rather than ions. Intermolecular forces between molecules bond them together as a solid. Generally there is an unequal sharing of electrons within the molecules. This gives rise to regions of electrical charge (polar molecules) and attractive forces (Fig. 8.8). The strength of the intermolecular forces depends on several factors that will not be discussed here. (We'll leave this for an introductory chemistry course.)

The intermolecular forces of molecular solids are not as strong as ionic bonds. As a result, molecular solids in general have low melting points, usually below 300°C. Also, they tend to be volatile and lose molecules as a vapor.

Many molecular solids can sublime, or pass directly from a solid to a vapor upon heating. Some

(a)

Figure 8.9
Ice lattice. (a) When water freezes, an open hexagonal (six-sided) structure results. (b) This six-sided structure is evident in snowflakes.

(b)

sublime at room temperature, for example, solid air fresheners and mothballs. Solid carbon dioxide (Dry Ice) is another molecular solid that sublimes readily. At room temperature, Dry Ice "disappears" by vaporizing.

Probably the most common molecular solid is ice. In one sense, ice is very unusual. It is almost unique in having a density less than that of the liquid from which it solidifies (density of ice = 0.918 g/cm³, density of water = 1.000 g/cm³). This is why ice floats in water (see Chapter 9).

When water freezes to ice, an open hexagonal (six-sided) lattice pattern results (Fig. 8.9). This structure is externally evident in snowflakes. The large empty spaces in the ice structure explain why ice is less dense than water. Also, the variation of the density of water with temperature explains why water freezes from the top down rather than from the bottom up. See Special Feature 8.2.

In some solids, each atom in a structure is covalently bonded to its neighbors. The resulting crystals are compact, interlocking, covalent network structures. Substances of this type are called macromolecular solids. In effect, the entire solid consists of one huge molecule. Common examples of mac-romolecular solids are diamond and silicon dioxide (SiO_2, the mineral quartz and the main component of sand). Macromolecular solids have relatively high melting points, often above 1000°C.

Polymers

Unlike macromolecular solids, which are in effect one large molecule, polymer substances are made up of giant or macromolecules. These are formed by the combination of small molecular units called **monomers** (Greek *mono,* meaning "one"). The monomer units form giant molecules, which make up a **polymer** (Greek *poly,* "many," and *meros,* "parts"). A linear polyethylene molecule is illustrated in Figure 8.10. A typical polymer molecular may contain a chain of monomers several thousand units long.

Polymers occur naturally and can be made synthetically. Polyethylene, as illustrated in Figure 8.10, is a synthetic polymer. That is, it is made in the laboratory and does not occur in nature. Hardly a day goes by when you don't use a dozen or more items made of synthetic polymers or plastics. Such items include dishes and cups, combs, telephones,

Water Density and Freezing at the Top

The densities of most liquids increase as the liquids are cooled and their temperatures reduced. This is the case for water as it is cooled to 4°C. However, when water is cooled below 4°C to its freezing point, its density decreases. This implies that the formation of an open lattice structure occurs over the temperature range of 4°C to 0°C, rather than taking place solely at the freezing point.

This unique property accounts for the fact that open containers of water freeze at the top first. Most of the cooling takes place at the open surface. As the temperature of the top layer of water is lowered toward 4°C, this cooler, denser water sinks to the bottom. However, below 4°C, the water at the top is less dense than the water below and remains at the top, where it freezes when the freezing point is reached.

Think of the environmental effects if this were not the case. Otherwise, lakes, ponds, and rivers would freeze from the bottom up, and much of the aquatic animal and plant life would be destroyed—not to mention what it would do for ice skating.

The density of water versus temperature. The maximum density of water occurs near 4°C. Below 4°C, water is less dense because of the formation of an open hexagonal structure of the molecular units.

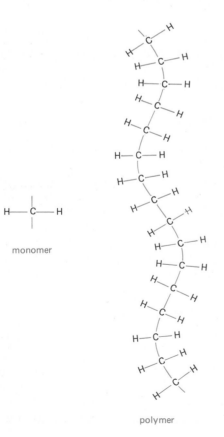

monomer

polymer

pens, false eyelashes, and fibers in clothing. A **plastic** is a polymer substance that will flow under heat and pressure and hence can be molded into various shapes. All plastics are polymers, but not all polymers are plastics.

Many natural polymers produced by plants and animals are essential to life. Many of these fall into the class of organic polymers known as proteins. Natural polymers are common in daily life, for example, wood, paper, and clothes made of cotton, wool, silk, and leather.

Some of our most useful synthetic polymers have resulted from copying giant molecules found in nature. For example, synthetic rubber is copied from natural latex rubber. Some synthetic polymers, however, do not have duplicates in nature, such as nylon, Dacron (polyester), and Teflon.

It is interesting to note that a typical radial auto tire is only 39 percent rubber, of which about 70

Figure 8.10
Polymers. Polymer molecules are long chains made up of a series of monomer units.

Figure 8.11
Polymer degradation. The ultraviolet portion of sunlight has sufficient energy to break the chemical bonds in polymers. If enough bonds are broken, the polymer becomes brittle and will break under pressure.

percent is synthetic rubber. Only in special-use tires that get very hot does the percentage of natural rubber increase because of its superior heat resistance. The rubber in a high-performance aircraft tire, for example, is almost all natural for this reason. However, the tread of a regular aircraft tire is about 50 percent synthetic rubber because of its better resistance to wear and abrasion. This characteristic is mainly responsible for synthetic rubber's wide use in auto tires.

The properties of a polymer depend on its molecular structures. Polymers in which the monomer units are arranged in nearly parallel chains form strong and flexible fibers. If the chains are tangled or linked, the polymer may form a strong film or a rigid solid. The melting points of polymers vary over a range of several hundred degrees.

Additives and chemical processes are used to give polymers desired properties. For example, many plastics are used where they are exposed to sunlight. The ultraviolet (uv) portion of the sunlight has sufficient energy to break the chemical bonds found in polymers. If enough bonds are broken, the polymer becomes brittle and will break under a force (Fig. 8.11). Compounds called uv stabilizers, which absorb uv light, can be added to reduce the degradation.

Most of the raw materials for synthetic polymers come from petroleum and some from coal. Only a

small amount of our petroleum consumption goes into making polymer products. See Special Feature 8.3.

Metallic Solids

Of the known elements, about 80 can be classified as metals. Of the four major types of solids, only metals are good electrical conductors in the solid state. All metals are solids at 25°C (77°F), with the exception of mercury.

The metallic lattice consists of positive ions surrounded by an electron "gas" or "sea." The electrons are donated by the atoms of the metal and belong to the crystal as a whole. The positive ions are anchored in position like bell buoys in a mobile "sea" of electrons. The electrons are free to migrate or wander throughout the lattice, somewhat like gas molecules in a closed container.

This simple model of metallic bonding explains why metals are good electrical conductors. However, metallic solids vary widely in hardness and melting-point temperature.

Alloys

While on the topic of metals, we should mention alloys. **Alloys** are blends of two or more metallic elements or metallic and nonmetallic elements that give materials with properties different from those of the individual elements. For example, bronze is stronger and harder than either of its constituents, copper and tin. Some common alloys are listed in Table 8.2.

Alloys have been known for a long time. The rise of some civilizations is credited to their knowledge of and work with metal alloys (e.g., those of the Bronze Age). We are familiar with alloys in the form of coins. Light aluminum and magnesium alloys are used in aircraft. However, the strengths of

Table 8.2
Some Common Alloys

Alloy	Constituents (Percentages by Weight)
Brass	Copper (60–86%) and zinc
Bronze	Copper and tin (5–10%)
Cast iron	Iron and carbon (2–5%)
Steel	Iron and carbon (<1.3%)
Stainless steel	Steel, chromium (18%), and nickel (8%)

Petrochemicals and Plastics*

Many people are unaware of how many of the products we use every day come from petroleum and natural gas. In the United States each family of four uses more than *two tons* of petroleum products annually. That's almost 1200 lb of chemicals each year for every man, woman, and child in the United States — a staggering total of 225 billion pounds of chemicals from petroleum, and to a lesser extent, natural gas.

Of the vast amount of petroleum and natural gas we consume, more than 90 percent is burned as fuels. Only about 5.5 percent is used for the manufacture of petrochemicals by the chemical industry. These petrochemicals vary widely in their functions and include such products as drugs, detergents, rubber, paints, fertilizers, dyes, perfumes, explosives, food preservatives, artificial sweeteners, and agricultural chemicals. Finally, about 1.5 percent of the oil and natural gas is used as raw material for plastics. This small percentage translates into the production of billions of pounds of polymers that yield many different and useful products.

In the post–World War II years, the United States was flooded with domestic and imported items of extremely low cost, low quality, and limited lifetime. This led to the image of "cheap plastics" with low durability. Today, however, the image of plastics has changed. Plastics perform an extremely broad range of functions, from heart valves and artificial kidneys to ski boots, nonstick surfaces, super glues, and spacecraft parts, and they compete with natural products in durability. No other materials except plastics could perform all these different functions.

Plastics are replacing more and more parts of your car. The use of 1 lb of plastic can replace an average of 3.5 lb of metal in an automobile. An automobile with 400 lb of plastic substituted for metal will weigh about 1000 lb less, which increases its gas mileage by about 3 mi/gal. The fuel savings are estimated to be about 160 million barrels of oil annually. That's more than the total amount used by the chemical industry as raw materials to make the polymers. As another example, synthetic polymer fibers are commonly used in fabrics, for both economical and practical reasons. If the world's synthetic fibers were replaced by cotton, this would require an additional 40 million acres of farmland.

Certainly the use of polymer plastics will increase. One can expect to find more applications in home construction and furniture because of the unlimited design freedom of plastics. Plastics will be used more in drink containers and food packaging. The 700 billion gallons of liquids consumed each year in the United States will find their way to the consumer more and more in plastic bottles. (Beer and champagne in plastic bottles?) Diseased or malfunctioning parts of the body will be replaced by specialized plastic components to a greater degree. We are indeed becoming a plastic society.

* Courtesy of Dr. Peter A. Vahjen, Professor of Chemistry, Lander College.

---

these alloys decrease rapidly between 300 and 400°C. Air friction on very-high-speed jet aircraft necessitates the use of titanium alloys. (The melting point of titanium is 1660°C, as compared with 660°C and 650°C for aluminum and magnesium, respectively.)

Alloys may be thought of as solid solutions. They are generally prepared by melting and mixing the constituent metals. Not all metals will form alloys, however. The fitting together of different metallic atoms (ions) in a common lattice is possible only if their sizes do not differ greatly. Some alloys are difficult to make because the heavier element tends to settle out in the melt before solidification. An example is aluminum antimonide, which is an alloy of aluminum and antimony. Antimony is more than three times denser and tends to settle faster than aluminum when mixed in the molten state.

One attribute of aluminum antimonide is that it may be 30 to 50 percent more efficient than silicon in solar cells and computer-circuit chips. The settling effect causes a lack of uniformity in the alloy, which severely limits its use in these applications. However, this effect does not take place in the "microgravity" of space (see Chapter 5). Experiments in *Skylab, Apollo-Soyuz,* and *Space Shuttle* orbital missions have achieved largely uniform mixtures in alloys that are used extensively in computers and other electronic products.

Also, metallic oxide alloys are now causing quite a stir in electrical superconductivity. See Chapter 17.

MECHANICAL PROPERTIES OF SOLIDS

A wide variety of materials are used in various applications. In many cases, the use of a particular material depends on its mechanical properties. In some instances it is desirable for a material to withstand an applied force or torque, for example, as on the axle of an automobile. In other instances, a material that can be stretched or easily deformed is required, for example, rubber in a rubber band or automobile tire.

The mechanical properties of solids depend on the composition and internal forces of the materials, as you might expect. The internal processes that occur in materials when they are subjected to forces are complex. However, these processes are manifested as changes in a material's external properties, which are described by common terms.

Elasticity

Solids have definite shapes because the forces between the particles resist changes. However, some solids, such as rubber, are easily deformed but return fairly well to their original shapes when external forces are removed. In such cases, the interparticle forces allow for stretching and twisting. When the distorting force is removed, the attractive forces

between the particles bring them back to their original positions and the solid as a whole back to its original shape. This property is called **elasticity**.

All materials are elastic to some extent, even steel. However, some materials are easily deformed and have little tendency to return to their original shapes. Such materials as putty and clay are generally said to be inelastic.

To better understand the terms *elastic* and *inelastic,* consider an "elastic" spring, as illustrated in Figure 8.12. When force (weight) is applied to a spring, it is stretched or deformed. In general, it is found that the stretch distance d is proportional to the applied force F, or $F \propto d$. For example, if a force of 4 N stretches a spring 2 cm, an additional force of 4 N will stretch it another 2 cm (total force 8 N and total stretch distance 4 cm). This relationship is known as **Hooke's law**.* Hooke's law also applies to a compressive force and compressed spring distance.

If the force is not too great, the spring will return to its original shape and length when the weights are removed. However, if too much force is applied, the elastic limit of the spring will be exceeded. When this happens, the spring will be stretched or bent out of shape. The deformation is then permanent, and the spring will not return to its original shape.

A straight steel wire also has elasticity, but this is not generally observed visually. Large forces and sensitive detectors are usually required to observe this behavior. Occasionally this elastic behavior can be observed for what are usually considered inelastic materials, as shown in Figure 8.13. This is the famous Tacoma Narrows bridge, which collapsed in 1940. More will be said about what caused the vibrations in the bridge in Chapter 14.

After the elastic limit is reached, metals and most polymers exhibit **plasticity** or **plastic deformation**. Unlike elastic deformation, plastic deformation is permanent; the material does not recover when the force of stress is removed. For example, a dent in an automobile fender is the result of a force that produces a plastic deformation. Plastic deformation was also required in forming the fender from sheet metal, but elastic deformations are usually desirable thereafter.

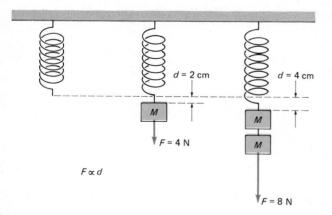

Figure 8.12
Hooke's law, $F \propto d$. If a force of 4 N stretches a spring a length of 2 cm, an 8-N force will stretch it 4 cm.

* Not all elastic materials follow Hooke's law (named in honor of Robert Hooke, an English physicist who first described the relationship). A rubber band is a good example.

Figure 8.13
Not a stress test. Winds caused this bridge to oscillate and be stressed until it collapsed. (See figure in Special Feature 14.2.)

Question: When a piece of wire, such as a straightened paper clip, is bent, it shows permanent plastic deformation. Yet when a bending force of the same magnitude is applied back and forth, the wire eventually breaks. Why is this?

Answer: Some metals and polymers, especially crystalline polymers, strengthen as a result of plastic deformation. This is called work-hardening or stress-hardening. When a wire is bent or flexed (stressed) repeatedly, it becomes harder and more brittle as a result of work-hardening in the bending area. The more brittle a material, the less stress it can stand before fracturing.

When metals are worked, a metal may have to be heated and cooled slowly (annealed) to remove the excessive hardening caused by work-hardening.

If enough force is applied to a spring or a steel wire (or any material), it will eventually break or fracture. Some materials, such as ceramics, fracture without noticeable plastic deformation. Breaking or fracture is a case of the external applied force exceeding the internal particle forces, which causes the bonds to break.

In applying a force to a solid, we commonly speak of applying stress and the resulting strain or deformational change in the object's dimensions. Let's illustrate these terms with an example that

(a)

(b)

Figure 8.14
Stress and strain. (a) The karate blow produces a compression strain on the top of the board and a tension (elongation) strain on the bottom. Both wood and concrete are weaker under tension, so the board cracks on the lower surface. Also, bone is much stronger than wood or concrete. (b) It is important to apply the impact force over a small area so as to produce a large stress.

takes into account the mechanical properties of materials as well as some of the principles of previous chapters. How about the breaking of a wooden board with a karate blow? See Figure 8.14.

The main purpose of a karate blow is to apply an impact force on a small area. This produces a large stress, which is the force/area. (A small area for a given force produces a large stress or concentrated force.) In the collision, the wooden board (or concrete block) is bent or strained. The upper surface is under a compression, and the lower surface is under a tension (elongation) strain. Both wood and concrete are weaker under tension, so with sufficient stress the board will begin to crack at its lower surface. The hand follows through and the board breaks in what appears to be an amazing demonstra-

tion of strength. Muscular strength is somewhat involved, but physical principles and the strength of materials are major considerations.

The blow is designed to make contact when the hand reaches its maximum speed so as to give a large transfer of momentum and energy to the board. In the collision, the bones of the hand receive a large compressive force, but bone is much stronger than wood or concrete (as you might guess and a karate expert knows or would hope). Of course, there are other considerations, such as the width and thickness of the board and the distance between its support points. Striking the board parallel to the grain is a good idea, as well as hitting the board properly so as to apply the impact force to a small area and produce a large stress [Fig. 8.14(b)].

SUMMARY OF KEY TERMS

Crystalline solid: a solid with a regular or orderly particle arrangement in a lattice structure.

Lattice: the geometric pattern or arrangement of particles in a crystalline solid.

Amorphous solid: a solid with random particle arrangement.

Liquid crystal: an in-between state in which a substance with liquid properties shows some degree of molecular order as in a crystalline solid.

Density: a measure of the compactness of the particles in a material, which is expressed as mass per unit volume: density = mass/volume.

Ionic solid: a solid consisting of oppositely charged ions, e.g., NaCl.

Molecular solid: a solid consisting of molecules with intermolecular bonding, e.g., ice.

Macromolecular solid: A solid consisting of covalently bonded atoms, such that in effect the solid consists of one huge molecule.

Polymer: a substance consisting of giant or macromolecules made up of repeating monomer units.

Plastic: a polymer substance that will flow under heat and pressure and hence can be molded into various shapes. All plastics are polymers, but not all polymers are plastics.

Metallic solid: a solid consisting of a lattice of positive ions surrounded by a "sea" of electrons, which gives rise to good electrical conduction.

Alloy: a blend of two or more metallic elements or metallic and nonmetallic elements.

Elasticity: the property of a solid whereby it returns to its original shape after a distorting force or stress is removed.

Plasticity: the property of a solid characterized by permanent deformation when a force or stress is removed.

[For more on density and volume, see An Extended View (Optional) in the Appendix.]

QUESTIONS

Solids

1. Describe how an amorphous "spring" solid might look.
*2. The simple cubic lattice of two different particles (ions) is illustrated in Figure 8.7. Two other types of cubic lattices are (a) face-centered cubic and (b) body-centered cubic. Using light and darkened circles to indicate different

particles, sketch how these lattices would look. (*Hint:* The names are descriptive of the lattices.)
3. Wood is a cellulose polymer. Explain why wood can be split "with the grain" but has to be sawed into pieces going "across the grain."
4. Diamond and graphite are both carbon. Why are the properties of two carbon substances so different?

Density

5. The density of uranium is 18.7 g/cm³, and the density of gold is 19.3 g/cm³. Yet an atom of uranium is more massive (heavier) than an atom of gold (238 vs. 197; see periodic table, Chapter 7). Why does uranium have a smaller density?

•6. When you squeeze and compress something, such as a sponge or a balloon, do you change its density? Explain.

7. The density of liquid mercury is 13.6 g/cm³, and the density of solid mercury is 14.2 g/cm³ (freezing point of −39°C). What does this tell you? How about floating?

8. Explain why sea water has a greater density (1.03 g/cm³ or 64.3 lb/ft³) than fresh water (1.00 g/cm³ or 62.4 lb/ft³).

•9. Density = mass/volume, so mass = density × volume. (a) What are the masses of gold and lead bricks with dimensions of 5 cm × 10 cm × 20 cm? (b) How much would these bricks weigh in pounds?

•10. Why isn't the density of water in SI units 1.000 kg/m³?

11. Other than hardness, what is a basic difference between "hard" wood and "soft" wood?

•12. To help understand the empty space in a lattice structure, consider the two-dimensional close-packed circles in Figure 8.15 and compute the percentage of empty space in the square not occupied by the circles. *Hint:* (a) calculate the area of the square, (b) calculate the total area of the circles (area of a circle = πr^2), (c) calculate the area of empty space = (a) − (b), and find percentage.†

Types of Solids

13. Compare the similarities and differences between ionic and macromolecular solids.

14. Compare the general melting-point temperatures of various solids, and explain the differences.

15. Why are mothballs and air fresheners volatile solids? Is there a practical advantage to this? Explain.

•16. Ice cubes in a freezer or freezer tray are observed to be smaller after having been in the freezer for a long time. Explain why.

17. Why is ice less dense than water? Water below 4°C is less dense than water at 4°C. What does this imply?

18. Many synthetic polymers are ultraviolet (uv) degradable but are not biodegradable (not

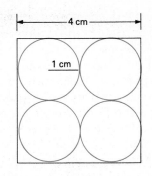

Figure 8.15
How much empty space? See Question 12.

broken down by bacterial action, as are paper and wood). What does this mean in terms of environmental litter, for example, plastic soda bottles and carryout soft drink lids? Discuss how this might be helped by the addition of uv *un*stabilizers to these products.

19. Suppose all the synthetic polymers suddenly disappeared from your classroom. How would things (and you) look?

20. Why are some metals better electrical conductors than others?

21. Magnesium and sodium are both metals, but magnesium has a much higher melting point than sodium (650°C vs. 98°C). Explain why.

•22. Is the quarter (25¢ piece) made with "sandwiched" copper an alloy? Explain.

•23. Explain why the density of brass is less than the density of copper.

Mechanical Properties of Solids

24. A long spring is stretched 4 cm when a mass is suspended from it (Fig. 8.16). If the spring is cut in half and the same mass is suspended from one of the halves, how far would the mass descend?

•25. Two identical springs stretch the same length (2 cm) when a 0.5-kg mass is suspended from either of them. If the springs are hooked together and the mass is suspended from the lower spring, what will be the stretch of each spring?

26. A wire is not very elastic, but if the wire is formed into a coil spring, there is a great deal of elasticity. What does this tell you about the interparticle forces of the wire?

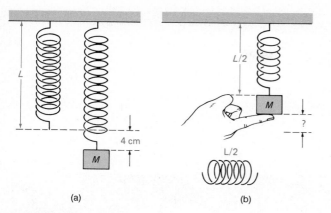

Figure 8.16
How much stretch? See Question 24.

	harder than iron	softer than iron
denser than iron		
less dense than iron		

Figure 8.17

*27. Why do hammers generally not have metal handles, in particular, sledge hammers? (Some carpenter's hammers do have metal handles, but with rubber grips.) Why do we use rubber mallets in some instances of hammering?

28. Automobile tires were once made of solid rubber. Why did we switch to inflated tires?

*29. Enter each of the following four solids into the appropriate boxes shown in Figure 8.17: lead, aluminum, diamond, and iridium (a metal used for penpoints). If you are not sure of the properties of these substances, make your best guess, or ask someone who works in a machine shop. What factors account for the hardness and densities of solids?†

30. Compare the advantages and disadvantages of ceramic and plastic (dinner) plates in terms of mechanical properties.

31. Discuss the various factors that are involved in the breaking of a board or concrete block with the hand (or foot) without injury.

† Questions 12 and 29 from Turk, J., and A. Turk, *Physical Science,* Third Edition, Saunders College Publishing, Philadelphia, 1987.

Liquids

On the average, about a million and a half gallons of water go over Niagara Falls each second.

FLUIDS

Liquids have definite volumes, but no definite shapes. The molecules of a liquid can move around, allowing the liquid to take on the shape of its container. We often use the term *fluid* to mean liquid. A **fluid** is a substance that can flow. Hence, gases as well as liquids are fluids (Fig. 9.1). Because of this common property of being able to flow, gases and liquids have many common properties, and they are often treated together in a chapter on fluids. However, there are many important differences, so we will treat them separately in different chapters (see Chapter 10, Gases).

A liquid can flow, but its molecules stay about the same distance apart. This means that liquids are essentially (but not perfectly) *incompressible*. As a result, a liquid keeps the same volume except for minute changes. The molecules of a gas, on the other hand, do not stay the same distance apart. This allows a gas to fill up the complete volume of any container, so a gas has no definite shape or volume. A gas is also *compressible,* with its volume being affected by pressures on it. For example, air (a mixture of gases) is compressible, but water is relatively incompressible.

A basic mechanical difference between solids and fluids is that a solid can support a shear force or stress (Chapter 8). For example, a solid can support the flat side or even the cutting edge of a knife blade without being cut or sheared if the force on the knife is not too great. But a liquid such as water (or a gas) could not support the shearing stress of the knife and would flow. A fluid flows under the application of a shear stress instead of being elastically deformed.

Keep in mind that although gases and liquids are different, many of the (fluid) principles discussed in this chapter apply to both.

PRESSURE

It is easy to apply a force to a solid. In fact, we didn't think much about it in previous chapters. A vector force arrow could easily represent a force applied with the flat of the hand, a finger, or even a pencil point. But since a liquid (or a gas) cannot support a shear stress, an object supplying a force might cut or plow right through it.

Hence, we have to think of applying a force to a liquid in a different way—in terms of pressure. Pressure takes into account not only the force but also the *area* of application. Pressure is defined as the force divided by area, or force per unit area:

Pressure = force/area

$$p = F/A$$

The SI unit of pressure is newton/per square meter (N/m²), which is called a pascal (Pa) in honor of the French scientist Blaise Pascal (1623–1662). In the British system, the common unit is pound per square inch (lb/in.²). For example, we might inflate automobile tires to a pressure of 30 lb/in.² (30 "pounds" pressure), or about 200 kPa (kilopascals).

To get a feel for the difference between force and pressure, try holding a pin as shown in Figure 9.2. If you push on the pin, the thumb and finger experience the same force (Newton's laws), but the

Figure 9.1
A fluid is a substance that can flow. Liquids and gases are fluids, but solids are not.

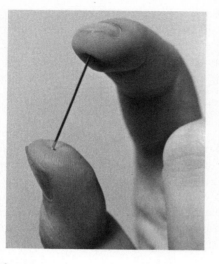

Figure 9.2
Pressure versus force. When holding a pin as shown, the thumb and finger experience equal forces (Newton's third law). However, the pressure on the finger with the smaller area of the pinpoint is greater, and it hurts. (Ouch!)

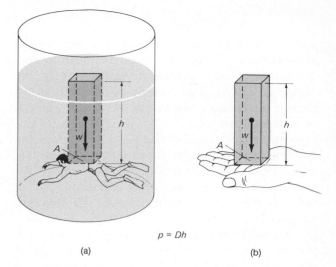

$$p = Dh$$

(a)   (b)

Figure 9.3
Pressure in a liquid. The pressure at any depth h in a liquid is due to the weight of the column of liquid above and $p = Dh$, where D is the weight density of the liquid ($D = w/V$).

finger holding the pinpoint hurts! This is because there is a smaller area of contact, or greater pressure. The force is the same, but the point concentrates it over a much smaller area:

$$p = \mathrm{F}/A \qquad p = \mathrm{F}/A$$
Head end   Point end

In order to apply pressure on a liquid, we commonly use a surface area of its container. For example, when you squeeze a liquid-filled plastic bottle, you apply pressure. Another common method is to put the liquid in a cylinder and apply a force to a piston or plunger, which puts pressure on the liquid. A medical syringe used to give shots is a good example.

Question: Why do spike heels on women's shoes dent tile floors?

Answer: This was a problem in the 1960's when spike heels were popular. A woman weighing 500 N (112 lb) applies the same weight force to the floor when wearing regular shoes or even when barefoot. The difference is pressure.

When spike heels are worn, a large part of the person's weight is distributed over the small heel area, and there is a greater pressure. This can cause the floor to be dented (plastic deformation).

Pressure in a Liquid

Inside a liquid without any externally applied pressure there is pressure due to the weight of the liquid itself. Most of us have experienced this when swimming underwater. The deeper one goes, the more pressure there is from the weight of the water above. For an open container of liquid, there is an additional pressure due to the weight of the atmospheric gases above the container. We'll ignore this for now and consider the effects of atmospheric pressure in the next chapter.

Consider a column of liquid, as shown in Figure 9.3. The pressure in the liquid at the bottom of the column is the weight force of the liquid column divided by its area. This can conveniently be expressed in terms of the weight density D of the liquid and the height or depth h of the column (sometimes called the **pressure-depth relationship**)*:

$$p = Dh$$

Pressure = weight density × depth

For example, the weight density of water is 62.4 lb/ft³. So, if you were swimming and dove down to a

---

* Weight density = weight/volume, $D = w/V$. The weight is given by $w = DV$. The volume of the column is $V = Ah$ (area times height). Hence, $p = F/A = w/A = DV/A = DAh/A = Dh$, or $p = Dh$.

depth of 5 ft, you would experience a pressure of $p = Dh = (62.4 \text{ lb/ft}^3)(5 \text{ ft}) = 312 \text{ lb/ft}^2$ as a result of the water above you. (This is 2.2 lb/in.² or about 15 percent of atmospheric pressure.)

The pressure on the bottom of the container due to the liquid may be found by using the same expression, taking h to be the total depth of the liquid. This is like considering the whole liquid as a column. The liquid pressure on the bottom of a container does not depend on the shape of the container, only on the depth of the liquid (and its weight density).

PASCAL'S PRINCIPLE

When you apply a force to one end of a rigid rod with the other end against something, the same pressure (force/area) is transmitted by the rod to the other end. This is because the rod is incompressible. The same is true for a liquid, since liquids are practically incompressible. However, there is an important difference. In a confined liquid, the pressure is transmitted undiminished to all the walls of its containers.

This fact was first stated by Blaise Pascal, after whom the SI unit of pressure is named, and is known as **Pascal's principle**:

> **Pressure applied to an enclosed liquid (fluid*) is transmitted undiminished to every other part of the liquid and to the walls of its container.**

As a demonstration of Pascal's principle, suppose some small holes are poked in a water-filled balloon—small enough so the water would not run out. Then, if pressure were applied to the balloon, the water would be forced out of all the holes. This demonstrates that the pressure is transmitted (equally) throughout the liquid.

A common application of Pascal's principle is shown in Figure 9.4. When you push on the brake pedal of your car, the brake fluid transmits the pressure to the brake cylinders that apply the brake shoes to the drums—then friction slows you down.

Pascal's principle is used in the hydraulic press, which is a simple machine. Recall from Chapter 2 that a machine is basically a force multiplier that

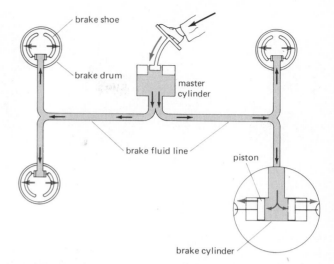

Figure 9.4
Pascal's principle. A practical application in the hydraulic (fluid) brake system of an automobile. The pressure applied by the person pushing on the brake pedal is transmitted to the brake cyclinders.

gives a mechanical advantage. Consider the systems shown in Figure 9.5. With equal forces (weights) on pistons of equal area, the pressure of one piston is transmitted undiminished to the other by the liquid (Pascal's principle). Since the areas are equal, the input force F_i is equal to the output force F_o, and there is no force multiplication.

Suppose the areas of the pistons aren't equal. Let the area of one piston be 100 times that of the other and the weight on the larger piston be 100 times that on the smaller piston. The pressures on the pistons are equal, as can be easily shown, $p_i = F/A$ and $p_o = 100 F/100 A = F/A$. But note that there is a force multiplication. A weight or force of 1 N is supporting a weight or force of 100 N, so there is a force multiplication factor of 100, i.e., $F_o = 100 F_i$. This means that an input force of 9.8 N could produce an output force of 980 N.

In general, the force multiplication factor is given by how many times greater the area of the larger piston is that of the smaller piston, or the area ratio. For example, if the area of the larger piston were 10 times larger than that of the smaller piston ($A_o/A_i = 10$), then there would be a force multiplication factor of 10.* See Special Feature 9.1 for an application of this hydraulic force multiplication.

---

* Pascal's principle holds for a gas, but the pressure in a gas is not transmitted as quickly as in a liquid because a gas is compressible whereas most liquids are not.

* Since $p = F/A$, we may write $F = pA$. Then forming the force ratio, $F_o/F_i = A_o/A_i$, since the equal p's cancel. For $A_o = 10 A_i$, we have $F_o = 10 F_i$.

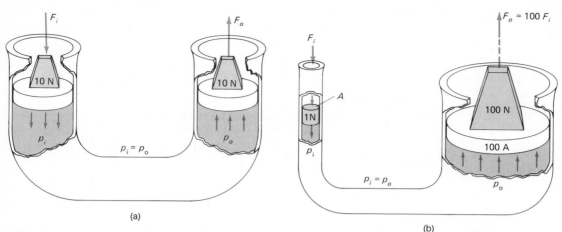

(a)

(b)

Figure 9.5
Mechanical advantage and Pascal's principle. (a) If the areas of the pistons are equal,
there is no force multiplication or a mechanical advantage of one. (b) The pressure on the
small piston is transmitted to the larger one (Pascal's principle), and with the large piston's
area 100 times greater, the output force is 100 times greater than the input force.

* Conceptual Physics, 8th Ed. Hewitt 226

The Hydraulic Garage Lift

A common application of the hydraulic press is in the
garage lift used to raise heavy automobiles and trucks.
Compressed air on the surface of an oil reservoir transmits pressure to the lift piston. The compressed air
pressure doesn't have to be too great. Suppose it were
200 kPa, which is about the pressure in your automobile tires (30 lb/in.^{2}). The pressure on the liquid in the
reservoir is transmitted to the lift piston. If the diameter
of the piston is 0.30 m, then its area is 0.07 m^2, and the
lift force would be 14,000 N ($F = pA$), or about
3150 lb. So the next time you see a car on a lift, you'll
know that Pascal's principle is in operation.

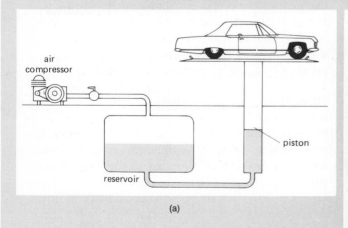

(a)

The hydraulic lift—an application of Pascal's principle.
The air pressure on the surface of the oil in the reservoir is
transmitted undiminished to the lift piston.

(b)

BUOYANCY AND ARCHIMEDES' PRINCIPLE

When lying in a tub of water or standing neck-deep in a swimming pool, you have no doubt noticed that your submerged arms (or legs) feel lighter. What causes this? Certainly your arms have their regular mass and weight. When you pick up something, you have to exert an upward force slightly greater than the object's weight. If someone helped you and supplied part of the force without your knowing it, you might think the object to be "lighter." So, there must be an upward "helping" force with you in the bathtub.

This phenomenon is called **buoyancy**, and the upward-acting force is called the **buoyant force**. To help you understand this buoyant force, consider an imaginary "block" of liquid, as shown in Figure 9.6. The water block exerts a downward force equal to its weight on the liquid surface at the bottom of the block, and the liquid exerts an equal upward force to keep it there. If the water block is removed and replaced with an object of the same weight and rectangular base, as shown in the figure, it would float in place.

Hence, the upward "buoyant" force must be equal to the weight of the block of water removed. This is the same as the volume of water the object would displace when normally placed in the water. The upward buoyant force then depends only on the weight of the volume of liquid an object displaces. This idea was expressed by the Greek scientist Archimedes (287–212 B.C.) and is known as **Archimedes' principle**:

> An object in a liquid (fluid*) is buoyed up by a force equal to the weight of the volume of liquid it displaces.

Archimedes' principle can be easily demonstrated, as shown in Fig. 9.7(a). When a heavy object suspended on a scale is submerged in a liquid, it weighs less and displaces a volume of liquid equal to its own volume. The weight of the displaced liquid is equal to the decrease in the measured weight of the submerged object, or the buoyant force.

You may be wondering what causes the buoyant force. Think in terms of the pressure in a liquid or fluid. As illustrated in Fig. 9.7(b), the liquid pres-

* This works for gases, too.

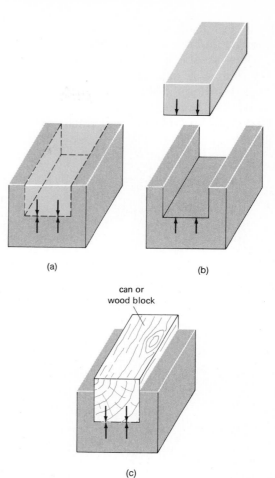

Figure 9.6
Buoyancy. If a "block" of liquid is removed and replaced by a material block of the same weight, it would float because of the upward buoyant force, which must be equal to the weight of the liquid displaced (the removed imaginary "block" of water in this case).

sure on the bottom of the block is greater than that on the top of the block, since the bottom portion is at a greater depth. The pressures on the sides of the block are equal and opposite, so they cancel each other. Hence, there is a net upward pressure or net upward (buoyant) force on the object. (Notice that the buoyant force depends on the pressure difference or the height of the object, and hence is independent of depth.)

Floating and Sinking

In terms of buoyancy we see that a submerged object will sink if the buoyant force is less than the object's weight. If the buoyant force is greater than a submerged object's weight, the object will be buoyed up and rise to the surface.

When the object reaches the surface and some of its volume is exposed, the buoyant force decreases (less liquid displaced). The object will float partially submerged when the weight of the displaced liquid or buoyant force equals the object's weight.

Rather than talk about "heavy" and "light" objects and liquids, it is convenient to consider densities. Recall that density is the amount of mass or weight in a given (unit) volume. It easily follows that

1. An object will float in a fluid if the density of the object is less than the density of the fluid.
2. An object will sink in a fluid if the density of the object is greater than the density of the fluid.
3. An object will be in equilibrium at any submerged depth if the density of the object and the density of the fluid are equal.

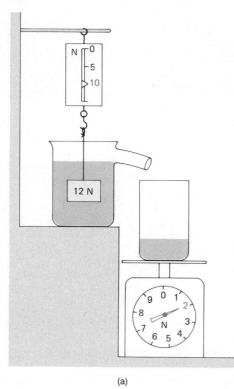
(a)

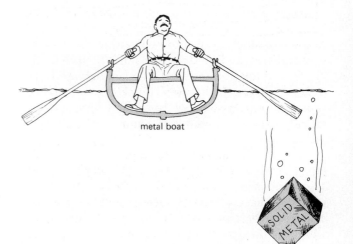
(a)

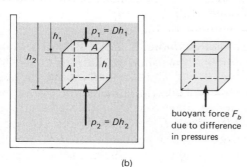
(b)

Figure 9.7
Archimedes' principle. (a) A demonstration of the principle. The weight of the volume of liquid displaced by the submerged block is equal to buoyant force, which causes the block to weigh less. (b) The source of the buoyant force is the difference in pressures — more pressure on the bottom of the block than on the top.

(b)

Figure 9.8
It's a matter of displacement. (a) The metal block and the boat weigh the same, but the boat floats since its overall density is less than that of water. (b) A submarine floats on the surface or submerges by changing its ballast or density.

The Tip of the Iceberg

Icebergs are huge floating masses of ice that have broken off the ends of glaciers or polar ice sheets. These floating "islands" of ice range in size up to 300 km (19 mi) long and 60 m (200 ft) thick. See figure at right.

The density of ice is 0.92 g/cm³, and the density of sea water is 1.03 g/cm³ (Table 8.1). Then, forming the density ratio, an iceberg floating in the ocean has 0.92/1.03 = 0.89, or 89 percent of its volume below the surface. We see only the "tip of the iceberg," or 11 percent (1/9) of its volume.

Icebergs pose a serious threat to shipping. After the *Titanic* struck an iceberg and sank in 1912, an international ice patrol was established by joint action of all nations with shipping interests in the North Atlantic. The ice patrol is the responsibility of the U.S. Coast Guard, which locates icebergs, predicts their drift, and issues warnings.

The tip of the iceberg. Almost 90 percent of an iceberg lies below the surface.

For example, the density of water is 1.00 g/cm³, and the density of wood is 0.70 g/cm³ (Table 8.1).* Thus, a block of wood will float in water. However, the density of iron is 7.9 g/cm³, so a block of iron would sink in water. If a block of some material had a density of 1.00 g/cm³ (the same as water), it would stay at any submerged depth at which it was placed.

Question: A block of iron will sink in water. But ships are made of iron. Why don't they sink?

Answer: In applying the preceding rules, you have to consider the density of an object as determined by its total volume. A block of iron will sink because in this form the density of the block is greater than that of water (Fig. 9.8).

But if the iron is fashioned into sheets and constructed so that the density of the structure as a whole is less than that of water, it will float. (The volume includes the hollow internal space.) In this case, water is displaced by the submerged hollow portion of the structure, and the buoyant force equals the weight force of an iron ship.

---

* The density units g/cm³ will be used instead of the SI units (kg/m³) because of the convenient smaller values. This is one of the advantages of the metric system and its submultiple units. (g/cm³ × 1000 = kg/m³)

A submarine can float or run submerged. For the sub to submerge, the ballast tanks of the sub are flooded with seawater until its total density is greater than that of seawater. The submerged depth can also be adjusted by means of engine power and fins. For the sub to surface, water is pumped out of the ballast tanks, thereby decreasing the sub's density.

It can be shown that the ratio of the density of a floating object and the density of the liquid (density of object/density of liquid) gives the fraction of the volume of the object that will be submerged. For example, if the density of a wooden block is 0.70 g/cm³, then in water the density ratio is (0.70 g/cm³)/(1.00 g/cm³) = 0.70, or 70 percent of the block's volume will be below the surface when floating in water.* See Special Feature 9.2 for another example.

---

* The relative density of a liquid is sometimes expressed in terms of specific gravity, which is the ratio of the weight of a volume of liquid and the weight of an equal volume of water. Since the volumes are equal, this is also the ratio of their densities: sp.gr. $= \rho_\ell/\rho_{H_2O}$ (where ρ is the mass density). The density of water is $\rho_{H_2O} = 1.0$ g/cm³, so the specific gravity is just equal to the magnitude of a liquid's mass density in these units (sp.gr. $= \rho_\ell/1.0$). Being a ratio, specific gravity has no units.

■ SURFACE TENSION

If you drop a razor blade or a needle into a glass of water, it will sink. But if you lay a razor blade very gently on the water surface with its flat side (long dimension of a needle) parallel to the surface, it will float (Fig. 9.9). Is there something wrong with Archimedes' principle? No, there is another effect here called **surface tension**.

Consider a molecule within a liquid (Fig. 9.10). It is attracted in every direction by the neighboring molecules surrounding it. However, a molecule on the surface of a liquid is attracted to the sides and downward, but not upward. These molecular attractions on a surface molecule tend to pull the molecule into the liquid. The net effect is that the surface becomes as small as possible. The surface is "stretched," or in a state of tension, hence the name *surface tension*. The surface behaves much like a thin elastic film. A razor blade or needle will sink if an edge penetrates this surface-tension "film."

The contracting surface force of a liquid accounts for our being able to carefully add more

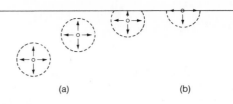
(a)   (b)

Figure 9.10
Intermolecular forces. (a) Within a liquid, a molecule has intermolecular forces in all directions because of surrounding molecules, so there is no net force. (b) At or near the surface, there is a net inward force that gives rise to surface tension.

water to a glass of water filled to the rim. The surface of the water can be raised above the rim of the glass because the tendency for the surface to contract keeps the water from spilling.

Surface tension is the reason why drops of liquids are almost spherical, for example, raindrops and falling drops of molten metal. (Lead shot is made by dropping molten metal from a tower, and the drops solidify into spherical shapes on the way down.) The contracting surface tension causes the free liquid to form into the shape with the smallest surface area. A sphere is the geometrical shape that has the smallest surface area for a given volume (Fig. 9.11).

Detergents

Back to the floating razor blade for a moment. Suppose you put a few drops of liquid detergent in the water. You'd find that the razor blade would sink,

Figure 9.9
Floating razor blade? Surface tension allows a razor blade to float on water.

Figure 9.11
Under tension. Surface tension causes a free liquid to form spherical drops.

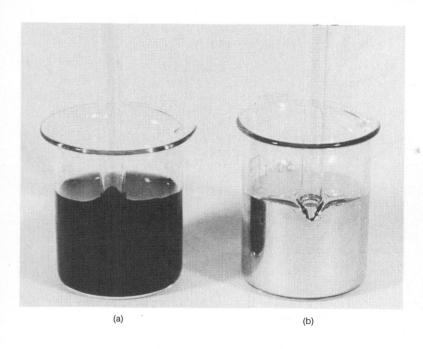
(a)   (b)

Figure 9.12
Adhesion and cohesion. (a) Because of adhesive forces between the water and glass, the water wets the glass. (The water was darkened with ink so that the effect could be seen.) Notice that the overall water surface is bowed in or concave. (b) The internal cohesive forces of mercury are greater than the adhesive forces with the glass, so it does not wet the glass. The overall surface of the mercury is bowed upward or convex.

because the detergent reduces the surface tension. This is also a detergent's job in cleaning. Surface tension keeps the water's surface "drawn in," and it does not "wet" particles of dirt or get under a particle on the fabric of cloth or on a dish. Detergents reduce the surface tension and make the water "wetter" so cleaning is more efficient. Ordinary soap is a detergent.

Question: Why does hot water clean better than cold water?

Answer: Hot water has less surface tension than cold water and so washes or "wets" better. This is because the molecules in hot water have more energy and move around more.

Modern detergents are more effective in reducing the surface tension in cold water than are ordinary soaps. So now you can do your washing with cold "power" and save energy.

■ CAPILLARY ACTION

Some liquids wet surfaces and others do not. For example, in Figure 9.11 the mercury drops do not wet the surface. This is because the contractive cohesive force (surface tension) of the mercury is greater than the adhesive force between the mercury and the surface.* A liquid that wets a solid has greater adhesion than cohesion.

If you examine the surface of water in a glass container, you will find it is not exactly level. It is slightly bowed downward at the center or has a concave shape (curved inward like a "cave"). The water in contact with the glass is lifted above the normal level because adhesion between the water and the glass is greater than the cohesion between the water molecules (Fig. 9.12).

If the container is filled with mercury instead of water, the surface of the mercury will be bowed upward at the center or will have a convex shape. In this case, the cohesion between the mercury molecules is greater than the adhesion between the mercury and glass.

The crescent shape (concave or convex) of the surface of a liquid is called the **meniscus** (pronounced "meh-NISS-kus"). This comes from the Greek word meaning "crescent moon."

When tubes of very small diameter, called capillary tubes, are inserted in water and in mercury, we find that the water rises in the tubes while the mer-

* Molecular attraction between *like* molecules is called a cohesive force. The molecular force of attraction between *unlike* molecules, for example, between a liquid and its container, is called an adhesive force.

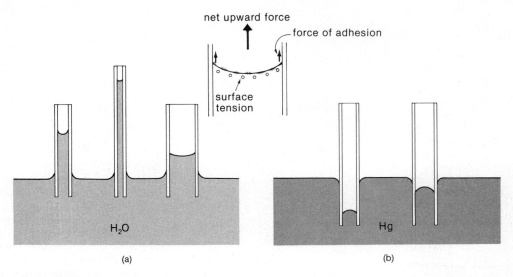

Figure 9.13
Capillary action depends on adhesion and surface tension. (a) Water wets the glass and rises in a capillary tube. (b) Mercury does not wet the glass and is depressed in a capillary tube.

cury level is depressed (Fig. 9.13). This is called **capillary action** or capillarity. The rising of a liquid in a capillary tube depends on both adhesion and surface tension.

Adhesion between the liquid and the tube causes the liquid to creep up the tube walls and produce a concave meniscus surface. The surface tension tends to flatten the surface by contraction. The combined action of these two forces raises the liquid column until the upward force is balanced by the weight of the liquid column. Notice that the liquid column is taller in a smaller capillary tube. Can you explain why in terms of forces?

For liquids that do not wet the tube (cohesion greater than adhesion), the surface tension causes the liquid in the tube to be depressed. The meniscus is convex in this case.

Capillary action causes oil to rise in lampwicks and liquids to be absorbed by paper towels. (A "quicker-picker-upper" has more capillary action.) An important environmental capillary action is the rise of water in the soil.

■ BERNOULLI'S PRINCIPLE

In the previous discussions, the liquids were generally at rest or in equilibrium. The study of liquid (fluid) flow is also an important topic, but it is a difficult one. Unlike in rigid-body solids, the molecules of a fluid can move around, and their motions within a fluid can get pretty mixed up.

To simplify matters for an introductory study of fluid flow, it is convenient to consider **streamline flow**. In streamline flow, the fluid particles follow parallel paths and don't get mixed up or cross each other. This idea is illustrated in Figure 9.14. The initial rising smoke approaches streamline flow (which is really an ideal case). Then mixing causes the flow of the smoke particles to become *turbulent.* You can see how complicated it would be to describe this motion.

Streamline flow is represented graphically by parallel streamlines. Notice that in streamline flow the streamlines never cross. If they do, this gives the particles the option of two or more paths, and the flow becomes mixed or turbulent. Turbulent flow is characterized by little whirlpools or eddies.

A group or bundle of parallel streamlines is often referred to as a tube of flow. In a uniform "tube" all the fluid particles have the same speed. But what would happen if the tube or pipe became constricted or decreased in diameter or cross-sectional area? You have no doubt experienced this effect when a nozzle is put on a garden hose (Fig. 9.15).

Without the nozzle, the water "particles" coming out of the hose have a low speed and are not "shot" very far. However, with a nozzle the water

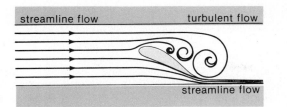

streamline flow   turbulent flow

streamline flow

(a)

Figure 9.15
Flow speed and area. Water comes out of a hose with a nozzle at a greater speed because the nozzle constricts or decreases the cross-sectional area. (The "loop of flow" doesn't really exist. Fluid flow would be something else if that happened.)

(b)

Figure 9.14
Streamline and turbulent flows. (a) When fluid particles flow in parallel paths, the flow is said to be streamlined. Mixing causes turbulent flow. (b) Smoke particle "flow," illustrating these conditions.

speed is increased, as evidenced by how far it is projected. (Recall from Chapter 4 that the range of a projectile depends on its initial speed.)

Streamline flow, along with some other physical principles, allows us to explain this and other general cases of fluid flow. In terms of streamlines (Fig. 9.16), a constriction in a pipe forces the streamlines closer together, which indicates that the speed of the liquid is greater.

This can be understood in a couple of ways. Since liquids are practically incompressible, if a certain mass of liquid goes into one end of the tube in a given time, an equal mass must come out of the other end (conservation of mass). The only way for this to occur is to have the liquid speed up in the smaller part of the tube.

Notice that the pressure in the liquid also changes, as indicated by the standpipes, which act

Figure 9.16
Velocity and pressure in streamline flow. The "bunching" together of streamlines increases the flow velocity and decreases the pressure as work is done in increasing the velocity.

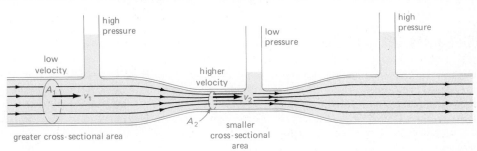

high pressure

low pressure

high pressure

low velocity

higher velocity

A_1 v_1 A_2 v_2

greater cross-sectional area

smaller cross-sectional area

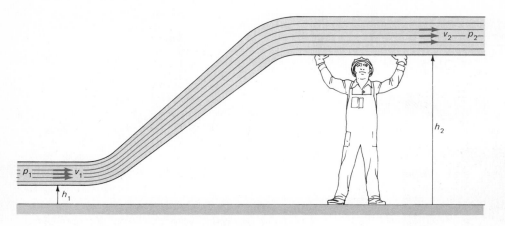

Figure 9.17
Bernoulli's principle takes into account changes in height or potential energy in fluid flow. See text for description.

as pressure gauges. The greater the pressure, the greater the height of the liquid in the standpipe. Why?

A change in pressure is a measure of the work done. As can be seen from Figure 9.16, the pressure decreases in the constricted portion of the tube, which means work was done on the liquid (in forcing it into the smaller tube). By the work-energy conservation ($W = \Delta KE$), the kinetic energy or speed of the liquid then increases.

Daniel Bernoulli (1700–1782), a Swiss mathematician, expressed these work-energy considerations for ideal fluid flow in what we call **Bernoulli's principle**. He also took into account potential energy, which involves a change in height along the length of fluid flow (Fig. 9.17).

In this case we have $W = \Delta KE + \Delta PE$, and work can go into gravitational potential energy of the fluid. In terms of work-energy conservation, this can be written

$$p + \tfrac{1}{2}\rho v^2 + \rho gh = \text{a constant}$$

where the work is proportional to a change in p, and the mass density ρ takes the place of mass in the kinetic and potential energy expressions.

From the equation we see that at a given height, where the speed is greatest, the pressure is least (and vice versa). We commonly use pressure to pump water into water towers (change in height), where it has potential energy. The fluid flow can be analyzed in terms of the conservation of energy, which is what Bernoulli's principle represents. Some common examples and applications of Bernoulli's principle will be given in the next chapter on gases (fluids).

■ VISCOSITY

Streamline flow is an ideal case but quite helpful. There are many other factors in real fluid flow. One of these is viscosity, which is the internal resistance of a fluid to flow—internal "friction." Basically, this depends on the attraction between molecules in the fluid or their ability to slide past each other. The greater the viscosity of a fluid, the greater its resistance or the slower its fluid flow (less fluidity).

Viscosity is temperature-dependent and generally decreases with increasing temperature. If a liquid is heated, its molecules move more rapidly, and the liquid becomes less viscous. Similarly, if a liquid is cooled, it becomes more viscous. Viscosity is the physical principle behind the old saying "as slow as molasses in January." It is also an important feature in proper lubrication. See Special Feature 9.3.

SUMMARY OF KEY TERMS

Fluid: a substance that can flow. Both liquids and gases are fluids.

Pressure: force divided by area: $p = F/A$.

Pressure in a liquid: $p = Dh$, where D = weight density and h = depth in liquid.

Pascal's principle: pressure applied to an enclosed fluid is transmitted undiminished to every other part of the fluid and to the walls of its container.

Buoyant force: the upward force on an object when submerged in a fluid.

SPECIAL FEATURE 9.3

Viscosity and Motor Oils

Viscosity is a prime consideration in the type of oil you use in your car. In the winter you want a low-viscosity or "thin" oil so it will flow more easily and lubricate the car engine properly, particularly when the engine is cold at startup. In the summer a higher-viscosity or "thick" oil is used because the temperature is generally higher and a "cold" motor is warmer.

Motor oils are rated in SAE (Society of Automotive Engineers) viscosity numbers. Common SAE numbers are 10, 20, 30, and 40. The greater the number, the greater the viscosity of the oil for a given temperature.

Oils of SAE 10W and 20W are used in automobiles in the winter. These oils are fairly fluid. The W denotes oils primarily for winter service — below 0°C (32°F). In summer, SAE 30 and 40 oils are used. These oils are fairly viscous. SAE numbers are commonly referred to as "weights," for example, "30-weight" oil.

You may have given all this up and now use a "multigrade" or multiviscosity oil year-round. They are sometimes called constant-grade oils, which is descriptive of their effective viscosities.

Such oils are made possible by a new group of polymer additives called viscosity improvers. These polymers have large coiled molecules. With an increase in temperature, the molecules become uncoiled and intertwine with each other. In this manner, the normal decrease in viscosity is counteracted. With the reverse action on cooling, the oil maintains a relatively constant viscosity over a temperature range.

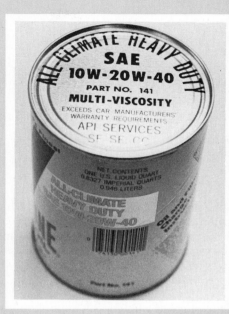

Multigrade or multiviscosity motor oil. Polymer additives or "viscosity improvers" facilitate oil flow at low temperatures and provide greater viscosity at high temperatures so the oil can be used year-round.

As shown in the figure, such oils are multirated, for example, SAE 10W–20W–40 (or 10W–40, for short). Multigrade oils maintain an intermediate viscosity that is functionally operable over the range of given SAE numbers, and you don't have to change oils seasonally.

Archimedes' principle: an object in a fluid is buoyed up by a force equal to the weight of the fluid it displaces.

Surface tension: the contracting force on the surface of a liquid due to unbalanced molecular attraction on the surface molecules.

Capillary action: the rise of a liquid in a small tube due to adhesion and surface tension.

Meniscus: the crescent-shaped surface of a liquid.

Streamline flow: flow in which fluid particles follow parallel paths.

Bernoulli's principle: the application of the conservation of work-energy to fluid flow.

Viscosity: the internal resistance of a fluid to flow.

QUESTIONS

Pressure

1. Given an ordinary brick, how many ways could you place the brick on a table so that the pressure on the table would be different? Which position would give the greatest pressure? Is the force on the table different in each case?

2. In case you didn't think of it, suppose you could place the brick in Question 1 on an edge or corner (using slight horizontal forces so that the vertical weight force of the brick is unaffected). How many ways could you position the brick in

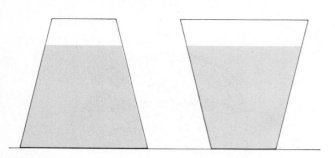

Figure 9.18
Difference in pressure? See Question 5.

this manner to get different pressures? Which position would give the greatest pressure?

*3. The formula for pressure in a liquid is sometimes written $p = \rho gh$, where ρ is the mass density and g the acceleration due to gravity. Is this the same as $p = Dh$? Explain.

*4. A cylindrical container of liquid has two holes in its side plugged with corks. One hole is near the base of the container and the other is midway down the side. If the corks are removed from the holes (same size), from which hole would the liquid flow faster, and why? (Can you apply Bernoulli's equation to this question? *Hint:* p in the equation is atmospheric pressure and doesn't vary between the holes.) Could you have the same flow rate from each hole by varying the size of the holes? Explain.

*5. Two containers, as shown in Figure 9.18, are filled to equal levels with a liquid. Which container has the greater pressure on its bottom?

6. Dams are built with wide bases. Why is this? Might it be necessary for a dam on a river to have a wider base than an ocean dam or a dike with water at the same height? (The little Dutch boy is said to have saved the day by sticking his finger in a leaking dike. How could he have held back the whole ocean with his finger?)

*7. A vertical cylinder 1 m in height is filled with water. If a piston applies a pressure of 2000 kPa at the top of the cylinder, what is the total pressure on the bottom of the cylinder? (*Hint:* The weight density of water is 9800 N/m³.)

Pascal's Principle

8. Suppose your water bed had a leak from a small hole. Would the water run out faster when you were "in" bed? Explain. (Water beds are used in some hospitals to help prevent bedsores, which bedridden patients can get from regular mattresses. How do waterbeds help?)

9. A hydraulic hand jack can be used to raise a car, a mobile home, or even a house. Explain how this can be done by hand. Why does the jack handle have to be moved many times to raise such a heavy object by only a short distance?

*10. In a hydraulic system such as that in Fig. 9.5(b), suppose that no energy is lost when the input piston is moved downward a distance of 50 cm. How far would the output piston be raised? (*Hint:* Explain this in terms of liquid volume displacement, $V = Ah$.)

*11. Suppose in the hydraulic system shown in Fig. 9.5(b) that the input force is the force applied to the larger piston. What would the mechanical advantage or force multiplication factor be in this case? How far would the large piston have to be pushed down to raise the output piston 10 cm?

Buoyancy and Archimedes' Principle

12. Cruise ships weigh on the order of 70,000 tons (600 million newtons). Explain how such heavy boats can float. What is the weight of the water displaced by the boat? Does a ship float higher or lower in the water when it is loaded? Why?

13. An ocean-going freighter loads up in a Great Lakes port and then sails out into the Atlantic. Will it float higher or lower in the ocean as compared with floating in the freshwater lake?

14. Why can humans float in water? Can you affect your floating depth in any way? Explain.

15. Why can people float easily when swimming in the Great Salt Lake in Utah?

Figure 9.19
A steel ball floating? See Question 17.

•16. If a Ping-Pong ball is held well below the surface of a body of water and then released, it rises and "pops" out of the water. What causes the ball to rise above the water surface?

17. A steel (iron) ball floats in a liquid (Fig. 9.19). What can you say about the density of the liquid? (See Table 8.1.) What do you think the liquid might be? It's a metal too.

18. Do ice cubes float higher in water than in mixed drinks? (Explain your answer. The density of alcohol is 0.79 g/cm³.) How about an ice float in a party punch that had fruit frozen into the ice? (*Hint:* Does fruit float in water?)

19. How could you use a plastic milk jug in a swimming pool to illustrate the density floating and sinking rules given in the chapter?

20. The setup shown in Figure 9.7 is used to illustrate Archimedes' principle. Could you also use it to determine the density of an irregularly shaped heavy (sinking) object like a rock? How about finding the density of an irregularly shaped floating object? (*Hint:* Use a sinker.)

•21. A hydrometer is a weighted glass instrument used to measure specific gravity (Fig. 9.20). The stem is calibrated to read 1.0000 when the hydrometer floats in water. (a) How do the markings on the stem run? That is, are the readings greater or smaller on the upper part of the stem (above the 1.0000 mark)? (b) Hydrometers are used to measure the "strengths" of battery acid and antifreeze in your car. How does the specific gravity indicate these strengths? (*Hint:* Battery acid is denser than water, and antifreeze is less dense.)

22. Would the buoyant force on an object floating in a container of water be increased if some of the water were displaced by being scooped out of the container?

•23. A container of water sitting on a scale has a certain weight. If a block of wood is placed in the water and it floats because of the upward buoyant force, would the scale reading change? Explain.

24. What is the principle of a life preserver?

•25. Suppose there is a hole in the bottom of a boat (Fig. 9.21). Explain why the boat sinks in terms of water displacement. How does bailing help?

Surface Tension

26. Why can water bugs "walk on water"?

27. Why are drops of dew on blades of grass or hanging from twigs semi-spherical? Why aren't they completely spherical?

•28. Why are soap bubbles easier to blow and longer-lasting than "water bubbles"?

29. Discuss the drops shown in Figure 9.11 and drops of dew hanging from a twig in terms of adhesive and cohesive forces.

30. Why is glue called an adhesive? Is its adhesive or cohesive force stronger? Why won't glue stick to some surfaces, for example, a greasy surface?

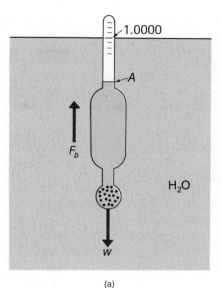

(a)

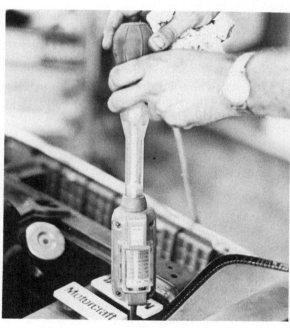

(b)

Figure 9.20
A hydrometer measures specific gravity. See Question 21.

31. What causes a "belly smacker" when you dive into a swimming pool improperly?
*32. To demonstrate the wetting action of modern detergents, an experimenter smeared the bottom side of a duck with detergent and then placed the duck in water; the duck sank. (It was saved, of course.) Why did the duck sink?

Capillary Action

33. At the beach, sand can be observed to be wet above the water line (even at high tide). Why is this? Why is it easier to walk on wet sand than on dry sand?
*34. A light liquid and a heavier (denser) liquid both have the same surface tension and adhesion to the walls of similar capillary tubes. Would both liquids rise to the same height in the tubes? Explain.
35. Why does a liquid rise higher in a capillary tube of a smaller diameter? (See Fig. 9.13.)
36. Would a detergent added to a liquid increase its capillary action?

Bernoulli's Principle

37. Why are you able to shoot water farther from a hose with a nozzle than from one without? How can you do this with a hose without a nozzle by using your finger?
*38. Some hospital patients receive fluids intravenously from bottles hung on tall stands. Why are the bottles elevated? What might happen if the bottles were hung too low?
39. While you are taking a shower and someone flushes a toilet (on the same floor), you may notice a decrease in the flow of the shower water, particularly the cold water. Why is this, when the water pressure coming into the house is still the same?
*40. Suppose the water pressure in your house was so great that it was too forceful when taking a shower. Would closing the valve part way reduce the pressure? Explain.
41. Explain what happens in terms of pressure and flow speed if the liquid flow in Figure 9.17 is reversed.

Figure 9.21
Help! See Question 25.

42. In Figure 9.17, the flow speed is decreased at the top of the elevation. Devise a way you could keep the flow speed constant.

Viscosity

43. What causes viscosity in fluids (liquids and gases)?
*44. Would the capillary action of a liquid be increased if you heated it so as to reduce its viscosity? Don't forget surface tension.
45. Does milk or cream have a greater viscosity? Which has the greater density?
*46. What would happen if you used SAE 40 oil in your automobile in the winter?

Gases

10

Hot-air balloons are buoyant in the surrounding cooler and denser air.

Now for a look at the other side of the fluid "coin"— gases. Gases are fluids since they can flow, and as such they can be described by many of the same principles that apply to liquids (Chapter 9). But an important difference between liquids and gases is that gases are easily compressible. The distances between gas molecules are relatively large, so molecules can be pushed or forced together.

Since little or no cohesive force exists between gas molecules, a gas has no definite shape. A gas can expand to fill the volume of any container and takes on the container's shape and size. On a big scale, this might be a "gravitational container." For example, the expansion of the Earth's atmosphere or that of a star is limited by gravitational forces, which determine the size and shape of a very large mass of gas.

◼ THE PERFECT GAS LAW

As with a liquid, pressure is used to express the force (per area) exerted on or by a gas. However, it may be difficult to see how pressure comes about in a confined gas. To get a picture of this, we use molecular theory.

Molecular theory views the molecules of a gas to be moving about at high speeds. They collide not only with each other, but also with the walls of the container (Fig. 10.1). Each collision with a wall exerts a tiny force on it. Taken alone, this would be insignificant. But the frequent collisions of billions of gas molecules exert a steady average force (per area) or pressure on the walls.

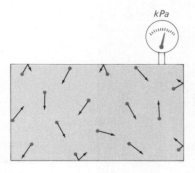

Figure 10.1
Gas pressure. Gas molecules colliding with the walls of a container exert, on the average, a steady force (per area) or pressure on the walls or on a pressure gauge.

The pressure of a gas depends on several factors. One is the volume of the gas or the size of its container. Consider a mass of gas in a cylinder, as shown in Figure 10.2. If the volume is decreased by one half, the pressure doubles. This is because the gas molecules occupy half the space and so make twice as many collisions with the walls. Similarly, if the volume is doubled, the pressure decreases (by one half, why?).

The relationship between the pressure and volume of a given quantity of gas (at constant mass and temperature) is given by

$$p_1 V_1 = p_2 V_2$$

This is known as Boyle's law, after Robert Boyle (1627–1691), an English chemist who is credited with its discovery. The relationship holds for all gases (except when the pressures are very large).

Using modern kinetic molecular theory, another relationship can be derived for the product pV. It is found that pV is proportional to the average kinetic energy of the gas molecules,

$pV \propto$ average kinetic energy of gas molecules

This makes sense because the more kinetic energy the gas molecules in a container have, the faster they move, giving rise to more collisions and a greater pressure.

The kinetic energy of a perfect gas is its internal energy. By adding or removing heat energy, we can therefore affect the pressure and/or volume (Fig. 10.3). When heat is added and the gas is kept at a constant volume, the pressure increases. However, there are cases in which both the pressure and the volume can change.

Temperature (T) is a relative indication of thermal energy, and in terms of temperature we may write

$$pV \propto T$$

For example, on a warm day an automobile tire may be properly inflated. But if it turns cold overnight, you may notice that the tire looks a little flat in the morning. The temperature decrease (reduction in internal energy) gives rise to the volume decrease.

Another factor that affects the pressure of a gas is its mass, or how many molecules there are. The more molecules, the more collisions and the greater the pressure. For example, suppose an automobile tire has a gauge pressure of 165 kPa (25

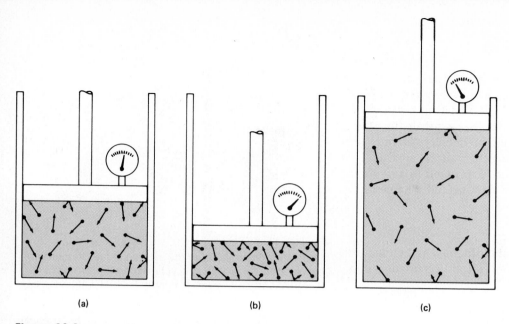

Figure 10.2
Pressure and volume. If the volume of gas (a) is decreased by one half, the pressure is doubled (b). In the confined space, the gas molecules collide with the container walls more often. If the volume is doubled (c), the pressure is decreased by one half. (Why?)

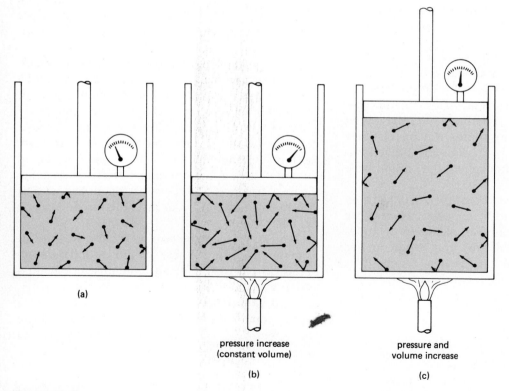

pressure increase
(constant volume)

(b)

pressure and
volume increase

(c)

Figure 10.3
Pressure, volume, and temperature. (a) A quantity of gas at a certain temperature has a particular volume and pressure. (b) When heat is added to the gas and the volume is kept constant, the pressure increases. (c) However, both the pressure and volume can change.

lb/in.²) and you *add* more air so that the pressure becomes 200 kPa (30 lb/in.²). The pressure increases because there are more molecules in the tire and more molecular collisions with the walls. Another way of looking at this is as a density increase. Notice in Fig. 10.2(b) that the density and pressure were increased by making the volume smaller.

All of these factors are expressed in a relationship called the perfect gas law (an older name is the ideal gas law):

$$pV = NkT$$

where N is the number of molecules in the quantity of gas and k is a constant (called Boltzmann's constant).*

■ THE ATMOSPHERE

Now that we know some of the factors that affect gases, let's take a look at some of the properties and phenomena of the mixture of gases in which we live—the atmosphere. Our atmosphere is the gaseous shell or envelope of air that surrounds the Earth. It is composed chiefly of two gases, nitrogen (78 percent) and oxygen (21 percent).

Just as certain sea creatures live at the bottom of the ocean, humans live at the bottom of a vast atmospheric sea of gases, which is on the order of 600 km (400 mi) deep. The Earth's gravitational attraction retains our atmosphere. The moon and the planet Mercury do not have atmospheres. Having relatively small masses and weak gravitational attractions, they lost any original atmosphere they might have had.

The downward attraction of the Earth's gravity causes our atmosphere to be concentrated or more dense near the Earth. One half of the entire atmosphere lies below 11 km (7 mi), and 99 percent lies below an altitude of 30 km (19 mi). So the air is quite "thin" in the upper part of the atmosphere.

Living at the bottom of the atmosphere, we support the weight of the air above us (Fig. 10.4). This

---

* The temperature in the perfect gas law is *absolute* temperature, which is measured in kelvins (SI unit). More will be said about this temperature scale in Chapter 11, where the perfect gas law is used to define *absolute* zero. The perfect gas law is sometimes written in another form. Since $pV/T = Nk =$ a constant,

$$p_1 V_1 / T_1 = p_2 V_2 / T_2$$

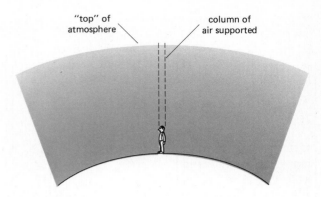

Figure 10.4
Atmospheric pressure. At the bottom of the atmosphere, we support the weight of a column of air, which gives a pressure of about 100 kPa (15 lb/in.²) at the Earth's surface.

gives rise to a pressure at sea level of about 100 kPa or 10^5 N/m², which in customary-system units is about 15 lb/in.² (actually 14.7 lb/in.²).

This means that the roof of a house with an area of 10 m × 25 m = 250 m² has a downward force of 25 million newtons (or about 2800 tons!) on it owing to the weight of the atmospheric column of air. ($p = F/A$ so $F = pA$.) This is balanced by the air pressure inside the house (Pascal's principle, Chapter 9).

An effective demonstration of atmospheric pressure can be made with a large metal can and a rubber stopper (Fig. 10.5). A small amount of water is placed in the open can, which is then heated over a Bunsen burner. When the water boils, as evidenced by water vapor seen coming out of the can, the fire is extinguished, and the stopper is pushed securely into the can opening.* After a short time, the can begins to make stressing noises and is slowly crushed as if by an invisible force. Can you explain why?

The steam from the boiling water drives the air from the can. After the heat is removed and the stoppered can cools, the steam condenses, leaving a partial vacuum in the can. The greater external atmospheric pressure on the can does the rest.

Of course, we don't notice the weight of the atmospheric gases because it is balanced by internal pressures. However, we sometimes notice varia-

---

* A rubber stopper is preferred to the usual gasketed screw cap that comes with the can, for safety reasons. Should one forget to turn off the burner, the pressure buildup will simply "pop" the stopper. With a screw cap in this case, the can might explode.

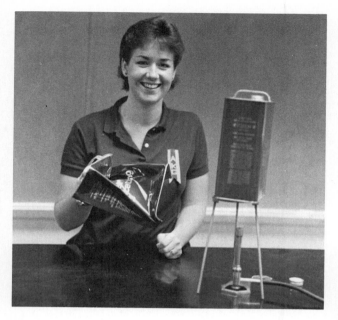

Figure 10.5
Demonstration of atmospheric pressure. A small amount of water in a metal can is boiled so that the steam drives most of the air from the can. When the heat is removed and the can is corked, the steam condenses, leaving a partial vacuum in the can, and atmospheric pressure crushes it.

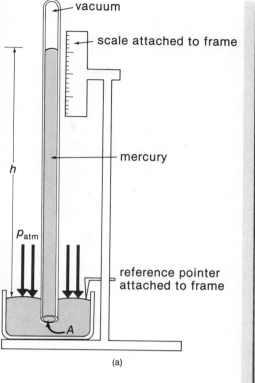

(a)

(b)

Figure 10.6
The mercury barometer. (a) The height of a mercury column depends on the atmospheric pressure on the mercury reservoir. (b) An actual mercury barometer.

tions or changes in pressure. A relatively small, quick change in altitude may cause one's ears to "pop." This is because the pressure in the inner ear does not equalize as quickly, which puts a force on the eardrum. When the pressure equalizes (swallowing helps), the ears pop.

Airplanes are equipped with pressurized cabins that maintain normal atmospheric pressure at high altitudes. Our bodies are accustomed to an external pressure of 100 kPa. Should this be substantially reduced, the excess internal pressure may be evidenced in the form of a nosebleed and other discomforts.

The Barometer

On TV weather reports the announcer gives you the temperature and so on, along with the barometric pressure (see Fig. 12.9). This is the atmospheric air pressure that is measured with a barometer.

The barometer was invented by Evangelista Torricelli (1608–1647), a contemporary of Galileo. The principle of the mercury barometer is shown in Figure 10.6, along with a modern instrument.

If a long tube closed at one end full of mercury is inverted in a container of mercury, some of it runs out, but a column of mercury remains in the tube. The atmospheric pressure on the mercury surface supports the column (via Pascal's principle). The pressure due to the weight of the mercury column is equal to the atmospheric pressure.

Figure 10.7
An aneroid barometer. The pressure is measured by an evacuated metal container sensitive to small pressure changes (seen in the background of the central opening with circular grooves). The lid or diaphragm of the container acts like a drumhead responding to pressure and moves an attached pointer.

The pressure of a liquid column is $p = Dh$ (Chapter 9). At sea level, the atmosphere supports a column of mercury 760 mm tall (30 in.). Instead of reporting the pressure in pascals or N/m^2, it is common practice to give the pressure in terms of the height of the mercury column. One millimeter of mercury (Hg) is called a torr, in honor of Torricelli. Hence,

$$1 \text{ atmosphere* } = 760 \text{ mm Hg} = 760 \text{ torr}$$
$$= 30 \text{ in. Hg}$$

The barometric pressure on weather reports is usually given in inches of mercury. The variations in pressure as high- and low-pressure air masses move across the country are important in weather forecasting.

Any liquid could be used in a barometer. Other liquids would be less expensive. Why not use water? That's a good question.

* A standard atmosphere (atm) is used as a unit for very large pressures. Another pressure unit used by meteorologists is the millibar. 1 millibar (mb) = 100 N/m^2 (pascals), so 1 atm ≈ 1000 mb.

Question: Why isn't water used in barometers?

Answer: One major reason is its relatively small density. The density of mercury (13.6 g/cm^3) is 13.6 times greater than that of water (1.00 g/cm^3). Therefore, a column of water with the same weight as a 760-mm column of mercury would be 13.6 times taller, or 10.3 meters (34 ft).

This would be a water column about as tall as a three-story building. Water barometers are a bit impractical.

Another type of barometer in common use is the aneroid barometer (aneroid means "without fluid"). This is a mechanical device having a sealed metal container with a diaphragm "lid" or cover that is sensitive to small pressure changes (Fig. 10.7). The diaphragm is much like a drumhead responding to pressure. A pointer is used to indicate changes in pressure.

Aneroid barometers with dial faces are common in homes. Notice in Figure 10.7 that fair weather is associated with high barometric pressures and rain with low barometric pressures. Do you know why?*

The altimeters used in airplanes and by sky divers to indicate altitude are really aneroid barometers. The pressure decreases rather uniformly with height in the lower atmosphere. If the barometer dial face is replaced by one calibrated inversely in height, the barometer becomes an altimeter. As you go up in altitude, the pressure becomes less, and this is read as increasing height on the barometer altimeter.

Atmospheric Effects

Many common effects are associated with the atmosphere and its pressure. You see or use some of them quite often. For example, a frequent use of atmospheric pressure is in drinking through a straw. It is generally believed that sucking on the straw

* When the atmospheric pressure is high, air heated near the Earth's surface does not expand greatly. Being only slightly buoyant (Archimedes' principle), it rises a relatively short distance before it cools off. When the pressure is low, the heated air rises higher. Clouds are formed when air rises high enough for water vapor to condense into droplets (Chapter 12). This does not generally occur with a high atmospheric pressure, and without clouds, there is no precipitation but rather fair weather. Barometric changes are associated with fronts, which are boundaries of high- and low-pressure air masses and generally regions of turbulent weather.

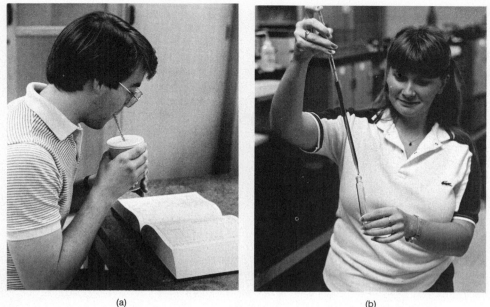

(a)   (b)

Figure 10.8

Atmospheric pressure in use. (a) When drinking with a straw, the person drawing on it reduces the pressure in the straw above the liquid, and atmospheric pressure on the surface of the liquid pushes it up the straw. (b) When a liquid is transferred with a pipette, the pressure on the liquid surface at the top is less than the atmospheric pressure on the liquid surface at the opening tip. (When the finger is first placed on the pipette, a slight amount of liquid runs out, increasing the empty space volume above the liquid and decreasing the pressure.) The net upward pressure supports the liquid column. When the finger is removed, the pressures are equalized, and the liquid drains from the pipette.

draws the liquid up the straw. Actually, the sucking action reduces the pressure in the straw. The atmospheric pressure on the liquid's surface then pushes the liquid up the straw (Fig. 10.8).

Liquids are often transferred and dispensed by medicine droppers and pipettes. A liquid goes up such glass or plastic tubes by the same principle as for a straw. The liquid remains in a pipette stoppered with a finger over one end because the external air pressure on the open end is greater than the pressure in the upper part of the tube. This pressure difference supports the weight of the column. When the finger is removed, the pressures are equalized, and the liquid flows from the pipette under the influence of gravity.

Another method of transferring a liquid is by siphoning action. A **siphon** is a tube filled with liquid, which is siphoned from one level over a small elevation to a lower level (Fig. 10.9). Atmospheric pressure is involved, but it really doesn't do the job as you might think.

Notice in the figure that the upward pressure at B, which is atmospheric pressure, p_a, and the up-

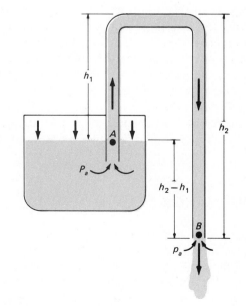

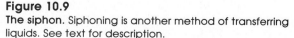

Figure 10.9

The siphon. Siphoning is another method of transferring liquids. See text for description.

(a)   (b)

Figure 10.10
Buoyancy. (a) When filled with enough hot air from a propane burner, (b) the balloons become buoyant and rise.

ward pressure at A are essentially the same. Thus, atmospheric pressure tends to push the liquid up each arm of the tube with equal force (uniform tube diameter).

However, the downward pressure at A due to the weight of the liquid in the shorter arm is less than the downward pressure at B, which is due to the weight of the liquid in the longer arm. As a result, there is a pressure difference between A and B, and the liquid flows through the tube. The pressure difference is given by $p = D(h_2 - h_1)$. The flow will stop if the liquid levels are equal, $h_2 = h_1$, or if the liquid level at A falls below the tube opening, allowing air to break the liquid column in the tube.

■ BUOYANCY

Being a fluid, the atmosphere also has buoyancy, with the buoyant force given by Archimedes' principle (Chapter 9). We don't feel this buoyant force because our bodies don't displace much air, so the force is quite small. Having a greater density than air, we "sink" in the atmosphere. However, if enough air is displaced, objects will float. This is observed for helium and hot-air balloons (Fig. 10.10).

Figure 10.11
A cylinder-piston mechanical force pump. See text for description. Variations in the air pressure in the chamber on the outlet pipe allow for a steadier flow between pumping strokes.

The Vacuum Cleaner

With more and more carpets in homes (wall-to-wall), the vacuum cleaner is an indispensable cleaning tool. Its operation involves a difference in pressure. Within a vacuum cleaner, the motor and blower create a partial vacuum (reduced pressure) in the cleaner chamber and hose. A vacuum cleaner "picks up" dirt because atmospheric pressure forces it into the partial vacuum.

Remember, it is pressure or a pressure difference that "pushes" a fluid. A vacuum does not "pull." Perhaps we should call a vacuum cleaner an "air-pressure" cleaner. (In some cases, we use the exhaust of a vacuum cleaner as an air blower.)

Pressure at work. Atmospheric pressure forces particles into a vacuum cleaner.

We commonly say that helium is "lighter" (less dense) than air and that "hot air rises." The perfect gas law shows why hot air is less dense than cooler air. Hot air has a greater temperature, and by $pV \propto T$, its volume increases. When heated air expands and increases in volume, its density is lowered ($\rho = m/V$). Hence, it is less dense than the surrounding cooler air, and it is buoyed up and rises.

PUMPS

Fluids (liquids and gases) can flow, but they must be given the energy to do so. This is the job of pumps. Basically, a pump is a machine in which mechanical energy is transferred to a fluid causing it to flow. For example, a fan may be thought of as a pump. Pumps are used to deliver water to homes, to pump oil and natural gas to furnaces and hot and cold air throughout our homes, to create partial vacuums, and to circulate blood in our bodies. There are many types of pumps. Let's consider here one common basic type—the force pump.

The Force Pump

A force pump mechanically displaces a volume of fluid by some means. For example, a cylinder-piston arrangement might be used, as shown in Figure 10.11. When the piston is brought forward, the liquid is forced from the cylinder. The air chamber on the outlet pipe is filled on the forward stroke, which compresses the air above and increases the pressure ($p \propto 1/V$, perfect gas law). The compressed air forces the water from the chamber during the backstroke of the piston and maintains a steadier flow.

During the piston backstroke, a partial vacuum is created in the cylinder, and the atmospheric pressure on the liquid in the reservoir forces more liquid into the cylinder. The cycle is then repeated.

Suppose the fluid in the force pump is air and it is pumped into a tank. The pressure in the tank is gradually built up, and the pump acts as a compressor. The compressed air can then be used in a garage lift, a spray painter, or some other application.

In some cases, gases are pumped to create partial vacuums—vacuum pumps. Here, the air or gas is pumped from a volume. Gases can be pumped mechanically and by moving vapors that carry the gas molecules along with them (diffusion). Mechanical methods can produce partial vacuums on the order of 10^{-3} torr (mm Hg). Diffusion pumps can produce partial vacuums on the order of 10^{-6} torr, which is about the atmospheric pressure at an altitude of 270 km (170 mi).

Space is a good partial vacuum. Outside a satellite in orbit at 500 km (300 mi) the pressure is about 10^{-10} torr. In the vacuum of "deep" space, there are only a few molecules per cubic meter and virtually no pressure. A more down-to-earth partial vacuum and its practical application are discussed in Special Feature 10.1.

175

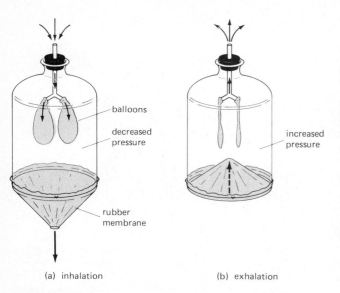

(a) inhalation   (b) exhalation

Figure 10.12
A demonstration to show the mechanics of breathing. A rubber membrane or sheet acts as the diaphragm and balloons as lungs.

Breathing

In a manner of speaking, air is pumped in and out of our lungs—at a rate of about 5 liters per minute. The intake and output of our respiratory pump are controlled by the diaphragm, the muscular partition between the chest and abdominal cavities. An apparatus used to demonstrate the breathing process is shown in Figure 10.12.

When we inhale, the diaphragm moves downward, thereby increasing the volume of the chest cavity. This decreases the pressure inside the chest ($p \propto 1/V$) to below the external atmospheric pressure. The atmosphere then forces air into the lungs. When we exhale, the diaphragm relaxes and moves upward. The internal pressure increases, and air is expelled from the lungs.

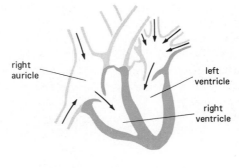
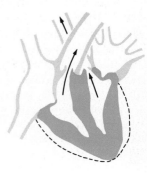

(a) intake   (b) output

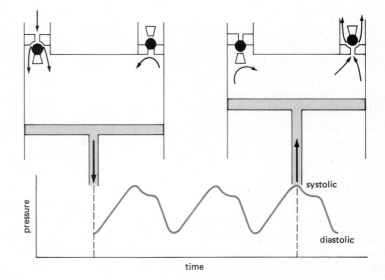

Figure 10.13
The heart as a force pump. The pumping action or "beating" of the heart causes the blood pressure to vary between a high (systolic) pressure and a low (diastolic) pressure.

Blood Pressure

Taking one's blood pressure monitors the pumping action of the heart. When the heart contracts, the blood pressure in the arteries increases. When the heart relaxes, the pressure goes down. The "upper" or maximum pressure is called the systolic pressure, and the "lower" or minimum pressure is called the diastolic pressure (Fig. 10.13).

When taking your blood pressure, a doctor or nurse (or even a coin-operated machine) is measuring the pressure of the blood on the arterial walls. This is done by an instrument called a sphygmomanometer (Greek *sphygmo,* meaning "pulse"). See Figure. This tongue twister is pronounced "ss-fig-mom-an-om-it-er." The dial pressure gauge is calibrated in mm Hg or torr. (Older types of meters use a mercury column to measure the pressure.)

An inflatable cuff is placed snugly around the upper arm. When inflated to a sufficient pressure, the cuff shuts off the blood flow in the large arteries of the arm. Air is then slowly released from the cuff, and the doctor or nurse listens with a stethoscope over the artery just below the cuff in the bend of the arm.

When the cuff pressure is slightly lower than the blood pressure, blood begins to flow through the artery with each beat of the heart. The rhythmic escape of blood beneath the cuff produces a distinct sound that can be heard through the stethoscope. As soon as the sound is heard, the pressure is noted, for example, 120 mm Hg (systolic or "upper" pressure).

As more air escapes from the cuff, blood flows more steadily through the artery. At a specific lower

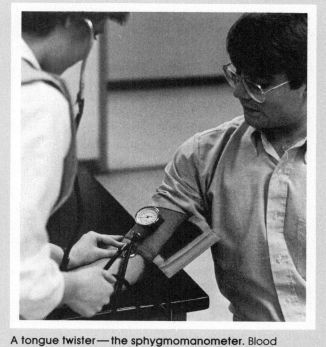

A tongue twister—the sphygmomanometer. Blood pressure is measured with a sphygmomanometer, as shown on the "patient's" arm. A student nurse listens with a stethoscope placed over a blood vessel.

pressure the distinct beating sound is no longer heard, for example, at 80 mm Hg (diastolic pressure). Blood pressure readings are commonly reported in a ratio form, such as 120/80 (a reading of "120 over 80"). Normally, the pressure in the arteries ranges between 100 and 140 systolic (contraction pressure) and between 70 and 90 diastolic (relaxation pressure).

The Heart

While reading about pumps you may have realized that there is an important pump within our bodies—the heart. The heart is basically a muscular force pump that pumps blood through the body's network of arteries, capillaries, and veins to bring nourishment and oxygen to the cells and to carry away wastes.

Rather than having a piston, the heart has muscular walls that contract and relax, thereby changing the volumes of the heart's chambers as it "beats," that is, as it forces blood through the circulatory network (Fig. 10.13).

The heart pumps or beats 60 to 90 times a minute in the average adult. With each pumping cycle, the heart distends and takes in freshly oxygenated blood arriving from the lungs. On contraction, the blood is forced through the aorta into the arterial network.

The working of the circulatory system is externally monitored by taking one's pulse and blood pressure. The pulse is commonly taken with one's fingers on a person's wrist. The pulsating beat is monitored at about 60 pulses per minute, on the average, for a calm, resting person. Blood "pressure" is measured with a special instrument. See Special Feature 10.2.

■ BERNOULLI EFFECTS

Bernoulli's principle, as discussed in Chapter 9, applies to fluids and so also to gases. There are many common effects of gas flow that can be explained by Bernoulli's principle, for example, why airplanes can fly and why baseballs can be made to curve.

In gaseous flow, the potential energy term in Bernoulli's equation is relatively unimportant. The elevation of the gas flow may change, but because of the relatively small mass or density of a gas (as compared to a liquid) the change in potential energy ($\rho g h$) is small. See Chapter 9.

Hence we need only to consider the work and kinetic energy terms in Bernoulli's principle ($p + \frac{1}{2}\rho v^2 = $ a constant) for a good approximation. This means that in regions where the gas speed is greater (streamlines closer together), the pressure is less, and vice versa. This is like the liquid flow in Figure 9.16. Keeping this in mind, let's look at airplanes and curve balls.

Airplanes are heavier (denser) than air, so by Archimedes' principle they would sink (or crash) or not even get off the ground by a buoyant force. So what upward force allows airplanes to take off and remain in flight? This "lift" force in terms of Bernoulli's principle is due to the airflow over the wings (Fig. 10.14).

Because the upper wing surface is curved, the speed of the air flowing over this surface is greater than the flow speed past the bottom surface. As a result, there is a greater pressure on the bottom of

Figure 10.15
Bernoulli effect. Because of the increased airflow over the top of a strip of paper, there is a pressure difference that causes the strip to rise. (Try it for yourself.)

the wing than on the top. This pressure difference provides the upward lift force. A lift force due to air passing over a surface can be demonstrated, as shown in Figure 10.15.

Some people prefer to analyze airplane wing lift in terms of momentum and Newton's second law. Notice in Figure 10.14 that the streamlines have a downward component or a downward component of momentum in that direction behind the airfoil. The rate of change of the vertical component of momentum (or upward force) on the airfoil passing through the air must equal the lifting force on the wing.

The airplane engines supply the forward motion that gets the air flowing over the wings. An airplane cannot "lift off" until it has gained sufficient air speed. Hang gliders must "take off" from heights and use gravitational potential energy to gain speed and lift. Updrafts may also provide some lifting force.

Another upward component of force also results from the air pressure on the leading edge of the wing when it is tilted at an angle to the horizontal — the so-called angle of attack. If the angle of attack is too great, turbulence develops over the upper wing

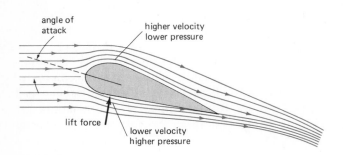

Figure 10.14
Airplane lift. Because of the curved surface of an airfoil or wing, there is an air speed and pressure difference between the upper and lower surfaces that gives a lift force. The lift can also be analyzed in terms of momentum and Newton's second law.

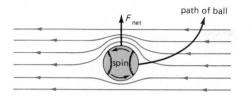

Figure 10.16
Bernoulli's curve ball. The difference in the air speeds on opposite sides of the ball gives rise to a pressure difference that causes the ball to curve.

surface (see Fig. 9.14), and the lift is lost. The airplane then "stalls" (loss of lift, not an engine failure) and falls.

A similar Bernoulli effect causes a baseball to curve. A pitcher causes the ball to curve by giving it an appropriate spin when thrown. If the air were an ideal nonviscous fluid, the spin would have no effect in changing the direction of the ball. However, since the air is viscous, friction between the ball and the surrounding air causes a thin layer of air to be dragged around the spinning ball (Fig. 10.16).

The actual velocity of the air at any point is then the resultant or vector sum of the velocities. Notice on one side of the ball that the spin direction and velocity of the air layer is opposite to that of the relative air velocity, and on the other side they are in the same direction. So the air speed is increased on one side of the ball and retarded on the other side (note the streamlines).

Then, by Bernoulli's principle, the low-speed side has a greater pressure than the high-speed side. The ball then experiences a net force toward the low-pressure (high-velocity) side and is deflected, or curves.

The effect of moving air and pressure is important in home fireplaces. Winds passing over a chimney reduce the pressure and cause the chimney to "draw" or to have a draft up the chimney. Tall chimneys get more airflow over the top and draw better than shorter chimneys. Without good draft, fireplaces often "smoke."

Another novel application of Bernoulli's principle is suspending a light ball, such as a Ping-Pong ball, in a stream of air (Fig. 10.17). You may have seen a similar demonstration in a store with a beach ball suspended above a vertical fan. The stream of

(a)

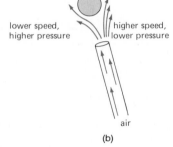

lower speed, higher pressure   higher speed, lower pressure

air

(b)

Figure 10.17
Bernoulli-trapped ball in an airstream. As the ball moves to the side, the air speed and pressure change, keeping the ball "trapped" in the airstream.

air pushes the ball upward, and it moves about. But what causes the ball to stay in the airstream?

Bernoulli's principle explains. As the ball moves to one side of the center of the diverging airstream, the air speed on the outer side of the ball is less than the air speed on the inner side nearer the center of the airstream. Hence, the pressure is higher on the outer side of the ball, and the pressure difference forces the ball back toward the center of the stream—a "Bernoulli-trapped" ball, so to speak.

A Bernoulli effect used in an application to harness the wind's energy is discussed in Special Feature 10.3.

Harnessing the Wind with the Bernoulli Effect

Jacques Cousteau, the famed environmentalist and explorer, has entertained and educated millions of people with sea explorations on his ship *Calypso*. The *Calypso* was built in 1942 as a mine sweeper, and after many years of service, much thought and effort have been given to finding its successor. In 1980 a prototype craft called the *Moulin à Vent* ("Windmill") was constructed. It was a wind-powered boat with no sails. Instead, this 20-m (65-ft) catamaran had a cylinder that was 13.5 m high (44 ft) and looked like a smokestack. However, the fuel was nonpolluting, abundant, renewable, and inexpensive — the wind.

Research on this concept led to the building of a two-cylinder craft named *Alcyone,* after the daughter of the Greek god of wind (a). The *Alcyone* was launched in 1985 and made a trans-Atlantic crossing the same year. Actually, the concept is a new application of an old idea. A stationary cylinder "sail" or column in the wind would produce no sideways "lift" or lateral force. This would be analogous to a cylindrical airplane wing, which would have similar streamline patterns on both sides and no pressure difference or lift force [compare (b) with Fig. 10.14].

However, if the cylinder were rotated, there would be a lateral propelling force, similar to that of a rotating baseball [(c); compare with Fig. 10.16]. Although this is a Bernoulli effect, it is sometimes called the Magnus effect, after the 19th century German scientist Heinrich Magnus, who explained it for a rotating cylinder. Anton Flettner, a German engineer, first applied the Magnus effect to ship propulsion in 1824. He replaced the masts of an old engine-converted sailing ship with two vertical cylinders that rotated at 100 to 150 rpm.

(a)

Taming the wind. (a) The *Alcyone,* named after the daughter of the Greek god of wind. The vessel uses smokestack-type "sails" to harness the wind.

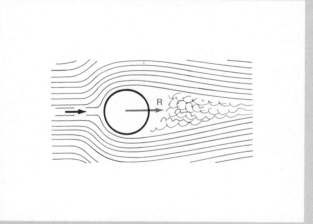

(b) Fixed Cylinder

(b) **Fixed cylinder.** A cylinder in a moving airstream (here flowing from left to right) develops resistance represented by the force *R*. Drag develops behind the cylinder, and the less streamlined the cylinder, the greater the drag.

SUMMARY OF KEY TERMS

Boyle's law: $p_1 V_1 = p_2 V_2$ or $pV =$ a constant (at a given temperature).

Perfect gas law: $p_1 V_1 / T_1 = p_2 V_2 / T_2$ or $pV = NkT$.

Barometer: an instrument used to measure atmospheric pressure.

1 atmosphere = 760 mm Hg = 760 torr
= 30 in. Hg
= 100 kPa = 14.7 lb/in.^{2}

Pump: a machine that transfers mechanical energy to a fluid, causing it to flow.

These cylinders worked as a complementary propulsion source, but there were disadvantages. The propelling force was sideways or lateral to the wind direction, in either one of opposite lateral directions, depending on the way the cylinder was rotated. But maybe you didn't want to go in either of these directions. In this case, you would have to "tack" into the wind (like a sailboat), which is slow. Flettner's large cylinders also offered a great deal of air resistance.

The *Alcyone* has nonrotating, but orientable cylinders. Running the length of each cylinder are two rows of vents, and each cylinder has a movable "shutter-flap" that can cover one row of vents. For propulsion, the flap always covers the windward row of vents, and a fan at the top of the cylinder "aspirates" or draws air into the column through the leeward vents. This causes a strong deflection of the air currents and creates a lateral "lift" force. [See (d), and note the wing-like taper of the cylinder in the photo as well a row of vents.]

Both the Turbosail* cylinder and the flap are oriented automatically by an on-board computer that monitors the wind direction. The orientation of the flap deflects the air so as to provide a force that propels the ship in the desired direction. The computer also adjusts the amount of additional diesel power that might be required to maintain a constant speed.

The *Alcyone* is 31.1 m (103 ft) long and 8.92 m (29 ft) wide and in a strong wind can maintain a cruising speed of 9 to 11 knots without assistance from its two diesel engines. So, when the wind blows, it can be harnessed by the Turbosails to give appreciable savings in fuel costs. The dream is the construction of an even larger *Calypso II* for future explorations.

* Registered trademark, Cousteau-Pechiney Turbosail system.

(c) Rotating Cylinder

(d) Aspirated Cylinder

(c) **Rotating cylinder.** When the cylinder rotates, the resistance force R changes directions and a "lift" force L is developed perpendicular to the airflow with a drag force D in the direction of the airflow. The strength of the lift force depends on the dimensions of the cylinder, the speed of rotation, and the speed of the wind.

(d) **Nonrotating, aspirated cylinder.** Using a fan to draw air into the column through the lateral vents causes a deflection of the air current, which gives rise to a lift force L similar to a rotating cylinder. The forward flap covering the other row of vents separates the airstream and provides an additional lift, as in the case of a horizontal airfoil of an airplane wing.

QUESTIONS

The Perfect Gas Law

1. (a) What happens to the volume of a quantity of gas if its temperature decreases without a change of pressure? Why? (b) What happens to the temperature of the gas if its volume is decreased without changing the pressure?

*2. If a quantity of gas is transferred from one container to another container half the size, how would this affect (a) the density and (b) the pressure of the gas? Would work be required to make the transfer? Explain.

3. When you pump up a tire with a hand pump, the lower part of the body of the pump becomes warm. Neglecting the friction of the plunger, why does this happen? Why does it get harder to pump as the tire becomes inflated?

4. If you inflated an automobile tire to proper pressure on a very cold day, what would happen if the weather suddenly turned warm?

The Atmosphere

*5. Water leaks from a cup with a hole, as shown in Figure 10.18. What would happen if you placed the palm of your hand tightly over the mouth of the cup, and why? (*Hint:* What happens to the pressure of the air above the water with respect to the atmospheric pressure outside the cup?)

6. When a soda bottle full of water (or soda) is turned over, the water does not run out steadily but comes out in spurts with a gurgling sound. Why is this? Why does the water flow out steadily when the bottle is slowly tilted?

*7. A glass is filled with water by submerging it, then lifting it up, as shown in Figure 10.19. How far could the glass be raised before the water runs out? The glass is in a saltwater aquarium. A curious sea urchin looks on in the background.

8. Gasoline cans with spouts commonly have plastic-capped vent holes on the top. What is the purpose of the vent, and what happens if you forget to remove the cap when pouring? (Some cans have a screw-top cap.)

9. Cakes and breads "rise" when baked because of the gas formed by the chemical reaction of baking soda in the batter. Cake mixes prepared for high-altitude locations have less baking soda. Why?

Figure 10.19
Why doesn't the water run out? The sea urchin at the right would like to know. See Question 7.

*10. Automobile tires are inflated to pressures on the order of 200 kPa (30 lb/in.²). Yet bicycle tires are inflated to pressures more than twice this. Why is there such a greater pressure for a lighter vehicle? *Hint:* Think in terms of pressure, force, and *area.* (How many times greater is the typical automobile pressure than atmospheric pressure?)

*11. Explain in terms of fluid principles (a) how a medicine dropper works, in both filling and dispensing, and (b) how a hypodermic syringe works. When a syringe is filled from a sealed, rubber-topped medicine vial, why is air first injected into the vial?

12. If a high-flying jet airliner lost a window, would things in the aircraft be blown out the window or toward the interior of the aircraft?

*13. A glass is filled with water and an index card is placed on the top. If you hold the card and turn the glass over, you can remove your hand and the card will not fall and the water run out (Fig. 10.20). Explain why. Could you turn the glass sideways or horizontally with the same result?

14. How do suction cups work? Would they work on the outside of a spaceship on the moon?

15. A "plumber's helper" or plunger is used to unplug drains. When a plunger is used on a sink, is the plug pushed or pulled in the drain?

*16. What would be the height of a barometer column if a liquid was used that was (a) one half as dense as mercury and (b) twice as dense as mercury?

17. Would an astronaut be able to drink with a straw on the moon? Explain. Would you take a job selling vacuum cleaners to "moon people"?

18. Would it be easier to drink with a straw on top of Mt. Everest or in Death Valley? Explain.

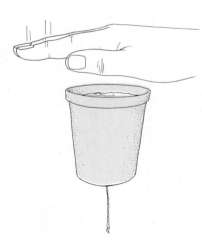

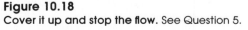

Figure 10.18
Cover it up and stop the flow. See Question 5.

Figure 10.20
The index card doesn't fall! See Question 13.

•19. Why is there a limit to how long (tall) a drinking straw could be practically? (*Big hint:* The height is 10.3 m.)

Buoyancy

•20. When a released helium-filled balloon rises in the atmosphere, it gets larger. Why? With increasing volume, does the buoyant force increase so the balloon will rise indefinitely?

21. Hot-air balloonists continually give the balloon a shot of hot air (from a propane burner) while in flight. Why is this necessary?

22. What is the limit of the gondola weight that can be lifted by a hot-air balloon?

•23. Why are hot-air balloons, and even more so the Goodyear blimp, so large? Wouldn't it be cheaper to make them smaller? What is a major difference between a hot-air balloon and a blimp in terms of navigation?

Pumps

24. Explain the principles involved in a "pitcher" pump or "lift" pump, as shown in Figure 10.21. It is common to have to "prime" such pumps by pouring water into the top if the pump has not been used for some time. Why is this necessary?

•25. When you open your mouth and breathe in quickly, are you sucking in air? Explain. (*Hint:* Think in terms of pressure difference.)

•26. Explain the difference between a pulse rate of 60 and a blood pressure of 120/80.

27. When an inflatable matress is blown up by mouth, what would limit the maximum pressure to which the matress could be inflated?

28. High blood pressure is associated with arteriosclerosis or "hardening of the arteries." The arterial walls are normally elastic in young people, but the elasticity diminishes with age. Fatty deposits can narrow the arterial passageways (which would decrease the pressure), but they also roughen the artery surfaces, which slows down the blood flow. Explain how the latter condition gives rise to high blood pressure. Would the heart have to work harder to maintain a normal blood flow?

Bernoulli Effects

29. Why do big aircraft require longer runways for takeoffs and landings?

30. Birds soar and glide using Bernoulli's principle, but how do they gain altitude? (a) What principle is involved here? (b) How do rockets fly without wings?

31. How do paper airplanes work? Is there any lift involved? (*Hint:* Think how they are thrown, i.e., orientation.)

•32. How do helicopters lift off vertically and fly without wings? (*Hint:* Think about sitting under a fan and one of Newton's laws.)

33. How does a Frisbee "fly"? Why does it fall?

34. Why does the top of a convertible (automobile) "pop up" or bulge when the car goes at a fast speed?

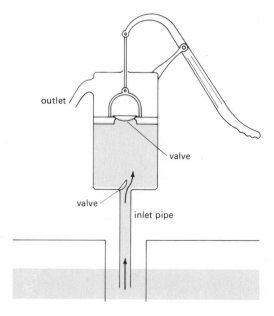

Figure 10.21
A "pitcher" or "lift" pump. See Question 24.

*35. When you drive your car on an interstate and a large tractor trailer with the "hammer down" goes by in the passing lane, the car sometimes seems to sway or veer toward the truck. What causes this?

36. What is the principle of an atomizer that is used to spray perfume? (An atomizer has a rubber bulb that is squeezed to cause the spray.)

37. Is the draft of a fireplace chimney good in terms of energy conservation? Explain.

38. A student in a class states that a nonspinning baseball will curve. Is this correct?

EXERCISES (Just for Fun)

1. On the average, how many times a minute do you breathe? If you take in 5 L of air per minute, how much air do you take in with each breath?

   Answer: assuming 10 breaths/min, 0.50 L or about 1 pint.

Some Exercises using $p = F/A$ that might surprise you.

2. Assuming the palm of your hand measures 6 in. by 4 in., when you hold it out face up, what is the weight of the air on the palm? If the amount seems unreasonable, explain how can this be so.

   Answer: about 350 lb.

3. What is the approximate total force due to the atmosphere on the body of a person with a total body area of 0.20 m²?

   Answer: 2.0×10^4 N (≈ 4500 lb).

4. A tabletop measures 1 m by 1 m. What is the force on the tabletop due to the atmosphere in (a) newtons and (b) pounds? How can such a table be picked up easily with such a force on its top?

   Answer: (a) $\approx 10^5$ N, (b) $\approx 22,500$ lb, (c) force also on bottom [Pascal's principle].

HEAT

Powder blown into the
flame of a Bunsen burner
explodes, giving a release
of energy.

PART III

The portion of physics having to do with heat may in a broad sense be called thermal physics. The effects of heat on the properties of matter are commonplace and have many practical applications in everyday life. With our present knowledge, it is quite natural to look toward the atomic structure of matter in describing these effects, and this provides great insight. However, it was not always so.

The major study of heat began with the start of the industrial revolution when it became important to understand the effects of heat and its conversion to mechanical energy. This study, sometimes called classical thermodynamics, concerned itself solely with the macroscopic properties of matter, such as temperature and pressure, without reference to the underlying atomic structure. Indeed, the atomic nature of matter was not fully understood at the time.

In the latter half of the 19th century, when the atomic nature of matter began to be understood, efforts were made to learn how the macroscopic or bulk properties of matter depended on the assumed behavior of atoms. One of the first successes of this study was with gases in terms of the average values of properties, such as the kinetic energy, of atoms and molecules. This part of the study of heat became known as kinetic theory.

In this section you will be treated to an overview of how macroscopic experimental observations are complemented by theoretical atomic theory in the understanding of the nature of heat and its effects.

Temperature
and Heat

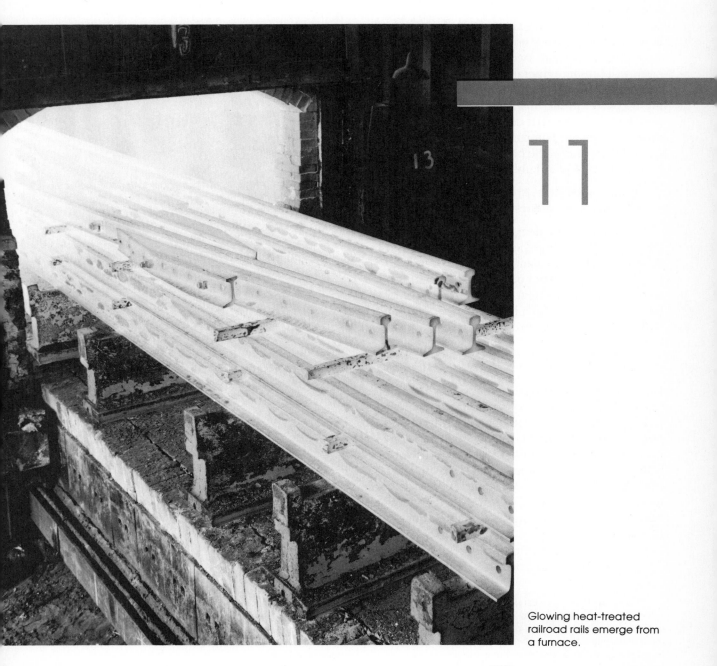

11

Glowing heat-treated
railroad rails emerge from
a furnace.

187

■ THE DIFFERENCE BETWEEN TEMPERATURE AND HEAT

The terms *temperature* and *heat* are used frequently, and we all understand their general meanings. We turn up the "heat" in the house to make it warmer or raise the temperature. We put things in the refrigerator to cool them off or lower the temperature. However, most people find it difficult to give precise definitions of temperature and heat. (Can you define them?)

It is evident from experience that temperature is associated with how hot or cold something is. For example, if you had two bowls of water at sufficiently different temperatures, you could tell which was hotter, or had a higher temperature, by comparing the water in the bowls with your hands.

Notice that this is a comparison, or *relative* measure. You are comparing the hotness and coldness of the water in the bowls relative to the (body) temperature of your hands. Hence, we can say

> Temperature is relative measure or indication of hotness or coldness.

With regard to heat, we know from previous chapters that heat is associated with energy transfer. When you put your hand in a bowl of water, your hand feels warmer or cooler because heat is transferred to or from it. Heat energy will always "flow" from a substance with a higher temperature to one with a lower temperature.* Thus, we can say

> Heat is energy transferred from one body to another because of a temperature difference.

It is a common observation that when a hot object (or fluid) is brought into contact with a colder object, the objects eventually come to the same temperature, or to thermal equilibrium in scientific jargon. For example, thinking of the two bowls of water again, if you dumped the water from one bowl into the other, all the water would eventually come to the same temperature. This would be somewhere between the temperatures of the individual bowls of water.

The preceding statements are really operational definitions. To obtain a better insight into tempera-

---

* An early theory of heat considered it to be a fluid-like substance called caloric (Latin *calor,* meaning "heat") that could be made to flow in and out of a body. Even though this theory has long since been abandoned, we still commonly say that heat "flows" from one object to another.

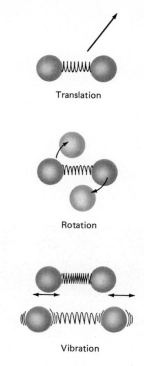

Translation

Rotation

Vibration

Figure 11.1
Molecular motions. In translational motion the molecule moves as a whole. There may also be intramolecular rotational and vibrational motions of the atoms.

ture and heat, let's look at molecular theory. Recall that all matter is made up of molecules—either single atoms or combinations of atoms. The molecules are continuously jiggling around in motion.

In a gas at ordinary pressures, the molecules move around freely, interacting through collisions. In liquids, the molecules also move around but are bound by relatively weak forces. In solids, the molecules move back and forth relative to each other, as if held together by tiny "springs" (see Chapter 8). The energy associated with these random translational motions is often called thermal energy.*

However, in diatomic and other complicated gases, liquids, and solids there may also be rotational and/or vibrational motions of the atoms within the molecules (Fig. 11.1). Then, too, there is potential energy associated with the molecular "springs." The *total* energy (kinetic plus potential) contained within a body is called its **internal energy**.

---

* *Thermal energy* is a general term rather than a scientific one. It is used in different contexts by different people. We will use it here in association with the *random* translational motion of molecules, which, as will be seen, has to do with temperature. An object moving as a whole, much as a thrown baseball, has translational motion, but this is ordered motion, *not* random motion.

In terms of these considerations, temperature is associated with the thermal energy or random motions of the molecules of a substance. Thermal energy is the "temperature" energy or the energy that gives rise to temperature. When heat is added to a substance, it can go into the thermal or random translational kinetic energy of the molecules, which increases temperature. It can also go into the internal vibrational and rotational energies, which do not raise the temperature. Generally, there is a combination of both. Because the molecules have different translational speeds or kinetic energies, we think in terms of the *average kinetic energy* of the molecules, and

> Temperature is a measure of the average random translational kinetic energy per molecule of a substance.

If a body has a high temperature, the average translational kinetic energy of each of its molecules is relatively high. If a body has a low temperature, the average translational kinetic energy of each of its molecules is relatively low.

You may have already guessed the molecular definition of heat:

> Heat is internal energy in transit.

That is, internal energy that is added to or removed from a body. For example, when heat is added to a body and its internal energy increases, some of the transferred energy generally goes into thermal energy or the translational kinetic energy of its molecules.* This increases the average kinetic energy of the molecules of the body, raising its temperature and making it hotter.

■ MEASURING TEMPERATURE

Now that we know what temperature is, how do we measure or express it? Humans have a temperature sense, but it is somewhat unreliable. It varies from person to person, and its range is too limited for scientific and even practical purposes. (Who wants to check the temperature of almost-boiling water?) Also, our temperature sense of touch does not provide a means for measuring temperature quantitatively.

*An exception occurs during a change of phase, for example, going from a solid to a liquid or from a liquid to a gas. This will be discussed in the next chapter.

Thermal Expansion

To measure temperature, we use some physical property of matter that depends on temperature. We construct a thermometer, a device that measures relative hotness or coldness by a change in some physical property. Fortunately, there are many physical properties that can be used (see Special Feature 11.3 at the end of this chapter).

Probably the most obvious and by far the most commonly used property is thermal expansion (and contraction, i.e., a negative expansion). Thermal expansion refers to the changes in the dimensions of substances that occur with changes in temperature.

Almost all substances expand with increasing temperature, but by different amounts. Conversely, most substances contract with decreasing temperature. Metals are a good example. The thermal expansion is small, but it can be made evident by using two metals in the form of a bimetallic strip (Fig. 11.2). Two metal strips, such as brass and iron, are bonded together. Because the metals expand differ-

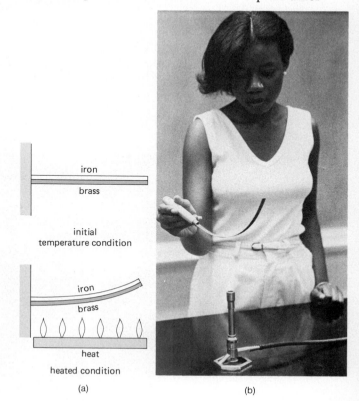

Figure 11.2
Bimetallic strip. Because the metal strips bonded together expand differently, the bimetallic strip bends toward the metal with the smaller linear expansion.

Figure 11.3
Bimetallic thermometer. A bimetallic strip wound in a coil or helix is used in a dial thermometer.

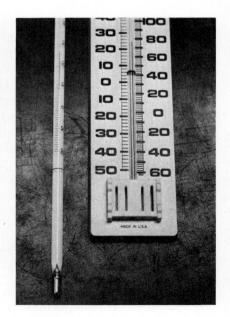

Figure 11.4
Liquid-in-glass thermometers. The liquid in a thermometer bulb expands and contracts, causing liquid in a capillary bore to go up and down and thereby indicating changes in temperature. Commonly used liquids are mercury *(left)* and colored alcohol *(right)*.

(a)   (b)

Figure 11.5
Allowances for thermal expansion. (a) Bridge expansion joint and (b) rocker support. Serious damage due to thermal expansion could occur without these.

The Thermostat

When we want to adjust the temperature in our homes, we usually turn the thermostat up or down. This common wall device controls the temperature by turning heating and cooling systems on and off.

If you pull the cover off a thermostat, you will ordinarily see a mechanism such as that shown in the figure. The temperature sensor is a bimetallic coil to which a glass vial containing mercury is attached. As the coil expands or contracts with temperature changes, the vial is tilted, and the mercury moves from one end to the other. Within the vial are electrical contacts. In moving to different ends of the vial, the mercury makes or breaks electrical contact to turn heating and cooling systems on or off, depending on the temperature for which the thermostat is set.

Setting a thermostat for a desired temperature tilts the vial to the appropriate position. Remove the cover of a thermostat set in the "off" position and observe the tilting of the vial and movement of the mercury when you adjust the temperature control.

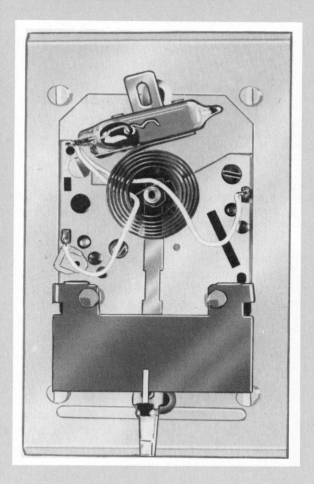

Thermostat. The expansion of a bimetallic coil causes a vial of mercury to tip and make or break electrical contact, which turns heating or cooling systems on and off.

ently, the bimetallic strip bends toward the strip with the smaller linear expansion.

A bimetallic strip could be used as a thermometer by calibrating its deflection, since each position of the deflecting end corresponds to a different temperature. A more convenient form of a bimetallic thermometer is a strip wound in a coil or helix. The deflection is indicated by a dial on a calibrated scale. Such dial thermometers are commonly used as cooking thermometers in ovens (Fig. 11.3). Some have long probes that can be stuck into meat roasts. A common application of a bimetallic coil for temperature control is given in Special Feature 11.1.

By far the most common temperature-measuring device is the liquid-in-glass thermometer (Fig. 11.4). This makes use of the thermal volume expansion of a liquid. It consists of a glass bulb, usually containing mercury or alcohol (colored with a dye to make it more visible), which is connected to a glass tube with a small capillary bore. An increase in the temperature of the liquid in the bulb causes it to expand up the bore, and the change in the height of the liquid in the tube provides a means of measuring a change in temperature.

Thermal expansion is an important consideration in many applications. It can cause bonded or glued joints to break because of "mismatch" expansion between the adhesive and bonded pieces with temperature changes. The different expansions of the materials can produce large stresses that break the bond. Engineers are well aware of the expansion of materials over seasonal temperature ranges. For example, you may have noticed the expansion joints and rocker supports on bridges (Fig. 11.5). Without these to allow for thermal expansion, damage could occur.

Question: If a piece of metal sheeting with a hole in it, as shown in Figure 11.6, were put in an oven and heated, would the hole get larger or smaller?

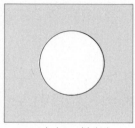 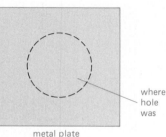

metal plate with hole   metal plate

where
hole
was

Figure 11.6
Area thermal expansion. When a metal plate with a hole
in it is heated, does the hole get larger or smaller? See
Question and Answer.

Answer: Most people think that metal would
expand so as to make the hole smaller—but
not so. The metal expands, but the hole gets
larger. One way of seeing this is to think
about the piece of metal that was cut out to
make the hole. If it was heated, it would
certainly expand and get larger.

The metal in the sheet with the hole
behaves the same as if the cut-out piece were
still there, so the hole gets larger. (For a more
conclusive proof, see Question 13 at the end
of the chapter.) As a practical example, when a
cannon is repeatedly fired, it gets hot, and the
diameter of its bore increases.

Temperature Scales

Thermometers are calibrated so that a numerical
value may be assigned to a particular temperature.
This is done by defining a temperature scale and a
standard unit of temperature. As with any measurement or scale, we can use two reference points or
marks. (Recall that the meter and the yard are practically defined by two reference marks on metal
bars.)

To establish a temperature scale, we may take
two reference or "fixed" points defined by physical
phenomena that always occur at the same temperatures, for example, the ice point and steam point of
water. These fixed points are convenient and readily
available. More commonly called the freezing and
boiling points, the ice and steam points are the temperatures at which pure water freezes and boils
under a pressure of one atmosphere (standard pressure).

The Kelvin temperature scale and the kelvin
unit are the official SI system standards. But before

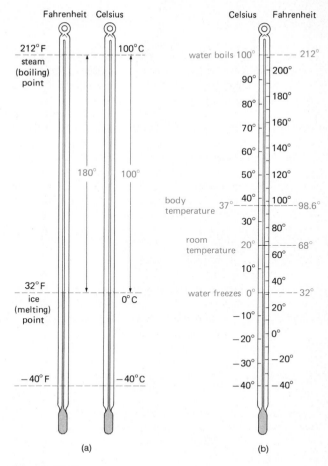

Figure 11.7
The Fahrenheit and Celsius temperature scales. The
Fahrenheit degree interval is larger than the Celsius
degree interval—almost twice (actually 1.8 or 9/5) as
large. See also Figure 11.4.

considering these, let's first discuss the more common Celsius and Fahrenheit temperature scales
(Fig. 11.7).

The **Celsius temperature scale** is named after its
originator, A. C. Celsius (1701–1744), a Swedish
astronomer. On this scale, the interval between the
ice and steam points is divided into 100 equal sub-
intervals or degrees (much like dividing a meter
into centimeters). The temperature is read in degrees Celsius (°C), with the ice point having a temperature of 0°C and the steam point a temperature of
100°C.* Room temperature is 20°C.

* This scale is commonly referred to as the centigrade
scale (Latin *centi,* meaning "one hundred," and German
grade meaning "degrees").

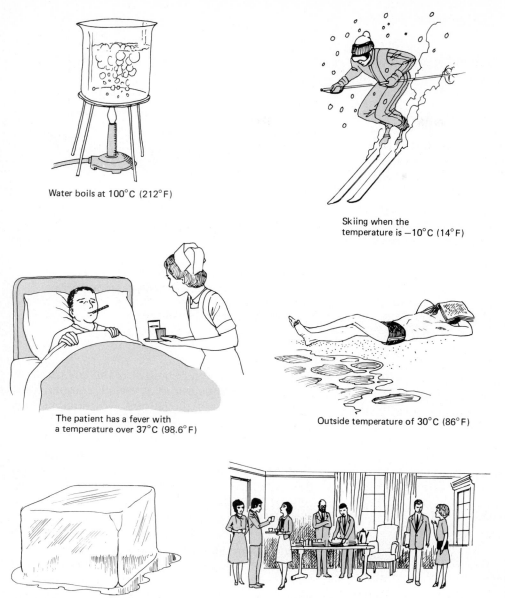

Water boils at 100°C (212°F)

Skiing when the
temperature is −10°C (14°F)

The patient has a fever with
a temperature over 37°C (98.6°F)

Outside temperature of 30°C (86°F)

Water freezes at 0°C (32°F)

Room temperature of 20°C (68°F)

Figure 11.8
Some Celsius-Fahrenheit comparative temperatures.

The **Fahrenheit temperature scale** is named after G. D. Fahrenheit (1686–1736), a German physicist who originated it and who invented the mercury-in-glass thermometer. On this common scale, the interval between the ice and steam points is divided into 180 equal degrees, and the temperature is read in degrees Fahrenheit (°F). As you probably known, the ice (freezing) point of water is 32°F and the steam (boiling) point is 212°F. Room temperature is 68°F (20°C).

Notice that the temperature *units* on both scales are degrees. However, the degree unit on the Celsius scale is larger than the degree unit on the Fahrenheit scale—almost twice as large (1 Celsius degree = 1.8 Fahrenheit degrees, since 100 Celsius degrees equals 180 Fahrenheit degrees). These temperature scales are compared in Figure 11.7.

We now commonly hear or see the temperatures on weather reports given in both degrees Fahrenheit and degrees Celsius so we can become accustomed to Celsius temperatures. The Celsius scale is used throughout most of the world for general temperature measurements, the most notable exception being the United States. Some examples of Celsius temperature conditions are shown in Figure 11.8.

Question: Could the temperature reading given on the weather report ever be the same on both the Celsius and Fahrenheit scales?

Answer: The temperatures on the Celsius and Fahrenheit scales always seem to be numerically different, for example, 20°C and 68°F. However, it is possible to have the same reading on both scales. This occurs at −40°, that is, −40°C = −40°F. (This is proved in the Extended Views [Optional] in the Appendix.) Whether you ever hear this temperature reading on a local weather report depends on where you live.

You may wonder why Fahrenheit chose 32° and 212° for the temperatures of the ice and steam points of water. Actually, he didn't. Other fixed points were used on his original scale.

Fahrenheit chose for the zero reading on his scale the lowest temperature then obtainable in the laboratory—that of an ice-salt mixture. The upper fixed point was taken to be normal human body temperature, which was designated as 96°. (More exact measurements showed this to be 98.6°F.) When the ice and steam points of water were adopted as fixed points, these corresponded to 32° and 212° on the Fahrenheit scale.†

Absolute Zero and the Kelvin Temperature Scale

The Celsius and Fahrenheit scales are used for everyday temperature measurements, and negative temperatures (below zero) are common. Thinking about this, you may wonder if there is a lower limit of temperature or an absolute zero. The answer is yes. It is determined by using the perfect gas law.

Recall from Chapter 10 that the perfect gas law is $pV = NkT$. Then for a gas in a rigid container (constant volume V), the pressure is directly pro-

† ". . . the degree 48, which in my thermometers holds the middle place between the most intense cold artificially in a mixture of water, of ice and sal ammoniac (ammonium chloride, a chemical 'salt') or even sea-salt, and the limit of the heat which is found in the blood of a healthy man.'' From Fahrenheit, D. G., *Philosophical Transactions,* Vol. 33, 1724, from Magie, W., *A Source Book in Physics,* Harvard University Press, Cambridge, Mass., 1963. In this article Fahrenheit reports the temperature of boiling rainwater to be 212° on his temperature scale.

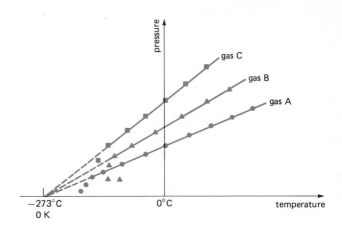

Figure 11.9
Absolute zero. The pressures of gases follow a linear relationship with temperature. Extrapolating to zero pressure gives the value for "absolute" zero temperature, 0 K (or −273°C).

portional to the temperature. If the pressures of some gases are measured at various temperatures and the data plotted, the points follow straight lines down to relatively low temperatures (Fig. 11.9).

At lower temperatures there are deviations from the straight-line relationship of the perfect gas law. This is because real gases eventually condense into liquids. Since there are no molecular interactions other than collisions in a "perfect" or "ideal" gas, it would theoretically remain a gas at any temperature. The pressure of a perfect gas would then be zero when the temperature is zero—absolute zero.

This absolute minimum temperature is implied by extrapolation of the straight lines in Figure 11.9 to zero pressure. It is found that they all converge to −273°C (actually −273.16°C or −459.69°F) on the temperature axis. This temperature is taken as absolute zero of the **Kelvin temperature scale**.* The Kelvin or absolute temperature (T_K) is related to the Celsius temperature by

$$T_K = T_C + 273$$

Notice that at $T_C = -273°C$, $T_K = 0$ K (see Fig. 11.10). The individual intervals on the Kelvin scale are the same size as those on the Celsius scale. This temperature unit is called a kelvin (*not* degree kelvin) and is abbreviated with the symbol K (*not* °K).

* Named after Lord Kelvin (William Thomson, 1824–1907), a British physicist.

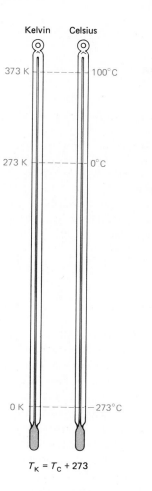

Kelvin   Celsius

373 K — — — — 100°C

273 K — — — — 0°C

0 K — — — — −273°C

$$T_K = T_C + 273$$

Figure 11.10
**A comparison of the Kelvin and Celsius temperature
scales.** The individual intervals on these scales are the
same size; a kelvin is the same size as a Celsius degree.

The kelvin is the SI base unit of temperature and is
used primarily in scientific work. When temperature intervals are involved, the kelvin may be used
interchangeably with the degree Celsius (°C).

However, in the perfect gas law, $pV = NkT$, the
absolute temperature must be used. Since the internal energy of an ideal gas is proportional to the
absolute temperature, this means that if the absolute
temperature is doubled, the internal energy is also
doubled, and vice versa. (The internal and thermal
energies are the same for a perfect gas. Why?) Such
is not the case for the Celsius (or Fahrenheit) temperature.

To illustrate this, suppose that a gas were at a
temperature of 0°C (or 0°F) and the internal energy
were doubled by adding heat. Certainly the Celsius

(or Fahrenheit) temperature is not doubled. Doubling zero (degrees) still gives zero. Also what if the
temperature of the gas were −10°C and you doubled the internal energy? These problems do not
occur with absolute temperatures.

Absolute zero (0 K) is believed to be the lower
limit of temperature, although we have never physically been able to obtain this temperature. (In fact,
there is a physical law that states that absolute zero
cannot be reached, as will be discussed in Chapter
13.)

No upper limit of temperature is known. In
terms of a measuring stick analogy, we have a temperature "measuring stick" with a zero end that extends indefinitely, or at least with no upper end
(limit) in sight. The interior temperatures of stars
are on the order of 100 million kelvins or degrees
Celsius. A difference of 273 degrees between the
temperature scales doesn't make much difference at
such temperatures. Various methods of measuring
temperatures here on Earth are discussed in Special
Feature 11.3 at the end of the chapter.

■ MEASURING HEAT

In the strict sense, a body does not contain heat but
contains internal energy. Heat is internal energy
transferred from one body to another as a result of a
temperature difference. The amount of heat or energy transferred is measured by some change that
occurs in the transfer process, usually a change in
temperature.

The SI unit of all forms of energy, transferred or
otherwise, is the joule. So in this system the unit of
heat is the joule. However, until the SI system is
universally adopted, you will often hear heat energy
expressed in a widely used unit called the **calorie**. It
is defined as follows (Fig. 11.11):

A calorie (cal) is the amount of heat required
to change the temperature of 1 g of water 1°C.

This is sometimes called a gram calorie to distinguish it from the larger Calorie associated with the
energy value of foods. See Special Feature 11.2.

Another common unit of heat energy is the **British thermal unit** (Btu). A Btu is the amount of heat
energy required to change the temperature of 1 lb of
water 1°F (1 Btu = 252 cal).

SPECIAL FEATURE 11.2

Counting Calories

If you are a weight watcher, you no doubt count Calories, which are used to specify the energy values of foods. The Calorie contents of food servings are now commonly listed on labels. However, this is not the same as the calorie (cal) defined as the amount of heat required to change the temperature of 1 g of water 1°C. The food Calorie (Cal) is really a kilocalorie (1 Cal = 1000 cal = 1 kcal). This amount of energy would change the temperature of 1 kg of water 1°C (see Fig. 11.11).

To distinguish between the kilogram Calorie and the gram calorie, the bigger unit is written with a capital C and is sometimes referred to as a "big calorie." For example, a piece of pie containing 300 Calories or "big calories" has 300 kcal or 300,000 "little calories" (gram calories). Perhaps if food values were listed in gram calories there would be a greater incentive for one to stay on a diet. At least it looks and sounds as though you get more.

Heat of Combustion

We burn fossil fuels to obtain energy for heating purposes. In a similar sense, foods are body fuels. These processes involve the conversion of chemical energy to heat.

The intrinsic energy value of a fuel or food is expressed in terms of **heat of combustion**. This is the heat produced per unit mass of a substance when burned in oxygen. The units of heat of combustion are calorie/gram (cal/g) or kilocalorie/kilogram (kcal/kg).

Some heats of combustion for typical substances are given in Table 11.1. Notice that gasoline

has a heat of combustion of 11,400 cal/g, so one gram of gasoline releases 11,400 calories of heat when burned. Compared with alcohol, which has a heat of combustion of 6400 cal/g, gasoline has a much greater intrinsic energy value.

From the table you can see that you would gain more weight from eating scrambled eggs, 2100 kcal/kg, than from eating boiled eggs, 1600 kcal/kg. Remember, the kilocalorie is the food "Calorie." Of course, you probably wouldn't eat a kilogram of eggs, at least not at one sitting. The Calories listed for foods in diet tables are given for average amounts or portions, for example, an average-sized egg.

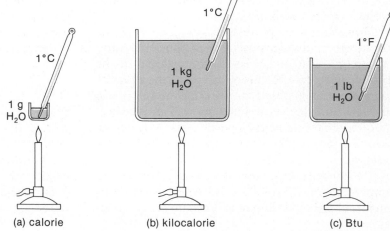

Figure 11.11
Units of thermal energy. (a) A calorie raises the temperature of 1 gram of water 1°C. (b) A kilocalorie raises the temperature of 1 kg of water 1°C. (c) A Btu raises the temperature of 1 lb of water 1°F.

(a) calorie   (b) kilocalorie   (c) Btu

Table 11.1

Typical Values of Heats of Combustion

Substance	cal/g or kcal/kg
Fuels	
Alcohol	6400
Coal	
Anthracite (hard coal)	8000
Bituminous (soft coal)	7500
Diesel oil	10,500
Fuel oil	10,300
Gasoline	11,400
Natural gas	10,000
Wood (pine)	4500
Foods	
Bread (white)	2000
Butter	8000
Eggs	
Boiled	1600
Scrambled	2100
Ice cream	2100
Meat (lean)	1200
Milk	700
Potatoes (white, boiled)	970
Sugar (white)	4000

■ SPECIFIC HEAT

Suppose you had equal masses of copper and aluminum at the same temperature, and you added equal amounts of heat to each. Would the temperature increase be the same for each? You might be surprised to find that if the temperature of the copper increased by 100°C, the corresponding temperature change in the aluminum would be only 44°C. To get the temperature of the aluminum to change 100°C, you'd have to add more than twice the amount of heat than was added to the copper.

If you tried the same experiment with mercury and water, the amount of heat required to give a 100°C temperature change in the mercury would produce only a 3.3°C temperature change in the water. The water would require more than 30 times as much heat for a 100°C temperature change. Why the big difference?

This shows that different substances have different capacities for storing internal energy. When heat is added to a substance, it can go into thermal or random kinetic energy that will increase the temperature. It can also go into the internal vibrational and rotational energies, which do not raise the temperature.

Generally, there is a combination of both. For example, it takes more heat to raise the temperature of a mass of hydrogen (a diatomic gas) by 1°C than for a degree change in an equal mass of helium (a monatomic gas). This is because in the hydrogen a portion of the internal energy is stored as atomic rotational or vibrational energy and is not effective in raising the temperature.

To express this difference, we say that hydrogen has a greater specific heat (or specific heat capacity) than helium. Each substance has a **specific heat** value, which is generally defined as follows:

> The specific heat is the amount of heat required to raise the temperature of a unit mass of a substance by one degree.

The specific heat is descriptive of how much heat energy a substance will "hold," or its "capacity," per unit mass for a given temperature change (one degree). For example, by definition of the calorie or kilocalorie, water has a specific heat of 1.0 cal/g/°C (calorie per gram per degree Celsius). By definition, 1 cal raises the temperature of 1 g of water by 1°C. The specific heat of water is also 1.0 kcal/kg/K. Why? The kelvin could be replaced with °C, since the temperature interval is the same. The specific heats of some substances are listed in Table 11.2.

In general, the greater the specific heat of a substance, the more heat it takes (per unit mass) to

Table 11.2

Specific Heats of Various Substances

Substance	cal/g-°C or kcal/kg-°C
Solids	
Aluminum	0.22
Brass	0.094
Copper	0.093
Glass (typical value)	0.16
Ice	0.50
Iron	0.11
Lead	0.031
Soil (typical value)	0.25
Wood (typical value)	0.40
Liquids	
Ethyl alcohol	0.60
Benzene	0.41
Gasoline	0.50
Mercury	0.033
Water	1.00
Gases	
Air	0.17
Steam	0.48

increase its temperature (or the more heat that must be removed to decrease its temperature). Notice in Table 11.2 that the specific heat of aluminum is twice that of iron. Hence, to raise the temperatures of equal masses of aluminum and iron by the same amount, you'd have to add twice as much heat to the aluminum.

We often say that certain materials "hold more heat." This is because such materials have relatively large specific heats. Since it takes more heat per unit mass to raise their temperatures or "heat them up," they have more stored energy. This is sometimes painfully evident when eating a baked potato or cheese on a pizza. A large portion of potato and

SPECIAL FEATURE 11.3

Thermometry — Temperature Measurement

Temperature is measured by reproducible changes in the physical properties of materials. As we have seen in this chapter, thermal expansion is a commonly used property in bimetallic and liquid-in-glass thermometers, but it is not the only one. For example, in addition to thermal expansion, electrical and radiation properties are used.

Various physical properties have different ranges over which they are applicable for temperature measurement or thermometry. Some of the common practical temperature-measuring instruments and their ranges are listed in Table 11.3.

In this feature, some of the not-so-common temperature-measuring devices will be discussed. Although some of the electrical and radiation principles involved will be discussed in later chapters, you should be able to grasp the general ideas from your practical experience.

Electrical Resistance Thermometers

The electrical resistance of most metals increases with temperature. (Electrical resistance is the opposition of a material to the flow of electric current. See Chapter 17.) These changes are rather large. For example, for

Table 11.3
Temperature-Measuring Instruments

Instrument	Approximate Range
Liquid-in-glass thermometer	
Alcohol	−80°C to 100°C
Mercury	−38°C to 350°C
Bimetallic thermometer	−40°C to 500°C
Electrical resistance thermometer	−272°C to 1600°C
Thermocouple	−260°C to 1600°C
Optical pyrometer	600°C upward
Infrared pyrometer	−20°C to 1700°C

platinum a 39 percent change occurs between 0°C and 100°C. Hence, changes in the electrical resistance of a metal can be calibrated to temperature and used as a resistance thermometer. Platinum, which can be used to measure over a temperature range of −260°C to 1100°C, is especially suitable.

If the metal in a circuit is connected to the appropriate meters, the temperature can be read. The metal temperature probe is convenient in many applications (Fig. 1).

Thermocouple Thermometers

The thermocouple thermometer is based on an electrical effect. If different metal wires, such as copper and iron, are joined to form junctions, a voltage difference exists between the wires if one junction is kept at a different temperature than the other (Fig. 2).

The small voltage is measured, and this is proportional to the difference in the temperatures of the junctions. If the temperature of one of the junctions (reference junction) is known, the temperature of the other junction can be determined. For example, a cold (reference) junction in ice and water is at 0°C, since ice and water co-exist at this temperature.

The thermocouple is particularly convenient for many applications because of the small size of the metal wire junction. It can be inserted wherever a temperature measurement is desired. The reference junction and electrical meter can be some distance away.

Pyrometers

Pyrometers make use of the radiation or light emitted by an object. The wavelength or color of the light is proportional to the temperature of the body. For exam-

(Text continues on page 200)

cheese is water, which has one of the highest specific heats.

The high specific heat of water arises because much of the added heat goes into the rotational and vibrational energies within the H_2O molecules. Because of its high specific heat, along with its availability and cheapness, water is used to store energy in solar homes. Energy stored in large bodies of water such as the oceans or the Great Lakes affects climates. Lakes warm up in the summer, and ocean currents, such as the Gulf Stream, bring warm water to various regions. In the winter the release of the stored energy helps moderate the climate in the surrounding regions.

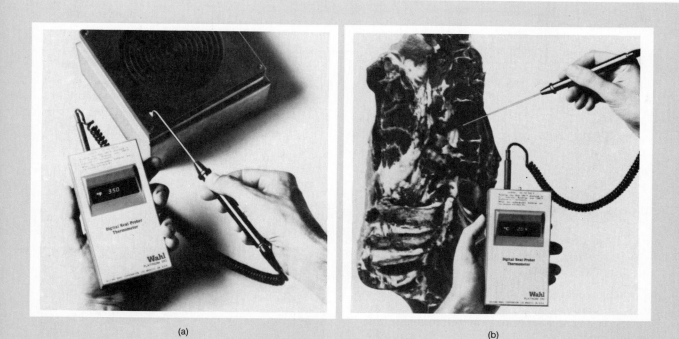

(a)

(b)

Figure 1
Resistance thermometers. Based on the change of electrical resistance of metals with temperature, resistance thermometers (with digital readouts here) are versatile temperature-takers. They measure (a) surface temperatures and (b) interior temperatures of solids and liquids.

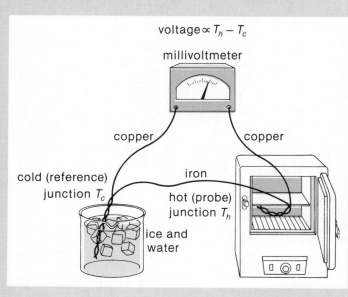

$$\text{voltage} \propto T_h - T_c$$

millivoltmeter

copper   copper

cold (reference)   iron
junction T_c
                    hot (probe)
                    junction T_h

ice and
water

Figure 2
The thermocouple thermometer. When the junctions of a thermocouple are at different temperatures, a voltage develops that is proportional to the temperature difference. One of the junctions is used for a reference at a known temperature.

Thermometry — Temperature Measurement

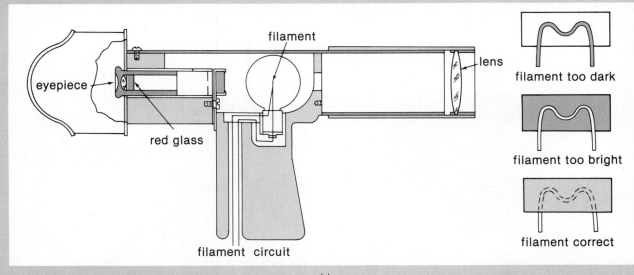

(a)

ple, when metals are heated to high temperatures they glow with different colors (emit different radiations) with increasing temperature. The approximate relationships are

Minimum visible red	475°C
Dull red	600°C
Cherry red	700°C
Light red	850°C
Orange	900°C
Yellow	1000°C
Blue-white	1150°C and higher

Thus, the temperatures of very hot bodies can be judged visually by their color and brightness (intensity). Direct visual estimation of temperature is subjective and may be inaccurate. An optical pyrometer is an instrument designed to improve visual estimates of temperature by providing the eye with a source of brightness for comparison (Fig. 3).

The comparison source is a lamp filament. Varying the current in the filament varies its brightness so as to match the brightness of the image of a hot body as seen through the pyrometer. The filament current is calibrated in terms of temperature.

(b)

Figure 3
Optical pyrometer. (a) The filament provides a visual brightness comparison of a temperature "background" that is matched. The filament current is proportional to the temperature. (b) An optical pyrometer being used to measure the temperature of molten steel.

(a)

(b)

All warm bodies emit infrared radiation. The characteristics of this radiation depend on a body's temperature. Infrared radiation cannot be seen with the human eye, so photocells and other special detectors must be used. An infrared pyrometer is used to detect the infrared radiation from a body, and the radiation is then related to its temperature.

These instruments have more precision and accuracy and a greater range than optical pyrometers. Infrared "thermometers" are now used quite extensively (Fig. 4). You don't even have to be close to take a body's temperature; you can just point the pyrometer at it.

Figure 4
Infrared thermometers. The characteristics of the infrared radiation emitted by a body are dependent on its temperature. (a) The infrared thermometer detects this radiation and reads out in temperature. (b) and (c) Some applications. The temperature of a surface can be determined from the emitted infrared radiation merely by pointing the instrument toward the surface.

(c)

SUMMARY OF KEY TERMS

Temperature: a relative measure or indication of hotness or coldness or, more specifically, a measure of the average random translational kinetic energy per molecule of a substance.

Thermal energy: the energy associated with the random translational motions of the molecules.

Internal energy: the *total* energy (kinetic and potential) contained within a body.

Heat: the internal energy that is added to or removed from a substance owing to a difference in temperature.

Thermal expansion: the changes in dimensions of substances that occur with changes in temperature.

Thermometer: a device that measures changes in temperature through a change in some physical property.

Celsius temperature scale: the scale with designations of 0° and 100°, respectively, for the ice and steam points of water. The temperature unit is the degree Celsius.

Fahrenheit temperature scale: the scale with designations of 32° and 212°, respectively, for the ice and steam points of water. The temperature unit is the degree Fahrenheit.

Absolute zero: the absolute minimum or lower limit of temperature, which is −273°C (−460°F).

Kelvin or absolute temperature scale: the scale based on absolute zero; its temperature unit is the kelvin. The ice and steam points of water are 273 K and 373 K, respectively.

calorie: the amount of heat required to change the temperature of 1 gram of water 1 degree Celsius.

Calorie: a kilocalorie (1000 cal) and the unit associated with the heat energies of foods.

Heat of combustion: the heat produced per unit mass of a substance when burned in oxygen.

Specific heat: the amount of heat required to raise the temperature of a unit mass of a substance by one degree.

[For more on how to change from the temperature on the Fahrenheit scale to the Celsius scale—and back, see an Extended View (Optional) in the Appendix.]

QUESTIONS

The Difference Between Temperature and Heat

1. Heat might be thought of as the "middle man" of energy. Why is this?

*2. Heat always flows from a body at a higher temperature into one with a lower temperature. Does it always flow from a body with more thermal energy into one with less thermal energy? How about internal energy? (*Hint:* Consider dropping a hot BB into a tub of cold water.)

3. When you leave an outside door open on a cold day, does the cold come in or the heat go out?

4. What is the difference between the thermal energy and internal energy of a perfect gas? How about a real substance?

*5. If equal masses of helium and oxygen gases are at the same temperature, do they have equal internal energies? Explain.

Thermal Expansion

*6. A piece of ice is placed on a bimetallic strip, as shown in Figure 11.12. What happens?

7. On a hot afternoon you may hear creaking noises in a house attic. Creaking may also be heard at night. Are there ghosts in the house?

8. Why are concrete highways poured in sections rather than in one long strip? Sidewalks are sometimes poured in strips, but groove joints are made. Is this for looks or a practical purpose?

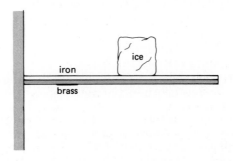

Figure 11.12
Ice on a bimetallic strip. See Question 6.

(a)

(b)

Figure 11.13
Thermal expansion. See Question 13.

9. On a hot afternoon you may observe gasoline dripping from a recently filled automobile tank. Is the owner trying to waste fuel and money? Explain.

•10. When cold water is run on a hot plate or a hot liquid is poured in a glass, the plate or glass often cracks. Why?

11. Certain glass and ceramic cooking dishes can be used in ovens for baking, and others cannot. What is the difference?

12. When one drinking glass is stuck inside another, an old trick to unstick them is to put water in one of them and run water at a different temperature on the outside of the other. Which water should be hot, and which should be cold?

•13. In the demonstration shown in Figure 11.13, at room temperature the ball fits easily through the ring (a). However, when the ball is heated, it will not fit through the ring (b). If both the ball and ring are heated, the ball fits through the ring as in the first figure. Explain why.

14. If you turn on a hot-water faucet or the hot water in a shower to a moderate flow, you may observe that the flow decreases after a while, but this is not observed for cold water. What causes the flow of hot water to change?

15. Most substances expand with increasing temperature. Explain this expansion in terms of molecular theory.

16. A thermometer is used to measure the temperature of a hot substance. Explain how this is done in terms of molecular theory. Is it possible for the thermometer to introduce error in the temperature measurement?

17. Water has unique thermal expansion properties (Fig. 11.14). As the temperature increases between 0°C and 4°C, the volume *decreases*. Compare this graph with the figure in Special Feature 8.2, and explain the differences. Explain why the volume of water increases in going from 4°C to 0°C. (*Hint:* See Chapter 8.)

•18. In light of the preceding question, would water be a good liquid for a liquid-in-glass thermometer? How would such a thermometer behave near the ice point?

•19. When a thermometer is inserted into hot water, the mercury or alcohol is sometimes observed to fall slightly before it rises. Why?

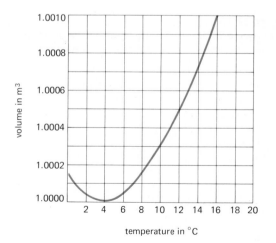

Figure 11.14
Volume versus temperature for water. See Question 17.

Temperature Scales

20. What is the most commonly used temperature scale in the world?

*21. Which unit is larger and by how much? (a) a degree Fahrenheit or a degree Celsius, (b) a degree Fahrenheit or a kelvin, and (c) a kelvin or a degree Celsius.

22. On what temperature scale would a temperature change of 10 units be the largest?

*23. What is the interval between the ice and steam points of water on the Kelvin scale? What is room temperature on the Kelvin scale?

24. A dilute (perfect) gas at constant pressure decreases in volume as its temperature is lowered. What does the perfect gas law predict the volume to be at absolute zero? Does this actually happen? Explain.

*25. A quantity of ideal gas is at room temperature (20°C). If enough heat is added to double its internal energy, what would be the temperature of the gas? (*Hint:* Change to kelvins and then change back.)

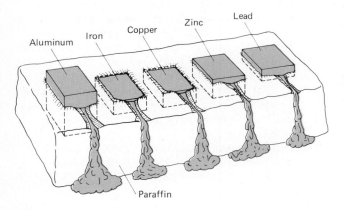

Figure 11.15
Different metal blocks with the same mass and same temperature melt to different depths. See Question 35.

Measuring Heat

26. The calorie is defined in terms of the Celsius temperature unit. How would the definition differ if it were defined in terms of the kelvin?

*27. One calorie is equal to 4.2 J. Give a definition of a joule in terms of heat measurement.

28. In Table 11.1, the heats of combustion are given in cal/g or kcal/kg. Are these ratios the same? Explain.

29. Why would scrambled eggs have a greater heat of combustion or more calories than boiled eggs? (See Table 11.1.)

*30. Gasohol is a mixture of gasoline and alcohol. Does gasohol have a greater intrinsic energy value than gasoline?

Specific Heat

31. When you eat a hot apple pie, you may find that the crust is only warm, but you may burn your mouth on the apple filling. Why is this?

32. Why is water used to store heat energy in solar homes?

*33. If equal amounts of heat are added to two containers of water and the temperature change of the water in one container is twice that of the water in the other container, what could you say about the quantities of water in the containers?

*34. Given equal masses of copper and lead at the same temperature, to which heat is added so each undergoes the same temperature change. To which metal was more heat added and how much more? (*Hint:* See Table 11.2.)

35. Blocks of metal with equal masses are heated to the same temperature and placed on a block of paraffin (Fig. 11.15). Explain why they melt to different depths when they are initially at the same temperature. Give an estimate of the specific heat of zinc.

Heat Transfer and Change of Phase

Heat transfer and a change of phase shown in action.

HEAT TRANSFER

Heat is the transfer of energy caused by a temperature difference. It goes on continuously, and in many instances we try to control and channel the flow of heat from one place to another for our benefit. For example, the Earth continuously receives heat from the Sun, and we use heat transfer in cooking foods and in heating (and cooling) our homes. Hence, heat transfer is an important practical consideration.

How is heat transferred? We know this requires a temperature difference, but what are the transfer mechanisms? There are three ways to transfer heat from one place to another. These methods are called conduction, convection, and radiation. Let's take a look at them.

Conduction

We can boil a pot of coffee on a stove because heat is *conducted* through the coffee pot from the hot stove burner. Heat transfer by conduction takes place through molecular collisions. To visualize heat conduction over a longer distance than through the bottom of a coffee pot, consider holding one end of a metal rod or poker in a fire. There is a temperature difference between the ends of the rod, so conditions are right for heat transfer.

The molecules and electrons in the heated end of the rod become agitated and move rapidly. These molecules and free electrons collide with their less energetic neighbors toward the cooler end of the rod and pass some of their extra energy to them. These molecules in turn collide with and pass energy on to their cooler neighbors, and so on.

As a result, energy is passed down the rod, in a "fire-bucket brigade" fashion, and heat is *conducted* through the rod. Energy arriving at the cooler end of the rod causes the molecules there to vibrate faster, and the temperature of this end rises. This may be painfully evident.

How well a substance conducts heat depends on the electrical bonding of its molecular structure. Solids are generally the best thermal conductors, with metals taking first prize. In addition to collisions between bound molecules, there are many electrons in a metal that are "free" to move around (not permanently bound to a particular molecule or atom, see Chapter 8). These free electrons contribute significantly to heat transfer or thermal conductivity. (As will be learned in Chapter 16, these free electrons of a metal are also responsible for electrical conductivity.)

Nonmetal solids have relatively few free electrons and are poorer conductors. Cooking utensils are made of metals because of their good thermal conductivity, but potholders are made of cloth and pot handles are made of wood or plastic because they are good *thermal insulators* (poor conductors).

Liquids and gases are, in general, poor conductors. Liquids are better thermal conductors than gases because their molecules are closer together. Gases are poor conductors because their molecules are relatively far apart and collision interactions are relatively infrequent. (The mobility of the molecules in liquids and gases gives rise to another form of heat transfer called convection, which will be discussed later.)

Many solids, such as cloth, wood, fiberglass, and Styrofoam, are porous and have many minute air pockets or spaces that add to their poor conductivity. We use such materials as thermal insulators. Examples include fiberglass insulation in our homes and Styrofoam "coolers" (Fig. 12.1). Insulating materials with air pockets are used instead of free air spaces because the air could move around in free spaces and promote heat transfer. (See the following discussion on convection.)

Figure 12.1
Poor thermal conductivity. Materials such as Styrofoam, which have a lot of air spaces, are poor thermal conductors. This property helps keep objects inside a cooler cold for long periods of time. Notice the 2-L (metric) bottles.

Question: People commonly say that an ice chest or cooler "keeps the cold in" or that home insulation "keeps the cold out." Is this correct?

Answer: No. Poor conducting materials or insulators are used in these cases to "keep heat out of the cooler" and to "keep heat in" the house. (In the case of air conditioning, home insulation helps keep the heat out.) Remember that heat is what is transferred, not *cold*.

The conductive heat-transfer capability of a substance is characterized by a quantity called **thermal conductivity**. Some typical values of thermal conductivities of materials are given in Table 12.1. Notice from the units of thermal conductivity that it gives the rate of energy flow per temperature difference ΔT. [W (watt) = joule/s, and W/K = J/s/K.] The length unit (m) arises from dimensional considerations of the material (area and thickness). Your instructor may wish to explain this more fully. We will use the values of the thermal conductivities only for comparison purposes.

Silver is the best conductor, with copper a close second (both are metals). Air is way down on the list, as you might expect. You might not expect snow to be a good insulator. In the far north, dogs and other animals often burrow in the snow to sleep. There are a lot of air spaces in the snow, as well as in their thick coats. As a poor conductor, the snow slows the loss of body heat. Also notice from the table that the conductivity of a vacuum is zero. Why?

Question: In a bedroom with a tile floor (common in dorms), when you rise and shine with both feet hitting the floor at the same time, you might remark, "Oh, that floor's cold!" A throw rug is usually obtained to avoid this discomfort (Fig. 12.2). How does this help when the rug and floor are in contact and in thermal equilibrium at the same temperature?

Answer: The rug and the floor are at the same temperature, but their thermal conductivities are different (see Table 12.1 and assume the rug is cotton). The tile floor only *feels* colder because it conducts heat from our feet faster. You should more correctly say, "Oh, that floor has a high thermal conductivity!"

Table 12.1
Some Typical Values of Thermal Conductivities

Material	Thermal Conductivity (W/m · K)
Good conductors	
Silver	425
Copper	390
Aluminum	235
Iron	80
Average conductors	
Ice	3.5
Brick	1.0
Concrete	0.8
Floor tile	0.7
Water	0.6
Glass	0.4
Poor conductors (good insulators)	
Wood	0.2
Snow	0.16
Fiberboard	0.1
Cotton	0.08
Glass wool	0.04
Styrofoam	0.033
Air	0.026
Vacuum	0

Figure 12.2
Is the tile floor colder than the rug? See Question and Answer.

Figure 12.3
Convection. Heat transfer by convection involves mass (particle) transfer. Rising hot air sets up a convection cycle that distributes heat around the room.

Convection

Compared to solids, liquids and gases are not good thermal conductors, but the mobility of the molecules in fluids permits heat transfer by another method—convection. Unlike conduction, convection involves mass transfer. Heat is not transferred by molecular collisions, but rather the heat is carried by the moving fluid en masse. For example, when you turn on the hot-water faucet, heat is transferred from the water heater to the kitchen or bathroom with or by the moving water.

Heat transfer can occur within a liquid or gas by convection (mass movement). When a fluid is heated at a surface or in one region, it expands. Becoming less dense, it rises or is buoyed upward in the surrounding cooler fluid (Archimedes' principle, Chapter 9). You may have noticed these rising "currents" in water being heated in a glass coffee pot or in air near a heating unit (Fig. 12.3).* When a portion of the fluid rises, colder, dense fluid descends to take its place, and a convection cycle is set up. This circulation mixes the fluid and distributes the heat.

* These currents in a fluid are seen because of the optical effect of the bending of light caused by temperature and density differences in the fluid. Another example is "seeing" hot air rising from a blacktop road in the summer. See Chapter 21.

On a larger scale, thermal convection cycles stir the atmosphere and result in surface winds, particularly near large bodies of water (Fig. 12.4). During the day, the land (soil) temperature increases faster than the water temperature. The land has a lower specific heat than water. The water also has mixing currents that distribute the solar energy. The air near the hot land surface is heated by conduction. The hot air rises and a convection cycle is set up. The bottom part of this cycle gives rise to a sea breeze. You may have noticed this continuous "on-shore" sea breeze at the beach on a sunny day, which makes it more pleasant.

At night, the situation is reversed. The water cools off more slowly than the land. As a result, the air over the water is warmer and the convection cycle is reversed, giving rise to a land or "off-shore" breeze.

A larger, but similar, effect occurs for summer and winter (compared to the heating effects of day and night), particularly in southeast Asia. The on-shore summer winds, called summer monsoons, bring moist air to the land, giving rise to heavy rains. In the winter (also warm in this area), the off-shore

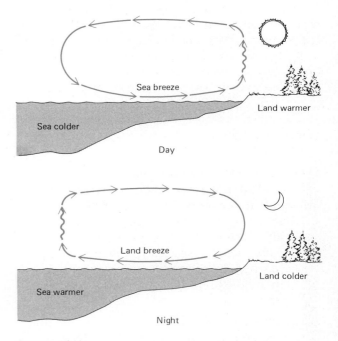

Figure 12.4
Natural convection cycles. (a) During the day, the land warms up more quickly than the water, and convection cycles are set up as illustrated, giving rise to a sea breeze. (b) At night, the land cools more quickly than the water, and the cycle is reversed and there is a land breeze.

winds or winter monsoons bring cool, dry air from the mountains.

Not all convection cycles are natural. In some cases the medium of heat transfer is moved by mechanical means. Common examples are the cooling systems of automobile engines and the forced-air heating of homes and buildings. In homes with forced-air heating, a blower fan aids the convectional transfer of heated air (Fig. 12.5). Air ducts allow the return of cooler air. Older homes used natural convection heating.

Convection currents are also important to the burning of a fire in a fireplace. The fire gets fresh air needed for combustion from convection currents, which are set up by the rising hot air in the chimney and aided by the "draft" of the Bernoulli effect (Chapter 10). The airflow can come from other por-

tions of the house and may take heated air up the chimney, or from the air in a closed room, that is replaced by colder air coming through cracks around windows or under doors. (A fireplace would not operate in a completely leakproof house.)

In either case, a fireplace is not very efficient. The efficiency is increased in fireplaces with "heat-olators," in which air pipes in the chimney are heated and the warmed air is circulated back into the room.

Radiation

Both conduction and convection require a medium or matter for energy transport. Yet energy is transmitted to us from the Sun through the void of space. Also, part of this solar energy is transmitted through

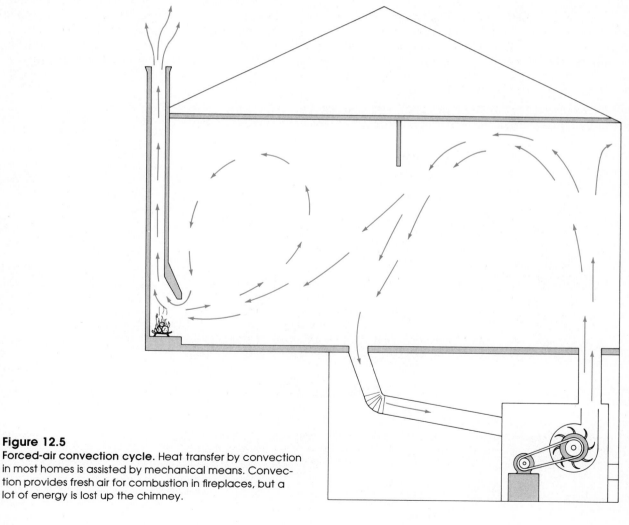

Figure 12.5
Forced-air convection cycle. Heat transfer by convection in most homes is assisted by mechanical means. Convection provides fresh air for combustion in fireplaces, but a lot of energy is lost up the chimney.

The Greenhouse Effect

If the Earth continually receives energy from the Sun, you may wonder why the Earth doesn't get hotter or why its average temperature does not increase. To maintain a relatively constant long-term average temperature, the Earth must lose energy, which on the average is equal to the amount it receives. This is accomplished through the re-radiation of energy back into space. Atmospheric effects on how this is done are important in preventing large variations in daily temperature. On the moon, which has no atmosphere, daily temperature variations range from about 100°C (212°F, the boiling point of water) on the day or Sun side to −173°C (−280°F) on the dark side.

Incoming solar radiation warms the atmosphere and the surface of the Earth, and the warm Earth re-radiates energy in the form of infrared radiation. The gases of the atmosphere, in particular water vapor and carbon dioxide (CO_2), are "selective absorbers." That is, they allow the visible incoming sunlight to pass through, but they absorb or trap certain infrared radiations. This atmospheric absorption helps to retain the Earth's energy, so we don't have the daily temperature fluctuations as on the moon. Clouds (water droplets) also assist in maintaining the Earth's warmth by absorbing infrared radiation. In the absence of cloud coverage, nights are "*cold and clear.*"

Hence, the atmospheric gases have a "thermostatic" effect in maintaining the Earth's daily temperature variations. We call this process the **greenhouse effect.** Glass has absorption properties similar to those of the atmospheric gases. As used in a greenhouse (figure below), the glass allows the visible sunlight to pass through, then blocks or absorbs the infrared radiation. Actually, in this case the warmth is primarily due to the prevention of the escape of warm air heated by the ground within the glass enclosure. The temperature of a greenhouse in the summer is controlled by painting the glass panels white, which reflects the sunlight, and opening panels to allow the hot air to escape.

The interior of a closed greenhouse is quite warm, even on a cold day. You have probably experienced the "greenhouse" effect in an automobile on a cold, sunny day.

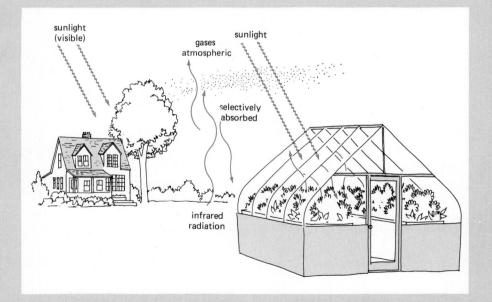

The greenhouse effect. The gases of the atmosphere, particularly water vapor and carbon dioxide, are selective absorbers of radiation with absorption properties similar to those of glass as used in a greenhouse. Visible light is transmitted and absorbed; infrared (heat) radiation is re-emitted.

the atmosphere directly to the Earth's surface. This couldn't be by conduction, since air is a very poor conductor. Convection is out too. Convection cycles begin at the Earth's warm surface and generally remove heat. Another heat transfer process, called radiation, is responsible here.

Radiant energy from the Sun is transmitted through space by means of electromagnetic waves, generally called **radiation**, which require no transport medium. These waves originate from accelerated charged particles (see Chapter 19).

Electrons vibrating in matter have acceleration, so all matter, whether hot or cold, radiates energy. The molecular vibration rate depends on the thermal energy and temperature of an object, so objects at different temperatures radiate different types of electromagnetic waves.

Waves and electromagnetic radiation will be discussed more fully in later chapters. For the present discussion, we will think generally of *radiant energy* that travels through space. One body emits radiant energy, and it may be absorbed by another body. The selective absorption by the atmospheric gases of the infrared radiation emitted by the Earth in the greenhouse effect is a good example. (See Special Feature 12.1.)

Another example is observed for visible radiation. A dark object is dark because it absorbs most of the visible radiation that hits it, whereas a light-colored object is light because it reflects radiation. You have no doubt noticed that a black asphalt driveway on a sunny summer day is much hotter on your bare feet than a concrete driveway or sidewalk. A good absorber is also a good emitter of radiation of the same type.

We have all experienced heat transfer by radiation when near an open fire. At an appreciable distance, you feel the heat on your face and hands from the fire. This cannot be by convection, since in general the air movement is toward the fire as part of the convection cycle initiated by it. Also, conduction is out. (Why?) Visible radiation is emitted by the burning material. But most of the heating effect comes from the invisible infrared radiation emitted by glowing embers or coals. We feel this radiation because it is absorbed by the water molecules of our skin. (Tissue is about 85 percent water.)

The water molecule has a natural vibration that is the same as an infrared radiation vibration, so the radiation is absorbed. This energy absorption causes the molecules to have larger vibrations.

Bumping into their neighbors, the molecules speed up as energy is transferred and the overall temperature increases.

Because of this, infrared radiation is commonly called "heat rays," and infrared lamps are used to keep foods warm in cafeterias and to warm aching joints in medical therapy. You may have noticed another example of this infrared radiation heating when the radiation was "turned off." Have you ever been "catching a few rays" (sunbathing) on a hot, sunny day when the Sun goes behind a cloud? It suddenly feels cooler. This is primarily because the infrared portion of the sunlight is absorbed by the cloud and not by you. A practical application of heat transfer by radiation is discussed in Special Feature 12.2.

■ CHANGE OF PHASE

Solid, liquid, and gas are said to be **phases (or states) of matter**. In general, when heat is added to or removed from a substance its temperature increases or decreases. However, at certain temperatures a substance will undergo changes of *phase,* for example, from solid to liquid or from liquid to gas (or vapor). While this is going on, say with heat being added, the temperature does not change. So where does the energy go?

Let's look at the common examples of the phase changes of water. Suppose you have a chunk of ice at $-10°C$. If heat is added, the temperature of the ice increases until it reaches $0°C$, the melting point. At this point, additional heat causes the ice to melt, and the ice and water co-exist at $0°C$ until all of the ice is melted.

Once the ice is melted, adding more heat causes the temperature of the water to increase. A similar situation occurs for the phase change from liquid to gas (water to steam) at the boiling point, $100°C$. Additional heat causes the temperature of the steam to rise. The graph in Figure 12.6 shows how the temperature of water varies with heat.

Note that in general when heat is added to a substance, its temperature changes. However, during a phase change, the temperature remains constant. This is because the heat energy goes into breaking up the lattice of a solid in becoming a liquid or in separating the liquid molecules in becoming a gas. We call the heat energy involved in a phase change **latent heat** (latent means "hidden").

The Microwave Oven

Heat transfer by radiation is a relatively new boon to cooking in the form of the microwave oven. Microwaves are a type of electromagnetic radiation that are also absorbed by water molecules. In a microwave oven (see figure), the microwaves are distributed by reflection from a metal stirrer and the metal walls. Because they reflect the radiation, the walls of the oven do not get hot.

Microwaves pass through plastic wrap and glass or ceramic cooking dishes but are absorbed by water molecules in the food. The water molecule has a rotational vibration that corresponds to the microwave vibration. The absorbed energy causes the food to be heated. The microwaves do not penetrate completely through the food but are absorbed in the outer layers. Heat is then conducted to their interior. You are advised to let large food items sit for a while after the oven has shut off so they will be heated or cooked throughout.

An important safety feature of a microwave oven is the automatic shut-off when the door is opened. If you

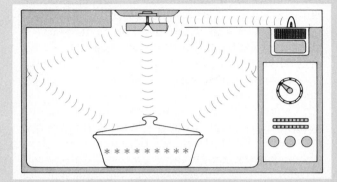

An example of heat transfer by radiation. In a microwave oven, microwaves (a type of electromagnetic radiation) are absorbed by water molecules in food, causing it to become hot.

were able to reach into the oven while it was running, you could be injured, since the water molecules in your hand would absorb the radiation.

The latent heat involved in the solid-liquid phase change is called the **latent heat of fusion**, and for the liquid-gas phase change it is called the **latent heat of vaporization**. For water, the latent heat of fusion at 0°C is 80 cal/g (or kcal/kg), and the latent heat of vaporization at 100°C is 540 cal/g. That is, 80 cal are required to melt each gram of ice at 0°C, and 540 cal are needed to convert each gram of water to steam at 100°C (Fig. 12.7).

The term *latent* or *hidden* heat may be more descriptively appreciated for the reverse processes of condensing and freezing. By the conservation of

Figure 12.6
Change of phase. A graph of how the temperature of water varies with heat. Phase changes or changes of state occur along the horizontal segments of the graph.

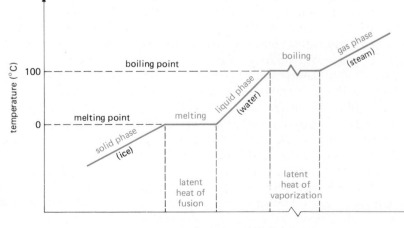

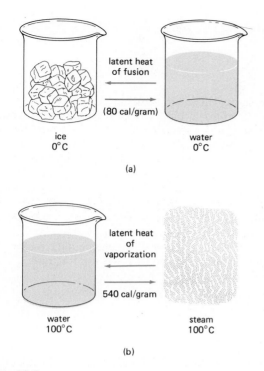

Figure 12.7
Latent heat. (a) The latent heat of fusion for water is 80 cal/g. (b) The latent heat of vaporization for water is 540 cal/g. These amounts of heat must be added or removed for each gram of water to change the phase at the respective phase change temperatures.

energy, when 1 g of steam condenses to water, 540 cal of heat are given up. This is why steam burns are more serious than burns from boiling water, even though both are at 100°C. When a gram of steam condenses on the cooler skin, an additional 540 cal is released that was seemingly "hidden."

When water freezes, for example, when rain changes to sleet or water vapor forms ice crystals, latent heat is given up. This is why sleeting and snowing are warming processes in the atmosphere. It was once common practice to keep large containers of water in cellars where canned goods were stored in the winter. If the temperature dropped below freezing (0°C), the water, which had a higher freezing point than the liquids in the canned goods, would freeze first and release latent heat. The released energy would moderate the cellar temperature and help prevent it from dropping to the point at which the jars of fruit and vegetables would freeze and burst.

There is another common change of state — solid to gas. Some solids, such as Dry Ice (solid carbon dioxide) and mothballs (*p*-dichlorobenzene), vaporize directly from the solid form. This direct change from the solid to gas phase is called **sublimation**. It requires latent heat of sublimation. A common example of the reverse situation (gas to solid) is frost. Frost is *not* frozen dew as some might think. Frost is formed by the direct change of water vapor to ice.

Evaporation

You don't have to be at the boiling point of a liquid to have a change of phase. The slow evaporation of water from wet clothes hanging on a line or from an open container of water at room temperature becomes evident after a time.

The molecules in a liquid are moving around at many speeds. (The temperature of the liquid is associated with an average of the speeds in terms of the average kinetic energy.) Some of the particularly fast molecules at or near the surface of the liquid may have enough energy to overcome the surface tension forces and break free of the liquid, or become "evaporated." Since the evaporated molecules take energy with them, the average kinetic energy of the liquid and its temperature are lowered. Hence, evaporation is a cooling process. See Special Feature 12.3.

Evaporation plays an important role in the cooling of our bodies. When hot, we perspire. The evaporation of perspiration takes heat from our bodies, and we feel cooler. How about rubbing alcohol? You really feel cool when it is rubbed on your skin. This is because alcohol evaporates more readily and cools more quickly. Air circulation also makes us feel cooler. Have you ever stood in front of a fan and thought how cool the circulating air feels? In a closed room, the air temperature is pretty much the same everywhere, and the fan is just blowing hot air from one side of the room to the other (Fig. 12.8). So why do we feel so much cooler?

There are a couple of reasons. Your body is usually warmer than the surrounding air, so it loses heat to the air by conduction. In still air, the air around your body is warmed, and the temperature difference becomes less. This reduces your cooling rate. Air circulated by a fan carries the warmed air away, and your cooling rate increases.

The air circulation also promotes cooling by evaporation. In still air, the evaporated water molecules from perspiration can collide with the mole-

SPECIAL FEATURE 12.3

Does Hot Water Freeze Before Cold Water?

This is a common question, and the answer depends on the conditions and principles discussed in this chapter. Suppose that two *covered* identical pans with equal amounts of water at temperatures of 20°C and 40°C, respectively, are placed in a freezer at 0°C. All other things being equal (except the cooling rates), the pan with the lower initial temperature (20°C) would freeze first. The pan at the higher temperature would lose heat at a faster rate (greater temperature difference), but it has more heat to lose. It would cool down to 20°C, then cool at the same rate as the other pan did during this time.

Think of it this way: Suppose you were going to run 20 m and another person was going to run 40 m on the same track. The other runner might run at a faster rate to your 20-m mark, but then if he slowed down and ran the last 20 m at the same rate as you, he would never be able to catch up, and you'd get to the finish line first.

It might be possible for the hot water to freeze first under some special conditions. Suppose the pans were *uncovered* and the hot water were very hot. In cooling down, more of the hot water would evaporate, which would leave less water (mass) in the pan to cool to freezing. Also, a hot pan may melt into the frost layer on the freezer shelf and with better contact would have greater conductivity and a faster cooling rate.

cules of the air. Being knocked backward, they may condense on you, which is a warming process. (You probably have had water condense on you in a moist bathroom after drying off from a shower and felt a bit warm.) Much of the cooling by evaporation can be canceled out by the warming of condensation. By moving the air with a fan, you remove the evaporated molecules and promote cooling by evaporation.

So, air circulation or wind plays an important part in how cool or cold we feel. At moderately low temperatures in a brisk wind, we may feel extremely cold. The wind promotes the loss of body heat, which adds to the chilling effect.

The effect of wind on how cold we feel is expressed in terms of a wind chill index or **wind chill factor.** This is the temperature in still air that would have the same cooling effect as that of wind on exposed flesh for a given air temperature. For example, if the air temperature is 10°F and the wind speed is 20 mi/h, the chill factor is −24°F; that's how could it would *feel* on bare skin (see Table 12.2, where customary units are used by the U.S. Weather Service).

Humidity

Another factor that affects our comfort is humidity. Humidity refers to the amount of water vapor in the air. The air always contains water vapor as a result of evaporation from oceans, lakes, the ground, and so on, and the amount varies.

The humidity may be expressed in terms of *absolute* humidity, which is simply the actual amount of water vapor in a given volume of air, say g/m³. The atmospheric water vapor can be thought of as being in solution with the air, similar to table salt being dissolved in or in solution with water. The amount of water vapor or table salt in solution depends on temperature. The greater the temperature,

Figure 12.8
Cooling effects. Air from a fan feels cool because it promotes evaporation and thus increases the cooling rate.

Table 12.2
Wind Chill Index (or Factor)*

Wind speed (mi/h)	Air Temperature (°F)						
Calm	30	20	10	0	−10	−20	−30
5	27	16	7	−6	−15	−26	−35
10	16	2	−9	−22	−31	−45	−58
Very cold							
15	11	−6	−18	−33	−45	−60	−70
20	3	−9	−24	−40	−52	−68	−81
Bitter cold							
25	0	−15	−29	−45	−58	−75	−89
30	−2	−18	−33	−49	−63	−78	−94
Extreme cold							
40	−4	−22	−36	−54	−69	−87	−101

* Find the air temperature in the top row and the wind speed in the left column. Read the chill factor where the corresponding row and column intersect. For example, a calm air temperature of 10°F and a wind speed of 20 mi/h together are equivalent in cooling effect to a temperature of −24°F.

the more vapor a given volume of air will hold. At a particular temperature, a volume of air or water can hold only so much water vapor or table salt. When this amount is reached, we say the solution or air is saturated.

Question: Why can warm air hold more water vapor than cooler air?

Answer: Consider a water surface and a volume of dry air above it with the water and air maintained at the same temperature (for example, a large body of water where evaporation would have a negligible effect on the temperature). When water evaporates, there is a small increase in pressure in the air. This results from the motion of the evaporated water molecules. Appropriately called vapor pressure, it is part of the total atmospheric pressure. As more molecules evaporate, the vapor pressure steadily increases and forces more and more molecules to return to the liquid.

Eventually, there are just as many molecules returning to the water as leaving. In this case, the air is said to be saturated or filled to capacity. However, if the temperature of the water and the air is increased, the molecules in the water would have a greater average kinetic energy and more would be able to

evaporate. Hence, at a higher temperature more moisture is needed in the air for saturation, or stated another way, warm air can hold more water vapor than cooler air.

It is more common to express humidity in terms of relative humidity. (You hear and see this on the weather report; see Fig. 12.9.) **Relative humidity** is the ratio of how much water vapor a volume of air actually holds to what it could hold (saturated) at that temperature. This fraction is usually expressed as a percentage. For example, if the relative humidity is 50 percent, a volume of air is "half full" or contains half the water vapor it could hold at that temperature. When the relative humidity is 100 percent, the volume of air is saturated, or it holds all the water vapor it can.

Since the amount of water vapor the air can hold depends on temperature, as the air temperature decreases the relative humidity increases (the same amount of water vapor in a volume of air that can hold less). The temperature may be decreased until the relative humidity is 100 percent. The temperature at which this occurs is called the **dew-point temperature**, or just the dew point for short.

At the dew point and below, the water vapor condenses. This is analogous to cooling a table salt–water solution to the temperature at which the

Figure 12.9
Relative humidity. The humidity on weather reports is given in terms of relative humidity. At 44 percent, this means the air holds 44 percent of the water vapor it could hold at the current temperature. (The barometer reading is in inches of mercury, and the winds are from the south at 8 mi/h.)

Figure 12.10
Cumulus cloud. Clouds are formed in rising air that cools below the dew point. Rising air commonly forms big, billowy cumulus clouds in the summer.

salt crystallizes out of solution. The water molecules do not coalesce or stick together in mid-air collisions but condense on blades of grass as dew or on particles in the air to form visible droplets. Masses of these visible droplets form clouds.

Clouds are formed in rising air that cools below the dew point. Convection cycles with rising air are a major method of cloud formation. You can watch big, billowy cumulus clouds form on a hot summer day (Fig. 12.10). Under the right conditions, some of the cloud droplets grow by condensing water vapor. These larger droplets then coalesce and form rain drops. Then, we may have a dark cumulonimbus cloud or "thunderhead."

If the relative humidity is high, along with the temperature, then it is hot and "muggy." Since the air has most of the water vapor it can hold, the evaporation from our bodies is small, and we feel hot. We feel most comfortable at room temperature (20°C) when the relative humidity is about 50 or 60 percent. At very low relative humidities, the rapid evaporation can cause drying of the nose and throat. We may feel cool but uncomfortable. To prevent these conditions by controlling the humidity, many people use humidifiers in their homes in the winter and dehumidifiers in the summer. (A humidifier may be a pan of water.)

■ BOILING AND FREEZING

The boiling and freezing (melting) points of substances vary widely. This is due to differences in the molecular bonding of the substances. For example, the boiling and freezing points of water are 100°C and 0°C, respectively; of silver, 2212°C and 960°C, respectively; and of nitrogen, −196°C and −210°C, respectively. All these phase-change temperatures are at a pressure of one atmosphere (standard pressure).

Pressure makes a difference on the boiling and freezing points. Let's first take a look at boiling. Why does a liquid boil?

Consider heating a liquid in a pan on a stove. As heat is added to the liquid and its temperature increases, the average kinetic energy of its molecules increases. Convection cycles are set up that carry the warmer, lower liquid to the surface, and the evaporation rate is increased.

As the temperature of the liquid near the bottom of the pan is further increased, a point is reached (the boiling point) at which the average molecular energy is equal to the work necessary to change the liquid into a gas (100°C at one atmosphere of pressure for water). Unless the molecules in the liquid are moving fast enough, the atmospheric pressure on the surface of the liquid keeps the gas (bubbles) from forming and the phase change from occurring.

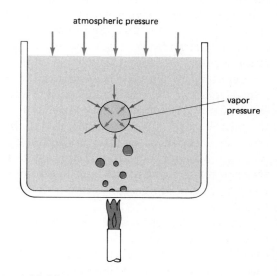

Figure 12.11
Boiling. When the vapor pressure in bubbles is greater than the external pressure, boiling starts. The bubbles rise and break through the surface, and the gas escapes.

The Pressure Cooker

If the pressure on a liquid is increased, its boiling point is raised. This is the principle of the pressure cooker, which is commonly used to cook foods (figure below).

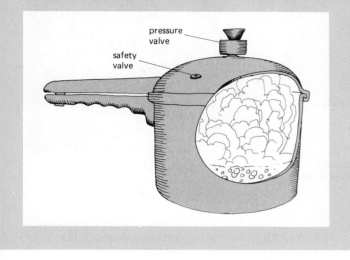

In a closed container (with an appropriate safety valve), the heating of water or some other liquid causes increased pressure on the water because of the evaporated gas above the water.

As a result, the boiling point is increased, and the water and food to be cooked can be heated to above 100°C. At a higher temperature, the food cooks in less time. For example, at a pressure of 1.4 atmospheres, the boiling point of water is 110°C; at 4.7 atmospheres, it is 150°C.

Pressure cooker. The boiling point of a liquid increases with increased pressure, and the liquid can be heated to higher-than-normal temperatures. This is the principle of the pressure cooker, in which foods cook more quickly as a result of higher temperatures.

More heat produces vapor bubbles in the liquid, usually near the bottom of the container where it is being heated and is hottest (Fig. 12.11). If the vapor (gas) pressure in the bubbles is greater than the pressure above because of the atmosphere (and the weight of the liquid*), boiling starts. The bubbles rise and break through the surface, and the gas escapes.

Thus, we can keep a liquid from boiling, or raise its boiling point, by increasing the pressure on the liquid. A practical application of this is given in Special Feature 12.4.

At reduced pressures, the boiling points of liquids are lower. For example, at high altitudes, where the atmospheric pressure is less (than at sea level), the boiling point of water is less than 100°C. At Pike's Peak the atmospheric pressure is about 600 torr (mm Hg), and water boils at about 94°C rather than at 100°C. This has an effect on cooking times.

By reducing the pressure on a liquid, you can have boiling without external heating. This is a common demonstration using water in a vacuum jar.

* This is negligible compared with atmospheric pressure, in most cases.

As air is evacuated from the jar, the pressure and hence the boiling point of the water are lowered. Eventually, the boiling point is lowered to the temperature of the water, and it boils "without heating." If this pressure is maintained, the boiling quickly stops because the latent heat of vaporization is taken from the internal energy of the water. This cools the water and lowers its temperature to below the boiling point.

If the pressure is reduced further, the boiling and cooling continue until the temperature is reduced to the freezing point and ice forms over the surface of the boiling water — boiling and freezing at the same time! Here we have water co-existing in all three phases — solid, liquid, and gas. This is known as the **triple point**, and it occurs at a pressure of 4.6 torr and a temperature of 0.01°C. This fixed point is used in the definition of the Kelvin temperature unit and scale.

The boiling points of all liquids generally increase with increasing pressure. But how about the freezing points? Most liquids contract on freezing or solidifying. Increased pressure helps this process along and thereby raises the freezing point. How-

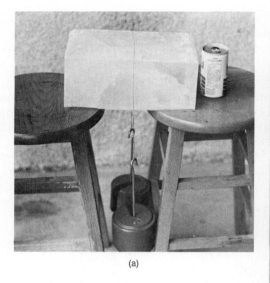

(a)

(b)

Figure 12.12
Regelation. (a) Weights suspended on a fine wire cause a large pressure on the ice under the wire, which lowers the freezing point. (b) The wire slowly "cuts" through the ice, which refreezes. (The soda was drunk while waiting.)

ever, there are some exceptions, most notably water. The open molecular structure of freezing water causes an expansion or increase in volume, so it is less dense as a solid (Chapter 8).*

Because the volume of a given amount of water is less than that of its ice in solid phase, it would seem reasonable that increased pressure on ice would tend to lower the freezing-point temperature and make it melt or have a smaller volume. This is the case, and increased pressure lowers the freezing (melting) point of ice. The melting point isn't lowered much, only about 0.0075°C per atmosphere of pressure increase.

It is sometimes believed that the melting-point reduction due to pressure is the principle of ice skating. An ice skater is able to skate or glide on the skates because of a lubricating thin film of water between the skate blades and the ice. The pressure ($p = F/A$) due to the weight of a person on the small area of a skate blade is large, perhaps 10 atmospheres. However, it would take a pressure increase of 100 atmospheres to reduce the melting point of

ice only 0.75°C. The primary cause of melting to produce the lubricating film is frictional heating between the skate blades and the ice.

Do you think it ever gets too cold to ice skate? It can, in the sense that you would not be able to glide on the skates. If it is really cold, the frictional heating between the skate blades and the ice may not be enough to lower the temperature below the melting point and provide a lubricating film of water. You would then "skate" as though you had left on your blade protectors — not in Olympic style.

The effect of pressure on the freezing point of ice can be observed by a demonstration, as shown in Figure 12.12. A fine wire with weights on each end is suspended over a block of ice. Because of the pressure of the wire, the freezing point is lowered, and the wire "cuts" slowly through the ice. But the block doesn't fall apart because the melted water refreezes.

This melting and refreezing due to pressure differences is called regelation. It takes awhile for the wire to cut through the block. The excitement is heightened if you have a game to guess when the cut will be complete, for example, in a Physics Olympics.*

---

* Other exceptions are gallium, bismuth, and certain alloys. These alloys are used in type metal that expands on solidifying, allowing the formation of raised type for printing. Another alloy, called gray iron, expands slightly on solidifying and is used to make metal castings against a mold, which reproduces its details. Many machine parts are made by casting iron (cast-iron parts).

* If you aren't familiar with Physics Olympics, check the index of *The Physics Teacher* in your library. They are exciting contests that use physical principles that you are learning. You may want to have some Olympic events of your own.

As a final topic, let's consider the boiling and freezing points of aqueous (water) solutions. For example, when you dissolve table salt in water, how does this affect the boiling and freezing points of the water? When you put salt in a pot of water when cooking, will the salt water boil above or below 100°C? You will find that the boiling point of the solution is a bit greater than 100°C. The salt ions are attracted to the polar water molecules, and more energy is required to vaporize them.

How about on the freezing side? Here, the dissolved salt lowers the freezing point. The salt ions get in the way of the water molecules forming their hexagonal ice-crystal structure. Only when the water molecules are slowed down sufficiently (lower temperature) are the attractive forces large enough to cause freezing. In general, adding anything to water causes this effect.

Question: Why is salt added to ice when making homemade ice cream?

Answer: Basically, to lower the temperature below 0°C so the ice cream mix will freeze. Heat is conducted from the mix through the inner metal "can" to the ice in the outer bucket. The metal can is rotated, and a stationary stirrer or "dasher" stirs the mix to keep it uniform and to promote the heat transfer (mechanical convection). The ice cream mix, being largely water, freezes at slightly below 0°C. If pure ice is used, the temperature of the ice cream mix would be lowered to 0°C, and then no more heat would be conducted, since it would be at the same temperature as the ice-water mixture outside.

However, when salt is added to the ice, it mixes with the outer layers of the ice. The freezing or melting point of the ice-salt mixture is well below 0°C and it melts, taking heat from the ice and the surrounding water. This melting lowers the temperature of the salt, ice, and water mixture to below 0°C, and more heat can be conducted from the ice cream mix, causing it to freeze—what we all wait for impatiently.

We put salt on icy sidewalks and roads in the winter for the same freezing-point effect. The temperature is cold enough to freeze water and produce ice, but if it is not below the freezing point of the ice-salt mixture, the ice will melt. If it was really cold and the temperature was below the freezing point of the ice-salt, then the salt would not "melt" the ice.

SUMMARY OF KEY TERMS

Conduction: heat transfer by molecular interaction with no net mass movement. Occurs chiefly in solids.

Thermal conductivity: how effective a substance is in conducting heat.

Thermal conductor: a material with good conductive heat transfer capability or thermal conductivity.

Thermal insulator: a material with poor thermal conductivity.

Convection: heat transfer by mass movement. Occurs in liquids and gases with movement of all or part of the fluid.

Convection cycle: the cyclic motion in a fluid due to localized heating and convective heat transfer.

Radiation: heat transfer by means of electromagnetic waves.

Change of phase: the transition from one phase or state of matter to another, for example, from solid to liquid.

Latent heat: the heat energy associated with a phase change that is involved in the work of changing the state of the material without a change in temperature.

Latent heat of fusion: the latent heat of a solid-liquid phase change. For water, this is 80 cal/g.

Latent heat of vaporization: the latent heat of a liquid-gas phase change. For water, this is 540 cal/g.

Evaporation: the vaporization of molecules from a solid or liquid.

Sublimation: the change of phase from a solid directly to a gas.

Humidity: the amount of water vapor in the air.

Absolute humidity: the actual amount (mass) of water vapor in a given volume of air.

Relative humidity: the ratio of how much water vapor a volume of air actually holds (absolute humidity) to what it could hold (saturated) at that temperature. This fraction is commonly expressed as a percentage.

Dew point: the temperature at which a volume of air is saturated or the relative humidity is 100 percent.

Boiling: the condition in a liquid when its vapor pressure equals the pressure above the liquid.

Triple point: the pressure and temperature at which a substance can co-exist in all three phases.

Regelation: melting and refreezing due to pressure differences.

QUESTIONS

Heat Transfer

1. On a molecular level, why is a material a good thermal insulator?
2. Good silverware (knives, forks, and spoons) has an actual silver coating. Is thermal conductivity a consideration here?
*3. Why do underground water pipes sometimes freeze only after it has been very cold for several days?
4. A medical emergency arises and you are told to boil some water as quickly as possible. In the kitchen you find two similar pots, one made of iron and the other of aluminum. Which one would you use?
*5. Is baked food more likely to burn on the bottom in an aluminum baking pan or a glass baking dish?
6. How would the rate of thermal conduction (conductivity) of the bottom of a metal cooking pan vary with (a) the area and (b) the thickness? (*Hint:* Think in terms of heat "flow.")

*7. Why do some metal pans (usually iron or steel) have a layer of copper on their bottoms? Is this only for looks?
8. Thermal underwear has a knitted structure with lots of holes (Fig. 12.13). Wouldn't a material without such holes be a better insulator?
*9. A bucket-brigade analogy is used to illustrate the molecular conductive process. Suppose a real bucket brigade passed buckets of hot water along the line. Is this actually a type of heat transfer process?
10. Thermopane windows have double panes of glass separated by a small air space. Why are these windows better for insulation than single-pane windows?
11. How do storm windows and doors help reduce your heating bill?
*12. Foam insulation is sometimes blown between the outer and inner walls of a house. If air is a poor conductor, why bother with insulation?
13. A big roaring fire in a fireplace is only about 10 percent efficient in heating a room. Why the low efficiency?

Figure 12.13
Holey long-johns. See Question 8.

Figure 12.14
Air-conditioner fins. See Question 16.

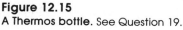

Figure 12.15
A Thermos bottle. See Question 19.

14. Discuss the energy balance and the average temperature of the Earth if its only heat-loss mechanism were conduction and convection.
15. Is a greenhouse kept warm by the greenhouse effect? Explain.
•16. The outside coils of window air conditioners have a fin network (Fig. 12.14). What is the purpose of these fins?
17. (a) Why do we generally wear dark clothes in the winter and light-colored clothes in the summer? (b) On a hot, sunny summer day, it is possible to cook an egg on the hood of a car. Could this be done faster on the hood of a black car or on the hood of a white car? (c) Why does dirty snow melt faster than clean snow?
18. Homemakers often complain that their pie crusts do not brown on the bottom in shiny aluminum pie pans as they do in the older metal pie pans. Why is this?
•19. A Thermos bottle is used to keep cold liquids cold and hot liquids hot. It consists of a double-walled, partially evacuated container with silvered walls (Fig. 12.15). Discuss how heat transfer is impaired in terms of conduction, convection, and radiation.
20. (a) When your skin is hot, the blood vessels in the skin dilate, or get larger in diameter. When the skin is cold (below 37°F), the blood vessels constrict. What is the purpose of this action? (b) Alcohol (taken internally) causes the blood vessels in the skin to dilate, and drinkers feel a warm "glow." Is this really a warming process for the body?
•21. Suppose you take cream in your coffee. If you have a hot cup of coffee that you want to drink later, which method would keep the coffee hotter longer: to put the cream in right away or wait until you are ready to drink the coffee?

(*Hint:* Think about the temperature differences and cooling [heat transfer]. Newton's law of cooling states that the rate of heat transfer is proportional to the temperature difference of objects or an object and its surroundings.)
22. The "radiator" of an automobile cooling system is where most of the heat is lost. This name imples that heat is lost mainly by radiation. Is this true?
23. Some questions about microwave ovens:
   a. In heating frozen foods in sealed pouches, why do you first poke holes in the pouch?
   b. Why are microwave ovens built so they will not operate with the doors open?
   c. What is the purpose of the metal grating on the inside of the glass in the door?

Change of Phase
•24. Why do various substances have different boiling and freezing points? Would you expect the latent heats to be different?
25. Explain why the latent heat of vaporization of water is almost seven times greater than the latent heat of fusion?
26. Why do some liquids evaporate more readily than others?
•27. People traveling in a hot region sometimes carry water in a porous canvas bag that stays wet. This is hung on the front bumper of a car or truck. What is the purpose of this?
28. (a) We sometimes blow on the surface of a hot cup of coffee or a spoon full of hot soup to cool it. How does this help? (b) In some states, highway signs warn: "Bridge freezes before road" (Fig. 12.16). Why is this?

Figure 12.16
Watch out for bridges! See Question 28.

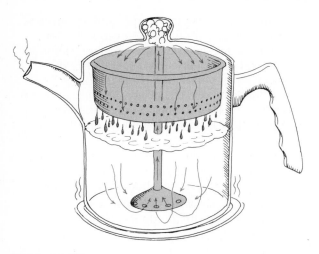

Figure 12.17
Perk your coffee. See Question 39.

29. To tell the wind direction, people sometimes wet a finger and hold it up. How does this help?

•30. What is the wind chill factor if the air temperature is 20°F and the wind speed is 15 mi/h? (See Table 12.2.)

•31. On a particular day there is a wind speed of 25 mi/h and a wind chill factor of 0°F. By how many degrees would the air temperature have to be lowered to have the same cooling effect in calm air?

32. Why do ice cubes in a refrigerator freezer get smaller with time, particular in a frost-free freezer?

•33. Fogs may be thought of as low-lying clouds. Why do fogs sometimes form in valleys overnight? Explain the phrase *the Sun burns off* a fog.

34. a. Why does the mirror in the bathroom fog up when you take a hot shower?

    b. Why does water condense on the outside of a glass containing an iced drink?

    c. Why can you "see" your breath on a cold day?

    d. When a teakettle boils, we say we can see the steam coming from the spout. Is this correct? Steam is invisible.

•35. Freeze-dried coffee crystals are made from frozen coffee. How are the crystals obtained?

36. Perfumes and colognes have an alcohol base. Why not use water since it would be cheaper?

Boiling and Freezing

37. A pan of water on a stove boils "faster" when the burner is on high heat and boils "slower" with the burner on low heat. Is the temperature of the water greater for a fast boil? Explain.

•38. Why does covering a pot of water with a lid help the water boil more quickly?

39. a. In a percolator coffee pot, water bubbles up through a central stem and "perks" down through the coffee grounds (Fig. 12.17). What causes the water to rise in the stem?

    b. Geysers also erupt for similar reasons. Water from underground streams fills long, vertical holes and is heated at the bottom by underground hot rocks. Explain why the boiling at the bottom starts at temperatures well over 100°C and why the spraying of water into the air at the surface is generally after the eruption gets under way. Why do geysers erupt periodically (like Old Faithful)?

40. Automobile cooling systems operate under pressure. (a) What is the purpose of this? (b) What would happen if you removed the radiator pressure cap immediately after turning off a hot engine, and why? (*Don't try this*—it's very dangerous.)

41. The antifreeze (ethylene glycol) used in automobile cooling systems has a freezing point of −17.4°C, which is lower than that of water. Why don't we replace the water completely with antifreeze?

42. At one time, you could purchase "temporary" antifreeze for a car, which contained alcohol, instead of the "permanent" type of antifreeze we now use. Why was the antifreeze "temporary"?

43. Is regelation an important factor in making snowballs? if so, why can't balls be made from dry powdered snow?

44. Where will water (a) boil and (b) freeze faster, on Pike's Peak or in Death Valley?

Thermodynamics, Heat Engines, and Heat Pumps

13

Professional race driver
Janet Gutherie at the
controls of a heat engine.

THERMODYNAMICS

As the name implies, thermodynamics deals with the transfer and actions (dynamics) of heat (Greek *therme,* meaning "heat"). In general, it is a broad and comprehensive branch of science that is concerned with all aspects of energy, but chiefly the relationship between heat and mechanical energy. The formal development of thermodynamics began less than 200 years ago, primarily growing out of efforts to produce heat engines—devices for converting heat energy into mechanical work. These include steam engines, gasoline engines, diesel engines, jet engines, and any device that converts heat into work.

Thus far our discussion of heat has not involved the practical conversion or transformation of heat to mechanical work or energy. But this is an important consideration. Where would we be without the internal combustion engine of the automobile? Still using horses perhaps, which are, in a sense, complicated physiological heat engines.

Some of the general aspects of thermodynamics have been presented in previous chapters. However, there are other important principles that govern the utilization of heat and work. These laws of thermodynamics are basic in the operation and design of heat engines and heat pumps, which are main topics of this chapter. A heat pump is a device that uses mechanical energy or work to transfer heat from a lower-temperature source to a higher-temperature region. Can you think of a common heat pump? How about a refrigerator or an air conditioner? An air conditioner takes heat from a cool room and transfers it to the hot outdoors. It takes work or energy to do this. Check your electric bills in the summer.

THE FIRST LAW OF THERMODYNAMICS

Since thermodynamics is concerned with energy transfer, it is important to keep track of or to account for the energy involved in a process. The first law of thermodynamics is simply the principle of the conservation of energy applied to a thermodynamic system. When heat is added to a system it doesn't disappear.

Let's look at a quantity or system of perfect gas, as illustrated in Figure 13.1. When heat is added to

Figure 13.1

Processes and the first law of thermodynamics. Heat (ΔQ) added to a gas system can go (a) totally into internal energy ΔU with no change in volume ($\Delta V = 0$) or (b) into doing work W with a change in volume and $\Delta U = 0$. (c) In the general case it goes into both, and by the conservation of energy, $\Delta Q = \Delta U + W$.

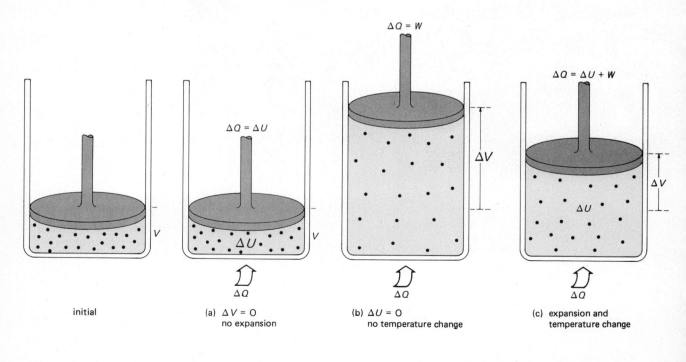

initial

(a) $\Delta V = 0$
no expansion

(b) $\Delta U = 0$
no temperature change

(c) expansion and
temperature change

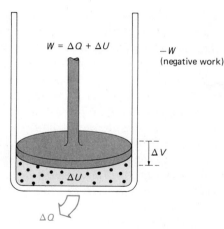

$W = \Delta Q + \Delta U$

$-W$
(negative work)

ΔV

ΔU

ΔQ

Figure 13.2
Work done *on* a system is taken to be negative. By the first law, the internal energy of the gas increases and/or heat is transferred from the system.

the system, several things can happen or are possible:

(a) If the volume is fixed (constant), the quantity of heat added (ΔQ) goes into increasing the internal energy (ΔU) of the gas, and $\Delta Q = \Delta U$. This is evidenced by an increase in the temperature (and pressure) of the gas. Recall that for an ideal gas $pV = NkT$ and $U \propto T$, in which T is the absolute temperature. No work is done in this case.

(b) Alternatively, with a movable piston the gas could expand and do work (W) in moving the piston, just as you would have to do work if you moved it manually. If all the added heat went into the work done *by* the gas as it expands, the internal energy of the gas would be the same afterward as before (same temperature), and $\Delta Q = W$.

Or (c), there could be a combination of these energy distributions, with heat going into increasing the internal energy *and* doing work, $\Delta Q = \Delta U + W$. This is the most general case, and the equation takes into account the other two cases with (a) $W = 0$ and (b) $\Delta U = 0$. This general conservation-of-energy principle is called the **first law of thermodynamics**.

Net heat transfer = change in internal
energy + work,

or

$$\Delta Q = \Delta U + W$$

Notice that this does not have to be a one-direction or a heat-added process. Suppose work was done on the system by an external force (Fig. 13.2).

By the energy balance of the first law, the internal energy of the gas increases and/or heat is transferred from the system.*

> *Question:* What would happen when work is done on a system of gas by compressing it (a) if no heat is removed from the system and (b) if the internal energy of the gas remains the same?
>
> *Answer:* The first law tells you right away. (a) When work is done on the system and no heat is removed ($\Delta Q = 0$), the internal energy is increased by an amount equal to the work done.
> (b) If the internal energy is unchanged ($\Delta U = 0$), then by the conservation of energy an amount of heat equal to the work done would have to be removed from the system.

In *any* case, the first law says in effect that you can't get more out of a system than you put in and/or than is already there. To see the last part of this statement, think about removing heat from a gas in a rigid container (no change in volume and no work). You could only remove a quantity of heat equal to the amount of internal energy the gas has. If you could do this (which you can't, in reality), the gas temperature would be at absolute zero. Why?

■ HEAT ENGINES

A **heat engine** is a device that converts heat energy to work. Many types of heat engines are available, and they come in many sizes and shapes — gasoline engines on lawn mowers, diesel engines in trucks, and steam turbines used in electrical generation. Basically, they all operate on the same principle of adding heat to a fluid that uses some of the energy to do mechanical work.

In thermodynamic processes we are not generally concerned with the component parts of an engine, such as pistons, cylinders, and gears. For theoretical purposes any heat engine can be conveniently represented by a diagram, as shown in Figure 13.3. Heat is taken from a high-temperature (T_{hot}) source or reservoir, some of which is used to do useful work (work out), and the remainder is

---

* To get the signs of the quantities in the equation of the first law to come out properly, work done *on* and heat removed *from* a system are taken to be negative ($-W$ and $-\Delta Q$).

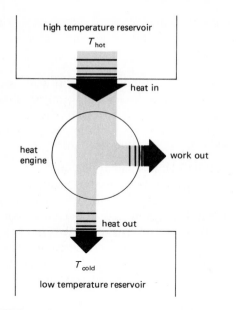

Figure 13.3
A heat engine diagram. In general, a heat engine takes heat from a high-temperature reservoir, converts some of it to useful work output, and rejects the remainder.

transferred to a low-temperature (T_{cold}) reservoir. Notice that the widths of the heat and work paths in the diagram are in keeping with the conservation of energy: heat in = work + heat out, or work = heat in − heat out.

In an actual engine, such as an internal-combustion gasoline engine, the high-temperature heat source is the exploding charge of vaporized gasoline and oxygen. Work is done by the expanding gases on a piston that is coupled to a crankshaft to give useful work output. The unused heat is "rejected" to the atmospheric surroundings (low temperature) through the engine exhaust system.

Most practical heat engines operate in a cycle. Heat could simply be added to a gas in a piston-cylinder arrangement, and we could have a one-shot work output in a single expansion process. However, in practical applications we want an engine to deliver work continuously. When an engine operates in a cycle, we have work output during each cycle. The idea of cyclic operation is illustrated for a special case in Special Feature 13.1.

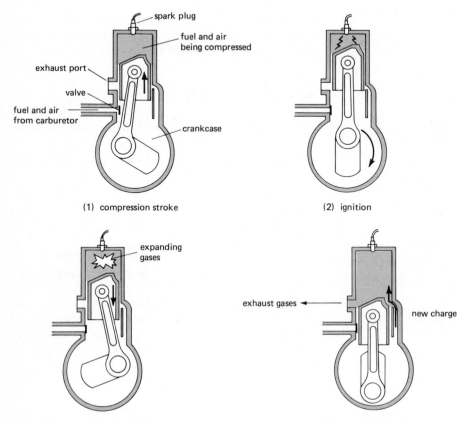

(1) compression stroke

(2) ignition

(3) power stroke

(4) exhaust and recharge

Figure 13.4
Two-stroke-cycle gasoline (heat) engine. The two strokes refer to the up and down motions of the piston (each a stroke), so there are two strokes per cycle. Notice some fuel is lost during the exhaust and recharge phase (4).

The Drinking Bird Heat Engine

A novel example of a cyclic heat engine is the toy drinking bird (figure below). It doesn't look much like a heat engine, but it falls under the definition of one.

To start the "engine," you wet the absorbent flock material on the head and beak of the bird with water. The liquid inside the body is ether, which has a low boiling point and readily vaporizes at room temperature. The evaporation of ether in the lower part of the body creates pressure above the liquid. The ether in the tube does not evaporate as readily because the head is cooled by the evaporation of water from the flock material covering, and there is less vapor pressure in the head. The pressure difference causes the ether to be forced up the tube into the head.

The rising liquid raises the center of gravity of the bird above the pivot point, and the bird pitches forward for a "drink" (to rewet the flocking). In this position, the pressures in the head and body are equalized, and the ether drains back into the body. The bird pivots back and the cycle begins again.

The bird could be hooked up so its motion could do some useful work. This heat engine is cheap to run, but it is doubtful if "birdmobiles" will ever replace piston-engine automobiles.

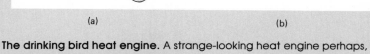

(a)   (b)

The drinking bird heat engine. A strange-looking heat engine perhaps, but it does the job with work output.

In general, we have two-stroke-cycle engines and four-stroke-cycle engines. Let's take a look at the difference. The cycle of a **two-stroke-cycle engine** is illustrated in Figure 13.4. Such internal combustion engines are used in some lawnmowers, motorcycles, outboard motors, and chain saws. This is the type of engine in which you have to mix oil with the gas for lubrication.

The two strokes refers to the up and down motions of the piston (each a stroke) during a cycle. Notice the compression and power strokes in Figure 13.4. Power (work output) is produced each time the piston goes up and down. This is an advantage for lightweight engine applications. However, there is a disadvantage. Notice that there is a waste or loss of fuel with the exhaust gases when the engine is recharged.

This is avoided in a **four-stroke-cycle engine**, which has two more piston strokes per cycle. (See Fig. 13.5.) There are separate intake and exhaust strokes that prevent fuel loss as in a two-stroke cycle. There are intake, compression, power, and exhaust phases in this cycle. Follow the cycle through in Figure 13.5 and check this out. All four strokes make up one engine cycle, and this requires two rotations of the crankshaft.

To keep engines running smoothly between power strokes, a flywheel attached to the crankshaft

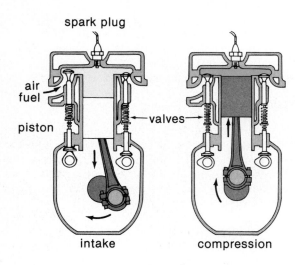

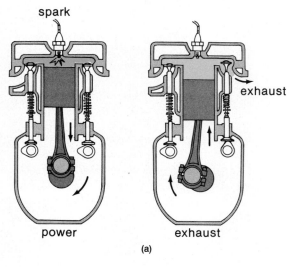

(a)

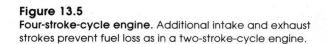

(b)

Figure 13.5
Four-stroke-cycle engine. Additional intake and exhaust strokes prevent fuel loss as in a two-stroke-cycle engine.

stores and supplies energy (see Chapter 6, Fig. 6.10). Most automobiles use four-stroke gasoline engines. To obtain greater work output, multicylinder engines—for example, those having four, six, or eight cylinders—are used. In each of these, all of the cylinders go through their four-stroke cycle in two rotations of the crankshaft (to which the piston rods are attached). The cylinder firing or ignition occurs in a regular sequence for smoother power output. For example, in a four-cylinder engine, the timing is such that one of the cylinders fires every half-revolution of the crankshaft.

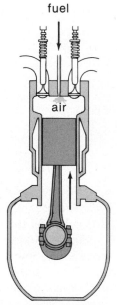

Figure 13.6
Diesel engine. In a diesel engine, high compression causes the temperature of the air to increase to the point that the fuel ignites spontaneously when injected into the cylinder. This eliminates the need for spark plugs as in a gasoline engine.

Question: What is the difference between a diesel engine and the common gasoline engine?

Answer: There are various differences, but fuel and ignition are two big ones. The diesel engine uses a lower-grade fuel. Diesel "oil" is not as highly refined as gasoline, so it is usually cheaper.

Probably the most notable difference is the absence of spark plugs to ignite the fuel in a diesel engine (Fig. 13.6). Air in the cylinder during the compression stroke is compressed three to four times more than the pressure in the cylinder of a gasoline engine. This compression raises the temperature of the air high enough so that the fuel ignites spontaneously when injected into the cylinder.

The diesel engine is named after its inventor, Rudolf Diesel (1858–1913), a German engineer.

THERMAL EFFICIENCY

Efficiency is what tells how "good" or how economical an engine is. It is common to measure cost efficiency in terms of fuel economy or how many miles per gallon (or kilometers per liter)—what we get out for what we put in. Recall from Chapter 2 that the mechanical efficiency of a machine is the ratio of the work output to the work input

$$\text{Efficiency} = \frac{\text{work out}}{\text{work in}}$$

which is the fraction (or percentage) of useful work done by a machine.

In heat engines, a similar thermal efficiency is used. Instead of work input, there is heat input. The work output is equal to the heat in minus the heat out (see Fig. 13.3). Hence, the **thermal efficiency** is defined as follows:

$$\text{Thermal efficiency} = \frac{\text{work out}}{\text{heat in}}$$
$$= \frac{\text{heat in} - \text{heat out}}{\text{heat in}}$$

For example, if a heat engine has a heat input of 1000 joules and rejects 600 joules while doing work in a cycle, its thermal efficiency is 40 percent:

$$(1000 \text{ J} - 600 \text{ J})/1000 \text{ J} = 400 \text{ J}/1000 \text{ J}$$
$$= 0.40 \ (\times 100\%)$$
$$= 40\%$$

An automobile has an overall efficiency of less than 15 percent. Here a lot of heat is lost because of friction as well as being rejected to the surroundings through the exhaust and cooling systems. In practical terms, this means that more than 85 percent of the energy from the gasoline used in your car is "wasted" or does not go into doing the useful work of propelling the vehicle. (See Special Feature 2.1.)

THE SECOND LAW OF THERMODYNAMICS

The first law of thermodynamics is concerned with energy balance or conservation. As long as the energy check sheet is balanced, the first law is satisfied. Suppose, however, that a heat engine operated so that *all* the heat input were converted into work (no heat out). This doesn't violate the first law, but something is wrong. In this case, the thermal efficiency of the engine would be 1 or 100 percent, and this just doesn't happen or has never been observed.

This fact is expressed in the **second law of thermodynamics**, which can be stated in several ways. One statement of the second law as it applies here is as follows:

No heat engine operating in a cycle can convert heat energy completely into work.

Or, put another way,

No heat engine operating in a cycle can have 100 percent efficiency.

If this were possible, the work output could be rechanneled into heat input, and the engine would operate indefinitely as a perpetual-motion machine. It is possible for all the heat to go into work in a single expansion process, but for an engine operating in a cycle, heat must be rejected or lost somewhere.

Since a heat engine must lose some heat, what is the best we can do or what is the maximum possible efficiency? A French engineer, Sadi Carnot (1796–1832), studied this question and came up with the answer. He found that the thermal efficiency of a heat engine is limited by the operating temperatures (temperatures of the hot and cold reservoirs, see Fig. 13.3), and the **ideal** or **Carnot efficiency** is given by

$$\text{Ideal efficiency} = \frac{T_{\text{hot}} - T_{\text{cold}}}{T_{\text{hot}}} = 1 - \frac{T_{\text{cold}}}{T_{\text{hot}}}$$

where the temperatures are absolute temperatures.

This is the maximum possible theoretical efficiency under *ideal* conditions. It sets an upper limit that can never be achieved. Actual thermodynamic conditions are never ideal, and friction is always present mechanically. But the ideal thermal efficiency does tell the engineer that a greater possible efficiency can be achieved by making T_{hot} higher and/or T_{cold} lower.

For example, if a steam turbine took in superheated steam at 500 K (227°C) from a high-temperature (T_{hot}) reservoir and rejected it to a cold-temperature (T_{cold}) reservoir at 400 K (127°C), the ideal efficiency is $1 - (400/500) = 0.20$, or 20 percent. But if the cold-temperature reservoir were at 300 K (27°C), the ideal efficiency would be $1 - (300/500) = 0.40$, or 40 percent. (In actual practice, the steam is condensed on the low-temperature side, which relieves the "back pressure" on the turbine to make it more efficient.)

You could also increase the high temperature to get increased efficiency. Diesel automobiles typically get better "gas" mileage (are more efficient) than their similar gasoline-powered counterparts, even though diesel oil has a lower heat of combustion than gasoline (see Table 11.1). This is mainly because diesel engines operate at higher temperatures than gasoline engines, which makes for greater efficiency.

■ HEAT PUMPS

A **heat pump** is a device that transfers heat from a low-temperature reservoir to a high-temperature reservoir. This is the reverse function of a heat engine. A heat pump is represented by a diagram in Figure 13.7. From experience you know that this heat transfer from a "colder" to a "hotter" reservoir will not happen spontaneously or by its own accord. This is expressed in another form of the second law:

> Heat will not flow spontaneously from a colder body to a hotter body.

Heat can flow spontaneously from a hotter body to a colder body. This is like heat flowing "down a tem-

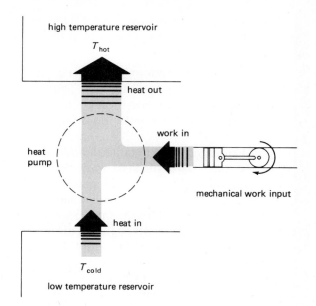

Figure 13.7
A heat pump diagram. In general, a heat pump takes heat from a low-temperature reservoir and transfers it to a high-temperature reservoir. This requires work input. A common practical heat pump is a refrigerator.

perature hill," as a ball would roll down a hill. The reverse, which would be like having a ball roll up a hill of its own accord, does not happen. Obviously, heat transfer "up a temperature hill" or a ball rolling up a hill requires something else. As you probably guessed or noticed from Figure 13.7, this is work (input).

One of the most common heat pumps is the kitchen refrigerator. Its function is to take heat from inside the refrigerator (low-temperature reservoir) and transfer it to the outside surroundings (high-temperature reservoir). The work is supplied by a compressor that uses electrical energy. A diagram of a refrigerator system is shown in Figure 13.8.

A working substance, called the refrigerant, is used to absorb and transport heat. It is a fluid that readily undergoes a liquid-gas phase change at the operating temperature. The most common refrigerant nowadays is Freon, a fluorocarbon compound with a boiling point of about −28°C. (Sulfur dioxide was once used, and ammonia is a common industrial refrigerant, e.g., in ice plants.)

Do you know how a refrigerator works? Let's take a look. It is convenient to divide the system into a high-pressure side and a low-pressure side (Fig. 13.8). Starting with the gaseous refrigerant in the

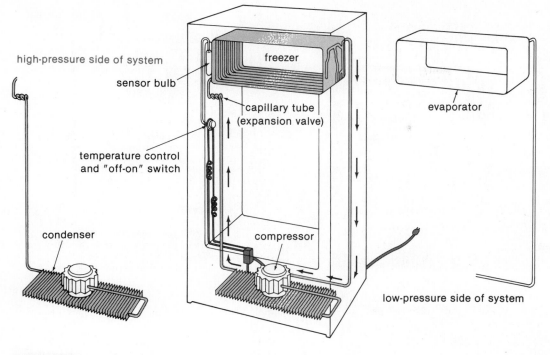

high-pressure side of system

sensor bulb

capillary tube
(expansion valve)

freezer

evaporator

temperature control
and "off-on" switch

condenser

compressor

low-pressure side of system

Figure 13.8
A refrigerator is a heat pump. A refrigeration process takes heat from inside the refrigerator (low temperature) and expels it to the outside environment (high temperature). Electrical energy provides the work to do this. See text for process description.

compressor, which supplies the necessary work, the gas is compressed and emerges at a high temperature and pressure. It then passes into the condenser, where it is cooled and liquefies, and the heat from the refrigerant is rejected to the surroundings (heat out into the room).

The liquid refrigerant then passes through an expansion valve into the low-pressure side of the system. Work is done by the liquid in getting through the valve at the expense of internal energy. This lowers the temperature of the liquid. The cooled liquid refrigerant then flows through the evaporation coils (low-temperature reservoir), where it absorbs heat and vaporizes (boils). The gaseous refrigerant carrying heat goes into the compressor and the cycle begins again.

Air conditioners are also heat pumps that have cycles similar to those of refrigerators. In this case, the low-temperature reservoir is usually the inside of a car or a building, and heat is pumped to the outside environment.

The term *heat pump* is now applied to the year-round heating and cooling systems that are becoming more and more common (Fig. 13.9). When

operating as a cooling system, the heat pump extracts heat from the air inside of the home and expels it to the outside, as an air conditioner does. In its other role as a heating system, the heat pump takes heat from the outside and delivers it to the inside of the home.

This is why heat pumps are more cost-efficient in the winter. No heating fuel is required. You may think that there would be little heat from the outside air on a cold day. But remember, the air has internal energy at any temperature, and there is a lot of air. When the outside temperature goes very low, the heating efficiency of a heat pump goes down too. To keep the house warm enough on very cold days, heat pumps are equipped with auxiliary electrical heating systems. Also, water-exchange heat pumps are now available. Heat is exchanged with an in-ground reservoir, which has smaller temperature variations than the air.

Energy-saving solar heating and cooling systems are now important considerations. Solar heating is readily understood, but how does one cool with solar energy or heat? One method is discussed in Special Feature 13.2.

(a)

(b)

Figure 13.9
Heat pump operations. (a) Cooling cycle. Heat is absorbed from inside circulating air (A) and transferred and expelled to the outside (D). The compressor (C) does the required work. (b) Heating cycle. Heat is absorbed from the outside air by the refrigerant in the outdoor coils (A). The compressor (B) pressurizes the refrigerant and sends it to the indoor coils, where it gives up heat to the circulating air. In both cycles, heat is pumped from a "colder" to a "hotter" region.

Matter, Heat, and Fluids

Temperature and heat. If finely divided powder is sprayed into a flame, combustion is rapid. Dust explosions can be very

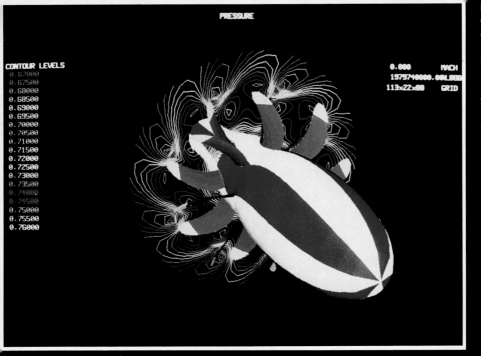

PRESSURE

CONTOUR LEVELS
0.67000
0.67500
0.68000
0.68500
0.69000
0.69500
0.70000
0.70500
0.71000
0.71500
0.72000
0.72500
0.73000
0.73500
0.74000
0.74500
0.75000
0.75500
0.76000

0.000   MACH
1979710000.00LBED
113x22x00   GRID

Not a balloon, but a p
Computer-predicted
contours for counterro
pellers. The blue area
lowest pressure conto
pink represent the hig
High-speed propeller
for cruise speeds at M
have considerable ge
differences from conv
propellers.

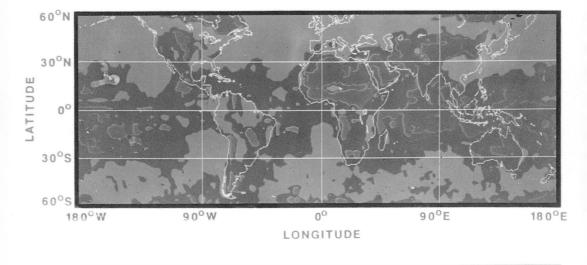

STRONG COOLING   MODERATE COOLING   MODERATE WARMING

Clouds cooling and heating the Earth as observed by the NASA Earth Radiation Budget Satellite, April 1985.

Cooling from Heat: The Absorption Refrigerator

There are solar cooling systems, and there are natural-gas refrigerators that have burning flames. Gas refrigerators were once quite common. These refrigerators had freezer compartments at the top, yet if you looked at the bottom you could see a gas flame. How can you get "cold" from "hot" or cooling from heat? The point to keep in mind is that heat energy, like all forms of energy, is capable of doing work—the essential ingredient in pumping heat. The trick is how to do it.

There are several solar cooling cycles or systems. Let's look at one of these, absorption cooling. This is also the principle of the natural-gas refrigerator. A basic absorption refrigeration cycle is illustrated in the figure below. It uses ammonia as a refrigerant and an ammonia-water solution as an *absorber*. To see how it works, start at the cooling unit or freezer [(1) in the figure]. Here, liquid ammonia absorbs heat from the cooling coils and evaporates (boils). The evaporation takes place in an atmosphere of hydrogen to speed up the process.

The ammonia–hydrogen gas mixture then goes to the absorber (2), where water absorbs the ammonia from the gas mixture. The hydrogen gas returns to the freezer, and the ammonia-water solution drains to the generator (3). Heat from a gas flame (or solar heat source) causes the liquid to bubble up through the tube, much as in a percolator coffee pot. Ammonia boils at a lower temperature ($-2\,°C$) than water, so the bubbles lifting the liquid to the rectifier (4) are practically pure ammonia vapor.

From the rectifier, the water drains back to the absorber, and the ammonia gas rises to the condenser (5). Here the ammonia gas is cooled and condensed back to liquid ammonia, which drains back to the freezer (1), and the cycle begins again.

There you have it—"cold" from "hot." The heat energy input merely supplies the necessary work

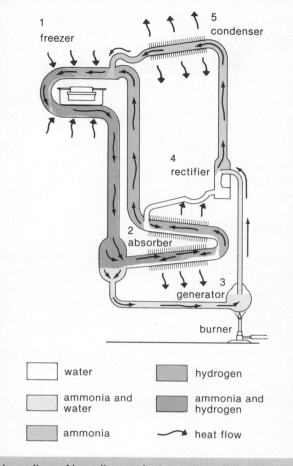

Absorption refrigeration cycle. In an absorption refrigerator, cooling is effected by heat, such as a gas flame (or solar energy). Cold from hot? See text for description of operation.

needed for the cycle as electrical energy does in more common refrigerators.

◼ THE THIRD LAW OF THERMODYNAMICS

How cold can we make it by pumping heat? We know there is a lower limit to temperature—absolute zero. Can we get that low? The answer is no. This is the **third law of thermodynamics**:

> It is impossible to obtain a temperature of absolute zero.

One way of seeing this is as follows: Suppose we could obtain absolute zero and have a cold-temperature reservoir at that temperature. Then, with $T_{cold} = 0$ K, we could have a heat engine with an ideal efficiency of 100 percent with any high-temperature reservoir.

$$\text{Ideal efficiency} = 1 - \frac{T_{cold}}{T_{hot}} = 1 - \frac{0}{T_{hot}}$$
$$= 1, \text{ or } 100\%$$

An ideal efficiency can never be practically attained, and moreover, the second law forbids a heat engine with 100 percent efficiency. So theoretically, absolute zero cannot be reached, as expressed in the third law.

The third law has never been violated experimentally, although scientists have come close to absolute zero — within 0.000001 (one-millionth) of a degree. Near absolute zero, each order of magnitude (factor of 10) reduction in temperature becomes more difficult with each step. For example, going from 0.1 K to 0.01 K is one step more difficult than going from 1 K to 0.1 K. That is, more work is required. To obtain absolute zero, an infinite amount of work would be required.

ENTROPY

As a final topic of thermodynamics, we'll consider entropy.

You may have noticed that the various statements of the second law tell what *does not happen.* In so doing, it indicates the *direction* in which a thermodynamic process can take place. For example, heat flows from a hot body to a colder body. Knowing there is a temperature difference, you know the direction of the process. A more general property that gives the direction of a process was suggested by Rudolf Clausius (1822–1888), a German physicist. This property is called *entropy,* a name coined by Clausius.

Although defined formally in terms of temperature and heat, **entropy** in effect is a measure of disorder. It is found that a system naturally or spontaneously moves from a more orderly state to a more disorderly state. This is expressed by an increase in entropy, and in this context the second law can be stated as follows:

The entropy of the universe increases in every *natural* process.

For example, although not a thermodynamic example, the natural tendency of an orderly desk or room is toward a state of disorder or increased entropy (Fig. 13.10). It's not very natural that a disorderly desk would spontaneously become orderly, which would be a decrease in entropy.

In the thermodynamic sense, entropy is a measure of the capability to do work or transfer heat. A system at a high temperature will naturally tend to do work on and/or transfer heat to its lower-temperature surroundings. In the process, the entropy of the system is increased, and the greater the entropy, the less *available* energy a system has. In terms of order, the heat energy is more "orderly" when it is

(a)

(b)

Figure 13.10
Entropy on the increase. Entropy in effect is a measure of disorder, which increases in every natural process. Straightening up or making a messy desk orderly (decrease in entropy) requires work or an expenditure of energy, which creates entropy such that there is always a net entropy increase.

more concentrated. When it is transferred or used (in a natural process), it is more "spread out" or "disorderly," and there is an increase in entropy.*

Since natural processes continually occur with heat transfer from "hotter" bodies to "colder" bodies, the entropy of the universe continually increases. In the limit, the universe should thermodynamically run down. The entropy would reach a maximum when everything is at the same temperature. This limit is called the **heat death of the universe**. The final temperature is estimated to be a few degrees above absolute zero. Not too warm, but this would occur billions of years from now, so don't worry.

---

* It should be noted that entropy can decrease in one part of a process, such as the transfer of heat by a heat pump. But this requires work or energy input from somewhere, which increases the entropy. Overall, there is a net entropy increase, and the entropy of the universe increases.

Thermodynamics is an empirical science, and its "laws," like other physical laws, are not based on absolute impossibilities of things happening but on the experience that they have never been observed. Because of this, we can say with a great deal of certainty, "Entropy can be created but not destroyed" and "Energy can be neither created nor destroyed," which are the essences of the second and first laws of thermodynamics, respectively.

SUMMARY OF KEY TERMS

Thermodynamics: the branch of science that deals with the relationship of heat and mechanical energy.

First law of thermodynamics: the conservation of energy applied to thermodynamic systems: net heat transfer = change in internal energy + work ($\Delta Q = \Delta U + W$).

Heat engine: a device that converts heat energy to work.

Thermal efficiency: the ratio of work output to the heat input of a heat machine:
work out/heat in = (heat in − heat out)/heat in.

Second law of thermodynamics: no heat engine operating in a cycle can convert heat energy completely to work, or no heat engine can have 100 percent efficiency.

[Optional Exercises are available in the Extended View section of the Appendix.]

Ideal or Carnot efficiency: the maximum possible theoretical efficiency of a heat engine under ideal conditions, and

$$\text{Ideal efficiency} = 1 - \frac{T_{\text{cold}}}{T_{\text{hot}}}.$$

Heat pump: a device that transfers heat from a low-temperature reservoir to a high-temperature reservoir.

Third law of thermodynamics: it is impossible to obtain a temperature of absolute zero.

Entropy: a quantity that indicates the direction in which a thermodynamic process can take place. The entropy of the universe increases in every natural process. Entropy is a measure of disorder or the unavailability of energy of a system.

QUESTIONS

Heat Engines

1. Draw a representative diagram of a heat engine with 100 percent efficiency.

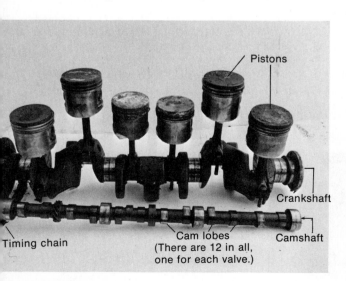

Figure 13.11
Inside an engine. See Question 7.

*2. Identify the heat input and heat output for the drinking bird engine (figure in Special Feature 13.1).

3. In Figure 13.5 the power rating (work/time) of the engine can be seen to be 8 hp. What will engines be rated in if we convert to the SI system?

4. Does a piston heat engine need more than one cylinder to be termed a heat engine? What is the purpose of multicylinder engines?

5. In an automobile engine, why does the engine run "rough" when its timing is off ("out of time") and the cylinders do not fire exactly at the right times? Does this have any effect on fuel efficiency or gas milage?

*6. What is the difference in the timings of a six-cylinder engine and an eight-cylinder engine? (See Question 5.)

7. An automobile crankshaft is shown in Figure 13.11. Explain the purpose of its somewhat eccentric shape. What type of multicylinder engine is this crankshaft from? The camshaft is also shown. As it rotates, the cam lobs lift the valves through lift rods. Why are there more lobs than the number of pistons?

8. During the compression stroke of a piston engine, how is the gas kept from leaking around the piston? (Cf. Fig. 13.11.)

*9. Diesel automobiles have starting heaters in their engines. What is the purpose of these?

Thermal Efficiency

10. Why does a car run better after the engine has "warmed up" on a cold morning?

*11. Show that in terms of the first law of thermodynamics, thermal efficiency is equal to $(\Delta Q - \Delta U)/$heat in. What is ΔQ? Is ΔU a consideration in a cyclic engine?

12. Given a choice between two heat engines with different operating temperatures, which would you choose? Why?

13. What would be the ideal efficiency of a heat engine if the temperature of the heat reservoirs were the same? How about the thermal efficiency?

14. Automobile engines can be water cooled or air cooled. Which type of engine would you expect to be more efficient and why?

*15. Suppose a temperature *below* absolute zero could be reached. What could the ideal efficiency of a heat engine be in this better-than-ideal (impossible) situation?

Laws of Thermodynamics

16. What is the underlying principle of the first law of thermodynamics?

*17. The heat output of a heat pump is greater than the heat input. Does this violate the first law of thermodynamics?

18. Give four statements of the second law of thermodynamics.

*19. In atmospheric convection cycles (Chapter 12), air is transferred from higher, colder altitudes to lower, warmer levels. Does this violate the second law of thermodynamics?

20. If the third law of thermodynamics were violated, would the other two laws also be violated? Explain.

21. Rudolf Clausius stated the second law of thermodynamics in terms of heat pumps, similar to the one in terms of heat engines. What would Clausius's statement say about heat pumps?

22. Some common sayings are, "You can't get something for nothing," "You can't even break even," and "I'll never sink that low." Are these expressions in any way descriptive of the laws of thermodynamics? Explain.

Heat Pumps

*23. Would leaving the refrigerator door open be a practical way to air-condition a kitchen? Explain.

24. Is it energy-efficient to put hot leftovers in a refrigerator? Explain.

25. Prior to refrigerators, "iceboxes" with blocks of ice in a top compartment were used to cool and keep food from perishing. Is the old icebox a heat pump? Why was the ice compartment on top?

*26. A water-exchange heat pump uses water, usually from a well or underground storage tank, as a working fluid. Such heat pumps are more efficient and more costly than the more common air-exchange heat pump. Why? When is the water the high-temperature reservoir?

27. A puddle of water is often observed under a car after it has been driven with the air-conditioner on. Where does this water come from?

*28. On hot, humid days the coils on home and auto air-conditioners sometimes frost and "freeze up." The conditioned air is then less cool. Why? Why does the compressor get hot (sometimes to the point that it burns out) when this happens?

Entropy

29. How does the entropy change in the process shown in Figure 13.12? Explain.

30. Does a quantity of water have more entropy when it is a liquid or when it is frozen? Explain.

*31. How does the entropy of a gas in a cylinder of a heat engine change during the compression stroke?

32. Is heat energy generally becoming less available? Explain.

Figure 13.12
A change in entropy?
See Question 29.

SOUND

IV

The Philadelphia
Orchestra makes pleasing
sounds.

PART IV

Don't make a sound! If you don't there will be no disturbance, no vibrations, and hence, no sound. In general, we consider silence to be a condition in which no sound waves come to our ears.

A discussion of sound usually leads to an old question: If a tree falls in the forest and no one is there to hear it, is there sound? You may have been puzzled by this question. However, in this part of our study, you'll learn that the answer is simply a matter of definition. As has been learned, definitions and basic principles are very important in science. Physically, sound is the propagation of energy or a wave in some material medium, most commonly, air. On the other hand (ear?), sound is generally described as speech, music, noise, and other effects perceived by the human ear. So, if a tree falls in the forest, sound as a wave or disturbance propagation exists, but must it be heard to be sound? As you can see, it depends on how you look at or define sound.

In this short part on waves and sound, we'll look at the fascinating physical and physiological mixture of one of our major sensory links to the environment.

Vibrations and Waves

14

Interfering water waves.

Vibrations and waves are quite common. We use the terms to mean various things. For example, some people talk about good vibrations ("vibes"), and we wave goodbye and wave flags. But these are different from the scientific meanings.

Examples of waves in scientific study are light waves and sound waves. Much of the information we receive about the world and our communications are through our senses of sight and hearing. Waves generally involve vibrations, but

> a **wave** is associated with a disturbance and energy transfer.

A disturbance requires energy, and the energy of the disturbance moves or propagates as a wave. When a motorboat travels on a lake, people on the shore are soon aware of it, even if it is unseen or unheard, because the disturbance made by the boat is propagated to shore as water waves. So when someone tells you not to make waves, you are being told not to make a disturbance.

■ VIBRATIONS

When something vibrates, it shakes or quivers. Automobiles usually have a few vibrations. The vibrational motion of a particle or a mass involves some type of back-and-forth motion or oscillation. Some examples of vibrating or oscillating objects are shown in Figure 14.1. They all repeat their motions periodically as a result of restoring forces. For the pendulum, the restoring force is not a mechanical one, but gravity.

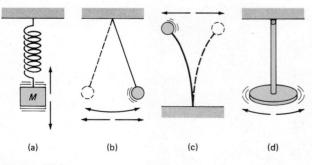

(a)   **(b)**   **(c)**   **(d)**

Figure 14.1
Vibrations and oscillations. (a) Mass on a spring. (b) Pendulum. (c) Mass on a flexible strip. (d) Disk on a rod (torsion oscillation).

To see how vibrational motion is described, let's consider a mass oscillating on a spring (Fig. 14.2). The restoring force of the spring is proportional to the displacement of the mass (Hooke's law, Chapter 8), and the mass is said to be in **simple harmonic motion**. A graphic picture of the motion is obtained by attaching a marker to the mass and moving a paper along behind the mass as it oscillates. This traces out a wiggly or wavy line called a sine curve.

The maximum displacement (distance up or down, $+A$ or $-A$, from the center or equilibrium position) of the mass is called the **amplitude** of the oscillation. This gives us a measure of the energy of the mass. If the mass had more energy, it would oscillate over a larger distance and have a greater amplitude.

Since the motion is repeated over and over, we say it is cyclic or goes through a cycle. The time for the mass to go through one complete cycle is called the **period** (T) of the oscillation. On the sine curve, this is the time for tracing out one complete "wiggle." For example, if the tracing took 0.5 s, the oscillating mass has a period of 0.5 s ($T = 0.5$ s). That's pretty simple.

Another term used to describe the motion is **frequency** (f). This is the number of cycles the mass goes through each second, or the number of wiggles per second on the sine curve. The frequency and the period are related by

$$f = 1/T$$

So if the period of the mass is 0.5 s, then it has a frequency of $f = 1/T = 1/0.5 = 2$ cycles per second (cps). The SI unit of frequency is the hertz (Hz),* so we say the frequency is 2 Hz (1 Hz = 1 cps). This unit isn't as descriptive as cps, but that's what it is.

If a stiffer spring were used, there would be more hertz or oscillations per second, and the wiggles of the sine curve would be more bunched up on the same time scale (Fig. 14.2). Since the frequency of oscillation is greater, the period would be smaller.

A sine curve can be used to describe any type of simple harmonic motion, for example, a mass oscillating horizontally on a spring (on a frictionless surface) or a pendulum for small oscillations, as shown in Figure 14.3. Notice here that the oscillations start

---

* After Heinrich Hertz (1857–1894), a German physicist and an early investigator of electromagnetic waves.

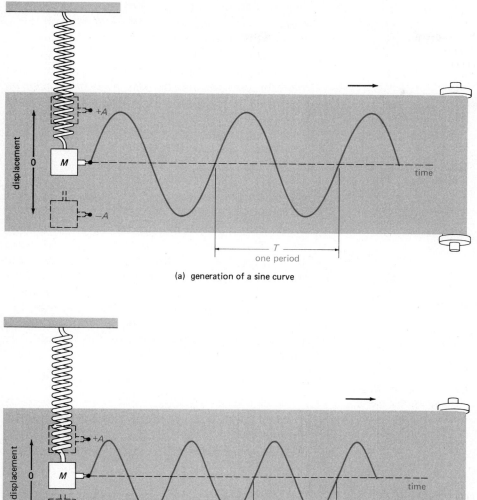

(a) generation of a sine curve

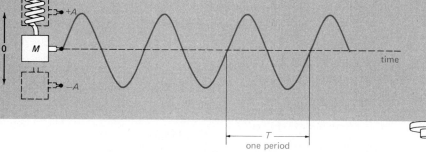

(b) shorter period, greater frequency

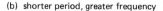

Figure 14.2
Simple harmonic motion (SHM). (a) The SHM of an oscillator can be described by a sinusoidal curve, such as that made by a mass oscillating on a spring. The period T is the time for one oscillation, and the frequency is the number of oscillations or cycles per second ($f = 1/T$). (b) The stiffer the spring, the shorter the period and the greater the frequency of the vibration. (The symbol A was used for the amplitudes in both oscillations, but the actual values are obviously different.)

at the amplitude positions and that the curve begins at a "crest." We say that the oscillation or sine curve is **out of phase** with the one in Figure 14.2 by a quarter of a cycle. If the masses oscillated together and the forms of their sine curves coincided, then they would be **in phase**.

Of course, in actual vibrations or oscillations, energy is lost because of friction. To keep something oscillating in a steady state with a constant amplitude, energy must be supplied by a driving force (Fig. 14.4). When this is removed, energy is lost, as evidenced by the decreasing amplitude, and

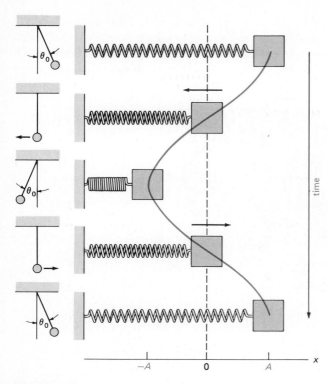

Figure 14.3
More simple harmonic motion. A sine curve also describes pendulum and horizontal-spring vibrations (neglecting friction).

the oscillation eventually dies out. This is called **damped harmonic motion.** Sometimes damping is desired and promoted by some means. For example, on dial bathroom scales the dial oscillates about your weight reading when you get on the scale. You want this oscillation to be "damped out" quickly so you don't have to wait for a long time to get an accurate reading.

So with these concepts under our belts, let's take a look at waves.

■ WAVES

When you drop a pebble into a still pond, what happens? The pebble disturbs the water, and water waves propagate outward. What we actually see in the rearrangement of the water's surface is the motion of the disturbance, with energy being transferred outwardly in the water. The water wave (or disturbance) moves outwardly, but water is not carried with it. (See, for example, the introductory photo in this chapter.) Hence,

> a **wave** is a disturbance that propagates through a medium or space.

Light and sound are both waves that propagate through space, but they are very different. Sound must propagate through a material medium—a gas

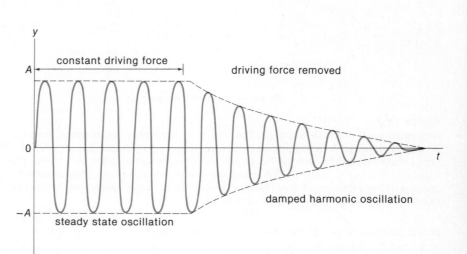

Figure 14.4
Steady-state and damped harmonic oscillations. Maintaining a steady state with a constant amplitude requires a constant driving force. The driving force adds the same amount of energy to the system that is lost. If the driving force is removed, the amplitude "decays" or is damped as energy is lost.

(commonly air), a liquid, or a solid. If there is nothing to propagate the disturbance, there would be no sound waves. Light, on the other hand, is the propagation of energy in the form of electric and magnetic fields. Light waves can travel through a transparent medium, but none is required. These waves can travel through empty space (vacuum).

But all waves originate from a disturbance. In the case of light, it is the motion of electrically charged particles. This kind of wave will be considered in more detail in a later chapter. For the time being, let's concentrate primarily on material waves.

Characteristics and Types of Waves

One of the simplest ways to "make waves" is to shake (disturb) the end of a stretched rope up and down (Fig. 14.5). The continual disturbance causes the rope "particles" to vibrate up and down, and the disturbances (waves) travel down the rope.

The waveform, which may be a sine curve, moves down the rope with a **wave velocity** (v). In general, this depends on the properties of the medium through which the disturbance travels. Wave velocity or speed is expressed in terms of how far the wave (disturbance) moves per time (length/time).

A **wavelength** (λ) is the distance between two crests of a wave (or any two adjacent points that behave identically or oscillate in phase, e.g., two

"trough" points, Fig. 14.5). Notice that a wave travels one wavelength during the time it takes for a particle to make one complete oscillation, which is the period (T) of oscillation. Since speed = distance/time, we may write

Wave speed = wavelength/period

$$v = \frac{\lambda}{T}$$

This means that the wave disturbance moves with a speed such that one "wave" or wavelength passes by an observer in a time T.

Recall that the period is related to the frequency (f) of oscillation by $f = 1/T$. The frequency is the number of cycles per second, and in terms of the moving disturbance, it is the number of "waves" (cycles) passing by each second. So in terms of the frequency we have

Wave speed = wavelength $\times$ frequency

$$v = \lambda f$$

or

wave speed = (length of a "wave") $\times$ (number of waves passing per second)

These equations give the distance a wave disturbance travels per time or its velocity (with direction added), and they hold for all kinds of periodic waves—in ropes and for light, sound, and water waves.

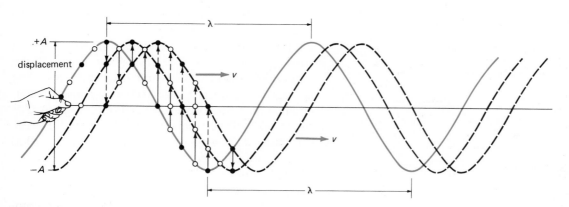

Figure 14.5
Wave characteristics. For a stretched-rope wave, the rope "particles" oscillate up and down with a maximum displacement (amplitude). The disturbance propagates down the rope with a wave speed of v. The distance between two crests, or two troughs, is called a wavelength (λ).

Figure 14.3
Longitudinal waves. In a longitudinal wave, the particles oscillate parallel to the direction of propagation. Examples are (a) waves in a long, horizontal spring with compressions and expansions, and (b) sound waves in air with compressions and rarefactions.

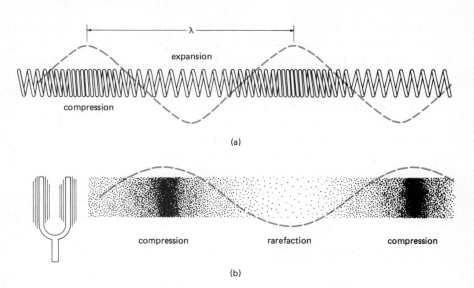

Transverse Waves

The directions of the disturbance (particle) oscillations and wave velocities are used to distinguish between different types of waves. In the case of a wave in a stretched rope (Fig. 14.5), the direction of the particle oscillations in the rope is perpendicular to the direction of the wave motion or velocity.

Such a wave is called a **transverse wave**. This type of wave is sometimes called a shear wave because the disturbance gives rise to shear forces.

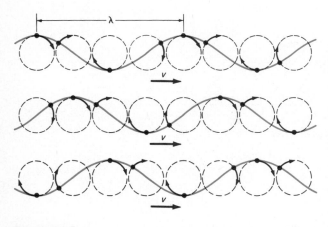

Figure 14.7
Water waves. Water waves are combinations of transverse and longitudinal motions. The water "particles" move in more or less circular paths. Starting at the top figure, notice how the crests move forward.

(You shear a candy bar when you hold it and pull up with one hand and push down with the other and snap it or twist it in opposite directions with a torsional shear.)

Pure transverse material waves occur only in solids. We say that liquids and gases cannot support a shear, meaning that there is no restoring force to a shear disturbance. A shear cuts through liquids and gases, so pure transverse or shear disturbances cannot propagate in these media.

Light waves are transverse waves. The electric and magnetic fields oscillate perpendicularly to the direction in which the wave travels through space (Chapter 19).

Longitudinal Waves

Suppose you moved the end of a horizontal spring back and forth in the direction of the length of the spring. The disturbance would propagate down the spring in the form of alternate compressions and expansions (Fig. 14.6). How do the spring particles move here? Notice that they oscillate parallel to the direction of the wave motion. Such a wave is called a **longitudinal wave**. (Remember *long*itudinal— *along*.) Notice in the figure how a sine curve can be used to describe the waveform.

Longitudinal waves are sometimes called compressional waves because of the compressions involved. Compressional waves can propagate through all types of media—solid, liquid, and gas. Sound is a longitudinal or compressional wave (Fig. 14.6). In a gas, the compressions are regions of

The Earth's Liquid Outer Core

You may have heard that the Earth has a liquid center or, more correctly, a liquid outer core (cf. figure below).* A good question to ask is, How do we know? We have never penetrated through the Earth's relatively thin crust. The deepest mine shafts and drillings of a few miles have only "scratched the surface." Hot molten material comes up from the upper mantle and

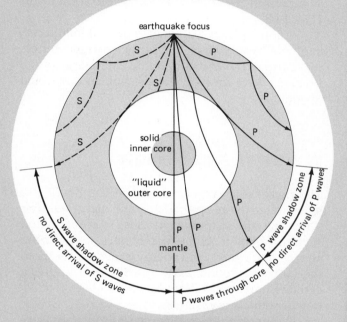

Evidence for the Earth's liquid outer core comes from waves generated by earthquakes. Transverse or shear waves cannot propagate through a liquid, but longitudinal or compressional waves can. Because of shadow zones where no waves are received, the Earth's outer core is believed to be a viscous liquid. (Note: Regions are not to scale, and outer crust is not shown.)

erupts, forming volcanoes, but this occurs only in certain areas, and overall the mantle is solid.

Waves are used to probe and investigate the interior of the Earth. We can produce minor disturbances through explosions, but nature supplies some big ones free of charge through earthquakes. Earthquakes are caused by the sudden release of built-up stress along faults or cracks in the Earth, such as the San Andreas Fault in California. The energy from these disturbances propagates outwardly in the form of seismic waves. There are two general types of seismic waves — surface waves and body waves. The surface waves move along the Earth's surface and cause most of the earthquake damage.

Body waves travel through the Earth. There are P-waves, which are compressional (longitudinal) waves, and S-waves, which are shear (transverse) waves. The *P* and *S* stand for *primary* and *secondary,* which indicate their relative arrival at a location. Primary waves travel faster than secondary waves through a particular material and reach a seismic detection station first.

The longitudinal P-waves can travel through a solid or liquid, but the transverse S-waves cannot travel through a liquid. When an earthquake occurs on one side of the Earth, P-waves are detected on the opposite side and S-waves are not. The absence of S-waves arriving in a "shadow zone" leads to the conclusion that the Earth must be liquid near the center. This is a highly viscous metallic liquid. When the P-waves enter and leave the liquid region, they are bent. This gives rise to a P-wave shadow zone, which tells us that only the outer part of the core is liquid.

* The Earth is divided into three regions: an outer crust about 24 to 30 km (15 to 20 mi) thick; the mantle, 2900 km (1800 mi) thick; and a core with a total radius of 3450 km (2150 mi). The solid inner core has a radius of about 1200 km (750 mi).

higher pressure and density. The intervening low-pressure and low-density regions are called rarefactions.

An application of the differences in the propagation properties of transverse and longitudinal waves is given in Special Feature 14.1.

Water Waves

When we see waves on the surface of water, it is tempting to think of water waves as being transverse

waves. Their profile looks like a transverse sine curve. However, these surface waves are combinations of transverse and longitudinal motions. The water "particles" move in more or less circular paths (Fig. 14.7). Notice that there is no net average motion of the water as a whole. You have probably noticed that a leaf or a fishing float or "bobber" just bobs up and down in place as water waves pass.

The diameter of the circular paths of the water particles decreases rapidly with depth (Fig. 14.8). In the open ocean a hundred meters or so below the

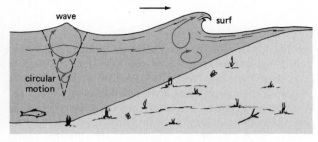

(a)

(b)

Figure 14.8
Surf. When water particles in a wave cannot complete the bottom parts of their paths, the wave "breaks" and the crest falls forward to form a surf.

surface, a submarine is undisturbed by the large waves due to a heavy storm on the ocean surface. However, as the wave approaches shallower water near a shore, the water particles have difficulty in completing their circular motion and are forced into more elliptical paths. The surface wave then grows higher and steeper.

Finally, when the depth becomes too shallow, the water particles can no longer move through the bottom part of their paths, and the wave "breaks." The crest of the wave falls forward to form a surf (Fig. 14.8).

At some places with the aid of tidal influences, surfs can be very large, to a surfer's delight. Surfers make use of the high amplitude and energy of the breaking wave for long rides — with an occasional "wipe-out."

ENERGY AND INTENSITY

The energy of a wave is related to its amplitude. But we are often interested in the rate of energy transfer. When you strike a tuning fork, it is given energy and set into vibration. As the fork vibrates, it causes the air near it to vibrate, and energy is given to the air in the form of a sound wave (see Fig. 14.6). After a certain time the fork "runs down," and the production of sound waves ceases.

To improve the energy transfer to the air, suppose that after you strike the fork, you hold the base of its handle against a tabletop. The sound is louder, and the fork stops vibrating sooner than if it is not touching anything. The loudness increases because the vibrating fork causes the tabletop to vibrate also. The large table surface is able to set more air in motion than the fork itself, and there is an area consideration in energy transfer.

To characterize the transfer of energy by a wave, we speak in terms of intensity. **Intensity** (I) is the rate of energy transfer through or to a cross-sectional area (A), or $I =$ (energy/time)/area (Fig. 14.9). Since energy per time is power, $I =$ power/area $= P/A,$ and the units of intensity are watts per square meter (W/m^2).

The intensity varies with the distance from the wave source. This can be seen by considering the intensities at various distances from a point source from which waves travel outwardly in all directions (Fig. 14.10). The area of a sphere around the

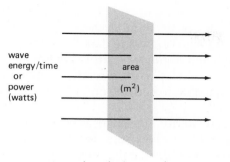

intensity *I* = power/area

Figure 14.9
Intensity. Intensity is wave energy per time or power passing through an area (*I = P/A*) and has units of watts/m² (W/m²).

sources is $A = 4\pi r^2$. So $I = P/A = P/4\pi r^2$, and the intensity from a small source "falls off" as $1/r^2$.

For example, if you double the radius of the sphere or distance from the source, the intensity is only one fourth as great. Farther from the source, the energy is "spread" thinner over a greater area, and the intensity is less. Imagine you have just so much paint to paint a house. Using all the paint, you'd have to spread it thinner on a larger house, so the paint "intensity" would be less.

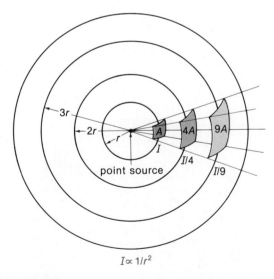

$I \propto 1/r^2$

Figure 14.10
The intensity decreases or "falls off" as 1/r². The energy propagating outward through a distance *r* from a point source passes through increasingly greater spherical areas (*A* ∝ *r*²), so the intensity is less and decreases as 1/r² (*I = P/A* ∝ 1/r².)

INTERFERENCE

Suppose two waves traveling in different directions run into each other. What do you think happens? As they pass through each other and interfere, the disturbances combine and add up. This is called the **principle of superposition**: When two or more waves interfere, the height or displacement of the combined wave at any point is equal to the vector sum of the individual wave displacements.

To illustrate this, consider two similar wave pulses traveling in opposite directions in a stretched rope [Fig. 14.11(a)]. As the waves interfere, the displacements add, and the waveform has a combined shape. Interference in which the adding of the displacements produces a taller combined waveform is called **constructive interference**. When the pulses are exactly superimposed, the combined waveform is twice as tall as the individual waves (*A + A = 2A,* assuming they have the same amplitude).

However, remember that the displacement can be negative, such as in a downward pulse [Fig. 14.11(b)]. Hence, when the two pulses interfere and the (vector) displacements are added, the height of the combined waveform is smaller. This is called **destructive interference**. In this case, when the pulses are exactly superimposed, the amplitude of the combined waveform is zero [*A + (−A) = 0*]. The upward pulse "fills in" the downward pulse, so to speak, in what is called total destructive interference. The energy of the disturbance is not "destroyed." It is still there, stored in the rope.

In either case, after the pulses interfere they pass on down the rope as though nothing had happened. Check out the interfering waves in the introductory photo of this chapter.

In some cases in which periodic waves continuously interfere, interference patterns are set up (Fig. 14.12). The waves spread out from each source and overlap or interfere. When a crest overlaps with a crest or a trough overlaps with a trough, constructive interference occurs. The waves are said to be in phase at these points. At other points where a crest overlaps a trough, destructive interference takes place. The waves are said to be completely out of phase here.

As the waves travel outward, these conditions occur at certain points and a pattern is formed, as shown in Fig. 14.12(b) for actual water waves. Along the dark lines radiating outward, crests overlap troughs, and destructive interference occurs. A

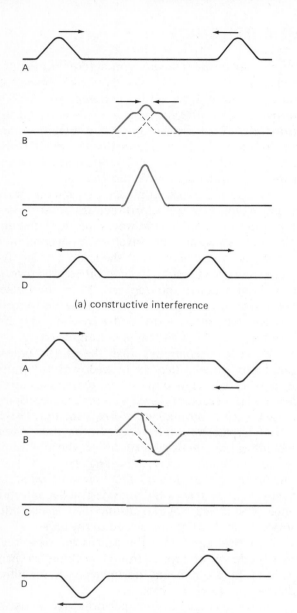

(a) constructive interference

(b) destructive interference

Figure 14.11
Wave interference. (a) Constructive interference occurs when the individual wave displacements add up or produce a greater combined waveform. (b) In destructive interference, the addition of displacements gives a smaller waveform.

Figure 14.12
Interference. (a) Continuous waves give rise to interference patterns. (b) Dark radial lines (Y) are regions of destructive interference. The bright radial lines (X) are regions of constructive interference.

small cork placed in one of these regions, say at Y, would be relatively motionless. The outward rays with alternate bright and dark lines are where constructive interference occurs (light is reflected from the tall crests). A cork placed in one of these regions, say at X, would bob up and down as the waves passed by.

Such interference patterns occur for all types of waves — water waves, sound waves, light waves, and so on — provided that the waves or periodic disturbances have the same frequency. When they don't, the waves still interfere, but the effect is different. We'll look at an example of this in the next chapter on sound.

STANDING WAVES

Interference also gives rise to standing waves. You have no doubt noticed that when you shake the end of a stretched rope up and down just right, the waveform appears to "stand" in the rope. This is due to the interference of the waves traveling down the

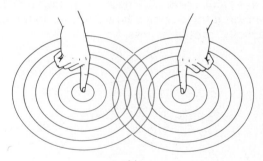

(a)

(b)

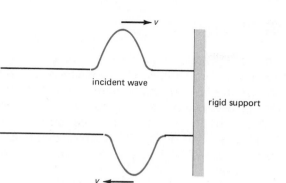

Figure 14.13
Reflection from rigid support. When a wave is reflected from a rigid support, the waveform is inverted.

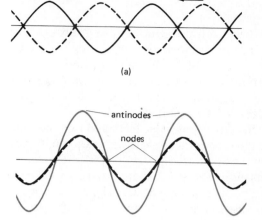

(a)

(b) waves move 1/4 λ

rope and the waves reflected back from the other end.

When a waveform is reflected from a rigid support, the reflected waveform is inverted. This is illustrated for a wave pulse in Figure 14.13. The incident wave exerts an upward force on the support, and the support exerts a downward reaction force on the rope, causing the reflected pulse to be inverted. For a continuous wave, the reflected wave is continuously inverted. (If the support can move freely, such as a light ring that can move up and down on a rod, the reflected wave is not inverted.)

When two waves of equal amplitude and wavelength traveling in opposite directions interfere, they are continuously in phase and out of phase. Suppose the traveling waves at a particular instant were as shown in Fig. 14.14(a). Their combined

(c) another 1/4 λ

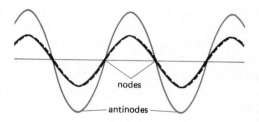

(d) another 1/4 λ

Figure 14.14
Traveling wave interference. Waves traveling in opposite directions in a string are continuously in phase and out of phase (a to d). Interfering waves of equal amplitude and wavelength give rise to standing waves (e and f).

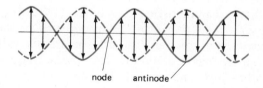

(e)

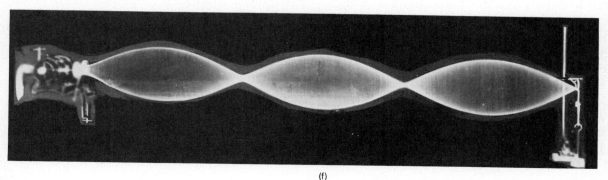

(f)

waveform would be a straight line (destructive interference).

When each wave moves one fourth of a wavelength (the distance of half a loop), they are in phase [Fig. 14.14(b)]. In another one-fourth wavelength movement the waves are completely out of phase, and in another one-fourth wavelength they are back in phase, and so on.

This action gives rise to a **standing wave** with the wave envelopes clearly visible, as shown in the photograph of a vibrating string. Notice that there are points along the string that do not move. These are called nodal points or **nodes**. The amplitude (maximum displacement) points of the envelopes are called **antinodes**.

Characteristic Frequencies

If you try to form standing waves in a rope, you'll find that they will occur only when you shake the rope just right—at the right frequencies. Otherwise, the waves will not interfere to form a standing wave. This is because of what we call boundary conditions.

Notice in Figure 14.15 that a node is required at each end of the rope (a *near* node at the shaking end). As a result, only a certain number of wave loops can "fit in" the rope. The first possibility is one loop, or half a wavelength (two loops per standing wavelength), then two loops (one wavelength), three loops (1½ wavelengths), and so on.

The speed of the waves in the rope is fixed by the tension in the rope and the mass along its length. So for particular wavelengths there can be only particular or **characteristic frequencies**. (By the equation $\lambda f = v$ or $f = v/\lambda$.) The waveforms of these characteristic frequencies are sometimes called normal modes of oscillation. The lowest frequency is called the **fundamental frequency** or first harmonic. The next frequency, which is twice the fundamental frequency, is called the second harmonic (or sometimes the first overtone), and so on. Each harmonic is a multiple of the fundamental frequency.

Get a rope and try this for yourself. It is difficult to get more than the lowest three or four harmonics in the rope. Standing waves occur in stringed and wind musical instruments. Wind instruments involve standing waves in air, as will be learned in the next chapter when we look at how musical instruments make sounds or tones.

■ RESONANCE

All things vibrate or oscillate at a characteristic frequency or frequencies. The characteristic frequencies of an object or material are *natural* frequencies of oscillation and depend on such factors as mass, elasticity or restoring force, and geometry (boundary conditions). In a sense, these are the frequencies at which an object or material "wants" to vibrate.

When an oscillator is driven at a frequency corresponding to a particular characteristic frequency, we say that the oscillator is driven in **resonance**. At a resonant frequency, there is maximum energy transfer from the driving source to the oscillator. This is not to say that energy will not be transferred at other driving frequencies. However, the oscillations may be erratic, as when you shake a rope at a frequency other than one of its characteristic frequencies.

A rope or string has many possible characteristic frequencies, as we have seen. But in some cases there is only one characteristic frequency, for example, in a child's swing or pendulum. A swing is basically a simple pendulum, which has only one characteristic frequency or just a fundamental frequency. This is determined by the length of the pendulum.

When a swing is pushed with this (resonant) frequency, it swings higher and higher (greater amplitude). You don't have to push hard, but you need the right rhythm or frequency so energy is transferred to the swing smoothly. Most of us have experienced what happens if we change the driving frequency slightly and push the swing before it reaches the return amplitude of its swing. Another example is bouncing a ball (Fig. 14.16). To have the ball bounce smoothly, you must drive it at its natural "bounce" frequency or in resonance.

A mass oscillating on a spring also has only one translational characteristic frequency. This depends on the spring constant, or how "stiff" the spring is, and the mass. However, there is another possibility. The mass may oscillate in a rotational "twisting" or torsional mode (see Fig. 14.1), and there is a characteristic frequency for this mode. Looking at the atoms of molecules as being held together by "springs," you might expect molecules to have certain natural oscillations, and they do.

As was learned in Chapter 12, water molecules absorb infrared and microwave radiations (heat

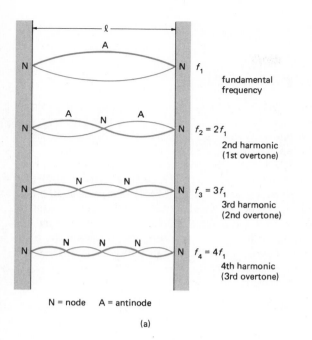

f_1 fundamental frequency

$f_2 = 2f_1$ 2nd harmonic (1st overtone)

$f_3 = 3f_1$ 3rd harmonic (2nd overtone)

$f_4 = 4f_1$ 4th harmonic (3rd overtone)

N = node   A = antinode

(a)

transfer). This is because the water molecule has a natural vibrational frequency in the infrared frequency region. The infrared radiation drives the molecule in resonance, and there is energy (heat) transfer. The water molecule also has a natural rotational oscillation at a lower characteristic frequency. This is in the microwave frequency region, and microwave resonance in water molecules is the principal energy absorption mechanism of microwave ovens (see Chapter 12).

The maximum energy transfer of the resonance condition is also important in electrical circuits, as will be learned in a later chapter. For example,

Figure 14.15
Characteristic frequencies. Boundary conditions permit only certain combinations of loops or half-wavelengths ($\lambda/2$). This allows only certain characteristic frequencies of vibration.

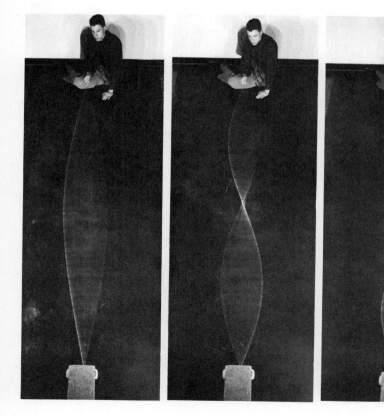

Figure 14.16
Resonance. Similar to pushing a swing, you must push or drive a bouncing basketball at its "bounce" frequency for maximum energy transfer. If this big fellow stooped over and shortened the bounce distance, he'd have to drive the ball at a greater frequency. Why?

Galloping Gertie: The Tacoma Narrows Bridge Collapse

The Tacoma Narrows Bridge near Tacoma, Washington, was opened to traffic on July 1, 1940. It was 2800 ft (855 m) long and 39 ft (12 m) wide. The steel girders used in its construction were 8 ft (2.4 m) tall.

During its first several months of use, many transverse modes of vibration were observed. On the morning of November 7, 1940, winds gusting up to 40 to 45 mi/h started the main span vibrating, with nodes, of course, at the main towers and at several places in between. The vibrational frequency was 36 Hz with an amplitude of 1.5 ft (figure below).

At 10 A.M. the main span of the bridge began to vibrate in a torsional (twisting) mode in two segments with a frequency of 14 Hz. This was apparently due to the loosening of a cable by which the roadway was suspended. Twisting in the wind, the bridge was nicknamed "Galloping Gertie."

The wind drove the structure near the critical velocity for the torsional mode, and this vibration built up by resonance. Shortly after 11 A.M., the main span broke and collapsed. The bridge was rebuilt using the original anchorages and tower foundations, but with a new design that stiffened the structure and increased the resonance frequency so that high winds would not set it into resonance vibrations — and it worked.

Tacoma Narrows Bridge collapse (1940). Gale winds set the bridge into resonance (torsional) vibration. The bridge, nicknamed "Galloping Gertie," oscillated for almost an hour before collapsing.

when you tune in a certain station frequency on your radio, you are adjusting a circuit so it will resonate electrically at that frequency. This gives you maximum energy transfer to the circuit by radio waves at that frequency, and you can pick up a particular station at its assigned frequency.

An example of an unwanted resonance is given in Special Feature 14.2.

SUMMARY OF KEY TERMS

Vibration: an oscillation or back-and-forth motion.

Simple harmonic motion: periodic motion in which the restoring force is proportional to the displacement (Hooke's law).

Sine curve: a graphic curve used to describe simple harmonic motion and waveforms.

Amplitude: the maximum displacement of an oscillation from the center or equilibrium position.

Period: the time (T) to complete one cycle in periodic motion.

Frequency: the number of cycles per second in periodic motion: $f = 1/T$. The unit of frequency is the hertz (Hz).

Damped harmonic motion: periodic motion in which energy is lost and the amplitude of the oscillation decreases with time.

Wave: a disturbance that is propagated through a medium or space.

Wavelength: the distance (λ) between two crests of a periodic wave (or any two adjacent points that behave identically, e.g., two troughs).

Wave speed: the speed of a wave, which is given by $v = \lambda f$.

Transverse wave: a wave in which the particle or oscillation motion is perpendicular to the direction of the wave motion or velocity. This is sometimes called a shear wave.

Longitudinal wave: a wave in which the particle motion is parallel to the direction of the wave motion or velocity. This is sometimes called a compressional wave.

Intensity: the rate of energy transfer through an area: $I = $ (energy/time)/area = power/area. The units of intensity are watts/m^2 (W/m^2).

Principle of superposition: when two or more waves interfere, the height or displacement of the combined wave at any point is equal to the vector sum of the individual wave displacements.

Constructive interference: interference in which the adding of the displacements produces a larger combined waveform.

Destructive interference: interference in which the addition of the displacements produces a smaller combined waveform.

Standing wave: a steady-state oscillating waveform due to the interference of two waves of equal amplitudes and wavelengths traveling in opposite directions. The motionless points of a standing wave are called nodes, and the amplitude points of the envelopes are called antinodes.

Characteristic frequencies: the particular frequency or frequencies at which an object or material naturally vibrates.

Resonance: driving an oscillator at a characteristic or natural frequency that gives maximum energy transfer from the driving source to the oscillator.

[Optional Exercises are available in the Extended View section of the Appendix.]

QUESTIONS

Vibrations

•1. In Figure 14.2 the sine curve motion of a mass oscillating on a spring is demonstrated. How could you show that an oscillating pendulum traces out a sine curve with time?

2. For a pendulum swinging in simple harmonic motion (small arcs), what positions of the swing correspond to the amplitude of the sine curve that describes its motion?

3. If a mass oscillating on a spring is given more energy by a driving force, how does this affect the (a) amplitude, (b) frequency, and (c) period of the oscillation? Explain.

•4. Two masses oscillate up and down on springs with the same amplitude and frequency. Compare the motions when they are (a) in phase, (b) one-fourth cycle out of phase, and (c) completely out of phase. (Draw some diagrams, and show where the masses are relative to each other at a particular instant.)

5. To maintain a steady-state oscillation, how much work must be done by the driving force? What happens if more or less than this amount of work is done?

6. Give some examples in which damped harmonic motion is undesirable and some in which it is desirable.

*7. An automobile has springs and shock absorbers to provide a smooth ride. Explain how they smooth the ride. Do you want damped or resonant oscillations? Why?

Waves

8. How does a bottle or twig bobbing up and down in a lake demonstrate that waves carry energy?

*9. How do electromagnetic waves carry energy? What disturbances cause electromagnetic waves?

10. When a motorboat moves across a lake, water waves continually lap the shore for some time after the boat has passed. But few or no waves reach the shore from a rowboat. Why is this?

11. In general, the wave speed in a medium is determined by the elasticity and particle mass of the medium. With the wave speed relatively constant in water, how is the distance between the crests of the surface waves affected when the disturbance frequency is increased?

*12. Sine curves with the same time scale representing two waves are shown in Figure 14.17. Which wave has the greater (a) amplitude, (b) frequency, (c) wavelength, and (d) period?

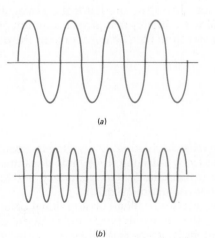

(a)

(b)

Figure 14.17
Frequency, wavelength, and period. See Question 12.

13. The wave velocities of some transverse and longitudinal waves are directed upward. What are the oscillation directions of the waves?

14. If earthquakes produced only S-waves, would we be able to tell whether the inner core of the Earth was solid? Explain.

*15. Neglecting tidal influences, why are surfs along some coastal areas greater than along others? Do tides influence surfs? Explain.

Interference

16. Two waves of equal amplitude interfere. What can you say about the amplitude of the combined waveform when the waves are (a) in phase, (b) out of phase, and (c) completely out of phase?

*17. Is energy destroyed in destructive interference? Explain. What is "destroyed"?

18. Explain why interference patterns are formed, that is, what causes a pattern?

*19. If a transverse wave and a longitudinal wave interfered in a medium, what would be the general waveform of the combined wave?

Characteristic Frequencies and Resonance

20. When a stretched rope is oscillating in its fourth harmonic frequency, how many standing wavelengths are formed in the rope?

*21. A long, limber rod is clamped at one end, leaving the other end free. The boundary conditions for a standing wave in the rod are a node and an antinode at the respective ends. In this case, only the odd harmonics are possible, e.g., f_o, $3f_o$, and $5f_o$. Draw a few possible standing waves, and demonstrate this in terms of wavelength.

22. When you drop a wrench or a metal bowl, it makes a particular sound if it hits a hard surface. Why is this? Why does a baseball bat make a different sound?

23. It is common for the "front end" of an automobile to shimmy and vibrate when it is out of alignment or when the tires are out of balance. However, this usually occurs only at a certain speed, and speeding up or slowing down causes the vibration to cease. Why is this?

*24. A swing has only one characteristic frequency, f_o. However, it can be pushed smoothly at frequencies of $2f_o$, $3f_o$, $4f_o$, and so on. Explain why, and describe the effect of the different driving frequencies.

25. When marching across small bridges, troops are ordered to break step or cadence. Why might marching in step be undesirable?

Sound
and
Music

15

What is it? Look at the
bottom for the keys—a
bird's-eye view of an
open piano.

■ THE NATURE OF SOUND

In the technical sense, sound waves are longitudinal waves that are propagated in solids, liquids, and gases. Without matter there are no sound waves. Most sound comes to us through the air. If we could live in a vacuum, there would be no normal sounds since there would be no medium to carry the waves.

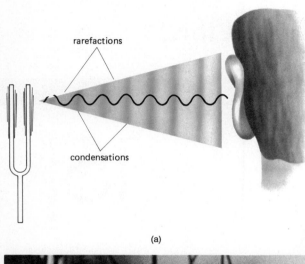

(a)

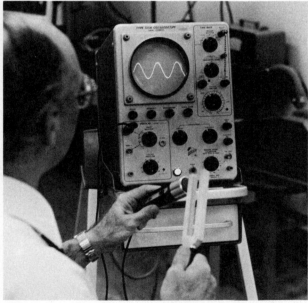

(b)

Figure 15.1
Tuning fork. (a) A tuning fork is essentially a bar bent in a U shape that vibrates at its fundamental frequency, causing sound disturbances in air. (b) The form of the wave from a tuning fork is displayed electronically on an oscilloscope.

Before there can be sound waves in air, some material body must vibrate. When the tires of an automobile "squeal" in a fast start, we hear the sound because something (tires) vibrates and disturbs the air. A sound pulse results from a single disturbance, such as a hand clap or explosion, and a continuous wave results from a continuous disturbance. An example of the latter is a tuning fork (Fig. 15.1). The fork is essentially a bar bent in a U shape that vibrates at its fundamental frequency (antinodes at each end). Higher harmonics are negligible, and a single tone or frequency is heard.

In the ear, the eardrum (a thin membrane) is set into vibration by the pressure compressions and rarefactions of the sound wave. These vibrations are transmitted by bone structures to the inner ear, where they are picked up by the auditory nerve.* When these vibrations have a frequency from about 20 Hz to 20 kHz (20,000 Hz), nerve impulses are initiated that are interpreted by the brain as sound.

The vibrational frequency range or spectrum is divided into three general regions (Fig. 15.2). Below the audible frequency region of human hearing is the infrasonic region (0 to 20 Hz). The longitudinal P-waves generated by earthquakes lie in this region (Chapter 14). Above the audible region is the ultrasonic region, which includes the elastic vibrations of crystals, such as quartz. The sound frequency spectrum does not continue indefinitely. There is a general upper limit of about one billion hertz, which is the upper limit set by material elasticity.

Humans cannot detect sound in the ultrasonic region, but animals can. For example, the hearing frequency response of dogs extends beyond that of humans. Whistles with ultrasonic frequencies, which do not disturb people, are sometimes used to call dogs. Some practical applications of ultrasound are given in Special Feature 15.1.

One of the most common audible sounds is the human voice. The energy for sounds associated with the voice originates from the muscle action of the diaphragm, which forces air from the lungs (Chapter 10). To produce variable sounds, this steady stream of air must be disturbed or "modulated."

---

* The inner ear contains a liquid, which is set into motion. This motion is picked up by hair cells, which initiate signals to the nervous system. The fluid and hair cells of the inner ear are also important in maintaining our balance, or equilibrium.

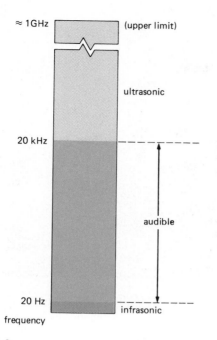

Figure 15.2
The sound frequency spectrum. We hear only sound disturbances with frequencies between about 20 Hz and 20 kHz (audible region). Below this is the infrasonic region. Above this is the ultrasonic region, which has an upper limit of about 1 billion hertz that is set by the limit of material elasticity.

The basic modulating organ is the larynx (the voice box), across which are stretched two membrane-like bands called vocal cords.

The vocal cords form a slit-like opening. As air is forced from the lungs, the flexible vocal cords allow the voice box slit to widen and narrow. Their vibrating and breaking up of the stream of air produces a variation in the pressure, and hence sound waves. The effect of the vocal cords is similar to that of the reed in some musical instruments, for example, the clarinet. The vibrating reed converts a steady air flow into a modulated one. The control of the vocal cords determines the frequency range of a person's singing voice.

The modulated wave from the larynx is further modulated in the numerous (resonance) cavities of the throat, mouth, and nose. Some vocal cavities have controllable parameters, such as the tongue and lips, that can be used to produce a wide variety of sounds. Sound out the vowel letters *a, e, i, o, u* (and sometimes *w* and *y*) and notice the changes in the position of your lips and tongue.

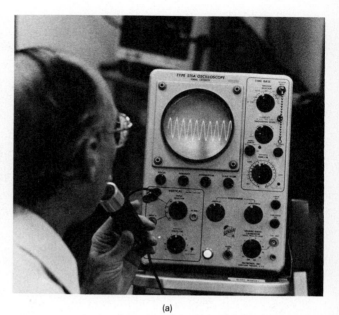

(a)

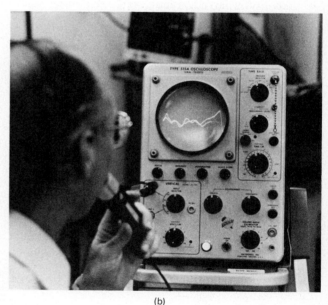

(b)

Figure 15.3
Human voice sounds displayed electronically. (a) Whistling a note gives a simple waveform. Compare with Figure 15.1(b). (b) Voice sounds have complex waveforms as a result of various overtones from the vocal cavities.

The waveforms of human sound can be complex, owing to various overtones from the vocal cavities (Fig. 15.3). Voice sounds are quite specific for individuals and provide "voice prints," similar to fingerprints. However, the use of voice prints for legal identification is still controversial.

Ultrasonics

Ultrasonic sound is in the frequency region above the audible region (> 20,000 Hz). This high-frequency sound finds many uses and applications. For example, ultrasonic waves can travel for kilometers in water and are highly directional or can be focused because of their short wavelengths. As a result, ultrasound is used in marine applications. For example, the depth of the ocean is determined by depth "sounding" techniques using a fathometer. A beam of ultrasound is directed downward from a ship and is reflected from the ocean floor. The depth is computed by knowing the ultrasound speed and elapsed time.

This detection and ranging technique is called sonar. Ultrasonic sonar is used to detect and determine not only the range of ships and submarines, but also that of schools of fish. In addition, combining or modulating audible sound waves with ultrasound makes underwater radio communication possible.

Ultrasound is used in many technical applications. One of the best known is ultrasonic cleaning [Fig. 1(a)]. Ultrasound in liquid baths is used to clean metal parts. Ultrasonic vibrations (small wavelengths) can loosen traces of foreign matter from otherwise inaccessible places. Jewelers make use of ultrasonic baths to clean rings and other jewelry.

Another important industrial application of ultrasound is ultrasonic drilling in the machining of very hard materials [Fig. 1(b)]. By means of an abrasive paste and an ultrasonic vibrator, the material is rapidly worn away. Since the ultrasonic drill does not rotate, the vibrator tip can be oriented to produce holes or surfaces of any shape. Ultrasonic soldering irons are also available. These are particularly useful in the soldering of aluminum. The ultrasound removes the aluminum oxide coating on the surface and eliminates the need for fluxes.

An ultrasonic beam or pulse traveling in metal can be used to detect flaws (Fig. 2). When the ultrasound strikes a flaw, which has different properties than the surrounding medium, reflection and refraction occur. The echo pattern is monitored, and an irregularity indicates the presence of a flaw in the metal. Such techniques provide a means of nondestructive testing of metal castings and other metal objects, such as airplane parts.

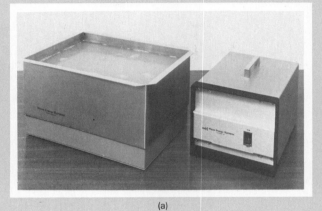

(a)

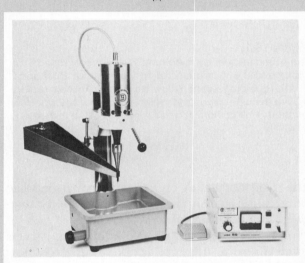

Figure 1
Ultrasonic applications. (a) An ultrasonic cleaning tank. (b) An ultrasonic drill makes holes in glass, ceramics, and gemstones by ultrasonic vibrations rather than rotary motion, as in ordinary drills.

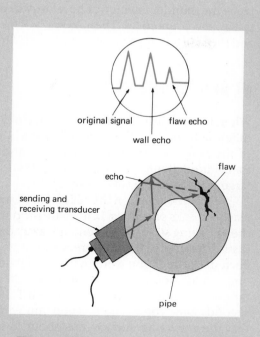

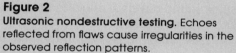

Figure 2
Ultrasonic nondestructive testing. Echoes reflected from flaws cause irregularities in the observed reflection patterns.

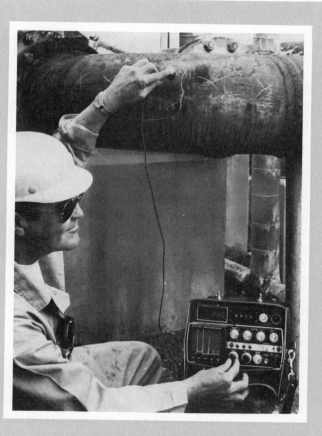

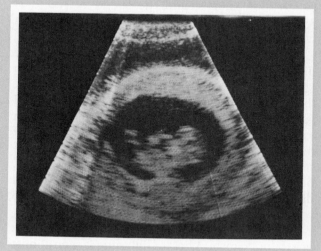

Figure 3
Medical ultrasonics. An ultrasonic scan of a human fetus.

In the medical field, ultrasonic "sonar" can be used to view soft internal tissues and organs, such as the liver or spleen, which are nearly invisible to X-rays. Ultrasound can also be used to "view" a fetus at different stages of development without the dangerous effects presented by X-rays. The different degrees of reflection of scanned areas are monitored and stored in a computer. The computer then reconstructs an "echogram" of the region (Fig. 3). Ultrasonic brain scans are used to detect tumors and cerebral hemorrhage. By applying the Doppler effect to reflected ultrasound, doctors can detect and monitor movements such as the actions of heart valves, the flow of blood, and the beating of fetal hearts.

SPEED OF SOUND

If you put your ear to a railroad rail and someone far down the track hits the rail with a hammer, you'll hear the disturbance in the rail before you hear it in the air. This is because sound travels faster in a solid than in a gas. The speed of sound in a liquid is intermediate.

This should come as no surprise, since there are differences in the stiffness or elasticity of different materials. In a highly elastic material, the restoring forces between the molecules cause a disturbance to propagate faster. Solids are more elastic than liquids, which in turn are more elastic than gases. Hence, the speed of sound is different in each. As a general comparison, sound travels about 15 times faster through a solid than through air and about four times faster through water than through air.

In air, the speed of sound is 331 m/s at a temperature of 0°C. The temperature affects the speed of sound in a medium because as the temperature increases, so does the speed of its molecules. As a result, the molecules collide more often, and a disturbance is transmitted more quickly to neighboring molecules or through the substance. In air, the speed of sound increases by about 0.6 m/s for each degree Celsius increase above 0°C. At room temperature (20°C) this would be an increase of 0.6 m/s × 20 = 12 m/s, and the speed of sound is 331 + 12 = 343 m/s. So sound travels faster on a warmer day, and you hear things sooner.

Notice that the speed of sound in air is quite different from the speed of light in air or vacuum (300 million meters per second). You may notice this difference sometimes when watching a ball game, say from the center-field stands. It is common to see a batter hit the ball, then hear the "crack" of the bat later. You see the ball being hit almost instantaneously because the speed of light is so fast, but if you are 150 m from home plate, the sound disturbance takes about half a second to travel to you.

This effect is also used to estimate the distance of a thunderstorm. A bolt of lightning is seen almost instantaneously. The resulting thunder comes rumbling along behind at about ⅓ km/s (33 m/s, or ⅕ mi/s)*. By timing the interval between seeing the lightning and hearing the thunder, for example, by counting seconds — "one-thousand-one, one-thousand-two," and so on — you can compute the distance the thunder traveled or how far away the storm is. Say you counted 6 s. Then, ⅓ km/s × 6 s = 2 km, and the storm is approximately 2 km (1.2 mi) away.

REFLECTION OF SOUND

When a sound wave strikes the surface of a material different from the one in which it is traveling, some of the wave disturbance is reflected and some is transmitted. If the surface is very rigid and smooth, a large portion of the wave is reflected.

When you mention the reflection of sound, one usually thinks of echoes. **Echoes** are produced when the reflecting surface is far enough away that we can distinguish between the direct sound and the reflected sound. If you were in the mountains and let out a shout, you might hear an echo reflected off a mountain surface. Suppose you heard the echo 2 seconds after you shouted. Then you would know that the reflecting surface was about ⅓ km away, since sound travels at about ⅓ km/s (1 second out and 1 second back).

This principle is used in the practical applications of sonar (Special Feature 15.1) and radar (Special Feature 15.2). Sonar makes use of ultrasonic waves to detect and determine the distance of a reflecting object in water, and radar uses radio waves in air. You can also tell if an object is moving, the direction, and how fast it's going using other wave principles we'll discuss shortly.

Sound reflections can cause problems in concert halls and auditoriums by giving rise to an "echo chamber" effect, caused by receiving multiple reflections at different times (Fig. 15.4). After a short time (depending on the size of the room), the multiple reflections are received so close together that they form a diffuse, continuous sound called **reverberant sound**.

If this reverberant sound persists too long, it interferes with subsequent sounds from the source — for example, a singer or an orchestra — and there is no clarity. We find this to be displeasing, and damping of the reflections is promoted by furnishing the hall with padded seats, drapes, and so on, that absorb a great deal of the sound rather than reflect it. On the other hand, if the reverberant sound is dampened too much, the fullness of the musical tones is lost.

---

* The speed of sound in air is on the order of 1100 ft/s or 750 mi/h (about ⅕ mi/s).

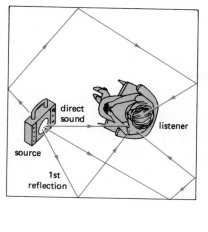

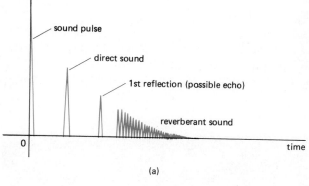

(a)

(b)

Figure 15.4
Acoustic reflections. (a) Reflections can give rise to an echo-chamber effect. Problems can also arise from reverberant sound that persists too long and interferes with subsequent sounds. (b) Concert halls are designed to prevent these problems.

■ REFRACTION OF SOUND

On a calm summer evening or night, it is possible to hear distant voices or sounds quite distinctly that ordinarily would not be heard. This is due to the refraction or "bending" of sound that occurs when a wave passes from one medium into another or into a different density region of the same medium.

For this to occur in air, the atmospheric conditions must be such that a cool air layer is near the Earth with a warm air layer higher up. These conditions often occur over bodies of water at or after sunset (Fig. 15.5).

Notice in the figure that the dashed lines represent the waves under regular conditions. They spread outward from the source and ordinarily would be too faint to be heard by someone on the opposite side of the lake. However, as the waves move through the warmer air they travel faster. As a result, the upper part of the wave is "bent" or refracted downward. This increases the sound intensity received by the person on the opposite side of

the lake, so the sound can be heard. Refraction will be considered in more detail in Chapter 21 for light waves.

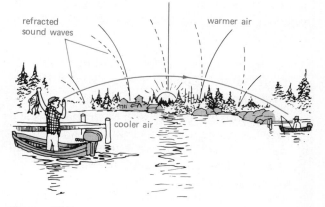

Figure 15.5
Refraction of sound. Because of different densities of air layers at different temperatures, sounds are refracted or bent and can be heard by someone at a distance or behind an object. Under uniform conditions, the sound waves *(dashed lines)* may not be intense enough to be heard.

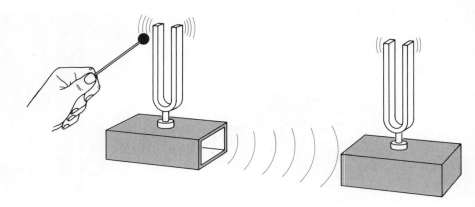

Figure 15.6
Sound resonance. When one tuning fork is set into oscillation, the sound amplified by the attached wooden box sets up vibrations in the other box, and the other tuning fork of the same frequency is set into resonance vibration.

RESONANCE

Resonance can be easily demonstrated with sound, in particular, with two tuning forks of the same frequency (Fig. 15.6). The "natural" frequencies of the tuning forks are usually stamped right on them. Suppose both tuning forks have the same frequency. If one is set into vibration with the other close by and then the vibrating fork is stopped after a short time, sound can still be heard. On investigation, you will find that the other tuning fork is vibrating and producing sound as a result of resonance energy transfer. (The wooden boxes shown in the figure amplify and direct the sound.)

If tuning forks of different frequencies are used, there is no observable transfer of sound energy. Just a small piece of putty or gum stuck on a receiving fork with the same frequency will change its natural frequency enough so that resonance is not observed.

Question: Opera singers are said to be able to shatter crystal glasses with their voices. How is this possible?

Answer: Opera singers with powerful voices have been able to do this (Fig. 15.7). If you wet your finger and move it around the rim of a crystal glass, you can get it to "sing" or vibrate at its resonance frequency. The wavelength of the sound is the same as the distance around the rim of the glass.

If a singer is able to sustain a tone with the proper frequency, the resonance energy transfer will increase the amplitude of the vibrations of the glass to the point that it shatters.

THE DOPPLER EFFECT

When you drop a pebble into a still pond, the water waves travel out in all directions on the surface, generally with the same wavelength and frequency.

Figure 15.7
Resonance in action. See Question and Answer. The glass isn't shattered by a singer, but the effect is the same.

The same is true for a stationary sound source. But what if the source is moving? You have probably heard a truck horn blowing as the truck approaches and passes by and noticed a distinctive change in the sound pitch or frequency. (The perceived sound frequency is commonly referred to as pitch.) The sound gets shriller (higher frequency or pitch) as the source approaches and becomes lower in pitch (lower frequency) as it recedes.

This change in frequency due to the relative motion between the source and the observer is called the **Doppler effect**.* (There is a similar effect if the source is stationary and the observer is moving.)

The effect of the moving source is to "bunch up" the waves in front of the source, and hence to increase the frequency (Fig. 15.8). Also, the waves behind the moving source are "spread out," which decreases the frequency. Notice that the Doppler "shift" or change in the frequency or wavelength (from when the sound is stationary) will depend on the velocity of the source.

The Doppler effect also applies to light waves. If we examine light coming to us from distant galaxies, we find that the frequencies of the light components (colors) are shifted as compared to those of stationary light sources here on Earth. This Doppler shift indicates that the galaxies are moving away from us and supports the theory of an expanding universe (see Chapter 27 on astrophysics).

A more "down-to-Earth" or terrestrial application of the Doppler effect is discussed in Special Feature 15.2.

Sonic Booms

A sonic boom results when a sound source travels faster than the speed of sound, and interference comes into effect. You may have experienced sonic booms from high-speed jet aircraft that rattle your windows and shake your doors. In fact, the jet aircraft are flying faster than the speed of sound, about 335 m/s, or 750 mi/h.

(Sonic booms are not commonly heard these days. Coming primarily from military aircraft, pilots are instructed to fly supersonic only at high altitudes and to avoid population centers.)

* After Christian Doppler (1803–1853), the Austrian scientist who first described the effect.

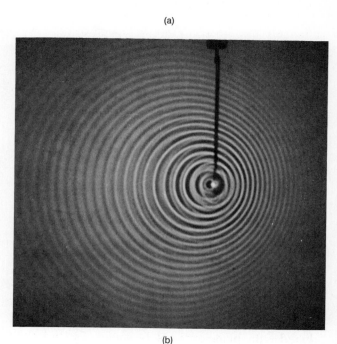

(a)

(b)

Figure 15.8
The Doppler effect. The waves in front of a moving source are "bunched up" and have a higher frequency, whereas the waves behind are spread out and have a lower frequency.

As a sound source approaches the speed of sound, the waves come closer together in the forward direction (Fig. 15.9). When a source travels at the speed of sound or faster, the waves interfere and build up in a V-shaped pressure ridge or shock wave, commonly called a bow wave. A similar bow wave is formed by the bow of a speedboat moving through water at a speed greater than the wave speed in water. The speeds of fast-flying aircraft or

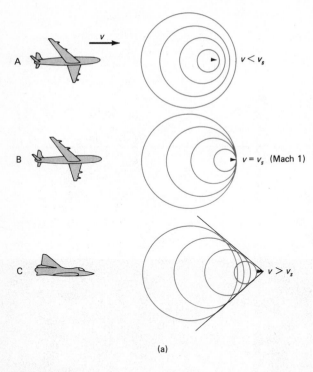

(a)

(b)

(c)

Figure 15.9
Bow waves. Bow waves are formed (a) by airplanes in air
similar to those (b) formed in water by boats or ducks. (c)
The shock wave boundaries are clearly seen for a
gun-launched projectile at Mach 7 in a wind tunnel.

objects are measured in Mach numbers. A jet aircraft
flying at the speed of sound is traveling at Mach 1. At
one and one-half times the speed of sound, the jet is
traveling at Mach 1.5; at Mach 2, twice the speed of
sound; and so on.

It might be thought that the boom is only produced as the plane exceeds the speed of sound, i.e.,
when "breaking the sound barrier." Actually, a
shock wave follows the aircraft at all times when
traveling at supersonic speeds. It tails out and
downward behind the airplane. When the shock
wave passes over an observer on the ground, a sharp
"sonic boom" is heard. In reality, there is a double
boom, since shock waves are formed at both the
front and the rear of a supersonic aircraft.

■ BEATS

Have you ever heard musicians tuning their instruments? A particular note (frequency) is sounded on
one instrument, for example, a piano, and the other
instruments are adjusted until they have the same
tone. As this is done, you can hear pulsating fluctua-

tions that vanish when things are in tune. These
fluctuations are called **beats** and result from an interference effect.

When two waves with the same amplitude and
frequency interfere in space, a combined wave pattern is formed (Chapter 14). Recall that this gives
regions in which a tone would be heard and others
in which there would be no sound (destructive interference).

But, when two sounds of equal intensity (amplitude) with slightly different frequencies interfere, a pulsed disturbance is formed. The amplitude
of the combined traveling wave varies, and a pulsating tone is heard, being louder when the regions of
constructive interference are received (Fig. 15.10).

The number of sound pulsations per second is
called the **beat frequency**, f_b. This is given by the
difference in the two tone frequencies $f_b = f_2 - f_1$.
For example, if the tones are from two tuning forks
with frequencies of 516 Hz and 512 Hz, the beat
frequency is $f_b = 516 - 512 = 4$ Hz, or four pulsations or beats each second. Musicians try to tune
their instruments together until no beats are heard,

Radar and the Doppler Effect

The Doppler effect is used in the determination of the velocity of a moving object by sonar (ultrasonic waves in water) and radar (radio waves in air). Let's discuss radar here, since it is more familiar. Radar stands for *ra*dio *d*etecting *a*nd *r*anging. The detecting and ranging (distance) come from reflection, the time delay, and wave speed, as discussed previously. Add the Doppler effect and there's more.

Suppose a law officer sitting along the side of the highway directs a beam of radio waves toward a car, and the waves are reflected back (figure below). If the car being observed is parked or not moving, the reflected waves have the same frequency. But if the car is moving toward the police car, the reflected waves have a higher frequency, or are Doppler shifted. In effect, the moving

car acts like a moving source, and there is a double Doppler shift — coming and going.

The magnitude of the shift depends on the speed of the car. A computer quickly calculates this and displays it for the officer to see. If this is above the speed limit, the next thing the driver might hear is the Doppler shift of a siren!

If the car is moving away from the patrol car, the reflected waves would have a lower frequency. The speed is still calculated from the magnitude of the frequency shift. It's a bit more complicated, but the speed of the car can be determined even when the patrol car is moving. The computer does this easily and can get you coming and going.

The Doppler effect in action. The Doppler frequency shift depends on velocity, and this effect is used in radar to catch speeders.

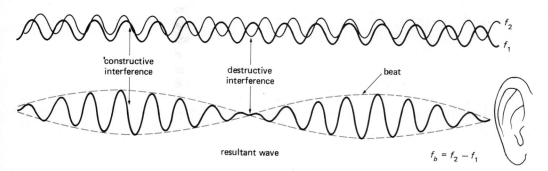

constructive interference

destructive interference

beat

resultant wave

f_2

f_1

$f_b = f_2 - f_1$

Figure 15.10
Beats. Traveling waves with equal amplitudes and slightly different frequencies give rise to pulsating fluctuations or beats.

which means $f_2 = f_1$, and they are in tune. The human ear can detect up to about seven beats per second.

■ SOUND AND HEARING

Sound waves are physical disturbances, and the physical properties that describe these waves are quantities that can be measured without reference to the ear or hearing. Basically, these properties are intensity, frequency, and waveform. The response of the human ear varies from person to person. The physiological terms used to describe the sensations of the ear are *loudness, pitch,* and *quality* (or *timbre*). These sensory effects and the physical wave properties have the following general correlations:

Sensory effect ↔ Physical wave property

Loudness ↔ Intensity
Pitch ↔ Frequency
Quality ↔ Waveform (overtones)

However, there is not a one-to-one correlation. The physical properties are objective and directly measurable. The sensory effects are subjective and vary with individuals. This is similar to our subjective temperature sense of touch and objective physical measurements with a thermometer.

Let's take a look at these similar, yet different, objective wave and subjective sensory effects.

Loudness and Intensity

How loud does a sound have to be for us to hear it? Another way of saying this is, What is the intensity threshold of hearing? We can just hear a 1000-Hz tone with an intensity of 10^{-12} W/m². Thus, for us to hear a sound, it must have not only a frequency in the audible range, but also a sufficient intensity.

As the intensity is increased, the perceived sound gets louder. At an intensity of about 1.0 W/m², the sound is uncomfortably loud and may be painful to the ear. This intensity is sometimes referred to as the threshold of feeling or pain.

Think a moment about this intensity range of hearing for our ears. The intensity at the threshold of feeling is a trillion times ($10^{12} = 1,000,000,000,000$) greater than the intensity at the threshold of hearing! The ear is sensitive to such an enormous range of intensities and the SI unit of watt/m² gives some pretty big numbers when we compare one intensity with another. Therefore, we commonly use another comparative measure of intensity called the **sound-intensity level**, which has units of **decibels** (dB).*

Here, sound intensities are compared to a standard level—the threshold of hearing. Using a logarithmic function, the large range of intensity of human hearing is compressed into a more manageable scale (Fig. 15.11). The threshold of hearing intensity has a sound level of 0 dB, and the threshold of feeling has a sound level of 120 dB.

The decibel scale is not linear. That is, if the intensity of a sound is doubled, the dB level is not doubled. In fact, if the sound intensity is doubled, the dB reading increases by three. For example, if a sound has an intensity of 50 dB, a sound with twice the intensity has a sound level of 53 dB.

Comparisons are conveniently made on the decibel scale in factors of tens of decibel differences:

An increase of 10 dB increases the sound intensity by a factor of 10.
An increase of 20 dB increases the sound intensity by a factor of 100.
An increase of 30 dB increases the sound intensity by a factor of 1000, and so on.

Thus, a sound with an intensity of 80 dB, for example, has an intensity 100 times greater than a sound with an intensity level of 60 dB. Too much sound intensity can be bad for the ear. See Special Feature 15.3.

Loudness is related to the sound intensity, but loudness is very subjective and ear-dependent. Here's the problem. The ear does not respond equally to all frequencies. For example, two tones with different frequencies, but with the *same* intensity levels (W/m²), are judged by the ear to have *different* loudnesses.

Pitch and Frequency

Pitch and frequency are often used interchangeably, but frequency is objective and pitch is subjective. For example, if a pure frequency tone is sounded at two intensity levels, nearly all listeners will agree that the more intense sound has a lower pitch (perceived "frequency"). This effect is most pro-

* The bel unit is in honor of Alexander Graham Bell, the inventor of the telephone. A decibel is one tenth of a bel.

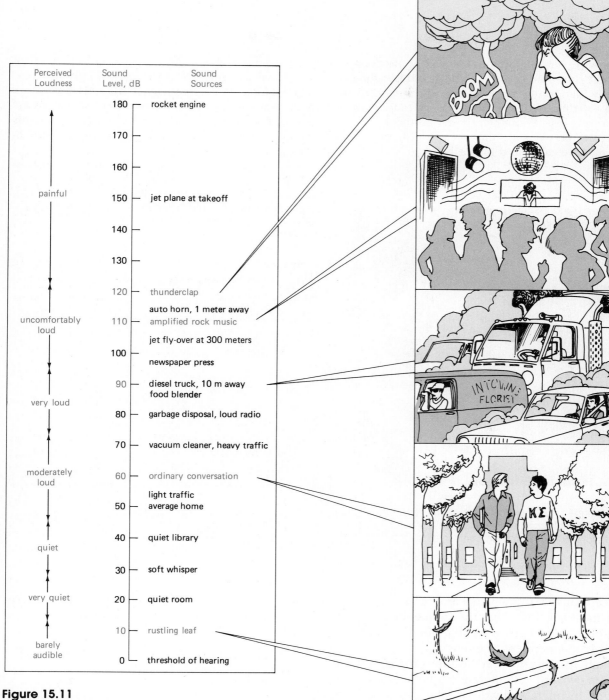

Perceived Loudness	Sound Level, dB	Sound Sources
	180	rocket engine
	170	
	160	
painful	150	jet plane at takeoff
	140	
	130	
	120	thunderclap
uncomfortably loud		auto horn, 1 meter away
	110	amplified rock music
		jet fly-over at 300 meters
	100	newspaper press
	90	diesel truck, 10 m away
		food blender
very loud	80	garbage disposal, loud radio
	70	vacuum cleaner, heavy traffic
moderately loud	60	ordinary conversation
		light traffic
	50	average home
	40	quiet library
quiet	30	soft whisper
very quiet	20	quiet room
	10	rustling leaf
barely audible	0	threshold of hearing

Figure 15.11
The decibel scale. An increase in the sound level intensity of 3 dB doubles the intensity. An intensity level change of 10 dB changes the intensity by a factor of 10; a change of 20 dB by a factor of 100; a change of 30 dB by a factor of 1000; and so on.

Noise-Exposure Limits

Sounds with very high intensities can be dangerous. Above the threshold of pain (about 120 dB), sound is painfully loud to the ear. Brief exposures to levels of 140 to 150 dB can rupture eardrums and cause permanent hearing loss. Consequently, ear protectors or ear valves must be worn in some occupations, and noise levels must be monitored (figure below).

Occupational loudness. The sound-level intensity of a machine is monitored with a sound-level meter. Federal regulations set limits for sound-level exposure times.

Longer exposure to lower sound (noise) levels can also damage hearing. For example, there may be a hearing loss for a certain frequency range. (Have you ever noticed a temporary hearing loss after listening to the music of a loud band for a long time?) Federal standards now set permissible noise-exposure limits, as listed in Table 15.1.

For example, 6 h is the maximum time that a worker can be exposed to sound-intensity levels of 92 dB. For a sound level of 95 dB, 4 h is the maximum exposure time. Notice that this is a doubling of intensity.

Table 15.1
Permissible Noise-Exposure Limits*

Maximum Duration per Day (hours)	Sound Level (dB)
8	90
6	92
4	95
3	97
2	100
1½	102
1	105
½	110
¼ or less	115

* When the daily noise exposure is composed of two or more periods of exposure of different intensity levels, their combined effect should be considered.

nounced for low frequencies around 100 Hz. The ear is most sensitive over the frequency range of about 500 to 5000 Hz. Here, the pitch of a tone is relatively independent of intensity.

Pitch discrimination is the ability to distinguish between two tones of different frequencies as being different in pitch. Individuals who can name the pitch of a tone on a musical (frequency) scale without comparison to an external standard are said to have perfect pitch. There is disagreement as to whether this is a special ability or comes with training. It does seem odd that most of us can remember

visual properties, such as color or shapes, exactly but not note pitches. Why do you suppose this is the case?

Quality (Timbre) and Waveform

The quality of a tone is the characteristic that enables us to distinguish it from another with the same intensity and frequency. Tone quality depends on the waveform or number of harmonics or overtones present (Fig. 15.12). You can objectively produce a

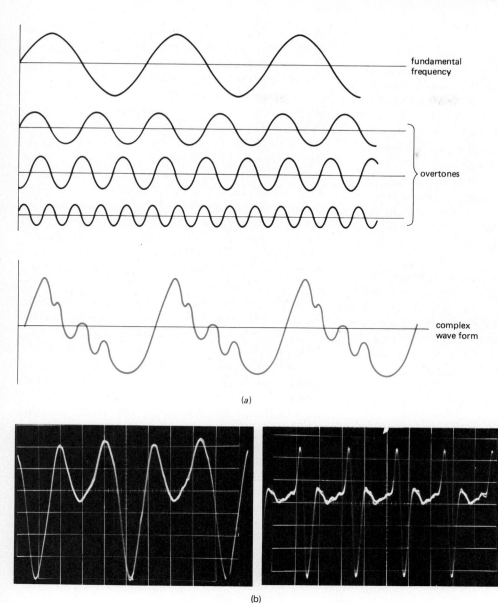

fundamental frequency

overtones

complex wave form

(a)

(b)

Figure 15.12
Quality. (a) The quality of a sound depends on the number and combinations of overtones. (b) Oscillograph displays of sound for a tone from a French horn *(left)* and trumpet *(right)*. The instruments have different qualities.

tone with the same intensity and frequency as your favorite singer, but a different combination of overtones gives the singer's voice a better quality or a more pleasing sound, along with a certain "richness."

The quality of a tone is highly subjective. Some sound qualities are pleasing to some people and cultures and not to others. Japanese music and in-struments may sound displeasing to some Americans, and vice versa. Some combinations of tones or chords generally sound pleasing, but other combinations are termed discordant or lacking in quality.

Such sounds may be classified as noise instead of music by some. Have you ever been told or thought yourself that hard rock music is noise? Different strokes (or quality) for different folks.

■ MUSIC

Music soothes (the savage beast), inspires, and delights. A Beethoven symphony can evoke a mood of sadness, whereas a lively march can get your foot tapping to the music. Music is very personalized. We all have our favorite songs. Such a wide variety of reactions and feelings make the psychological interpretation of music very complex.

When tones played together are judged by a listener to be pleasant, the combination is referred to as a **consonance** or consonant sound. On the other hand, if the combination is judged to be harsh or unpleasant, it is called **dissonance** or a dissonant sound.

Dissonance is important to music. A bit of dissonance breaks up the blandness of continual consonance, adding excitement or restlessness. But music has become increasingly dissonant. Modern music is much more dissonant than older classical music. Why do you think this is? Is it actual subjective preference or a desire to be different?

Musical Scales

We designate particular musical notes or the fundamental frequency of a tone by constructing musical scales. As with temperature scales, there are various ways to do this. Musical scale intervals are based on pleasing or consonant sounds when the frequencies of two notes form ratios of whole numbers. The simplest scale in Western-culture music is the *just*

diatonic scale. The international standard frequency of 440 Hz for the note A_4 is the starting point for constructing the consonant intervals defined by whole-number ratios, as shown in Figure 15.13.

Starting with a C note on a piano and playing successive white keys to the right from a C note to the higher C note, this gives the familiar "do–re–me–fa–sol–la–ti–do," where each note is represented by a letter (Fig. 15.13). This is the C-major scale. Notice that the frequency doubles from C_4 to C_5 (from 264 Hz to 528 Hz). This interval is called an **octave** (Latin, meaning "eight"), with eight steps or notes. A piano keyboard covers seven octaves, and the C_4 of the middle octave is called "middle C." The frequency of each note, D, E, and so on, doubles with each octave interval; for example, the frequency of D_5 is twice that of D_4.

If you look at the intervals or frequency ratios in Figure 15.13, you will quickly notice an oddity. There are relatively larger intervals (⅑⁄₈ and ¹⁰⁄₉), whereas others are smaller (¹⁶⁄₁₅). The larger intervals are called whole intervals, or tones, and the smaller intervals are called half intervals, or semitones.

On the piano, this means that some adjacent white keys differ by a tone and others by a semitone. The scale would be more uniform if we had a complete range of semitones. The black keys of the piano provide this and divide an octave into 12 intervals or steps.

The just diatonic scale has many consonances, but there are serious problems because the tone and

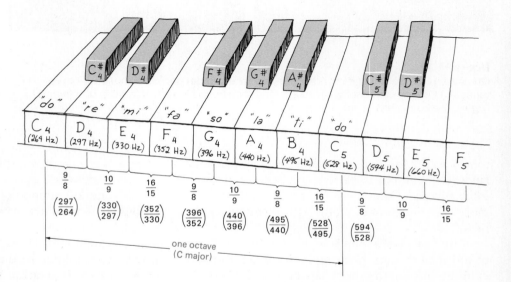

Figure 15.13
Notes and frequencies of a just diatonic music scale.

semitone intervals are not uniform along the scale, even with 12 steps. When played in another major scale or key, say D-major (do–re–me sequence starts with D), certain notes are not the same because of the nonuniform semitone interval; for example, an E note is different in these different major scales.

The problem could be alleviated if a scale had 12 equal semitone intervals in an octave. Such a scale is called the *equally tempered scale*. The standard frequency for this scale is still $A_4 = 440$ Hz (Table 15.2). The advantage of this scale is that all intervals are the same in any key. Also, the sharp ($\#$) of one note and the flat (b) of the next higher note are the same frequency, for example $C^\#$ and D^b. This is not generally the case on the just diatonic scale. However, the equal semitone interval ratios are no longer whole numbers, and some consonance is lost.

Pianos are usually tuned to the equally tempered scale, since they can be used to play music in many keys. Some quality may be lost, but if pianos were tuned to the just diatonic scale, compositions in some keys would sound fine and others would be particularly dissonant.

Musical Instruments

Do you play a musical instrument? If so, it falls into one of the three general instrument categories: string, wind, and percussion. All involve tones resulting from standing waves.

String instruments include the guitar, violin or fiddle, bass, and piano (see introductory photo in this chapter for the piano). Here, standing waves are formed in strings fixed at each end. When a string instrument such as a guitar or violin is being tuned, what does the musician do? He or she tightens or loosens the strings, i.e., adjusts the string tension.

This changes the pitch or frequency of the string sound until no beats are heard with the tuning source. Actually, the wave speed in a string is being adjusted. This increases with the tension or tightening. Adjusting the wave speed changes the frequency, since the length between the ends (between the bridge and neck) or the fundamental wavelength of the string is fixed, and $\lambda f = v$.

The wave speed also depends on the mass per unit length of the string. You have probably noticed that the high-frequency strings on a string instru-

Table 15.2
The Equally Tempered Scale for the Central Octave of a Piano

Note	Frequency (Hz)
C_4	261.6
$C_4^\#$ (or D_4^b)	277.2
D_4	293.7
$D_4^\#$ (or E_4^b)	311.1
E_4	329.6
F_4	349.2
$F_4^\#$ (or G_4^b)	370.0
G_4	392.0
$G_4^\#$ (or A_4^b)	415.3
A_4	440.0
$A_4^\#$ (or B_4^b)	466.2
B_4	493.9
C_5	523.2

ment are thin and light as compared with the low-frequency strings, which are thick and heavy.

Once the strings on a violin or guitar are tuned to certain notes, other notes of the musical scale are obtained by changing the effective lengths of the strings by fingering (Fig. 15.14). This shortens the

Figure 15.14
Changing notes. On stringed instruments, diferent notes on the musical scale are obtained by changing the effective lengths of the strings.

fundamental wavelength of a vibrating string and increases the frequency. A piano, on the other hand, has a string for each note or frequency, and tuning is a major operation.

The quality or overtone combination of a string instrument's sound depends on how the strings are excited. This may be done by plucking, bowing, or striking (as with a hammer in a piano). Where, as well as how, a string is excited affects the quality of the sound. The bodies and cavities of string instruments are important to the quality and amplification of sound. Vibrating strings themselves disturb little air and would produce little sound intensity, but they are coupled with the instrument body, for example, through the bridge on a guitar and violin. The body acts as a resonance cavity for amplification and the production of overtones that gives the instrument a particular quality. Recall how our different vocal cavities create different voice qualities.

The making of the bodies of string instruments is more of an art than a science. For example, for a fine violin, maple wood for the bottom plate is better when cut in December or January and taken from the south side of the tree. The spruce wood for the top plate is better when taken from a tree exposed to the sun. For the piano, a cast-iron frame introduced

in 1855 by Henry Steinway gives this instrument its brilliance over its ancestor, the clavichord.

In wind instruments, standing waves are set up in air columns. For instruments such as trumpets and trombones, the vibrations of the player's lips are the driving force. For woodwinds such as the clarinet and saxophone, a vibrating reed does the job. For the flute and piccolo, the fluttering airstream resulting from blowing against a hole sets the air column into vibration.

To visualize how standing waves in air columns are formed, consider open and closed pipes, as illustrated in Figure 15.15. A closed pipe is really open at one end (the antinode end) so that sound can get out. An open pipe has antinodes at each end. An antinode position is where the interference gives the standing wave in air the maximum vibrations and the combined amplitudes of the traveling waves give an amplification of the sound.

Different harmonics can fit in the pipes and give the sounded note quality. A pipe organ has a pipe for each note, similar to a piano having a string for each note. For instruments such as the clarinet, trumpet, or saxophone, the length of the "pipe" is varied, and different notes are obtained by means of holes or "stops" and valves (on a trumpet). For a trombone,

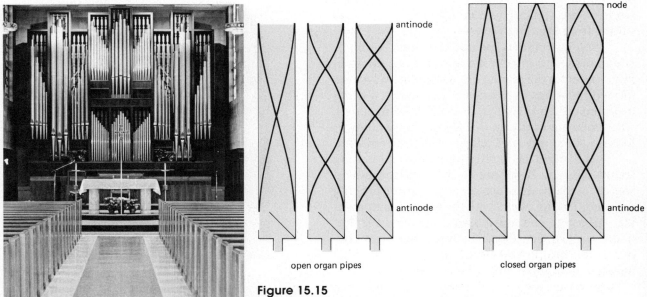

Figure 15.15
Notes in pipes. In a pipe organ, standing waves are set up in pipes, giving fundamental frequencies and overtones.

the length of the pipe is physically varied by moving the slide. Standing waves can be formed in cone-shaped and flared pipes as well as in cylindrical pipes. In percussion instruments, such as drums and cymbals, standing waves are set up in two-dimensional membranes or surfaces. This analysis is more complicated.

In summary, music is in the ear of the listener, just as beauty is in the eye of the beholder. The introduction of dissonant musical sounds by Igor Stravinsky in his works in the early 1900's made audiences furious. Today, we have musical tones produced by synthesizers, which produce sounds electronically. These look more like computers than musical instruments. Here the emphasis is more on pure tones than on quality, and some people find "Moog music" (named after Robert Moog, a pioneer in electronic sound) somewhat displeasing. Indeed, music is a very personal thing, and you have to "make your own kind of music."

■ SUMMARY OF KEY TERMS

Audible frequency region: region of human hearing frequency response, 20 Hz to 20 kHz. Lower than 20 Hz is the infrasonic region, and above 20 kHz is the ultrasonic region.

Speed of sound: (temperature dependence) the speed of sound v_s increases with temperature. In air,

$$v_s = 331 + 0.6 \, T_c \quad \text{m/s}$$

where T_c is the temperature in degrees Celsius.

Echo: reflected sound that can be distinguished from the direct sound because of the distance of the reflecting surface.

Refraction: the bending of waves as they pass from one medium into another or into a different-density region of the same medium.

Doppler effect: a change in the frequency of a wave due to relative motion between the source and observer.

Sonic boom: the shock wave associated with aircraft traveling at supersonic speeds.

Beats: the pulsating fluctuations due to the interference of waves with equal amplitudes but slightly different frequencies (f_1 and f_2). The beat frequency is $f_b = f_2 - f_1$.

Loudness: the subjective sensory effect related to the intensity of sound.

Decibel: a unit of sound-intensity level that gives a comparative measure of intensity.

Pitch: the subjective sensory effect related to wave frequency.

Quality: the subjective sensory effect related to the waveform or number of overtones in a wave.

Consonance: a combination of sounds judged pleasing to the ear.

Dissonance: a combination of sounds judged displeasing to the ear.

Musical scale: a scale of note frequencies with constant intervals defined by whole-number ratios of frequencies.

Octave: a musical scale interval defined by a doubling of frequency or a ratio of 2:1.

[Optional Exercises are available in the Extended View section of the Appendix.]

■ QUESTIONS

The Nature of Sound

*1. Some people say there are two things necessary to have sound, and others say there are three things. Why the difference?

2. How do flying insects produce a high-pitched buzzing sound?

3. An alarm clock is hung in a closed glass jar. When the alarm goes off, it can be heard. Why? What would happen if the air were pumped out of the jar? You may be able to hear the alarm faintly. Why?

*4. What happens to the energy when a sound "dies out"?

5. Children often make a "telephone" out of a tin can and string (Fig. 15.16). Explain the sound principles involved.

Speed of Sound

6. The speed of sound in air is given for still air. How would wind affect the propagation of sound? What happens when you shout into the wind?

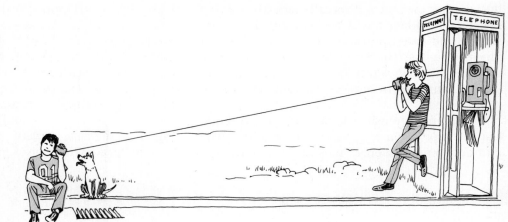

Figure 15.16
Cheap telephone. See Question 5.

7. Sound travels faster in moist or humid air. Why is this? (*Hint:* The major components of air are N_2 and O_2. In a volume of moist air, some of these molecules are replaced by H_2O molecules. Check out their masses, using the periodic table.) Could this be the basis of the weather saying, "Sound traveling far and wide, a stormy day will betide"?

•8. In a popular lecture demonstration, the instructor breathes in helium (He) gas. When the instructor talks, his or her voice has a high-pitched "Donald Duck" sound. What causes this? (*Hint:* The vocal cavities and wavelengths are the same.)

Sound Effects

9. Why do sounds in an empty building or room sound "hollow"?

•10. After a snowfall, it "sounds" particularly quiet. Why is this?

11. When you hear a jet aircraft flying overhead and look in the direction of the sound to see the plane, you may have trouble finding it there. Explain why.

•12. Singing in the shower is a popular pastime (Fig. 15.17). Why does your voice sound fuller in the shower?

13. Is there a resonance effect in bouncing a ball (for example, when dribbling a basketball)? Explain.

14. It is common to be able to hear music, particularly the bass part, through walls or ceilings (floors if you are on top). What does this tell you about the natural frequency of the walls?

The Doppler Effect

•15. Does the Doppler effect apply to all waves? Explain.

16. Is there a significant Doppler effect when a source moves at right angles toward or by a stationary observer at some distance?

•17. If a sound source and an observer are both moving with the same velocity, is there a Doppler effect? Explain.

18. Binary stars revolve about each other. How is this detected using the Doppler effect?

19. It is known that the Sun rotates. How might this be determined using the Doppler effect?

•20. Does a sonic boom occur only when a jet plane "breaks the sound barrier"?

21. How fast would a "jet" fish have to swim to create an "aquatic boom"? (Give a numerical answer.)

Figure 15.17
The singer. See Question 12.

22. Associate some Mach numbers with the adjectives *subsonic, sonic,* and *supersonic.*

Sound and Music

*23. Do beats have anything to do with the beat of the music? Explain.

24. Why is an echo not as loud as the direct sound?

*25. Can there be a negative dB level, for example, −10 dB?

26. Estimate the dB levels that might be heard from the sound sources shown in Figure 15.18.

27. On the low-frequency strings of instruments, a second wire is wrapped around the main wire. What is the purpose of this?

*28. What is the effect when the base of the handle of a vibrating tuning fork is placed against a tabletop? Is a similar effect used in musical instruments? Explain.

29. Explain how musicians tune instruments using beats.

*30. Suppose you break a string on a guitar and the only spare available is lighter than the broken string. Using this string in a pinch, how would you compensate so that the string might produce the proper tone?

31. What is the relationship between octaves and overtones on a string instrument?

32. Guitar keyboards have raised ridges or frets to show where the fingers should be placed, but violins do not. Why do you suppose this is the case?

33. Give some examples of subjective sensory effects or differences for the senses other than hearing.

*34. What is the difference between a pure tone and a musical note?

35. Normal voice sounds range in frequency from 500 to 5000 Hz. High-fidelity music requires a much greater frequency range. Why?

*36. Stereophonic music is recorded with two microphones and sound tracks and is played back through two speakers. How does this provide better quality or listening? Quadraphonic systems use four microphones, two forward and two back. How does this enhance the sound?

Figure 15.18
Intense sound levels? See Question 26.

ELECTRICITY
AND MAGNETISM

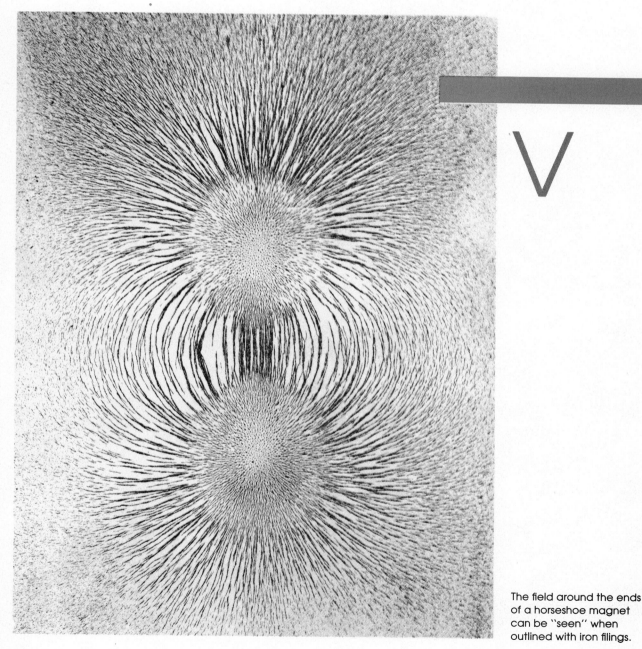

V

The field around the ends
of a horseshoe magnet
can be "seen" when
outlined with iron filings.

PART V

Gravity (Chapter 5) is one of the four fundamental forces that we experience daily. The other fundamental force that can be experienced outside of the nucleus is the electromagnetic force. We usually separate the electromagnetic force into an electric force and a magnetic force for convenience of study, but these forces are closely interrelated. Basically, you can't have a magnetic force without an electric force, as will be learned in this section.

In an overall sense, gravity holds the world together on a macrocosmic scale—the solar system and the universe (although the latter is expanding). As was learned in Part Two (Properties of Matter), the electric force is similarly important in holding matter together on a microcosmic scale—atoms and molecules. Even so, gravity and electromagnetism are used in many practical applications.

In this part of our study, we will concentrate on the aspects and applications of electricity and magnetism. Consider this: How would your life be without the use of electricity and magnetism? We get glimpse experiences of this when our domestic or business electrical service goes off for short periods of time. Indeed, we have become an electric and magnetic society. For example, our automobiles would not run, or even start, without these effects. Many people might consider this part of our study the most important part of practical physics. See if you do.

Electrostatics — Charges at Rest

16

Lightning flashes illuminate the Kitt Peak National Observatory in a 45-second exposure. (© 1972 Gary Ladd)

To kick off the study of electricity, we will start with the simplest case: charges at rest, or electrostatics. In this chapter you will find that many of the principles studied previously apply to electricity, for example, force, field, work, and energy. Electricity and gravity (Chapter 5) are the most common fundamental interactions we experience. There are many similarities in the principles used to describe them, but there are also differences, since they result from different fundamental properties. Let's take a look.

■ ELECTRIC CHARGE AND FORCE

What is electricity? Suppose you were asked this question. What would you say? We use the term *electricity* frequently, but many people would have trouble telling you what it is. A typical answer is that it is a current. But what's a current? Finally, they get around to saying that it has something to do with electric charge. If you ask what is electric charge, an appropriate answer would be that **electric charge is a fundamental property**.

Basically, we don't know what electric charge is, only what it does. (The term *fundamental* hides our ignorance.) Electric charge is involved in a fundamental interaction called the electric force, just as mass is involved in the fundamental gravitational interaction or force. But on a comparative scale, the electric force is billions upon billions, even trillions upon trillions, of times stronger than gravity.

Electric charge is associated with the atomic particles, the electron and proton. As was learned in previous chapters, it is the electric forces between these particles that hold solids and liquids together. The electron and the proton have the same magnitude of electric charge, which is measured in the standard unit of the **coulomb** (C). See Table 16.1. Even though the particles have the same size or magnitude of charge, there is a difference, as evidenced by the electric force interactions. Unlike the gravitational force, which is always attractive or

seeks to bring masses together, the electric force can be either attractive or repulsive. This is expressed in the **law of charges**:

Like charges repel, and unlike charges attract.

To designate this difference, we speak of positive (+) and negative (−) charge (Fig. 16.1). Negative charge is arbitrarily associated with that of the electron and positive charge with that of the proton. The charge of the electron (or proton) is the smallest unit of free charge we have ever found in nature. All multiple charges, commonly designated by the letter *q*, are whole-number or integer multiples of the charge of the electron.

■ COULOMB'S LAW

The magnitude of the electric force between two charges is given by an expression discovered by the French physicist Charles A. deCoulomb (1736–

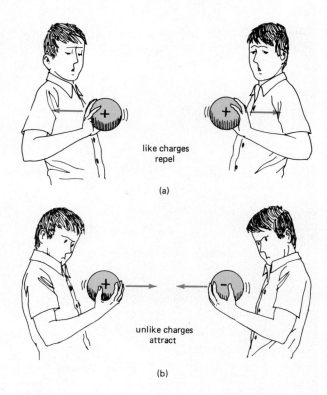

like charges
repel

(a)

unlike charges
attract

(b)

Figure 16.1
The law of charges. This law gives the direction of an electrostatic force acting on a charge or particle.

Table 16.1
Particles and Electric Charge

Particle	Electric Charge	(Mass of Particle)
Electron	$e^- = -1.6 \times 10^{-19}$ C	$(m_e = 9.11 \times 10^{-31}$ kg$)$
Proton	$p^+ = +1.6 \times 10^{-19}$ C	$(m_p = 1.67 \times 10^{-27}$ kg$)$

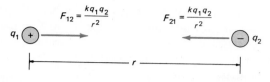

Figure 16.2
Coulomb's law. Coulomb's law gives the magnitude of the equal and opposite forces between two static electric charges.

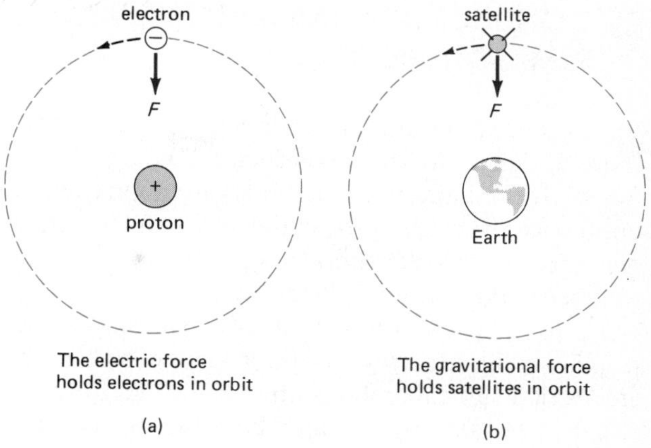

The electric force holds electrons in orbit

(a)

The gravitational force holds satellites in orbit

(b)

Figure 16.3
Electric and gravitational forces. The electric and gravitational forces may perform similar functions, such as supplying the centripetal acceleration for (a) an orbiting electron in an atom (an oversimplified model) and (b) a satellite orbiting the Earth.

1806), after whom the standard unit of charge is named:

$$F = \frac{k q_1 q_2}{r^2}$$

This expression is known as **Coulomb's law**. The q's are the charges (in coulombs), and r is the separation distance of the charges (Fig. 16.2).

Notice how similar this formula is to that of Newton's law of gravitation, $F = G m_1 m_2 / r^2$. (Is Nature trying to tell us something?) There is a big difference in the constants G and k. Recall that the universal gravitational constant G is 6.67×10^{-11} N-m²/kg². The Coulomb constant k is 9.0×10^9 N-m²/C². This is a much larger number and tends to indicate that the electric force is much stronger than the gravitational force. However, the comparative magnitudes of the masses (m's) and charges (q's) must also be considered.

Another obvious difference between these forces is that gravity only attracts, whereas electric forces may either attract or repel. The vector direction of the Coulomb force is given by the law of charges.

The gravitational and electric forces may perform similar functions (Fig. 16.3), but each has its own place in the scheme of things. The gravitational force provides the centripetal force to hold a satellite in orbit, and the electric force provides the centripetal force to hold an electron in orbit in an atom. There is also a gravitational force between the electron and proton (mass particles), but this is very small compared with the electric force between the particles and is usually neglected. We often refer simply to charges rather than to particles or masses with charges.

Question: Just how much stronger is the electric force than the gravitational force?

Answer: To get a comparison, consider the situation of the simple (hydrogen) atom, as illustrated in Figure 16.3. Then, the electrical and gravitational forces are:

$$F_e = \frac{k p^+ e^-}{r^2} \quad \text{and} \quad F_g = \frac{G m_p m_e}{r^2}$$

where the m's are the masses of the particles and r is their separation distance. When we form a ratio, the r's cancel, and

$$\frac{F_e}{F_g} = \frac{k p^+ e^-}{G m_p m_e}$$

Putting in the values of k and G and the magnitudes of the charges and masses from Table 16.1, we find that

$$\frac{F_e}{F_g} = 10^{40}$$

or

$$F_e = 10^{40} F_g$$

This means that the electrical force (F_e) is 10,000 trillion, trillion, trillion times the gravitational force (F_g)! That's even bigger than the national debt, which is measured in trillions (10^{12}) of dollars.

CONDUCTORS, INSULATORS, AND SEMICONDUCTORS

Just as there are thermal conductors and insulators, there are also electrical conductors and insulators. Recall from Chapter 12 that metals are good thermal conductors because some of the electrons in a metal are "free" to move around (not permanently bound to a particular molecule or atom).

The electrical conductivity of a material also depends on the electron bonding or how tightly the atoms of a substance hold on to their electrons. In some materials, such as metals, some of the outer electrons in the atoms are relatively free to move. Such materials are called **conductors**.

In other materials, such as wood or glass, there is little electron mobility. The electrons are firmly attached to their atoms. These materials are called **insulators** (poor conductors). Ionic solids, such as NaCl, are poor conductors because the transferred electrons are tightly bound to the ions, and the ions are "fixed" in their lattice positions (Chapter 8). However, they do become good conductors when melted or dissolved in water. In these cases, the ions (charged particles) are free to move in the liquid.

In some materials, the charge mobility is in between that of good conductors and poor conductors. Such materials are called **semiconductors**. In some instances we "dope" or add impurity atoms to poor crystalline conductors, such as pure germanium and silicon, so they become semiconducting.

These solid-state semiconductors have given rise to a whole new field of solid-state electronics with component names like *transistors* and *silicon chips*. You probably own a "transistor" radio, and *solid-state circuitry* is now a common term. For example, automobiles now have solid-state ignitions. Notice where they "fit in" in terms of the relative conductivity in Figure 16.4.

An important semiconductor application involving electrostatics is discussed in Special Feature 16.1.

ELECTROSTATIC CHARGING

What does it mean when something is electrically charged, and how do we do it? Being made up of atoms yourself, you contain a bunch of electric charges, but ordinarily you are not "charged." An atom is electrically *neutral*. That is, it has an equal number of protons and electrons, or just as many plus charges as negative charges. However, if we transfer a charge (an electron) from one atom to another, as in ionic bonding, we have a *charged* particle or ion.

Hence, an electrically charged object involves an imbalance of electrons and protons. If an object has more electrons than protons, it is negatively charged. If it has more protons than electrons, it is positively charged. The difference or imbalance of charge is called the *net charge*. For example, a so-

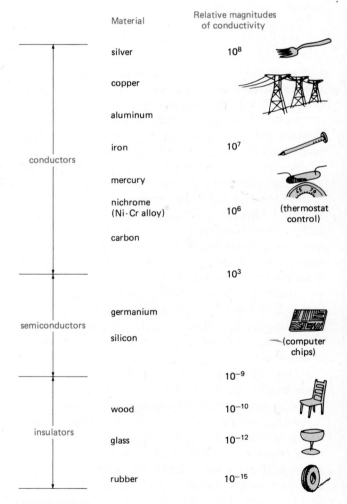

Figure 16.4
The relative conductivity of some materials. Materials are generally divided into conductors, semiconductors, and insulators.

dium ion (Na⁺), having lost an electron, has a net charge of $+1$ (one more proton than electrons). If the electron is transferred to a chlorine atom, the chlorine ion (Cl⁻) has a net charge of -1.

The total net charge of the neutral sodium and chlorine atoms together (or separately) is zero. The total net charge of the ions together is still zero $(+1 - 1 = 0)$. The charge was simply transferred or moved. The fact that the net charge of a system is constant is called the **conservation of charge**. (The net charge may be other than zero, but it remains constant.)

The conservation of charge means that charge can be neither created nor destroyed *or* that charge must be created or destroyed in pairs. It was once thought that only the first part of the statement was true, but the latter part also applies. Charge pairs can be created or destroyed. Oppositely charged particle-antiparticle pairs can be created from energy. When a charged particle and its oppositely charged antiparticle (Chapter 7) interact, they are both annihilated or destroyed, with the mass going into radiant energy or other particle pairs (conservation of mass-energy). So, on a big scale, the net charge of the universe is constant (but no one knows for sure what this net charge is).

Charging by Friction

Have you ever shuffled across a carpeted floor and been zapped by a spark when you reached for the doorknob? (Then maybe you tried it by bringing your finger close to the ear or nose of a friend or your little brother or sister.) This happens because you are electrically charged, and there is enough electric force to cause the air molecules to be ionized (electrons freed from molecules). There is then a flow of charge, giving rise to a spark. Or, you may have stroked a cat and heard the crackle (mini-thunder?) or seen the sparks in the dark. In both cases, electrons were transferred by friction from one object to another, giving each a net electric charge.

The outer electrons of the atoms of some materials are loosely bound and can be freed and transferred to another object. For example, when a hard rubber rod is rubbed with a piece of fur, electrons are transferred to the rod from the fur. With an excess of electrons, the rod is negatively charged. (How about the fur?)

Similarly, when a glass rod is rubbed with silk, the glass gives up electrons to the silk and becomes positively charged (deficiency of electrons). Since rubber and glass are poor conductors, the charge stays on the rods for some time before leaking off, if the air is dry.

Question: Why do static electricity effects, like sparks jumping to doorknobs, occur best on dry days?

Answer: Charging by friction occurs best in dry air. In moist, humid air an invisible film of water condenses on objects, which makes the surfaces conductive. Charges quickly leak off, and there is no buildup of charge.

You can demonstrate with an electroscope whether there is a net charge on the rods (Fig. 16.5). For example, if you bring a negatively charged rod

Figure 16.5
An electroscope. A common type of electroscope used to demonstrate electric charge. Here, a rubber rod that has been rubbed with fur is tested.

Xerography and Electrostatic Copiers

Xerography is a dry process by which anything written, typed, printed, or photographed can be copied. The word is coined from the Greek *xeros*, meaning "dry," and *graphein*, meaning "to write."

The heart of xerography is the photoconductor. Some semiconductors are light-sensitive, or photoconductors. In the dark, a photoconductor such as selenium is a good insulator and hence can be electrostatically charged. However, when light strikes a photoconductor, it becomes conductive and permits the electric charge to leak away from the illuminated part.

In transfer electrostatic copying, a photoconductor-coated plate, drum, or belt is electrostatically charged and then illuminated with a projected image of the writing or object to be copied (figure at right). The illuminated portions of the plate become conducting and discharge.

The plate then comes into contact with a negatively charged black powder called toner or "dry ink." The charge-retaining portions of the plate attract the toner, and it adheres to these regions. Paper is then placed over the plate, and a positive charge is delivered to the back of the paper. This attracts the toner from the plate surface to the paper. The image is then fused into the paper permanently by heat, and a copy of the original image is produced.

Electrostatic copying. Xerography process steps and an electrostatic copier. This Xerox copier/duplicator uses a photoconductive semiconductor to make copies.

How Xerography Works

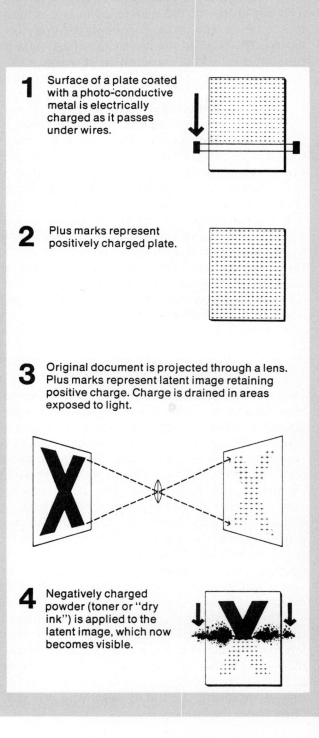

1 Surface of a plate coated with a photo-conductive metal is electrically charged as it passes under wires.

2 Plus marks represent positively charged plate.

3 Original document is projected through a lens. Plus marks represent latent image retaining positive charge. Charge is drained in areas exposed to light.

4 Negatively charged powder (toner or "dry ink") is applied to the latent image, which now becomes visible.

5 Paper (or other material) is placed over plate and given positive charge.

6 Positively charged paper attracts dry ink from plate, forming direct positive image.

7 Image is fused into the surface of the paper or other material by heat for permanency.

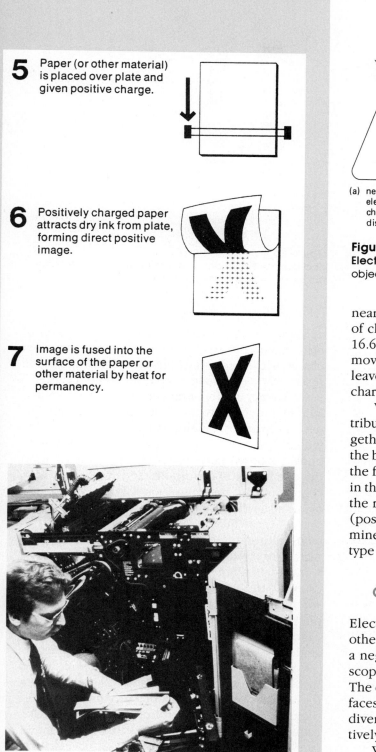

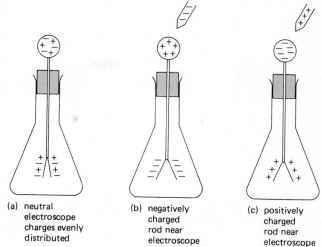

(a) neutral electroscope charges evenly distributed

(b) negatively charged rod near electroscope

(c) positively charged rod near electroscope

Figure 16.6
Electric charge. An electroscope can be used to tell if an object is electrically charged.

near the metal bulb of the electroscope, by the law of charges electrons in the bulb are repelled (Fig. 16.6). Since the metal is a conductor, the electrons move as far away as possible—to the metal foil leaves. The leaves, which now have excess negative charges, are repelled apart or diverge.

When the rod is removed, the electrons redistribute over the conductor, and the leaves come together. If a positively charged rod is brought near the bulb, electrons are attracted to the bulb, leaving the foil leaves positively charged, and they diverge in this case too. The electroscope tells you whether the rods are charged but not how they are charged (positive or negative). However, this can be determined by charging the electroscope with a known type of charge.

Charging by Contact

Electrons can be transferred from one object to another simply by touching or contact. For example, if a negatively charged rod is touched to an electroscope bulb, electrons are transferred to the bulb. The electrons spread to all parts of the metallic surfaces, and with an excess of electrons the leaves diverge (Fig. 16.7). The electroscope is then negatively charged.

With a charged electroscope you can now tell whether a rod is negatively or positively charged. For example, with a negatively charged electro-

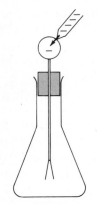 

(a) electroscope touched with negatively charged rod; charges transferred to bulb

(b) electroscope is negatively charged (net negative charge)

Figure 16.7
Charging by contact. Knowing the charge on the rod, the charge of the electroscope is known.

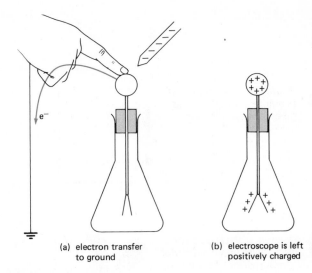

(a) electron transfer to ground

(b) electroscope is left positively charged

Figure 16.8
Charging by induction. A charge opposite to that on the rod is induced on the electroscope.

scope, if a negatively charged rod is brought near the bulb, the leaves will diverge further. If a positively charged rod is brought near the bulb, the leaves will come together. This demonstrates that there are two types of electric charges and repulsive and attractive forces.

Charging by Induction

An object can be charged without friction or contact of a charged rod but simply by bringing a charge close by. For example, suppose a charged rod is brought near an electroscope bulb, as shown in Figure 16.8. Touching the bulb with the finger "grounds" the electroscope.* This provides a path for the electrons to get farther away from the charged rod, and some of them are conducted to ground.

When the finger is removed, the electroscope is left with a deficiency of electrons, and it is positively charged. How could you test to see if it is positively charged using the negatively charged rod?

Charging by induction need not involve a removal of charge. An object can have regions of charges on different surfaces. For example, if a charged rod is brought near an insulated metal ball,

---

* Electrical ground refers to earth (ground) or some other large conductor that can receive or supply electrons without significantly affecting its own electrical condition. By analogy, this is like adding a cup of water to or taking a cup of water from the ocean.

as shown in Fig. 16.9(a), there is a separation of charge and a net electric force, even though the net charge of the ball as a whole is zero. The induction results in a *polarization* or separation of charge.

All materials, including insulators, can be polarized by induction to some degree. Some materials have permanent dipoles (Chapter 7, a molecule with different regions of charge), and molecular dipoles can be *induced* in others. Induction results in the orientation of the dipoles and opposite surface charges of equal magnitude on opposite sides of the material, as illustrated in Fig. 16.9(b).

This principle is used in a capacitor, which can be used to store electrical energy. When two metal plates are connected to a battery, the battery does work in charging the plates, and we say energy is "stored" in the capacitor [Fig. 16.9(c)]. However, when an insulating material, called a dielectric, is placed between the metal plates and the capacitor is charged, work also goes into inducing dipoles (separating charge) or orienting permanent dipoles.

When the charged capacitor is disconnected from the battery, the "work" is stored in the capacitor as electrical energy. Depending on the type of dielectric material, the amount of energy stored because of work done in inducing or orienting dipoles can be many times that of charging a capacitor without a dielectric.

Some common examples of combinations of charging by friction and induction are shown in Fig-

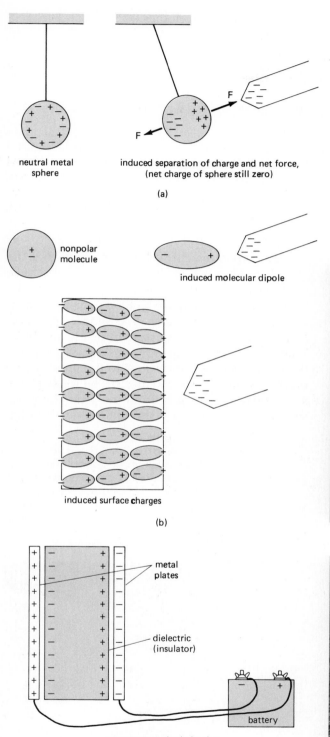

ure 16.10. When you rub an inflated balloon on your hair (or sweater), it will stick to a wall or ceiling. The balloon is charged by friction, and the charged balloon induces an opposite charge on the wall, which provides the attractive electric force. Similarly, after being run through the hair, a hard rubber comb attracts bits of paper.

"Static electricity" can be undesirable as well as used practically. Medical personnel in operating rooms wear shoe coverings with conducting strips to prevent electrostatic sparking in the presence of flammable gases. You can now buy products to pre-

Figure 16.9
Polarization or separation of charge by induction. (a) In a metal ball. (b) In molecular dipoles. (c) Energy storage by induction in a capacitor with a dielectric.

Figure 16.10
Examples of electrostatic forces. Through charging by induction, (a) balloons stick to walls and (b) bits of paper are picked up by a rubber comb.

Figure 16.11
Products to prevent static cling.

vent "static cling" in clothes (Fig. 16.11). The electrostatic precipitator discussed later in Special Feature 16.3 is an example of a practical use of electrostatics.

LIGHTNING AND LIGHTNING RODS

During the development of a storm cloud, there is a separation of charge, presumably from the frictional separation of water droplets. This causes regions of different charge in the cloud (Fig. 16.12). The bottom of the cloud is generally negatively charged, and an opposite charge is induced on the surface of the Earth.

Air is a poor electrical conductor, but when the charge buildup is great enough, the electric force ionizes the air, and lightning occurs. This can be from cloud to cloud or from cloud to Earth, primarily a tree or tall building—the shortest distance. The downward ionized path is nearly invisible. When this makes contact with positive charges near the Earth, many electrons flow to the ground. The bright lightning flash or return stroke we see actually propagates *upward* from the ground and results from electron flow to the cloud along the origi-

Figure 16.12
Lightning. Lightning results from a separation of charge in a storm cloud. The ionization of the air allows the conduction of electric charges between a cloud and Earth or from cloud to cloud. Ben Franklin was one of the first to demonstrate the electrical nature of lightning.

The Lightning Rod

One of the practical outcomes of Franklin's work with lightning was the lightning rod, which was described in *Poor Richard's Almanac* in 1753. It consists simply of a pointed metal rod that is attached to the top of a building; the rod is connected by a wire to a metal rod driven into the Earth (grounded). Franklin wrote that the rod "either prevents the stroke from the cloud or, if the stroke is made, conducts the stroke to Earth with the safety of the building" (see figure at right).

The idea that a lightning rod may prevent a lightning stroke comes from the fact that charges tend to accumulate at points or sharp edges, and charges would leak off or be attracted here. This could conceivably leak off the cloud-induced charge and prevent lightning from occurring, at least in the rod area.

However, the general view today is that the main function of a lightning-rod system depends on the principle of establishing contact with the downward ionized "leader" from a cloud before it reaches the structure to be protected and discharging the charge flow harmlessly to ground.

Shazam! Lightning strikes the Empire State Building, which has an elaborate lightning "arrestor" system.

nal ionized path. We visually perceive the stroke to be heading downward.

Ben Franklin is often said to have been the first to demonstrate the electrical nature of lightning. In 1750 he suggested an experiment using a metal rod on a tall building. But a French physicist named d'Alibard set up the experiment first and drew sparks from a rod during a thunderstorm. Franklin later performed a similar experiment using a kite. He drew sparks with his knuckles from a key hanging at the end of the kite cord (Fig. 16.12).

Ben was extremely lucky that he wasn't electrocuted. Under no circumstances should you try to duplicate this experiment. Others who have were killed. An insulator, such as a tree, is usually shattered when struck by lightning, whereas a conductor may melt and fuse. People near a lightning strike are often knocked unconscious by the electrical shock. If you come upon someone who has been "struck by lightning" and is still alive, give mouth-to-mouth resuscitation if the person is not breathing and keep him or her warm to treat physical shock.

Question: Is there any truth in the common sayings, "Lightning never strikes twice in the same place" and "Rubber tires on cars make them safe from lightning"?

Answer: These are general misconceptions. Lightning can and does strike in the same place. For example, the Empire State Building is struck, on the average, about 23 times a year. As you can imagine, this building has an elaborate lightning-rod system (Special Feature 16.2).

The National Weather Service tells us that there are *no* absolutely safe places from lightning, although enclosed automobiles and buildings are among the safest. Perhaps rubber tires, being insulators, make people feel safer (like wearing rubber-soled shoes). But lightning, which travels kilometers through insulating air, would not be halted by a few inches of rubber. The lightning would travel through the metal frame, then jump to ground through the air, or along a wet tire surface, or through the tire itself. The latter is a greater possibility now with steel-belted radial tires.

ELECTRIC FIELD

Recall from Chapter 5 that another way of looking at the gravitational interaction is to consider the possible effect on a mass rather than the cause. This was done in terms of gravitational field in space, which

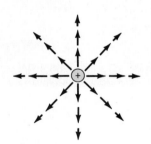

electric field vectors near positive charge

(a)

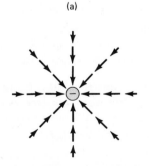

electric field vectors near negative charge

(b)

Figure 16.13
Electric field vectors. The electric field points away from a positive charge (a) and toward a negative charge (b). That is, it points in the direction of the force that would be experienced by a positive charge.

described the force properties around a mass or mass distribution. We can think of any other mass as interacting with the field rather than with the mass producing it. For example, the gravitational effects on a satellite can be described by the Earth's gravitational field without considering the Earth itself.

Similarly, an electric charge produces an **electric field** around it that interacts with any other charges in the field. Like a gravitational field, the electric field can be represented by vector arrows (Fig. 16.13). The electric-field arrows represent the electric force per unit charge, or

$$E = \frac{F}{q}$$

Electric field = force/charge

This is the force per charge brought into the field, *not* on the charge producing the field. The units of electric field are newton/coulomb (N/C).

Since the electric force can be either attractive or repulsive, the direction of the electric field is designated to be in the direction of the force that a positive charge would experience at that location. We say that an electric field is mapped out by placing a small, positive test charge at various positions. In Figure 16.13 the arrows are away from the positive charge, as prescribed by the law of charges for a positive test charge. If the charge producing the field is negative, the electric-field arrows are in the opposite direction, or toward the charge.

The field arrows can be joined to form "lines of force" (Fig. 16.14). The closer together the lines of force are, the stronger the electric field in that region. Thus, a positive charge in an electric field would experience an acceleration in the direction indicated by the arrowheads. A negative charge in an electric field would experience force in the opposite direction. Why?

The electric fields for an electric dipole and charged parallel plates are shown in Figure 16.14. Notice that the electric field between charged parallel plates is uniform (the same everywhere, except in the fringe areas near the edges). This is a capacitor without a dielectric. We say that the energy resulting from the work done in charging the plates is stored in the electric field. Remember that we talked about electromagnetic radiation propagating and carrying energy through space. This is a wave combination of an electric field and a magnetic field (discussed in a later chapter) and is another example of energy being stored in a field.

An important environmental application of electric fields is given in Special Feature 16.3.

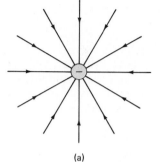

(a)

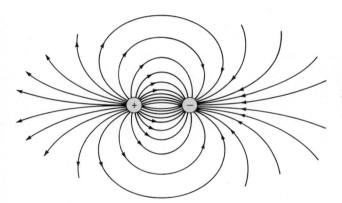

(b)

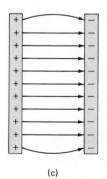

(c)

Figure 16.14
Lines of force. When the electric field vectors are joined, lines of force are formed for (a) a point charge, (b) an electric dipole, and (c) charged parallel plates.

■ ELECTRIC POTENTIAL (VOLTAGE)

We use electrical energy every day to do work. This energy is available when we need it, so it must be stored as potential energy. To visualize how this energy is stored by charges, consider the situation in Figure 16.15. Remember, potential energy is the energy of position.

When a charge is moved toward a like charge or away from an unlike charge, work is required against the electric forces. This is like doing work in compressing and stretching electrical "springs." As a result, the electrical charges have potential energy. If a charge is released, it moves or accelerates as its potential energy is converted into kinetic energy.

Rather than dealing with total potential energy in practical electrical cases in which there are numerous charges involved, it is more convenient to

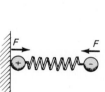

(a)

(b)

Figure 16.15
Electric potential energy. (a) When like charges are brought closer together, or (b) unlike charges are separated, work is done against the electric force, which is "stored" as electric potential energy. This is analogous to doing work in compressing or stretching electrical "springs", as illustrated by the fellow applying an external force.

The Electrostatic Precipitator

A practical application that uses an electric field and electric force is the Cottrell or electrostatic precipitator, which is used to remove particulate matter from flue gases. The developmental work on this type of precipitator was done by F. G. Cottrell, an American physical chemist, in the early 1900's and was used in the smelting industry in 1912.

The basic principle of an electrostatic precipitator is illustrated in Figure 1. The positively charged plates are called discharge plates, and the negatively charged plates are called collection plates. As a result of a large electric field between the plates, a corona discharge occurs around the needle projections on the positive plates. (Some precipitator designs use wires as positive discharge electrodes.)

The particulate matter, such as soot, in flue gases passing through the precipitator chamber is ionized. The positively charged particles then move toward and collect on the negative collection plates. The particulate matter accumulates and falls to the bottom of the chamber, where it is removed.

Electrostatic precipitators are capable of removing more than 90 percent of the particulate matter from flue gases (Fig. 2). They are commonly used in fossil-fuel electrical generating plants and industrial operations in which there is a large amount of particulate matter in the flue gases resulting from incomplete combustion and impurities.

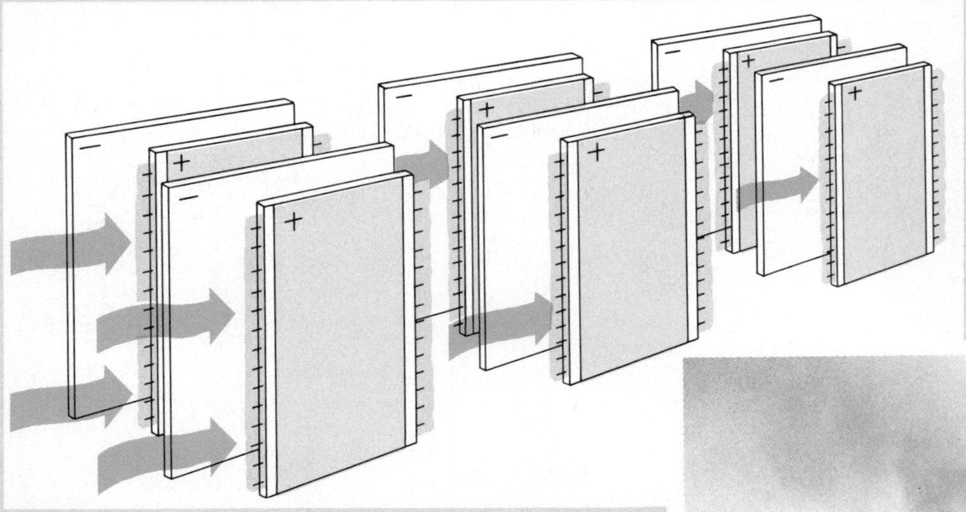

Figure 1
Electrostatic precipitation. The large electric field between oppositely charged plates gives rise to corona discharge at the needle projections on the positive plates. Particles in the flue gases passing through between the plates are ionized, attracted toward, and collected on the negative plates. Other designs use wires instead of positive plates.

Figure 2
They're good! Electrostatic precipitators are extremely efficient. The two smokestacks on the left are equipped with precipitators; the two smokestacks on the right are not.

consider the electric potential energy per charge, or the **electric potential,**

$$\text{Electric potential} = \frac{\text{potential energy}}{\text{charge}}$$

Keep in mind that the electric potential is not the same as electric potential energy, but rather the (electric) potential energy per charge. Perhaps to avoid any confusion, the electric potential is commonly called **voltage.** This is easy to remember, as the unit of electric potential is the **volt** (V).*

You are no doubt familiar with these common terms. We have 12-V batteries in our cars, and lamps and other appliances in our homes operate on a voltage of 110 V. More about this later. Here, we're getting the basic fundamentals and electrical jargon.

Incidentally, it is interesting to note that the volt and other common electrical units are the same in the SI metric and British systems. So, electrically, we have already converted to SI units.

THE VAN DE GRAAFF GENERATOR

An application that uses the principles discussed in this chapter is the Van de Graaff electrostatic generator, which was invented by the American physicist Robert Van de Graaff around 1930. It generates electrostatic potentials of *millions* of volts. The high voltages are used to accelerate charged particles that can be used to probe the nucleus and to produce X-rays for medical and industrial applications.

The basic design of the Van de Graaff generator is shown in Figure 16.16. Electric charge is "sprayed" on a motor-driven insulating belt by a corona or spark discharge from a row of metal points. The charges are deposited on the inside of a metal sphere and, being repelled from each other, are distributed over the outside of the sphere. This leaves the inside uncharged and able to receive more charge from the belt. The operation results in the buildup of a huge charge on the sphere and the "stored" electric potential of millions of volts.

When the sphere is charged to a high enough potential, corona discharge can occur through the ionization of air (Fig. 16.17). At STP (standard temperature and pressure, 0°C and 1 atm), the electric

---

* The volt is named in honor of Alessandro Volta (1745–1827), an Italian scientist who invented one of the first batteries.

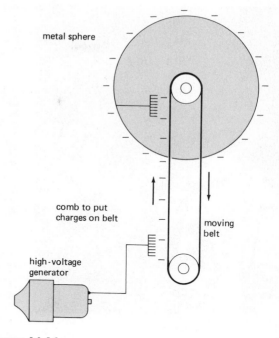

Figure 16.16
A diagram of a Van de Graaff generator. The accumulated charge on the sphere can result in an electric potential of millions of volts.

Figure 16.17
Big volts! Ionization discharge from a Van de Graaff generator. The ionization of air requires an electric field of about 3 million volts per meter.

(a)

(b)

Figure 16.18
Effects of high Van de Graaff voltages. (a) A human electroscope.
This young fellow is insulated, and when touching the Van de Graaff
sphere, he is charged to the same potential. (b) An electric field
strong enough to light a fluorescent bulb. The lamp only lights out to
the hand. The person can "turn down the lights" simply by moving
the hand along the tube toward the generator.

field required for the ionization of air is about 3
million volts per meter (V/m).*

Although the corona discharge is rather spectacular, it is usually undesirable, since it limits the voltage or amount of charge that can be accumulated on

---

* In addition to N/C as a unit of electric field, there is
volt/meter (V/m). This comes from the relation $E = V/d$
(volt/meter).

the sphere. The discharge can be suppressed and
higher voltages achieved by surrounding the
charged sphere with a gas of high ionization potential under high pressure. Electric fields greater than
10 million volts per meter can then be achieved.

Some other effects of the high voltage of a Van
de Graaff generator are shown in Figure 16.18. Notice the human electroscope with the hair taking the
place of the leaves.

SUMMARY OF KEY TERMS

Electric charge: the fundamental property associated
with electric force.

Law of charges: like charges repel and unlike
charges attract.

Coulomb's law: the relationship that gives the
electrostatic force between two charges,
$F = k\, q_1 q_2 / r^2$.

Conductor: a material in which electrons are free to
move.

Insulator (nonconductor): a material in which there
is little electron mobility.

Semiconductor: a material with charge mobility
between that of a good conductor and a poor
conductor (insulator).

Conservation of charge: the net charge of a system is
constant.

Electrical induction: the redistribution of charge in
an object due to the influence of a charged body
nearby but not in contact.

Electrical ground: the Earth (ground) or some other
large conductor that can receive or supply
electrons without significantly affecting its own
electrical condition.

Electric field: the force field surrounding a charged
body. Electric field = force/charge, in the
direction a positive charge would experience a
force.

Electric potential: the electric potential energy per
unit charge. This is commonly called **voltage**
and is measured in volts.

QUESTIONS

Electric Charge and Force

1. Two negative charges of −1 C each are placed at opposite ends of a meterstick. (a) Could a free electron be placed somewhere on the meterstick so it would be in static equilibrium (zero net force)? How about a proton? (b) Could an electron or proton be on a line with but not on the meterstick so it would be in static equilibrium? Explain.

•2. A charge of +1 C is placed at one end of a meterstick and a charge of −1 C at the other end. (a) Could a free electron be placed somewhere on the meterstick so it would be in static equilibrium (zero net force)? How about a proton? (b) Could an electron or a proton be placed on a line with but not on the meterstick so it would be in static equilibrium? Explain.

•3. About how many electrons would it take to make up a charge of 1 C? How many protons would it take?

4. The gravitational force is weaker than the electric force. It is easy to feel or experience the gravitational force, for example, by picking up a heavy object. There are many electrons and protons around. Why don't we generally feel the electric force?

5. A large charge of $+Q$ and a small charge of $-q$ are a short distance apart. How do the electric forces on each charge compare? Is this comparison described by another physical law (cf. Chapter 1)?

6. Is the Coulomb constant k a universal constant like G? Explain.

•7. An electron is a certain distance from a positive charge. If the electron is moved twice as far from the other charge, how is the electric force affected?

8. An electrically neutral object (like yourself) can be given a net charge by several means. Does this violate the conservation of charge? Explain.

Electrostatic Charging

•9. Given an object and an electroscope, how could you tell if the object were charged? If it were, how could you tell what type of net charge it had?

10. Another type of electroscope, called a Braun electroscope, is shown in Figure 16.19. Explain the principle of its operation. Does this type of electroscope have any practical advantages over the foil-leaf electroscope?

11. When thin plastic food wrap is pulled from its roll in a box and cut off, it often sticks together. Explain why.

•12. An electroscope is negatively charged, and its leaves diverge. What would happen to the leaves (and why) if you touched the bulb with (a) your finger and (b) a glass rod that had been rubbed with silk?

13. Will a charged electroscope remain charged indefinitely? Explain.

14. How could you charge an electroscope negatively by induction? How could you prove it was negatively charged?

15. What causes static cling in clothing? Why is this more of a problem on a dry day? Why isn't it a problem with all clothing?

16. Why is dust so difficult to get off a phonograph record?

17. We can rub balloons on our hair to charge them electrostatically. How could a bald-headed person charge a balloon so that it would stick to the wall?

•18. An inflated balloon is charged and placed against a smooth ceiling. What determines if the balloon will stick to the ceiling?

19. Getting zapped by a spark when you are about to touch a doorknob involves charging by friction and induction. Explain why.

•20. What would happen in Fig. 16.9(a) if the rod is touched to the metal sphere?

21. Why must work be done in charging a pair of capacitor plates even without a dielectric

(a)   (b)

Figure 16.19
Another type of electroscope. (a) Electroscope uncharged. (b) Electroscope charged. See Question 10.

between them? Why is more work required when there is a dielectric between the plates?

22. Why must a capacitor dielectric be an insulating material?

•23. Two metal spheres on insulating rods are in contact, as illustrated in Figure 16.20. How could you charge both spheres by induction without directly touching the metal spheres? How would the spheres be charged, positively or negatively?

Lightning and Lightning Rods

24. Why is it not a good idea to seek shelter under a tree in a thunderstorm?

25. When it starts to rain, and sometimes before, lifeguards make everyone get out of a swimming pool. Is this a wise precaution, or do the lifeguards just want to take a break and are using the rain as an excuse? Explain.

26. Does lightning ever strike the same place twice? If it does, why is it likely to do so?

27. It is said that there is absolutely no safe place from lightning. Does this apply to astronauts in an orbiting spacecraft or on the moon?

•28. Do lightning rods repel lightning? Explain.

Electric Field

29. How is the direction of the force acting on a charged particle due to an electric field determined?

30. Sketch the electric field in the vicinity of two isolated (a) positive charges and (b) negative charges. (Don't forget the directional arrowheads.)

31. A net charge on a metal conductor resides on the outside of the conductor. Why is this the case?

•32. Why do electric field lines always begin on positive charges and end on negative charges?

33. How can you tell from an electric field diagram where the field is stronger or weaker?

•34. When a closed, hollow metal object, such as a metal box, is charged by induction by outside charges or is given a net charge by contact, the electric force or electric field inside the object is zero. Why? (*Hint:* See Questions 31 and 32.)

35. Question 34 is an example of electrostatic shielding. If there are charges inside the shield, these will set up their own electric field. But the

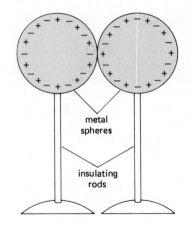

Figure 16.20
Everything's neutral. See Question 23.

field inside will be unaffected by charges outside and will experience no force because of them. Would it be possible to construct a gravitational shield? Explain. (Electrostatic shielding is common in electronic equipment where it is desired to keep the electric fields originating in one part from interfering with components in another part.)

36. Would it be safe to get inside a Van de Graaff generator sphere and then have it charged? (*Hint:* See Question 31.)

•37. Explain why the boy's hair in Figure 16.18 stands on end. Why is he insulated from ground?

38. We say that energy is stored in an electric field. How might this be described by a three-dimensional field representation? (*Hint:* See Chapter 5.)

Electric Potential (Voltage)

39. Distinguish between electric potential energy and electric potential.

•40. One charged object has twice the electric potential energy as another. Does it necessarily have twice the electric potential? Can it have?

41. Knowing the charge and voltage of an object, how could you determine its electric potential energy?

42. What are the British units for (a) electric charge and (b) electric potential?

Electric Current —
Charges in Motion

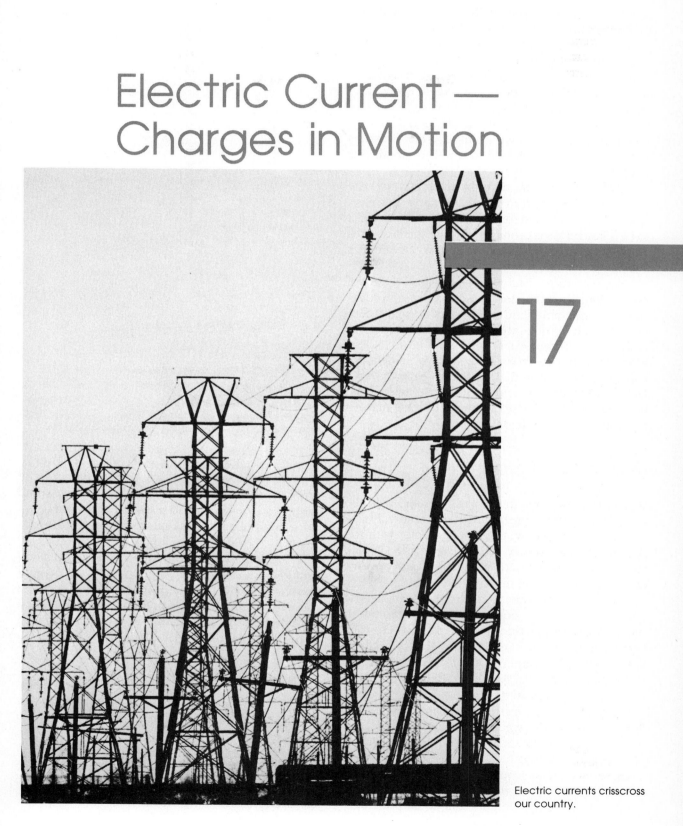

17

Electric currents crisscross
our country.

Static electricity is important, but the vast majority of electrical applications involve electric currents or charges in motion. When you turn on a flashlight or a car's headlights, there is a flow of electrons in a wire circuit. The lamp filament (a wire too) is part of the circuit, and here electrical energy is turned into heat and light (radiant energy). Similarly, when you switch on a light in the home there is a flow of electrons in a circuit, but in this case the electrons move alternately back and forth. More about this later. First, a little history.

Early investigators thought electricity was the result of different fluids—positive and negative. Recall that the early theory of heat also considered it to be a fluid. This is probably why we speak about a "flow" of charges and about heat flow. (A fluid is a substance that can flow.)

Ben Franklin advanced a single-fluid theory of electricity. He postulated the existence of a tenuous, imponderable fluid. All bodies normally contained a certain quantity of the fluid. A surplus or deficit created electrical properties. With an excess, a body was positively excited, and with a deficit it was negatively excited. This theory gave rise to a later idea that it was the positive electric charges that flowed or moved. Today, we know it's the electrons that move in a conductor.

▮ ELECTRIC CURRENT

In solid conductors, particularly metals, it is the outer atomic electrons that are free or relatively free to move. As we know, the atomic nuclei with the positively charged protons are fixed in the lattice structure. (In ionic liquid conductors, both positive and negative *ions* can move.)

But what causes electrons to move or an electric current to flow in a conductor, such as a common metal wire? Stop and think for a minute (maybe for a few seconds). What causes heat to flow or water to flow downhill? A temperature *difference* or a height (potential energy) *difference* (Fig. 17.1). In the electrical case, it is an electric potential or voltage (energy/charge) *difference*. In a sense, the electron flow or current is down an electrical "hill," from a higher to a lower potential. This could be due to an electrostatic charging.

Of course, the "flows" in Figure 17.1 would be only temporary. When thermal equilibrium (equal temperatures), equal heights, and equal potentials

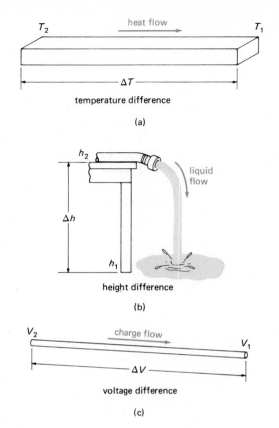

Figure 17.1
Flow. "Flows" generally require differences in conditions.

are reached, or when there are no differences, the flows cease. To have a sustained heat flow would require a heat pump and a system of thermal conductors or a circuit to maintain the temperature difference.

The same is true for water flow and electric current (Fig. 17.2).* In the water "circuit" the path for the water flow is complete, and the pump supplies the work to maintain the potential difference for a continuous flow. The water flow can do work, such as turning a paddle wheel. Similarly,

A sustained electric current requires a complete circuit and a voltage source.

Conducting wires usually complete an electric circuit, and the voltage source, which does work to

* Because of familiarity, water circuit analogies are instructive for some aspects of electrical circuits. An analogy is an explanation based on similarities or resemblances between two things. Obviously, there is not a one-to-one correlation in the circuits, and many things are different.

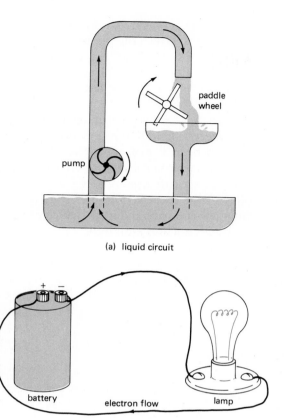

(a) liquid circuit

(b) electric circuit

Figure 17.2
Circuit analogies. For a sustained flow, a completed circuit and a method to maintain a potential difference are required for both liquid and electrical circuits.

maintain the electric potential difference, may be a battery. The electric current can light a bulb in the circuit.

In terms of charge movement, the **electric current** is the net amount of charge q that passes through an area of a conductor per time:

$$\text{Electric current} = \frac{\text{charge}}{\text{time}}$$

or
$$I = \frac{q}{t}$$

The units of electric current are then coulomb per second. This is given the name **ampere** (A), in honor of the French physicist Andre Ampere (1775–1836), an early investigator of electricity. It is common to speak of current in "amps," for short. Small currents are expressed in milliamps (mA, 10^{-3} A) and microamps (μA, 10^{-6} A).

Voltage Sources

A voltage source or electrical "pump" is required to maintain a potential difference for the flow of current. There are many types of voltage sources, which are often called "seats" of electromotive force (emf). This is an unfortunate name because electromotive force is not a force, but rather the *energy per charge* (voltage) given by the source. The most common types of voltage sources are batteries and generators.

A battery produces a voltage by chemical means. There is a certain "terminal" voltage between or *across* the battery terminals. For example, most automobile batteries are 12-V sources. When you attach the battery cables to the positive (+) and negative (−) terminals, the circuit has a "pumping potential" of 12 V. (The positive terminal has a potential of +12 V and the negative terminal is grounded or at 0 V. The voltage difference between the terminals is then $\Delta V = V_2 - V_1 = 12 - 0 = 12$ V.)

The electron flow in the circuit is away from the negative terminal toward the positive terminal. We call this **direct current (dc)** because the flow of charge is direct, or in *one direction.*

As long as the proper chemical action of the battery is maintained, the voltage is constant with time (Fig. 17.3). We often say that a car battery is a 12-VDC (12 volts–direct current or 12 volts-dc) source.

The voltage of a generator, on the other hand, fluctuates with time (see Chapter 19). It can alternate between positive and negative values (Fig. 17.3). As the polarity of the voltage alternates back and forth (positive and negative), so does the direction of the electron current. We call this **alternating current (ac)**. Common household voltage is 120 VAC (volts–alternating current or volts-ac).* The voltage frequency is 60 Hz. This means the electrons in the conductor move back and forth 60 times per second.

Alternating voltage has advantages, and its production will be considered in a later chapter. For the time being, we'll take a closer look at direct-current battery sources. Before doing that, however, let's consider current in a bit more detail.

---

* You may hear household voltage given as 110 VAC, 115 VAC, or 120 VAC. This is because the voltage varies between 110 V and 120 V. (When the abbreviation volts-ac is used, it is common to capitalize the *ac*, i.e., VAC.)

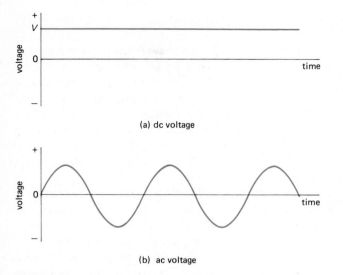

(a) dc voltage

(b) ac voltage

Figure 17.3
Voltages. (a) A dc voltage, such as that maintained by a battery, is relatively constant with time. The magnitude of a dc voltage may vary, but the polarity (+ or −) does not change, so the current is always direct or in one direction. (b) The polarity of an ac voltage, such as common household voltage, alternates.

Charge Flow

The idea of electron *flow* is convenient in describing electrical current, but don't get the idea that the electrons flow around the circuit like water through pipes. Water circuits are helpful analogies, but they are different and one difference is the "flow."

In the absence of a voltage difference, the electrons in a metal wire move around randomly and chaotically at very high speeds—on the order of 1000 km/s (about 2 million miles per hour). With the electrons moving around in all directions, there is no net flow one way or the other. When a potential difference from a battery is applied across the wire, the electrons still collide and move chaotically. However, there is a general charge movement along the wire toward the lower potential. This electron movement has an average or **drift velocity** on the order of 0.1 cm/s—not much of a flow. In the case of alternating current, the electrons move periodically back and forth and there is *no* net current flow.

Yet when you switch on a light, it comes on instantaneously. Telephone conversations in the form of electric signals travel hundreds of kilometers through telephone lines. If this depended on actual electron flow, telephone conversations would be pretty slow. What then travels through a circuit? It is the electric field. The circuit wires act as guides for the electric force field that drives the conducting electrons.

A battery or a generator connected to a circuit causes an electric field to travel through the circuit with a speed near that of light. The energy in the field does the work in the circuit. For example, the alternating electric field associated with an alternating voltage source causes the electrons in a lamp filament to vibrate back and forth, which causes the filament to get hot and emit light.

So it's really energy that flows in a circuit, producing only a slight motion of the electrons. But we still talk in terms of current and electron flow because they conveniently describe the overall electrical effects.

■ BATTERIES

A battery converts chemical energy into electrical energy in a chemical cell. The term *battery* refers to a "battery" of cells, although we often call a single cell a battery.

The Italian scientist Alessandro Volta, after whom the volt unit is named, is credited with constructing one of the first practical batteries. Basically, a battery consists of two electrodes in an electrolyte (Fig. 17.4). With the appropriate materials (Volta used zinc and copper electrodes in a dilute electrolyte solution of sulfuric acid), one electrode, called the **cathode**, becomes negatively charged, and the other electrode, the **anode**, becomes positively charged as a result of chemical processes. That is, a potential difference develops across the battery electrodes or terminals.

When a battery is connected to a circuit, the potential difference causes a flow of electrons that is maintained by chemical reactions and ionic conduction in the battery. We commonly say that the battery "delivers" current to the circuit or that the circuit "draws" current from the battery.

A wide variety of batteries are available for all the battery-operated devices in existence today. Probably the most common batteries are the single-cell 1.5-V D-cell flashlight battery and the multicell 12-V storage battery used in automobiles (Fig. 17.5). The flashlight battery is called a "dry" cell. It

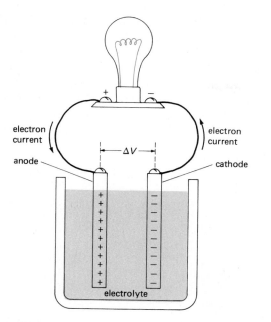

Figure 17.4
A battery. A simple battery consists of two electrodes and an electrolyte. Chemical action gives a potential difference across the terminals, and in a completed circuit a current is maintained by the chemical reactions and ionic conduction in the battery.

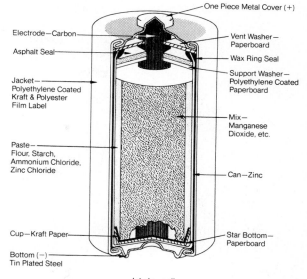

(a) dry cell

(b) storage battery

Figure 17.5
Common batteries. (a) A dry cell, such as the D-cell battery, is commonly used in flashlights. (b) An automobile storage battery. This 12-V battery is a combination of six 2-V cells.

is not actually dry, only relatively so compared with an automobile battery. The electrolyte (ammonium chloride and other additives) is in the form of a paste. A dry cell becomes "dead" when its active materials have been depleted. Most dry cells cannot be recharged.

A storage battery, on the other hand, can be recharged. In the common lead storage battery, there are many plate electrodes. The positive electrode is lead oxide (PbO_2), and the negative electrode is spongy (porous) metallic lead. The electrolyte is a water solution of sulfuric acid (H_2SO_4). When current is drawn from the battery, both electrodes change chemically to lead sulfate ($PbSO_4$), while the electrolyte is converted to water and the acid solution becomes more dilute.

The "strength" of a battery may be tested by determining the specific gravity of the electrolyte solution with a hydrometer (Chapter 9). If a current is passed through a storage battery in the opposite direction to that of the normal current flow, the battery is recharged, with the electrodes and electrolyte restored to their original conditions (reverse chemical action).

The 12-V auto battery actually consists of six 2-V cells connected in series—the positive terminal of each cell connected to the negative terminal of the next cell. When connected in this fashion, the voltages of the cells add. This is analogous to a series of pumps raising water to successively higher gravitational potentials (Fig. 17.6).

Batteries may also be connected in parallel. In this case, all the positive terminals have a common connection, as do the negative terminals (Fig. 17.7). When batteries with the same voltage are connected in parallel, the potential difference across the combination is the same as the voltage of the individual batteries. However, each battery supplies an equal fraction of the current delivered to the circuit, similar to each pump supplying a fraction of the water in the parallel water analogy. The lamp in Figure 17.7

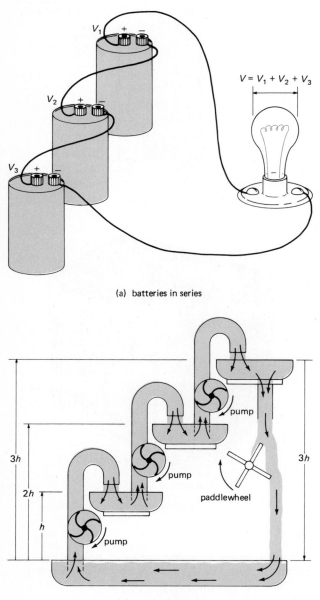

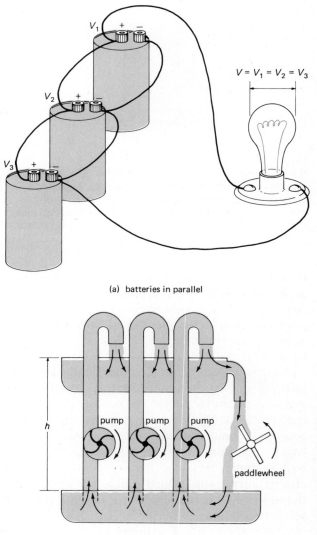

(a) batteries in series

(a) batteries in parallel

(b) pumps in series

(b) pumps in parallel

Figure 17.6
Batteries and pumps in series. Each battery or pump adds to the potential difference in the respective circuits. Notice a completed circuit is required for a sustained flow in each case. In series, the voltages of the batteries add, as do the heights or gravitational potential of the liquid. The liquid does work in turning a paddle wheel, and the electric potential lights a lamp.

Figure 17.7
Batteries and pumps in parallel. Each battery or pump contributes a fraction of the "flow" in the respective circuits. In parallel, the voltage across the lamp is the same as that of an individual battery (assumed equal). By analogy, the liquid gravitational potential is the same for each pump as delivered to the same height.

could be lighted using only one battery. An advantage of connecting three batteries in parallel would be that you could go three times as long before you had to change batteries, since each battery would supply one third of the current.

Question: Are the batteries in an ordinary flashlight connected in series or parallel?

Answer: They are connected in series by contact (Fig. 17.8). The rounded, raised piece of metal on the top of the battery is the positive terminal, and the flat metal bottom is the negative terminal (see Fig. 17.5). The batteries are put in the flashlight with plus (+) to minus (−) terminal contact, so they are in series. The spring on the base end of the flashlight (which is part of the circuit) provides pressure so there is good contact.

When you switch on a flashlight, you are making or "closing" (completing) the circuit. When you switch it off, you break or "open" the circuit. This is the same for a light switch in a house.

ELECTRICAL RESISTANCE

Motion is generally accompanied by resistance, for example, frictional resistance between objects and internal resistance or viscosity in fluid flow. Similarly, when there is charge flow, there is electrical resistance. On the atomic level, this opposition to charge flow arises from collisions between electrons and lattice atoms or ions of a material. Hence, **electrical resistance** is a material property.

But resistance also depends on such things as the size of a conductor and its temperature. For example, the resistance of a wire conductor depends on its length and cross-sectional area. Think about a pipe with fluid flow. The longer the pipe, the more resistance there is. However, the greater the diameter or cross-sectional area of the pipe, the more fluid flow there would be, indicating less fluid resistance. Similarly, for an electrical wire conductor, the longer the wire, the greater the resistance, and the greater its diameter or cross-sectional area, the smaller the resistance.

Temperature also affects electrical resistance. As a general rule, the resistance of most metallic conductors increases as the temperature increases, whereas the resistance of most nonmetallic conductors decreases as the temperature increases. The increased atomic or ion motion in metals generally hinders the flow of charge through more collisions, whereas in nonmetals (e.g., carbon and semiconductors) there is a greater flow of charge or decreased resistance. Hence, cooling a metal conductor decreases its resistance. The extreme case of superconductivity is discussed in Special Feature 17.1.

OHM'S LAW

How much current flows in a circuit depends not only on how much resistance there is in the circuit, but also on the voltage source. The greater the electric potential, the greater energy per charge there is. The relationship of the current (I), the voltage (V), and the resistance (R) for many conductors is expressed by **Ohm's law**, which was formulated by Georg Ohm (1789–1854), a German physicist:

$$\text{Current} = \frac{\text{voltage}}{\text{resistance}}$$

or $$I = \frac{V}{R}$$

Figure 17.8
Series or parallel battery connections in flashlights? See Question and Answer.

Superconductivity

The electrical resistance of a metal conductor can be decreased by cooling. You may wonder how far one can go with this decrease. Strangely enough, for certain materials you can go all the way. At relatively low temperatures some materials exhibit **superconductivity,** and the electrical resistance vanishes or goes to zero!

In this state, an electric current established in a superconducting loop would persist indefinitely, with no resistive losses. Currents of several hundred amps introduced in a superconducting ring have been observed to remain constant for several years. In a sense, the conducting electrons of a superconductor never collide with the material lattice, while a conductor at normal temperatures always has some resistance and energy loss. The trick or requirement for superconduction is maintaining materials at very low temperatures.

Superconductivity was discovered by a Dutch physicist named Heike Onnes in 1911 and was first observed in solid mercury at a temperature of about 4 K (4 kelvins above *absolute* zero, $-269\,°C$ or $-452\,°F$ — pretty cold!). The mercury was cooled to this temperature using liquid helium. The boiling point or the temperature at which helium condenses to a liquid is about $-267\,°C$ at 1 atm of pressure. (Liquefaction or an increase in the boiling point is accomplished at higher temperatures by increased pressure, cf. Chapter 12.) Lead also exhibits superconductivity when cooled to liquid helium temperatures.

Liquid helium is relatively expensive, $3 to $6 a liter, depending on the amount you buy. So there has been a search for other materials that become superconducting at higher temperatures. Over the years, other superconducting metals and alloys were found, and the critical temperature crept upward to about 18 K ($-225\,°C$ or $-427\,°F$). In 1973, this went up to 23 K ($-250\,°C$ or $-418\,°F$).

In 1986 there was a major breakthrough and a new class of superconductors was discovered with higher critical temperatures. These were ceramic "alloys" or mixtures of rare earth elements such as lanthanum and yttrium. The new superconductors were prepared by grinding a blend of metallic elements and heating them at a high temperature, which produces a ceramic material. For example, one of the new superconductors was a mixture of yttrium, barium, and copper oxide. In 1986 the critical temperature got up to 57 K ($-216\,°C$ or $-357\,°F$), and in 1987 a critical temperature of 98 K ($-175\,°C$ or $-283\,°F$) was obtained. In the first part of 1988, during the time this book was being readied for press, a new thallium compound was reported that achieves superconductivity at 125 K ($-148\,°C$ or $-234\,°F$). There have also been reports of bits of materials becoming superconducting at room temperature (293 K), but these are questioned.

Superconducting levitation. A piece of "high temperature" material with a small permanent magnet on top is cooled with liquid nitrogen. When the material becomes cold enough to superconduct, the magnet is repelled, rises, and levitates.

The 98 K material was a major breakthrough. Such "high temperature" superconductors can be cooled using liquid nitrogen. (See figure above.) Liquid nitrogen has a boiling point of 77 K ($-196\,°C$ or $-321\,°F$) and costs only about 25 to 40 cents per liter. (Nitrogen is readily plentiful as the major constituent of the air.) This is an important scientific discovery that will probably revolutionize many things. However, it's going to take a while, which many reports fail to point out.

One of the main applications of superconductors is in superconducting magnets. Magnets are used in motors (Chapter 18), and the greater the strength of the magnet, the more powerful the motor. In such electromagnets, the strength of the magnet depends on the current in the windings (wires). Without resistance, no loss occurs and there is a greater current. (Superconducting magnets have been used on ship motors for some years but with liquid helium.)

With superconducting magnets, things are more efficient and larger currents can be conducted. Potential applications are magnetically levitated trains, which are repelled off the track and ride on cushions of air. Superconducting magnets could also be used for propulsion. (Experimental "MagLev" trains have already been built using low-temperature superconducting magnets; see Chapter 19.) Other applications might be underground electrical transmission cables or electric cars.

However, it is generally thought that such applications are some distance in the future. A more immediate application will probably be in computer interconnects. Interconnects are the metallic connections between computer chips by which they "talk" to each other. Superconducting interconnects would decrease power dissipation and possibly speed up the signal transfer, which makes for faster computers. The absence of electrical resistance opens up many possibilities.

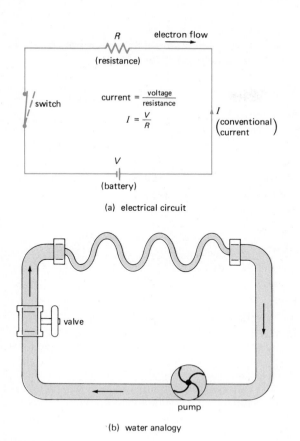

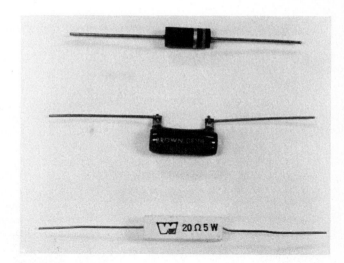

Figure 17.10
Commercial resistors. The value of the top resistor is given by the color-coded bands, whereas the values of the others are stamped directly on the resistors.

Figure 17.9
Ohm's law. (a) For a given voltage (V), the current (I) is determined by the resistance (R) in the circuit $I = V/R$. (b) A liquid-circuit analogy. The bent tube offers resistance to the liquid flow, and the pump supplies the potential. Certainly there are differences in the circuits, but the analogies may be helpful. Notice how an electrical switch is analogous to a valve.

Ohm's law shows that the greater the voltage, the greater the current flow, and the greater the resistance, the smaller the current flow.

An electrical circuit and diagram, along with a water analogy, illustrating the elements of Ohm's law are shown in Figure 17.9. Notice how the circuit elements are represented by symbols in the circuit diagram. It is customary to designate the *conventional current I* in the direction that positive charges would flow, even though the actual electron net flow is in the opposite direction—perhaps a leftover from Franklin's single-fluid theory.

The unit of resistance is the ohm (Ω, a capital Greek omega, used to avoid confusion with a capital O or zero). If the battery in Figure 17.9 were 12 volts and the lamp resistance were 6 Ω, then by Ohm's

law the current in the circuit would be $I = V/R = 12V/6\ \Omega = 2$ amps. (The connecting wires are considered to have negligible resistance.) Commercially available resistors (Fig. 17.10) are often put in circuits to affect the current flow.

It should be kept in mind that not all conductors are "ohmic" or follow Ohm's law. In general, metals are ohmic conductors, but some other materials, such as carbon, are not.

ELECTRIC POWER

When an electric current flows, the electric charge moves from a higher to a lower potential, and work is done. This results in the conversion of energy from one form (electrical energy) to another. Examples of these other forms include heat energy in an electric toaster, radiant energy in a light bulb, and mechanical energy in turning a motor.

Recall that the rate at which work is done (or electrical energy is dissipated) is power—work/time (or energy/time). **Electric power** is given by the product of the current and the voltage:

$$\text{Electric power} = \text{current} \times \text{voltage}$$

or*

$$P = IV$$

* IV = (charge/time) $\times$ (work or energy/charge)
= work or energy/time = power (P).

(a)

(b)

Figure 17.11
Power ratings. (a) The light bulb uses 60 W (60 J/s) at 120 V. (b) The power requirement of this appliance is found by taking the product of the current and voltage ($P = IV$).

Table 17.1
Typical Power and Current Requirements of Some Common Household Appliances

Appliance	Power	Current ($I = P/V$) $V = 120$ V
Air conditioner		
room	1500 W	12.5 A
central	4500 W	37.5 A
Blanket, electric	180 W	1.5 A
Blender	800 W	6.7 A
Coffee maker	1625 W	13.5 A
Dishwasher	1200 W	10.0 A
Food processor	330 W	2.75 A
Heater		
portable	1400 W	11.7 A
water	4500 W	37.5 A
Iron	1100 W	9.2 A
Microwave oven	625 W	5.2 A
Refrigerator		
regular	400 W	3.3 A
frost-free	500 W	4.2 A
Stove		
range-top	6000 W	50.0 A
oven	4500 W	37.5 A
Television		
black-and-white	50 W	0.42 A
color	100 W	0.83 A
Toaster	950 W	7.9 A

Electric power is expressed in watts (W) or kilowatts (kW). The power requirements or wattage ratings are usually indicated on electrical components (Fig. 17.11). For example, a 60-W light bulb uses or dissipates 60 J of electrical energy per second. Unfortunately, not all of the electrical energy is converted to visible light—only about 5 percent. Most of the radiation is in the infrared region. The filament temperature of a common light bulb is on the order of 2500 to 3000°C.

The electrical heating effect is sometimes called "joule heat" or I^2R losses.* The heat-energy conversion comes about from electrons colliding with the lattice atoms or ions of a conductor and the transfer of energy, which increases the internal (heat) energy of the conductor (Chapter 11).

* Ohm's law is $I = V/R$ or $V = IR$, and since $P = IV$ we can write the equation for power as (a) $P = IV = I(IR) = I^2R$ or (b) $P = IV = (V/R)V = V^2/R$.

In applications in which heating effects are wanted, as in hair dryers and electric heaters, joule heating is promoted. This is done by having a heating element of relatively low resistance (but much higher than that of the connecting wires), so there will be a large current to take advantage of the square of the current in the I^2R losses. Another way of seeing this is through the equation for power in the form of $P = V^2/R$.

For a fixed voltage, when the resistance is decreased, the power increases. A typical hair dryer may have a wattage or power rating of 1200 W. This is the energy consumption (per time) of the hair dryer. Most of this energy goes into joule heat in the dryer heating element. Some is used to turn the motor of the hair-dryer blower. The connecting wires in the hair-dryer cord also dissipate joule heat, but the resistance of the copper wire is only a very small fraction of an ohm. Even with a large current flowing in the hair-dryer circuit, the I^2/R loss of the cord is a relatively small percentage.

Typical power requirements for some common household appliances are given in Table 17.1. Simply dividing the wattage rating by 120 V gives the current an appliance draws or uses ($I = P/V$).

■ ELECTRIC CIRCUITS

As we know, a sustained current requires a complete path or circuit (along with a voltage source). When there is a break in the path, we say that it's an open circuit. Switches are convenient means of opening and closing circuits and are commonly used to turn circuit components, like appliances, on and off.

It is also important to know how components are connected in a circuit. There are two basic ways of doing this, in *series* and in *parallel*. Each has different voltage and current characteristics and different applications. For example, are the lights in a circuit in your home wired in series or in parallel? Let's take a look and see.

Series Circuits

As the name implies, the components in a series circuit are in an in-line series, so the same current flows through all the components. This is shown for a light bulb circuit in Figure 17.12. If you trace around the circuit with your finger, you will note that it goes through all the bulbs, just as the current flow does.

Suppose there was a 12-V battery in the circuit "pumping" the current or "raising" the charge to a higher potential difference or voltage. The individual voltage "drops" across the lamp resistances would not be 12 V, but they would add up to 12 V: $V = V_1 + V_2 + V_3 = 12$ V. We express this by saying that the voltage drops around a circuit are equal to the voltage rise (of the battery).

This is really just a statement of the conservation of energy. If all the lamps were the same, there would be a 4-V voltage drop across each. This is analogous to the potential energy or height drop shown in the water circuit in Figure 17.13. The total resistance of a series circuit is the sum of the individual resistances. For example, if R_1, R_2, and R_3 were 10 Ω, 15 Ω, and 30 Ω, respectively, the total (equivalent) resistance would be 55 Ω.* This means that

*In a series circuit the total resistance is the sum of the individual resistances, $R_s = R_1 + R_2 + R_3$ in this case.

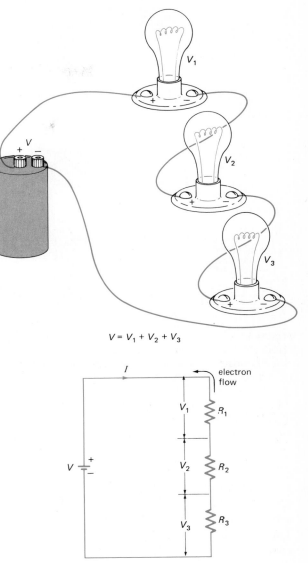

$$V = V_1 + V_2 + V_3$$

circuit diagram

Figure 17.12
Series circuit. The current flowing through each light bulb (resistance) is the same, and the voltage "drops" across the light bulbs add up to the voltage "rise" of the battery ($V = V_1 + V_2 + V_3$).

the three resistances could be replaced with a single 55 Ω resistance—the battery wouldn't know the difference.

Now suppose you hooked light bulbs in series in a household circuit with the 120-V source provided by the electric company. You'd have a problem. The light bulbs are rated for a certain wattage at 120 V, but there would be only 40 V across each, so they would glow very dimly.

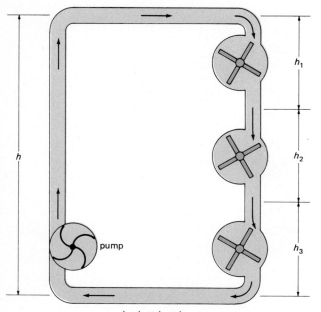

$$h = h_1 + h_2 + h_3$$

Figure 17.13
Paddle wheels in series. A liquid circuit with paddle wheels (resistances) in series. There are many differences from an electrical series circuit, but notice that the analogy of the gravitational potential (height) "rise" of the liquid, as supplied by the pump, is equal to the potential "drops" across the wheels ($h = h_1 + h_2 + h_3$).

What would happen if one of the light bulbs blew out? The circuit would be opened and no current would flow, so all of the lights would go out. This wouldn't be very practical in a house circuit, even if the voltage were adequate. How would you like to have a light bulb in a lamp blow out, and all the other lights on that circuit go dead, along with a radio you may have plugged into the circuit? Such problems are avoidd by wiring circuits in parallel.*

Parallel Circuits

In a parallel circuit, all the components have a common connection to the high-potential side of the voltage source and a common connection to the low-potential side (Fig. 17.14). In this manner, the light bulbs are connected in "parallel" branches in-

* Do not get the idea that series circuits are totally impractical. There are many electrical applications other than household circuits in which series-circuit characteristics are wanted.

stead of being in a series. The voltage drop across each bulb is the same and is equal to the voltage rise of the battery, $V = V_1 = V_2 = V_3$.

However, the current from the voltage source divides at the common junction, and a fraction of the current flows through each bulb. If the bulbs are the same and have equal resistances, each conducts one third of the total current (conservation of charge). If the bulbs are different, the greatest current is through the bulb with the least resistance — following the path of least resistance. Not all the current follows the path of least resistance; it divides proportionately.

A parallel water circuit is shown in Fig. 17.15. Think of what happens to the water flow from the pump when it reaches the common parallel pipes. It divides proportionately among the pipes, but the potential drop for each is the same.

In a parallel circuit with a 120-V source, we don't have the previous problem of a (household) series circuit. The voltage across all the bulbs in parallel is 120 V, which is what they are rated for. If one bulb blows out, the circuit paths through the other bulbs are not disrupted, and the bulbs remain lighted. Because of these characteristics, household circuits are wired in parallel. When you plug an appliance into a wall socket, you are connecting it in parallel into a circuit.

A household circuit is illustrated in Figure 17.16. Most appliances operate on 120 VAC. The two power lines coming into a house are at voltages of +120 V and −120 V. Connecting between one of these "hot" lines and ground (0 V) gives a potential or voltage difference of 120 V. Connecting across both lines gives a voltage difference of 240 V, which is needed for the power requirements of some air conditioners and electric stoves.*

Notice the fuse in the circuit in Figure 17.16. This is an important safety feature, as will be learned in the next section. (A circuit breaker serves the same purpose.) You might think that the more appliances or resistances you put in a parallel circuit, the lower the current flow would be. But, oddly enough, the more individual resistances added to the circuit, the less the total resistance — sort of "opposite" to a series circuit. This can cause problems with large currents and joule heating, as discussed in the Electrical Safety section that follows.

* When regular household voltage is considered to be 110 V, this difference is 220 V.

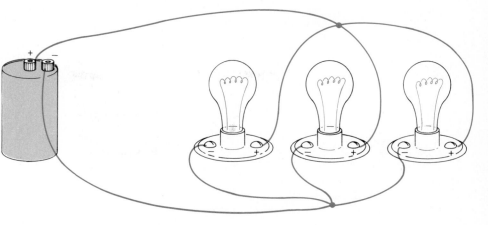

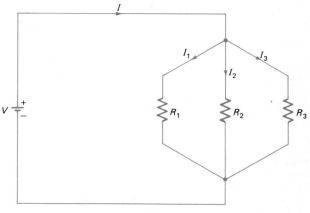

$$I = I_1 + I_2 + I_3$$

circuit diagram

Figure 17.14
Parallel circuit. The voltage drop across each light bulb (resistance) is the same, and the current supplied by the battery divides among the bulbs according to their resistances. By the conservation of charge, $I = I_1 + I_2 + I_3$.

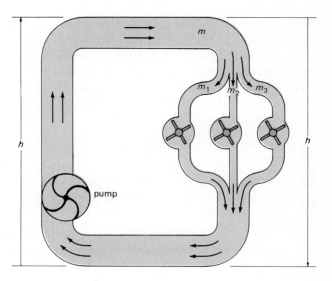

Figure 17.15
Paddle wheels in parallel. A liquid circuit with paddle wheels (resistances) in parallel. There are many differences to that of an electrical parallel circuit, but notice the analogy of the gravitational potential (height) "rise" of the liquid as supplied by the pump is equal to the potential "drop" across each wheel. The liquid divides among the parallel wheels, and by the conservation of mass, $m = m_1 + m_2 + m_3$.

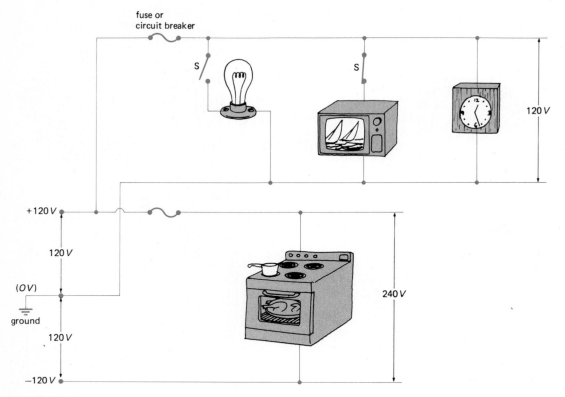

Figure 17.16
Household circuits. For a 120-V circuit, the connection is made between ground and a
potential of 120 V (either +120 or −120 V). For a 240-V circuit, the connection is made between +120 V and −120 V. Keep in mind that it is the voltage or potential *difference* that is
important.

In fact, the total resistance of the circuit is
always less than the smallest appliance resistance.
For example, suppose the parallel resistances R_1,
R_2, and R_3 in Figure 17.14 were 10 Ω, 15 Ω, and
30 Ω, respectively. The total resistance of the arrangement is 5 Ω.* That is, the three resistances
could be replaced with a 5-Ω resistance without the
battery knowing the difference. Recall that these
three resistances connected in series had a total resistance of 55 Ω, so there is more current in the
parallel circuit ($I = V/R$).

Question: Are Christmas-tree lights wired in
series or in parallel?

Answer: Most of the newer strings of
Christmas-tree lights have a shunt resistance
in the bulb wired in parallel with the bulb
filament. Each of these parallel-bulb combina-

* See example in An Extended View (Optional) in the
Appendix.

tions is wired in series along the string. In
this manner, when one bulb blows out, the
others remain lit because the shunt resistance
still completes the circuit at that point.

At one time, Christmas-tree lights without
shunt resistors were wired directly in series.
When one bulb blew out on these older
strings of bulbs, they all went out. Then you
had to hunt for the bad bulb to replace it.
There was big trouble when more than one
bulb blew at once.

ELECTRICAL SAFETY

The importance of electrical safety cannot be over-
stressed. Electrical accidents can result in property
damage, personal injury, and death. This does not
mean that you should fear electricity, just that you
should use it properly and safely.

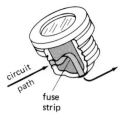

Figure 17.17
Fuses. A fuse "blows" or opens the circuit when the joule heat from a current larger than the rated value melts the fuse strip. Edison-base fuses (one shown in lower left of photo) are interchangeable, and any value of fuse can be put into a socket. Type-S fuses (*top of photo*) have an adapter installed in the socket specifically for a rated fuse so that they cannot be interchanged. Notice the different threads on the fuses.

The common methods of protecting against property damage through circuit overloads (too much current) and overheating are fuses and circuit breakers. All electrical circuits in the home are required to be protected by these means. When too much current flows in a circuit, it becomes hot (joule heat) and perhaps melts the insulation and starts a fire. An overload may also burn out and damage appliances. Electronic equipment commonly has fuses to protect the component parts from overloads. Even your automobile circuits are fused for overload protection.

A **fuse** is essentially a short piece or strip of metal with a low melting point. When the current in a fused circuit exceeds the fuse rating, for example 15 or 20 amps, the joule heat melts or vaporizes the fuse strip. The fuse "blows" and the circuit is opened.

Another common problem is that the insulation on wires may become worn, for example, on an extension or appliance cord. If the bare wires touch each other or if a high-voltage or "hot" wire touches ground, this is called a short circuit, since the path of the circuit is effectively shortened. This provides a low-resistance path, and there is a large current, which also blows the protecting fuse.

There are various types of fuses. The common type in older homes is the Edison-base fuse (Fig. 17.17). Its base is the same as that of an ordinary light bulb. Edison-base fuses are interchangeable in the fuse socket: for example, a 30-amp fuse can be placed in a 15-amp circuit. Because of this, Edison-base fuses are no longer permitted in new installations.

To prevent interchanging, a nontamperable fuse is used. This fuse, called a "Type-S" fuse (trade name "Fustat") will not fit or screw into an Edison-base socket. A nonremovable adapter is installed in the socket. Each differently rated Type-S fuse has an adapter with different-sized threads. As a result, a 30-amp Type-S fuse will not screw into a 15-amp adapter (Fig. 17.17).

It is more common now to use **circuit breakers** in place of fuses (Fig. 17.18). If the current in a circuit exceeds a certain value, the breaker is activated, and a magnetic relay (switch) breaks or opens the circuit. The circuit breaker can be "reset" or closed by a manual switch (see Chapter 18, Fig. 18.11).

In either case—whether a circuit is opened when a fuse blows or when a circuit breaker "trips" —steps should be taken to find out why and to remedy the problem. Remember, these are safety devices, and when they open a circuit they are telling you something—namely, that the circuit is overloaded or shorted.

Switches, fuses, and circuit breakers are always placed in the hot (high-voltage) side of the line, so as to interrupt power flow to the circuit element. However, fuses and circuit breakers may not always

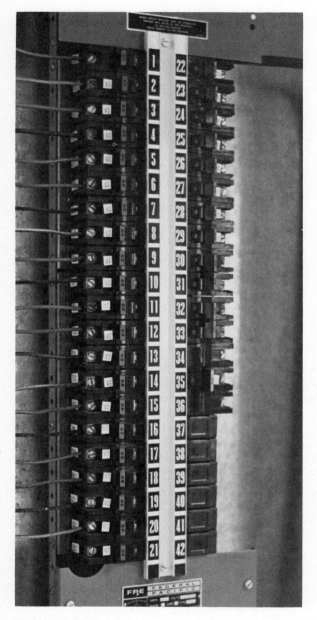

Figure 17.18
Circuit breaker. A circuit breaker panel. If the current in a circuit exceeds a certain value, the breaker is activated, and a magnetic relay (switch) "breaks" or opens the circuit. The circuit breaker can be "reset" and the circuit closed by a manual switch. See Chapter 18 for detailed operation (cf. Fig. 18.11).

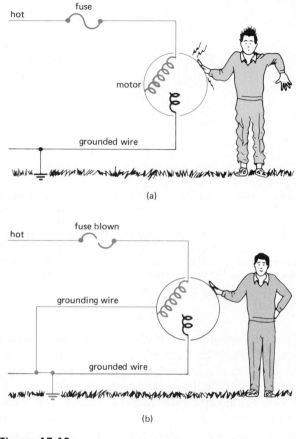

Figure 17.19
Grounding. (a) A hot wire in contact with the metal casing of a motor or tool, which is not connected to ground, presents a potentially dangerous situation, since a person could provide a path to ground. (b) If the casing is grounded, the circuit would be completed to ground and the fuse blown. This is the purpose of the grounding prong of a three-prong plug.

protect you from electrical shock. An example is given in Figure 17.19. A hot wire inside a tool (e.g., an electric drill) or appliance—for example, from the winding of a motor—may come into contact with the metal frame or housing. If you touch the frame, watch out. See Special Feature 17.2 for some possible effects.

To prevent this, a grounding wire is used. The circuit is then completed (shorted) to ground, and the fuse in the circuit is blown. This is why many electrical tools and appliances have three-prong plugs (Fig. 17.20).

The grounding prong is usually U-shaped. In the wall receptacle, this connection runs to ground. The use of a three-prong plug in an older-type "two-hole" socket requires an adapter. The grounding lug, or in some cases a wire, of the adapter should be connected to the grounded receptacle by means of

Personal Safety and Electrical Effects

Personal safety in working with electricity involves common sense and a fundamental knowledge of electricity. The main thing to remember is not to become part of a circuit yourself. As with any other circuit component, the current drawn by the human body depends on its resistance and the voltage source, $I = V/R_{body}$.

If the skin is dry, there is usually a high resistance, and an electrical shock may be only an uncomfortable and surprising tingle. However, if the skin is wet — for example, with perspiration — the resistance may be low enough to allow injurious and fatal current flows. Table 17.2 shows that only milliamps are needed.

If you come into contact with a "hot" wire, say at 120 V, current could flow through parts of your body. If the circuit is completed through your hand (finger to finger), a shock and perhaps a burn can result. However, if the circuit is completed through the body (hand to hand or hand to foot), there can be serious results, depending on the amount of current.

One may lose muscle control and not be able to let go. (Muscles are controlled by nerves, which are activated by electrical impulses.) Larger amounts of current can cause breathing difficulties, and the heart may have uncontrolled contractions (ventricular fibrillation). If the current is greater than 100 mA (0.1 A), death can result. This isn't much current on a relative basis. A 60-W light bulb operating on 120 V draws $I = P/V = \frac{60}{120} = 0.5$ amp of current. Keep in mind that it is the current that kills, not the high voltage itself.

Table 17.2
The Effects of Electric Currents on Humans*

Current (mA)	Effect
1	Barely perceptible
5 – 10	Mild shock
10 – 15	Difficult to let go
15 – 25	"Muscular freeze"; cannot let go
25 – 50	Breathing difficulty
50 – 100	Breathing may stop; ventricular fibrillation
>100	Death

* Varies with individuals

Figure 17.20
A three-prong plug and adapter. An adapter has a grounding lug or wire that should be connected to the grounded receptacle by means of the plate-fastening screw. Otherwise, the whole purpose of the grounding system is defeated.

the plate-fastening screw. Many times people do not ground the grounding lug, which defeats the whole purpose of the safety grounding system.*

Have you ever tried to plug something in and the plug wouldn't go, but when you turned it over, it fit? If you take a look at a wall receptacle, you'll find that one of the slits is bigger than the other. If you look at the plug, you will find that one of the prongs is bigger than the other (Fig. 17.21). This is called a polarized plug.

Polarizing in the electrical sense refers to a method or identification by which proper connections can be made. The original intent of this type of plug was as a safety feature. The small slit in the receptacle is the hot side and the large slit the neutral or ground side, if properly connected. The hous-

* When the grounding lug is connected to the plate-fastening screw, it is assumed that the receptacle box is grounded by means of a third dedicated grounding wire (not the zero potential *ground* wire, which is a current-carrying wire in the circuit). If this is not done, the three-prong plug does not serve its intended purpose.

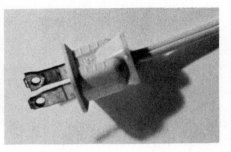

Figure 17.21
A "polarized" plug. This plug was an early attempt at electrical safety that is still with us. It's a good backup system *if* the receptacle and appliance are properly wired.

ing of an appliance could then be connected to the ground side all the time by means of the polarized plug. The effect would be similar to that of a three-prong plug. Then, if a condition such as that in Figure 17.19 occurs, the hot wire is shorted to ground, which blows a fuse or trips a circuit breaker.

However, this system left too much to human error. If the receptacle or the appliance is not wired (polarized) properly, a dangerous situation can exist. The polarization is ensured with a "dedicated" third grounding wire as in a three-prong plug system, which is the accepted safety system. The original two-prong polarized plug system remains as a general backup safety system, *provided* it is wired properly.

SUMMARY OF KEY TERMS

Electric current: a movement or flow of electric charges that does work through energy transport. Units of electric current are coulomb per second or ampere (A, commonly stated as amps for short).

Voltage source: an electrical "pump" that supplies a potential or voltage difference for the flow of current, commonly a battery or generator. The unit of voltage (difference) is the volt.

Direct current (dc): electric current that flows in only one direction because of the fixed polarity of the voltage source, e.g., a battery.

Alternating current (ac): electric current in which the charge motion alternates back and forth owing to a changing polarity of the voltage source.

Battery: a single cell or series of cells that converts chemical energy into electrical energy.

Electrical resistance: the opposition of a material to electric current flow, which is measured in ohms (Ω).

Ohm's law: the relationship of the current, potential difference (voltage), and resistance in many conductors.

$$\text{Current} = \frac{\text{voltage}}{\text{resistance}}$$

or

$$I = \frac{V}{R}$$

Electric power: the time rate of electrical energy transfer or rate of doing work, which is given by the product of the current and voltage. The unit of power is the watt (W).

$$\text{Electric power} = \text{current} \times \text{voltage}$$

or

$$P = IV$$

Series circuit: a circuit in which the components (such as resistances) are arranged in a series such that the same current flows through each.

Parallel circuit: a circuit in which the components (such as resistances) are arranged in parallel branches with common junctions such that the current divides among the branches.

Electrical safety: important procedures, practices, and devices that prevent electrical accidents that could result in property damage and personal injury.

(For more on resistances in series and parallel, see An Extended View (Optional) in the Appendix.)

QUESTIONS

Electric Current

1. People often say that they left the house lights "burning." Is this a correct statement? How do you think it might have originated?
2. If a wire is connected to two objects and there is a current in the wire, what do you know about the electric potentials of the objects? In which direction is the current?
3. If the drift velocity of electrons in a conductor is so small, why does an auto battery influence the starter as soon as you turn on the ignition switch?
•4. What is "direct" in direct current?
5. Since the charges move back and forth in alternating current, there is no net current. What "flows" in an ac circuit, and what does this do to give electric-current effects, say, in a light bulb?
6. If you switch the lead wires to a circuit rapidly back and forth between the terminals of a battery, would there be an "alternating" current in the circuit? Explain.
7. An electric wall receptacle is sometimes called an outlet. Does it let something out?
8. When you plug in a lamp in the home and turn it on, do electrons flow through the plug to the lamp? Explain.
•9. If a net charge of 10 C passes through a cross-sectional area of a conductor in 2 s, how much current is in the conductor?

Voltage Sources

10. What is electromotive force? What is a "seat" of electromotive force?
11. How does a battery complete a circuit with regard to current?
•12. Suppose that two batteries in series are both placed backwards in a flashlight so that the base of one battery makes contact with the bulb. Would the flashlight work?
13. Suppose you got one of the two batteries in a flashlight reversed. Would the flashlight work? Explain.
14. On a very cold day, a car battery is very sluggish in starting the engine because the current output of the battery is reduced. Why is this?
15. If you had two 1.5-V D cells and another 9-V dry cell, what would be the maximum voltage you could get by connecting the batteries, and how would you connect them?
•16. For the circuit shown in Figure 17.22, in which case would the bulb(s) be (a) the brightest and (b) the dimmest, and why? (All the batteries and bulbs are the same.)

Ohm's Law and Electric Power

17. Voltage sources "pump" energy into a circuit. Where does the energy go?
18. We are told not to waste electricity. How is electricity "wasted"?

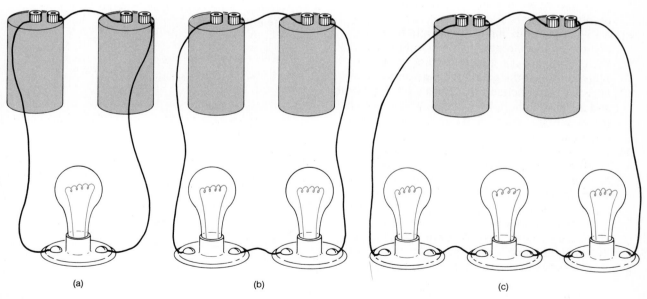

(a)   (b)   (c)

Figure 17.22
Let there be light. See Question 16.

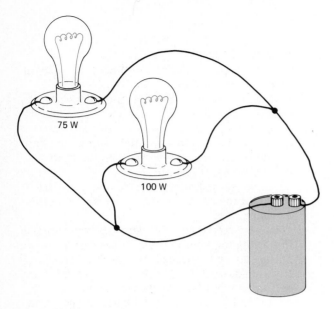

Figure 17.23
Current and voltage. See Question 21.

19. Dim lights set certain moods. One way of dimming lights is by using a variable-resistance dimmer switch, which usually has a rotating knob. What causes the lights to dim (and brighten) when the knob is rotated? (*Hint:* Think in terms of Ohm's law.)

*20. Which has the greater resistance, a 75-W or a 100-W light bulb? If the tungsten filament in each bulb were the same length, how would they otherwise differ?

*21. A 75-W light bulb and a 100-W light bulb are connected in a circuit, as shown in Figure 17.23. Which bulb has the greater current through it? Which bulb has the greater voltage drop across it?

22. Sketch a battery circuit that has a 75-W light bulb and a 100-W light bulb in series. Which bulb has the greater current through it? Which bulb has the greater voltage drop across it?

23. Although the connecting wires of a circuit have small resistances, they still have I^2R losses and heat up. How does this affect the resistance of the wire? How does the change in resistance affect the joule heat?

24. How much more power does an electric can opener use (rating shown in Fig. 17.11) than a 100-W light bulb?

25. Which six items are the biggest users of electrical energy in a typical home? (See Table 17.1, but take into account hours of use.)

*26. A hot dog can be cooked by "plugging it in," so to speak (see Fig. 17.24). (This is a lecture demonstration to be done by the teacher and should *not* be attempted at home. It can be dangerous.) It takes only a minute or so for the hot dog to be heated. There are now commercially available hot dog cookers that operate on the same principle and cook several hot dogs at one time. What causes a hot dog to cook this way? Would the hot dogs be "wired" in series or parallel in the cooker?

*27. Some light bulb questions (several billion bulbs are used in the United States each year):
   a. The gas in a light bulb is not air but rather is a mixture of argon and nitrogen at low pressure. Why isn't air used? Wouldn't it be cheaper?
   b. After long periods of use, a gray spot develops on the inside of the bulb. (Check this out on a used bulb.) What is the gray spot? (*Hint:* It's metallic.)

28. What do you think would happen if a 120-V appliance were plugged into a 240-V outlet? (The plugs are different, so this can't be done normally.) How about a 240-V appliance plugged into a 120-V outlet?

Electric Circuits

29. Give a common example of both a series circuit and a parallel circuit.

30. How are the circuits in a home wired? What would happen if they were wired the other way?

*31. Are auto headlights wired in series or parallel? How could you find out if you weren't sure?

32. Are the ceiling lights in a room all that is in that circuit, or are there usually some wall plugs in the circuit too?

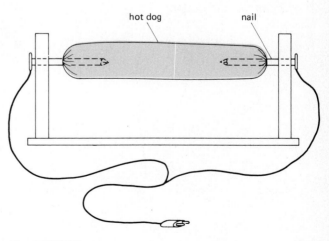

Figure 17.24
Hot dog! See Question 26.

Figure 17.25
High voltage or high current? See Question 38.

*33. If two 6-V batteries connected in series are in a circuit with a resistance of 4 Ω, (a) how much current flows in the circuit? (b) If the batteries were connected in parallel, how much current would be in the circuit? (c) If another 2-Ω resistance were placed in the circuit in series with the other resistance, how much current would be in the circuit with the batteries in parallel and series connections? (d) What is the voltage drop across the resistances in each case?

Electrical Safety

34. Why is a switch placed in the high-voltage side of a circuit?

35. Why are Edison-base fuse boxes no longer used in home installations?

36. A foolhardy and dangerous thing to do when no spare Edison-base fuses are available is to put a penny in the socket behind the blown fuse. Why is this dangerous?

*37. Why can birds sit on high-voltage power lines (which are bare and have no insulation) and not be electrocuted?

38. Warning signs commonly state, "Danger—High Voltage" (Fig. 17.25). Wouldn't it be more appropriate to say, "Danger—High Current"? Explain.

39. If a power line happens to fall on a parked automobile in which you are sitting, you should sit still and not touch any metal parts of the car. Why? Wouldn't the power line be shorted to ground and trip some protective device in the line?

40. Some people cut off the grounding plug of a three-prong plug so it can be used in a two-prong receptacle. Is this a wise thing to do? Explain.

41. What's wrong with the wiring arrangement shown in Figure 17.26, and how should it be properly wired?

42. A warning sign on the handle of a hair dryer reads, "Danger. Electrocution possible if used or dropped in tub. Unplug after each use." Explain the reason for this warning.

*43. An old saying used by people who work around high electrical voltages is, "You should keep one hand in your pocket." What is the logic behind this saying?

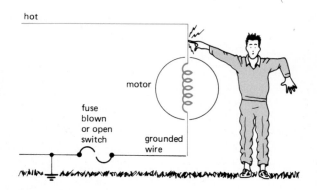

Figure 17.26
Shocking! See Question 41.

Magnetism

18

Iron filings show the natural magnetism of a lodestone.

Electricity and magnetism are closely related. In fact, we will find they are inseparable, although we talk about electric forces and magnetic forces. Electric forces arise from stationary or moving electric charges, and magnetic forces involve moving electric charges. Since both forces ultimately arise from electric charges, the general term *electromagnetic force* is sometimes applied to both.

Electromagnetic forces are basic in the operations of motors, generators, and electric meters and in many other practical applications. The basic principles and applications of electromagnetism are the subjects of this and the next chapter.

■ MAGNETS AND MAGNETIC FIELDS

Like most people, you have probably been fascinated by the attractive and repulsive forces between magnets—a hands-on example of force at a distance. Magnets are readily available today because we know how to make them. But they were once quite scarce and existed only as natural magnets or rocks that were found in nature.

Natural magnets, called lodestones, were found as early as the sixth century B.C. in the province of Magnesia in ancient Greece, from which magnetism derives its name. Lodestones could attract pieces of iron and other lodestones. For centuries, the attractive properties of natural magnets were attributed to supernatural forces. Early Greek philosophers believed that a magnet had a ''soul'' that caused it to attract pieces of iron (see introductory photo in this chapter).

Sometime around the first century A.D., the Chinese learned to make artificial magnets by stroking pieces of iron with a natural magnet. This led to one of the first practical applications of magnets, the compass, which implied that the Earth itself has magnetic properties. (The Chinese are said to have developed the compass, but several other peoples claim this invention, too.)

The most familiar magnets are the common bar magnet and the horseshoe magnet (a bar magnet bent in the form of a horseshoe). If a bar magnet is dipped into iron filings, the filings cling to the magnet in concentrations near the ends of the bar (Fig. 18.1). These regions of apparently concentrated magnetic strengths are called **magnetic poles**. The poles are distinguished as the north (seeking) pole and the south (seeking) pole. This comes from the properties of a compass. The north magnetic pole ''seeks'' and points north, and the south magnetic pole points south.

We have never been able to isolate a single magnetic pole, as we have done with electric charges. That is, we have never found a magnetic ''monopole'' to occur by itself—the poles always occur together. If you break a magnet in two, you get two smaller dipole magnets. The reason for this is discussed later in the chapter.

When magnets are brought close together, it is quickly observed that the magnets attract each other in some cases and repel in others. This action is described by the **law of poles**, which is similar to the law of charges:

Like magnetic poles repel, and unlike magnetic poles attract.

That is, N-N and S-S repel, and N-S (or S-N) attract, as illustrated in Figure 18.2.

Also as in the electric case, it is convenient to describe the magnetic force in terms of a magnetic field. The **magnetic field**, commonly called a B

Figure 18.1
Magnetic poles. Regions of concentrated magnetic strength, as shown by attracted iron filings, are called magnetic poles.

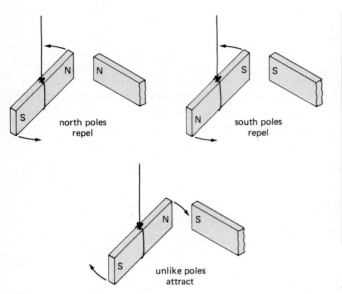

Figure 18.2
The law of poles. Like magnetic poles repel, and unlike poles attract.

field, is the magnetic force per *pole.** (Recall that the electric field is the electric force per *charge.*) Its direction is determined by the direction of the force experienced by a north magnetic pole. For example, if a compass is placed in the vicinity of a magnet, the magnetic field can be mapped out and is in the direction of the north pole of the compass (Fig. 18.3).

The magnetic field can be "seen" by iron filing patterns. When iron filings are sprinkled on a piece of paper over a magnet, they become induced magnets and line up with the field, as shown in Figure 18.3.

Question: The "north-seeking" pole of a compass needle is actually the north pole of the magnetic needle. According to the law of poles, this north pole should be attracted toward a south magnetic pole. Why does it point north?

Answer: This is because the Earth's south *magnetic* pole is near its *geographic* north

---

* This is an older designation used here for illustration. Technically, the magnetic field is defined as the magnetic force "per moving charge," since magnetic fields are produced by moving electrical charges, as will be learned. The unit of the magnetic field in the SI system is the tesla (T), in honor of Nikola Tesla, a Yugoslav scientist who worked in the United States.

pole. The Earth's magnetic field is similar to that of a bar magnet located in the Earth (Fig. 18.4). When a compass needle lines up with the Earth's magnetic field, its north pole points north (magnetic north) but toward a south magnetic pole.

The Earth's magnetic poles do not coincide with the rotational geographic poles. The south magnetic pole is located north of Prince Wales Island, Canada, at about 75°N, 101°W. This is some 1600 km (1000 mi) from the geographic north pole (true north). The other magnetic pole is even farther removed from the south geographic pole, being located in Wilkes Land, Antarctica (69°S, 14°E).

The Earth does not have a big magnetized chunk of iron or a bar magnet as used in Figure 18.4 for illustration. It is believed that the Earth's magnetic field is associated with its liquid outer core and rotation. The magnetic poles also "wander" slowly and are not always in the same place.

There is also evidence that the Earth's magnetic poles have switched polarity at various times in the past, most recently about 700,000 years ago. During a period of pole reversal, the south magnetic pole is near the south geographic pole. We are not certain why these changes occurred, or if it will happen again. More about this later in Special Feature 18.1.

ELECTROMAGNETISM

Now we can produce magnetic fields at will — even turn them on and off. In electrical experiments carried out in 1820, the Danish physicist Hans Christian Oersted discovered a relationship between electricity and magnetism. It was found that

An electric current produces a magnetic field.

This can be shown by an arrangement illustrated in Figure 18.5. When the circuit is open and there is no current, the compass needle points in its normal north-south direction. However, when the switch is closed and there is an electric current in the circuit, the compass needle is deflected, indicating the presence of a magnetic field other than that of the Earth.

Another important feature is that the strength or magnitude of the magnetic field depends on the magnitude of the current — the greater the current, the greater the magnetic field ($B \propto I$).

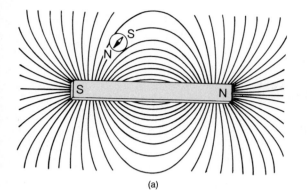

(a)

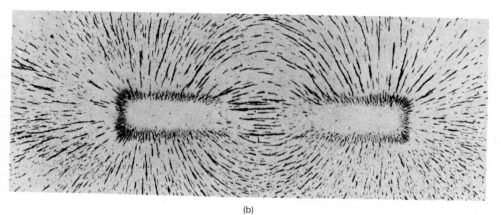

(b)

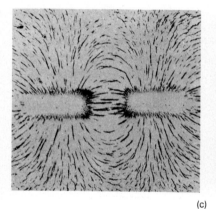 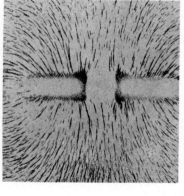

(c)

Figure 18.3
Magnetic fields. (a) The magnetic field at a particular point is the
magnetic force per pole in the direction of the force experienced by a
north magnetic pole. (b and c) Iron filings line up with the magnetic field
and show its pattern.

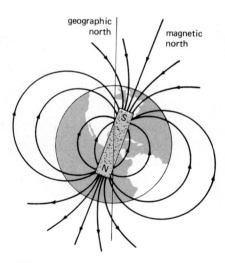

Figure 18.4
The Earth's magnetic field. The
magnetic field of the Earth is similar to
that of an imaginary bar magnet inside
the Earth, with the south magnetic pole
near the geographic north pole. The
north pole of a compass then points
toward geographic north (law of
poles—unlike poles attract).

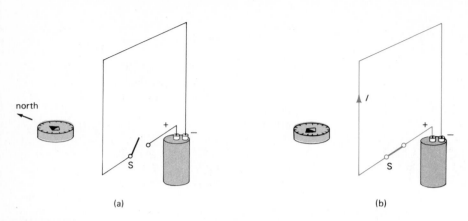

Figure 18.5
An electric current produces a magnetic field. When the switch is closed and current flows in the circuit, the compass needle is deflected, indicating the presence of a magnetic field.

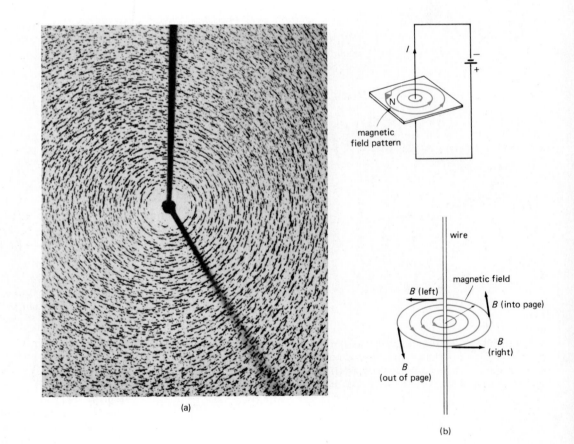

Figure 18.6
The magnetic field around a straight wire. (a) The magnetic field is in circles around a current-carrying straight wire, as shown here by iron filings. (b) The circular magnetic field lines may be mapped out using a compass. The magnetic field vector B is tangential to the circle at any point.

The Earth's Magnetic Field

We don't really know what causes the Earth's magnetic field. But the fact that a current produces a magnetic field leads scientists to speculate that there may be a flow of charge in the Earth's liquid outer core, which is believed to be chiefly iron, that gives rise to the magnetic field. This is thought to be associated with the Earth's rotation.

Some of the other planets have magnetic fields, whereas others have none or very weak magnetic fields. For example, the magnetic fields of Saturn and Jupiter are much larger than that of the Earth. These planets are largely gaseous and have fast periods of rotations of 10 to 11 hours. Venus and Mercury, on the other hand, have very weak magnetic fields. These planets are about as dense as Earth but rotate relatively slowly (243 and 58 days, respectively).

Another theory is that the Earth's magnetic field arises from charge in motion in thermal convection cycles in the liquid outer core as a result of heat coming up from the central core. If the heat flow subsided and the directions of the convection cycles were reversed with subsequent heat flow, this could account for the Earth's magnetic pole reversals.

Using a compass to map out the magnetic field around a straight-wire segment, we find that the field lines are in concentric circles around the wire (Fig. 18.6). At any point on a field-line circle, the field direction is tangential to the circle.

The magnetic field around a current-carrying loop of wire is somewhat like that of a bar magnet (Fig. 18.7). If a number of loops of wire are wound in a tight coil, called a solenoid, the fields of the wire loops add, such that the field outside a current-carrying solenoid is very much like that of a bar magnet, as well as having a relatively uniform field inside the coil.

There are important and practical differences, however. The strength of the magnetic field of a solenoid can be varied by varying the current, as well as by being turned on and off. You can also reverse the magnetic field by reversing the direction of the current in the coil.

Electromagnetism gives us some clues to the Earth's magnetic field. See Special Feature 18.1.

Question: If magnetism is produced by moving electric charges, what produces the magnetic field of a bar magnet?

Figure 18.7
Magnetic fields. Around (a) a current-carrying loop and (b) a current-carrying solenoid. The effects of the loops of wire in a solenoid with many loops give a relatively uniform magnetic field inside the coil. Notice that the fields of the loop and the solenoid are similar to that of a bar magnet.

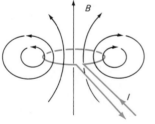

(a)

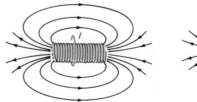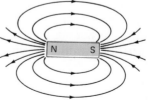

(b)

Answer: Recall that the electrons in an atom or ion revolve in orbits about the nucleus. This orbital motion of an electronic charge is a current and gives rise to a magnetic field. Also, electrons have intrinsic "spin" contributions to the magnetic field. This cannot be accounted for by classical mechanics, such as for a particle in orbit, and its origin is not known. We assume it derives from the spinning of the electron on its own axis, hence the term *spin.*

All materials contain atomic electrons. Whether a material is magnetic or can be made into a permanent magnet depends on whether the electron magnetic fields add or cancel. This is the topic of the next section. Read on.

MAGNETIC MATERIALS

A bar magnet easily picks up nails, paper clips, and iron filings, and we say that these are magnetic materials. On the other hand, a magnet has no observable effect on nonmagnetic materials such as wood or aluminum. The magnetic properties of a material depend on the magnetic fields of its electrons, as discussed in the previous Question and Answer. The magnetic properties of the most common magnetic materials depend on the electron spin, so we'll concentrate on this.

Because of the electron spin, each electron acts as a tiny magnet. If two electron spins are in the same direction, their magnetic fields add. However, if a pair of electron spins are in opposite directions, their magnetic fields cancel. In most atoms, the electron magnetic fields cancel each other, and most materials are nonmagnetic.

Common magnetic materials, called **ferromagnetic materials**, are iron, nickel, and cobalt. In these materials there is not a complete field cancellation; for example, an iron atom has four uncanceled electron spin fields. As a result, each atom acts like a little magnet, and we say the atom has a magnetic moment (similar to an electric dipole moment). There is an interaction among adjacent atoms, and large groups of them line up with each other. These groups of aligned atoms are called **magnetic domains.**

However, an ordinary piece of iron by itself is not a magnet. This is because the magnetic domains are randomly oriented, and their effects cancel (Fig. 18.8). In the presence of a magnetic field, the domains are induced into alignment, and the iron becomes magnetized. The degree of magnetism depends on the degree of alignment. When the magnetic field is removed, thermal motion generally causes the domains to go back into a random orientation, and the magnetism is lost. However, after being in a strong magnetic field, the domains may retain some alignment and magnetism. You may have noticed this in a paper clip that has been around a permanent magnet.

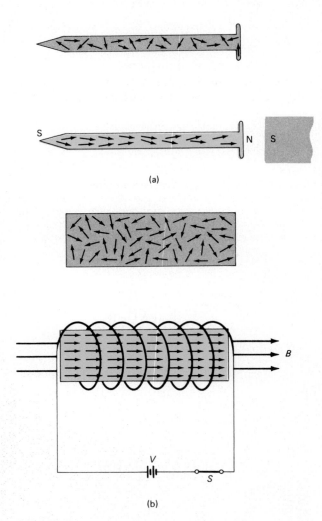

(a)

(b)

Figure 18.8
Magnetic domains. Ordinarily, the magnetic domains in a magnetic material such as iron are randomly oriented, and their effects cancel. A magnetic field from a magnet (a) or from a current-carrying wire (b) causes the domains to become aligned, and the iron is magnetized.

"Permanent" magnets are made by placing a piece of iron or some other ferromagnetic material or alloy in a solenoid in an oven. The iron is heated up, and the magnetic field of the solenoid is turned on. This magnetic field is maintained as the iron cools and the domains are "locked in" their alignment. Relatively lightweight magnets are made of alnico, an aluminum-nickel-cobalt alloy. The magnetism of a permanent magnet can be destroyed by heating it or striking it on a hard surface. Why? Also, you can now understand why breaking a magnet in two gives two smaller magnets.

The magnetism of natural magnets arises from iron minerals in the rocks. When molten material containing these minerals is extruded upward, as in volcanic eruptions, and solidifies in the Earth's magnetic field, the rock becomes magnetized. The orientation of the magnetization is in the direction of the Earth's magnetic field. These solidified rocks, called igneous rocks, are worn down by erosion. The rock fragments are carried away by water, and they eventually settle in bodies of water, where they become the layers of future sedimentary rocks.

In the settling process, the magnetic fragments become generally aligned with the Earth's magnetic field. By studying this remanent magnetization in geologic rock formations, scientists have learned about changes that have taken place in the Earth's magnetic field, such as pole reversals.

Question: Are coins (money) magnetic?

Answer: Some are and some are not. It depends on their metal content. In general, the U.S. "silver" coins aren't magnetic, but Canadian coins are. This is because Canadian coins still contain nickel (a ferromagnetic material). Get a magnet and a few coins and see for yourself.

Have you ever put a Canadian coin in a coin-operated machine and have it not work or "go down"? This is because the machines are equipped with magnets to stop metal slugs or washers (usually iron) from entering the coin mechanism so as to prevent damage and economic losses (Fig. 18.9). Coins containing ferromagnetic nickel are likewise attracted to the magnets.

ELECTROMAGNETS

A solenoid with an iron core is the principle of an electromagnet used to pick up magnetic metal objects, for example, in a junkyard or in the application shown in Figure 18.10. When the solenoid current is turned on, the iron core becomes magnetized and

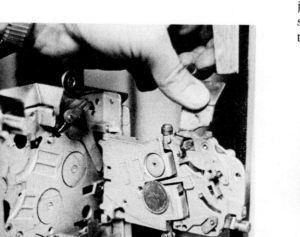

Figure 18.9
Magnetic money. A Canadian coin sticks to a magnet in the mechanism of a coin-operated machine. Canadian coins contain the ferromagnetic element nickel. U.S. coins are not magnetic.

Figure 18.10
An electromagnet at work. Pellets of iron oxide used in a steel-making process are lifted by an electromagnet.

attracts magnetic materials. When the current is turned off, they can be dropped or placed where wanted.

Electromagnets can be used to produce very strong magnetic fields — the greater the current, the greater the magnetic field. But there are limitations. With very large currents there are also large I^2R losses in the magnet coils. Large electromagnets used in scientific work are often water-cooled to carry away the joule heat.

There are also superconducting magnets. If a superconducting material is used for the coils and they are cooled below a certain critical temperature, the resistance and I^2R losses drop to zero. (See Special Feature 17.1. An application of superconducting magnets will be given in the next chapter.)

You have probably used electromagnets without knowing it. A common application is the relay, which is an electromagnetic switch. One application is shown in Fig. 18.11(a). When the coil is en-

(a) relay

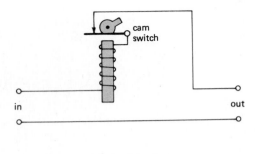

(b) circuit breaker

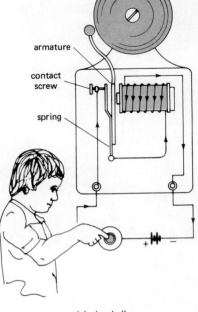

(c) doorbell

Figure 18.11
Applications of electromagnets. (a) A magnetic relay is an electromagnetic switch. (b) Diagram of a circuit breaker used for electrical safety (see Figure 17.18). (c) A doorbell circuit.

The Auto Starting System

You may have heard of (or had trouble with) the "solenoid" associated with the starter of a car. This solenoid is similar to a relay, except that the electromagnet contains a movable core (figure below). A solenoid is an electromagnetic device that converts electrical energy into linear motion (of the core). When the electromagnet is energized, the core is drawn into the coil, providing mechanical action.* A spring may be used to return the core when the coil circuit is opened (not shown in the figure).

* The core becomes a magnet with the opposite polarity to that of the coil, and the core is "drawn" (attracted) into the coil.

In the starting system of an automobile, when the starter switch is closed, a low-current starter solenoid is activated. The movement of the core or plunger mechanically engages the starter gear and closes a switch that connects the cranking motor to the automobile (high-current) battery. When the starter switch is released, the solenoid is de-energized, and a spring within the solenoid assembly pulls the gear out of mesh and interrupts the current flow to the cranking motor.

Solenoids are also used in home door chimes. Here you usually just get a "ding-dong" sound rather than continuous ringing, as with a doorbell (see Fig. 18.20 and Question 26 at the end of the chapter).

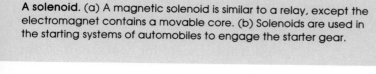

(a)

(b)

A solenoid. (a) A magnetic solenoid is similar to a relay, except the electromagnet contains a movable core. (b) Solenoids are used in the starting systems of automobiles to engage the starter gear.

ergized, the electromagnet attracts the contact arm and opens the high-voltage circuit. Notice that relay switching allows low-voltage control of high-voltage circuits. High-voltage contact switches tend to become pitted and burned, which causes poor contact. The low-voltage control of lights in homes is coming into increasing use. Also, large-current motors and machines can be controlled from a distance without having to run heavy wires to the controlling switches. This provides economy and safety.

A simplified diagram of an electromagnetic circuit breaker is also shown in the figure. With a safe amount of current in the circuit, the electromagnet is not strong enough to attract the contact arm and

break the circuit. But when too much current flows in the circuit, the breaker switch is tripped. See if you can use your knowledge of circuits and electromagnets to explain why a doorbell rings continuously when the button is pushed [Fig. 18.11(c)]. Another common use of a magnetic device is described in Special Feature 18.2.

■ MAGNETIC FORCES ON MOVING ELECTRIC CHARGES

As we have learned, a current produces a magnetic field. This can cause a magnet (compass) to move as a result of a force interaction, as was illustrated in

Figure 18.5—a case of a stationary current-carrying wire giving rise to a force on a movable magnet. Nature tends to show symmetry, and we can have a stationary magnet giving rise to a force on a movable current-carrying wire (Fig. 18.12). Another way of saying this is that

> A moving electric charge or current in a magnetic field experiences a force.

Notice in the figure how the force is reversed when the direction of the current is reversed. This could also be done by turning the magnet over, thereby reversing the direction of the magnetic field.

This force is experienced by a single moving charged particle or a beam of charged particles, as well as a current in a wire, in which case there is a force on the wire. Positive and negative charges experience forces in opposite directions. In any case, there must be charge in motion.

The force is greatest when the charge moves perpendicular to the magnetic field. It becomes less as the charge motion becomes nearly parallel to the field and is zero when the charges move parallel to the field. The size of the force also depends on the magnitudes of the magnetic field and the electric charge and on the speed of the moving charge. The direction of the force is always perpendicular to

both the direction of the charge motion and the magnetic field direction.

It is interesting to note that the ampere is defined by the mutual force interaction of two parallel, current-carrying wires. One wire produces a magnetic field, and the other wire experiences a force (and vice versa).

■ MOTORS

A force being available, we can usually do something useful with it. In the electromagnetic case with a force on a wire, we can build an electric motor, which converts electrical energy into mechanical energy. Think of how life would be without electric motors. We'd still be winding up clocks and phonographs, carrying water, and doing a lot of other manual work.

The basic principle of a motor is illustrated in Figure 18.13. Practical motors have many windings or loops of wire on the rotating armature. As can be seen in the figure, the electromagnetic forces on the current-carrying wire loop in the magnetic field of a permanent magnet are perpendicular to the wire and give rise to a rotating magnetic field and torque.

The split-ring commutator is essential to the operation of this dc motor. In starting from rest, if the commutator were not split or in two pieces, the loop would go through only one half-cycle, since it would be in static equilibrium and stop when vertical (see figure), when the torque would be zero.

But just at that moment, the end of the loop that was on the negative part of the commutator makes contact with the positive part (and vice versa for the other end of the loop), and the current and forces are reversed. The rotational inertia of the armature carries it through this position, and the loop rotates through another half-cycle. This process repeats, and the motor armature rotates continuously— pretty clever.

Ac motors of this type (there are many designs) have two separate rings that aren't split for continuous contact. Since the current alternates or changes direction itself in this case, there is no need to change contacts.

Small dc motors are made with permanent magnets, but large dc and ac motors usually have electromagnets that give larger magnetic fields and larger forces on the armature.

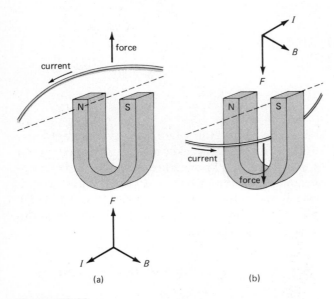

Figure 18.12
A moving electric charge or current in a magnetic field experiences a force when field lines are crossed. The current shown here is the conventional or positive-charge current.

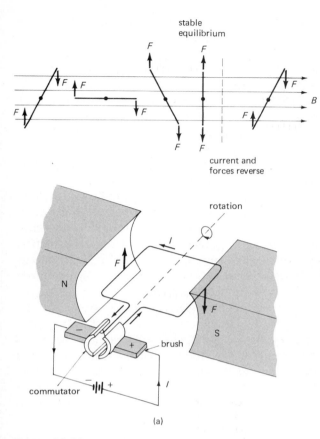

(a)

(b)

Figure 18.13
Motors. (a) The principle of a motor. A current-carrying loop in a magnetic field experiences a force or a torque and rotates. A dc motor has a split-ring commutator to change the current direction in the loop. This is done automatically in the windings of an ac motor. (b) A universal motor of the type used in vacuum cleaners. Note the many windings.

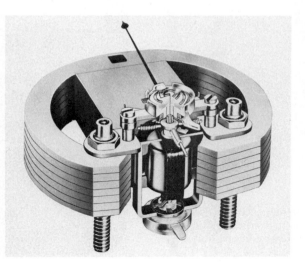

■ THE GALVANOMETER

The electromagnetic force on a current-carrying wire is the principle of the galvanometer, which is used in electrical meters that measure current and voltage (ammeters and voltmeters). A galvanometer movement is shown in Figure 18.14. A small current in the pivoted coil between the pole faces of a permanent magnet produces a force on the coil, caus-

Figure 18.14
A galvanometer movement. When current passes through the pivoted coil between the pole faces of the magnet, a torque causes the coil to rotate. The deflection, as indicated by a pointer, is proportional to the current.

ing it to rotate, and the attached needle is deflected. The greater the current, the greater are the force and the needle deflection. A restoring spring returns the needle (and coil) to its zero position when there is no current in the coil.

A large current through the fine-wire coil would burn it out, so in an ammeter a low "shunt" resistance is connected in parallel with the coil (Fig. 18.15). The shunt resistance then carries most of the current. The meter is calibrated to read the total current. By having a selection of shunt resistances, an ammeter can read various ranges of current.

A voltmeter also has a galvanometer, but it has a large resistance in series with it to limit the current when placed across a high voltage. The meter is calibrated in volts instead of amps by using Ohm's law ($V = IR$).

■ THE EARTH'S MAGNETOSPHERE AND IONOSPHERE

The forces on charged particles in a magnetic field give rise to many interesting geophysical effects associated with the Earth's magnetic field. The Earth is continually being bombarded by charged particles and radiation from the Sun and by cosmic rays. Cosmic rays (mostly protons) come from outside the solar system, perhaps from distant stars, or maybe they are vagabonds left over from the Big Bang. We're not sure. The Sun continually hurls off clouds of high-velocity ionized particles (electrons and protons) and radiation. We call this stream of emissions the solar wind.

The Earth's magnetic field is deformed slightly, owing to the "pressure" of the solar wind (Fig. 18.16). On the daylight side of the Earth, the field lines are compressed slightly, whereas on the night side they are stretched out in a long tail. This confines the geomagnetic field to a region scientists call the **magnetosphere**. Notice how the Earth and its magnetosphere look like an object in a wind tunnel with a bow shock wave (Chapter 15).

The magnetosphere acts as an obstacle to the solar wind and deflects a great deal of it. But some charged particles enter the geomagnetic field and experience forces perpendicular to the direction of the field lines. As for any charged particle in a magnetic field, it moves in a corkscrew path around a field line (either north or south from the equator).

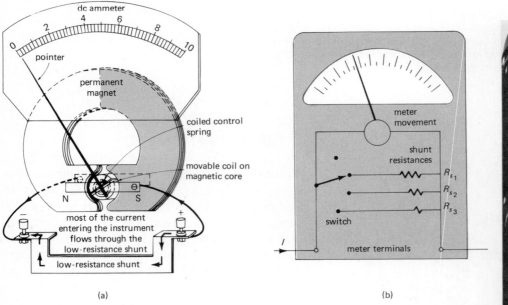

Figure 18.15
The ammeter. (a) Exposed view. The parallel shunt resistance carries most of the current and prevents the galvanometer coil from being burned out. (b) With a selection of shunt resistances, an ammeter can read various ranges of current. Some ammeters now have digital readouts.

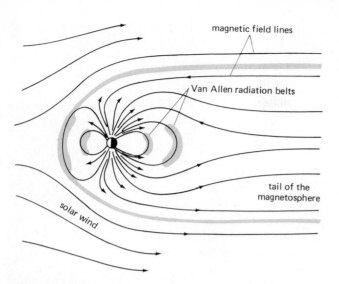

Figure 18.16
The Earth's magnetosphere. The region of the geomagnetic field is called the magnetosphere. This is shaped by solar winds. Two large, donut-shaped regions of magnetically confined charged particles are called the Van Allen radiation belts.

Thus, the magnetic dipole field of the Earth acts as a trap (or "magnetic bottle") and confines many particles at certain altitudes.

The particles bounce back and forth and drift around the Earth. Two large, donut-shaped regions

at altitudes of several thousand kilometers are called the Van Allen radiation belts (Fig. 18.16).[*] Several other radiation belts were made in the lower Van Allen belt region in the late 1950's and early 1960's as a result of high-altitude nuclear explosions. These involved experiments to determine the behavior of charged particles in the Earth's magnetic field.

At lower altitudes (100 to 400 km) there are noticeable concentrations or layers of ions. Hence, this region is called the ionosphere. Here, where the air is more dense, nitrogen and oxygen molecules are ionized by energetic particles and radiation from the Sun. The layers vary in ion density and are called the D, E, and F layers (Fig. 18.17).

Since the production of ions requires direct solar radiation, the concentration of charged particles varies from day to night, particularly in the D and E layers, which weaken at night (the D layer virtually disappears). The upper F layer, however, is present both night and day because of the low density of the atmosphere in this region, and the ions and electrons do not recombine as rapidly as they do at lower altitudes.

[*] After James A. Van Allen, an American scientist who identified the radiation belts in 1958 from data from instruments he had on board the first U.S. satellite, *Explorer I.*

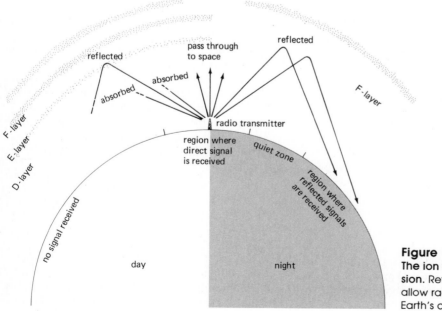

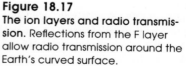

Figure 18.17
The ion layers and radio transmission. Reflections from the F layer allow radio transmission around the Earth's curved surface.

You may have had some contact with the ion layers through AM radio reception. Have you ever turned on a radio at night and been surprised to receive a station hundreds of kilometers away in another part of the country? This is because the F layer bends and reflects medium- and high-frequency radio waves (Fig. 18.17). Ham-radio operators make good use of this in reflecting the straight-line radio waves around the curvature of the earth for long-distance transmissions.

During the day, when it is present, the D layer absorbs most radio waves as they pass through and on the return reflection from the higher E and F layers. However, at night the absorbing D layer all but disappears, and the E layer is very weak. Of course, you may be in a "quiet zone" and not receive the station. In this zone, you are too far from the station to receive the direct signals and not in position to receive the reflected radio waves.

Radio and other communications are also disturbed during periods of increased solar activity, such as periods of maximum sunspots and solar flares. Solar flares are violent magnetic storms on the Sun that spew out enormous quantities of charged particles and radiation. One spectacular sight that apparently results from these solar storms is aurora—aurora borealis (northern lights) in the northern hemisphere and aurora australis (southern lights) in the southern hemisphere (Fig. 18.18). These eerie, flickering lights generally occur in the

Figure 18.18
Aurora. The aurora borealis, or northern lights, as seen in Alaska.

Earth's polar regions and follow a timetable of being a day or so after a solar disturbance.

It is believed that the stream of charged solar particles is trapped in the Earth's magnetic field, and the particles are guided toward the poles. They excite or ionize oxygen and nitrogen molecules, which emit light on de-excitation or recombination —the glow of the aurora.

SUMMARY OF KEY TERMS

Magnetic poles: regions of concentrated magnetic strength, designated as the north pole and south pole.

Law of poles: like magnetic poles repel, and unlike magnetic poles attract.

Magnetic field: the magnetic force field surrounding a magnet or electric charge in motion. At a particular location, the magnetic force or pole in the direction of the force experienced by a north pole.

Magnetic domain: a group of aligned magnetic atoms whose interaction gives the domain region magnetic properties as a whole.

Ferromagnetic materials: strong magnetic materials, such as iron, nickel, and cobalt.

Magnetic force: a force experienced by an electric charge (or current) when moving in a magnetic field in a direction other than parallel to the field lines.

QUESTIONS

Magnets and Magnetic Fields

1. Which way would a compass located at the geographic north pole point? How about when located at the nearby south magnetic pole?

*2. Are nails attracted to each end of a bar magnet? Why or why not?

3. Nails hanging on a magnet are induced magnets (Fig. 18.19). Identify the poles on each of the nails.

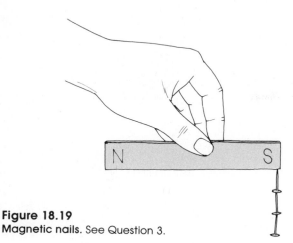

Figure 18.19
Magnetic nails. See Question 3.

4. Why do airplane pilots and ship navigators following a compass course have to make corrections to stay on a course charted on a map?

5. If the Earth's magnetic field underwent another pole reversal, what effect would this have on such things as navigation and the Van Allen belts? Think of the magnetic field weakening, going to zero, and building up again with reverse polarity.

6. Why do iron filings show magnetic field patterns?

Electromagnetism

*7. A long, straight horizontal wire carries a direct current to the north. In what general directions would a compass needle point if the compass were placed (a) above the wire, (b) below the wire, and (c) on either side of the wire? What would happen in each case if the wire carried alternating current?

8. Why does stroking a needle on a permanent magnet magnetize the needle? Should the needle be stroked back and forth if this is a desired effect? Explain.

9. Why does striking a magnet with something hard or hitting it on a hard surface weaken its magnetism? How about heating the magnet?

*10. A fellow student (who hasn't yet taken physics) thinks that a bar magnet results from a flow and separation of north and south magnetic poles, similar to electrostatic charging by induction. How could you disprove this theory?

11. Why doesn't breaking a magnet in two give single magnetic poles? How about if you kept breaking it into smaller pieces?

12. Do the poles of a magnet have definite points of location? Discuss the concept of poles in terms of electromagnetism.

13. Could you make a compass out of an electrically charged needle?

*14. You are given four items—a piece of string, a needle, and two identical iron bars, one of which is a permanent magnet. Using the string, needle, and unmagnetized bar only one time each, give three ways you could identify which bar is the magnet. (*Hint:* Two ways are pretty easy. If you get stuck on the third, keep in mind that the unmagnetized bar becomes an induced magnet when the magnetic field lines are generally along the length of the bar.) What would happen if you put the end of one bar at the mid-length point of the other?

15. If a current-carrying solenoid were suspended on a string so it could rotate freely, what would happen? Consider both dc and ac currents.

16. Could an electromagnet be made without a metal core? What are the advantages of an iron core? Does it add to the field?

*17. An electromagnet attracts magnetic materials. Does it make any difference if the magnet is operated on ac or dc? (Consider the frequency of the polarity change for the ac.)

Magnetic Forces on Moving Electric Charges

*18. What would happen in Figure 18.12(a) if the current were reversed and the magnet rotated around the wire so the pole ends pointed downward?

19. A charged nonmetallic ball is suspended on a string. Would it be possible to get the charged ball swinging by using a stationary magnet (without touching it, of course)? What would happen if you set the ball swinging with the stationary magnet nearby?

20. An electric field can be used to distinguish between positively and negatively charged particles. Could you do this with a magnetic field? Explain.

*21. An electron and a proton are traveling along the same straight line, say up the middle of this book. (a) What would happen if a uniform magnetic field were turned on in a direction perpendicular to and out of the page? (b) Parallel to the page in a bottom-to-top direction? (c) Parallel to the page in a side-to-side direction from right to left?

Magnetic Applications and Effects

22. There are both dc and ac motors, but inside a motor there must always be an alternating current. Why is this, and how is this done in a dc motor?

23. What is the difference between an electromagnetic relay and a solenoid?
24. Distinguish between electromagnetic and electromechanical devices.
‡25. Explain how the doorbell in Figure 18.11 works.
26. Explain how the door-chime mechanism shown in Figure 18.20 works.
‡27. An ammeter is a low-resistance instrument and is connected in-line or in series in a circuit. A voltmeter is a high-resistance instrument and is connected across or in parallel with a circuit component to determine the voltage drop across it. Why are high and low resistances critical for these instruments? (*Hint:* Think in terms of voltage drops and distribution of current.)
28. Why do most of the energetic charged particles in cosmic rays and the solar wind not reach the Earth's surface?
‡29. Why do ham-radio operators generally make long-distance transmissions at night or early in the morning?

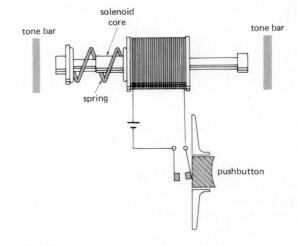

Figure 18.20
Ding dong. See Question 26.

Electromagnetic Induction

19

Using the mechanical
energy of water and
electromagnetic
induction, hydroelectricity
is generated.

As we learned in the last chapter, a current produces a magnetic field. With the similarities and symmetries in electromagnetic interactions, you may have thought about the reverse situation: Can a magnetic field produce a current? The answer is yes, under certain conditions.

After it was found in 1820 that a current produces a magnetic field, several scientists of the day asked themselves the same question. In about a decade, electromagnetic induction was discovered, and voltages and currents were induced in circuits using magnetic fields. This important discovery forms the basis for the production of electricity we use in our homes and at work. Have you ever wondered how electricity is generated? You'll find out in this chapter.

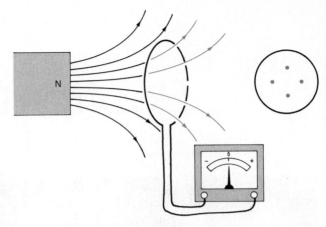

(a) no motion

FARADAY'S LAW OF INDUCTION

The trick in inducing a voltage or current in a circuit, say a wire loop, with a magnetic field is that the number of field lines through the loop must vary with time. There are several ways of doing this. For example, a magnet could be brought toward a loop of wire, as illustrated in Figure 19.1. When the magnet is stationary, there is no current in the loop. If the magnet is brought quickly toward the loop, the number of field lines through the loop increases, and a current is induced in the wire, as indicated by the deflection of the galvanometer.

If the magnet is pulled away quickly, the number of field lines through the loop decreases. The galvanometer then shows another deflection, but in the *opposite* direction, indicating that the polarity of the induced voltage and the direction of the current have changed. The magnitude of the induced voltage depends on how fast the change in the number of field lines through the loop takes place. If you moved the magnet very slowly, the induced voltage would be too small to detect. The current in the loop depends on the induced voltage and the resistance of the loop (Ohm's law, $I = V/R$).

In a practical demonstration of electromagnetic induction in this manner, a coil with several loops of wire would be used to increase the magnitude of the induced voltage. The change in the field lines through each loop contributes to the voltage. For example, a circuit with a coil of ten loops would have ten times the induced voltage of a single loop.

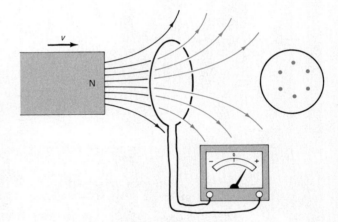

(b) magnet moved toward loop

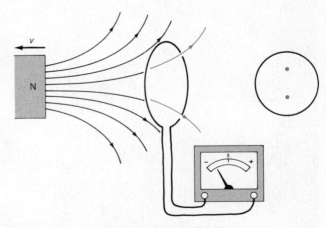

(c) magnet moved away from loop

Figure 19.1
Magnetic induction. A change in the number of magnetic field lines through a wire loop by moving a magnet induces a current in the wire.

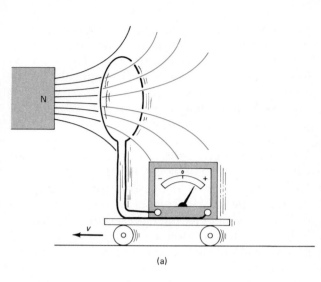

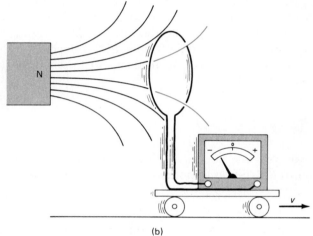

Figure 19.2
Magnetic induction with a stationary magnet and moving loop. (a) Loop moving toward magnet and number of field lines through loop increasing. (b) Loop moving away from magnet and number of field lines through loop decreasing. Note that the change in the number of field lines depends on the relative motion of a magnet or loop, so the motion of either or both could induce a current.

The same effects would be observed if the magnet were held stationary and the loop moved (Fig. 19.2). The change in the number of the field lines depends on the relative motion of the magnet and loop, so either (or both) could be moved. In one case, the field lines cut through the wire, and in the other, the wire cuts through the field lines.

Such experiments were done independently by Michael Faraday in England in 1831 and by Joseph Henry in the United States at about the same time.

The result was what we call **Faraday's law of induction**, which is generally stated:

> A voltage is induced in a conducting loop when there is a change in the magnetic field through the loop.

Question: Would there be an induced voltage or current in a loop when the plane of the loop is moved perpendicularly to a uniform magnetic field?

Answer: No. Although the wire of the loop cuts through field lines, the number of field lines through the loop on the average does not change, since the field is uniform. For every field line coming into the loop, another one leaves it, so the number of field lines through the loop stays constant.

You may have already thought of how a voltage could be induced in a loop in a uniform magnetic field. If a loop is rotated in the field, the number of field lines through the loop changes (Fig. 19.3) The field lines through the loop go from a maximum number when the plane of the loop is perpendicular to the field to zero when it is parallel to the field. On the next half-cycle or rotation, this goes from zero to a maximum again. This idea will be important later, when we discuss generators.

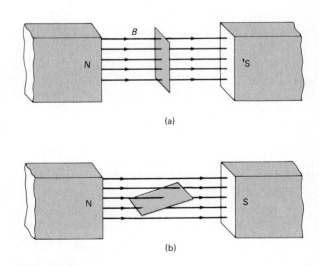

Figure 19.3
Magnetic induction by rotating a loop in a magnetic field. The number of field lines through the loop goes from some maximum value (a) to zero when the plane of the loop is parallel to the field.

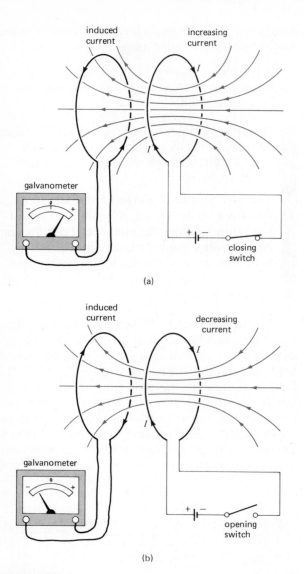

(a)

(b)

Figure 19.4
Magnetic induction by a time-varying magnetic field.
(a) When the switch is closed, a current builds up in the
right loop and a magnetic field builds up (changes)
through the other loop, inducing a current in the loop.
(b) The current in and magnetic field of the right loop decrease when the switch is opened, and the changing field
through the other loop induces a current in the opposite
direction to that in (a).

Another way to induce a voltage without having
any moving parts is to have the magnetic field
change with time. Consider the circuits with parallel loops shown in Figure 19.4. When the switch in
the battery circuit is closed, the current goes from
zero to some constant value. This occurs in a very

short time, and the increasing current produces an
increasing magnetic field around the loop. The field
goes through the second loop, the field through this
loop changes (increases, since originally it was
zero), and a voltage is induced in the loop. This
process is called **mutual induction** because it depends on the interaction of two circuits.

When the current in the battery circuit is steady,
so is the magnetic field, and there is no induced
voltage in the second loop. (Why?) However, when
the switch is opened, the current goes to zero, and
the magnetic field decreases or "collapses." The
decreasing change of the field through the second
loop gives rise to an induced voltage with a reversed
polarity and the current in the opposite direction.

You might think that there is a "moving part,"
since the switch has to be opened and closed. But
suppose the first loop were hooked to an ac source.
Would the effect not be the same? We'll return to
this idea later in the discussion of transformers.

Several effects of electromagnetic induction are
shown in Figure 19.5.

■ LENZ'S LAW

Knowing the direction of the induced voltage and
current is often important. This is given by **Lenz's
law**[*]:

> An induced current is in such a direction that
> its effects oppose the change that produces it.

For example, suppose a magnet is brought
toward a coil (Fig. 19.6). The change producing the
induced current is an increase of field through the
coil. By Lenz's law, the induced current is in the
direction that produces a magnetic field in the opposite direction to that of the magnet, so as to oppose the change or keep the field from increasing
(field lines in opposite directions cancel).

In effect, the induced magnetic field in the coil
is like that of a magnet with a like pole toward the
pole of the permanent magnet so as to oppose it.
Similarly, if the magnet is moved away from the coil,
the induced current is in the opposite direction, so
the induced magnetic field adds to the decreasing
number of field lines. The coil behaves an an attractive magnet in this case.

[*]After Heinrich Lenz, the German physicist who deduced
this law in 1834.

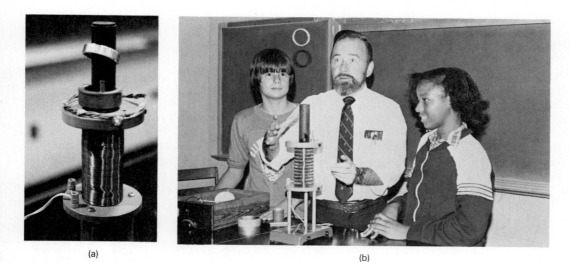

Figure 19.5
Effects of electromagnetic induction. Alternating current in the coil produces an alternating magnetic field that is concentrated by the vertical iron rod (a). This produces a current in the loops of wire, as shown by the glowing bulb just above the coil. A current gives I^2R losses, and water can be quickly boiled in the hollow metal container by induction heating. (Although difficult to see in the photo, the slight variation on the dark rod is condensed water vapor.) A current in magnetic fields experiences a force. This is upward in this case, and a metal loop "levitates" on the rod. (The hollow metal container is fastened down behind the rod.) (b) If a ring is pushed down over the rod where the magnetic field and the upward force are stronger and is then quickly released, the force causes it to jump into the air.

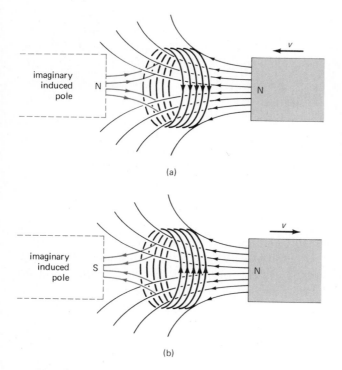

Figure 19.6
Lenz's law. The effect of Lenz's law in opposing the change of a moving magnet is as though there were an imaginary induced magnetic pole to oppose the change.

Lenz's law is really a rather subtle statement of the conservation of energy. To see this, consider a dc motor (Chapter 18, Fig. 18.13) and suppose that an induced current did not oppose the change producing it. In a dc motor with a current from an external source in its armature loops, the loops experience a torque and turn in a magnetic field. This turning changes the number of field lines through the loops (Fig. 19.3), so there is also an *induced* voltage and current in the loops.

If the field of the induced current added to that of the permanent magnet, then there would be a greater change in the number of field lines, more induced current, and so on—a runaway condition. This would be a violation of the conservation of energy, which doesn't happen because the effects of the induced current oppose the change that produces it.

So, in a dc motor the induced voltage and current are in the opposite direction and oppose the voltage and current from the external source. This gives rise to what is called a **counter or back emf,** which is an opposing voltage.* The net voltage to drive current through the resistance of the motor

*emf = electromotive force. See Chapter 17.

MagLev: The Train of the Future

A train that levitates a centimeter or two off the track and travels at 480 km/h (about 300 mi/h)? This may well be the train of the future using electromagnetic induction—the MagLev train (Fig. 1). Conventional trains have wheels that ride directly on the rails with wheel flanges against the rails for guidance, which gives large frictional losses. The maximum practical speed for such trains is about 185 mi/h (300 km/h).

The electromagnetic principles of the experimental MagLev train are involved in levitation, guidance, and propulsion. The on-board magnets are powerful superconducting magnets. At low temperatures, some materials show the property of superconductivity—a condition of zero electrical resistance (see Special Feature 17.1). In the absence of resistance, very large currents and magnetic fields are possible ($B \propto I$). Simplified diagrams of the operating principles of the MagLev train are shown in Figure 2 to assist in understanding the following descriptions.

1. Levitation. When a magnet on board the train approaches and passes over coils on the guideway, a current is induced in a coil and it becomes an electromagnet with the same polarity as the on-board magnet (Lenz's law). The repulsion of the magnets causes the train to "levitate" off the guideway.

2. Guidance. When the train deviates to either the right or left, a current is induced in coils along the sides of the guideways (which are also used for propulsion). The magnetic forces bring the train back to the center of the guideway.

3. Propulsion. The train is propelled by a "linear" synchronous motor (most motors rotate). This operates on the interactions of on-board magnets and the changing polarities of magnetic coils along the sides of the guideway [Fig. 2(c)]. The on-board magnets are attracted by the coil magnets of different polarity immediately ahead of them and repelled by those immediately behind. The polarity of the

Figure 1
MagLev train. The MagLev train levitates one-half inch off the guideway and is guided and propelled by electromagnetic principles.

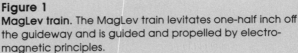

coils is less than the source voltage by the amount of the back emf (voltage). The back emf depends on the rotational speed of the armature—the faster the rotational speed, the greater the change of field lines and the greater the back emf. But the back emf tends to reduce the armature current, and hence the torque on the armature, which may allow the rotation speed to decrease. Thus, the back emf controls the rotational speed of the armature to some extent.

A back emf can be induced in a stationary coil. For example, if a coil is in a battery circuit and the circuit switch is closed, the current and magnetic field build up in the coil. The field lines of any one of the coil loops cut through the other loops, and a back emf opposes the current buildup. This is called **self-induction**. When the switch is opened, the back emf is in the opposite direction, so the induced current opposes the current decrease. In an ac circuit, there is a continuing back emf or opposition to current flow. A coil or *inductor* in an ac circuit acts somewhat like a resistance, except that the inductor opposition to current flow depends on the frequency of the driving voltage source. Why?

An application involving electromagnetic induction and Lenz's law is given in Special Feature 19.1.

guideway magnets is reversed at regular, synchronous intervals as the train moves along so that this attraction-repulsion relationship is maintained. The time interval between the guideway coil reversals establishes the speed of the train.

Causing the polarities of the guideway coils to be "out of sync" would slow the train, giving a braking action, which may be supplemented by frictional skids.

American and Japanese companies have built prototype MagLev trains. Speeds in excess of 300 mi/h have been achieved on experimental guideways. The MagLev train's greatest potential is currently in heavy, short-distance commuter transit.

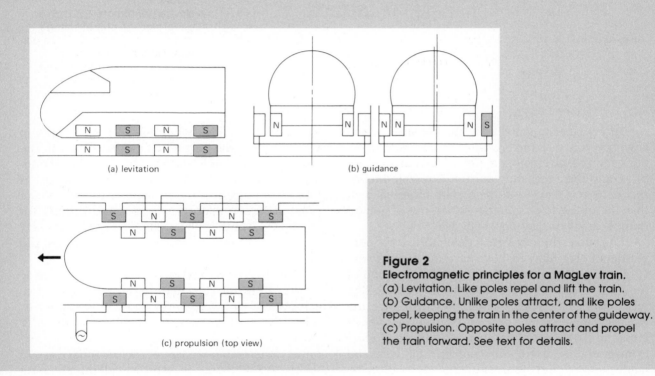

Figure 2
Electromagnetic principles for a MagLev train.
(a) Levitation. Like poles repel and lift the train.
(b) Guidance. Unlike poles attract, and like poles repel, keeping the train in the center of the guideway.
(c) Propulsion. Opposite poles attract and propel the train forward. See text for details.

■ GENERATORS

You have probably suspected that electromagnetic induction is used to generate electricity. Recall that a motor is a device that converts electrical energy into mechanical energy. A generator has the opposite function.

A generator is a device that converts mechanical energy into electrical energy.

Diagrams of simple ac and dc generators are shown in Figure 19.7. In the ac generator, the polarity of the induced voltage changes each half-cycle (by Lenz's law), and there is an ac voltage output. An ac generator is sometimes called an alternator. The ac electricity we commonly use has an alternating frequency of 60 Hz, and the 110- to 120-V voltage is a special time-averaged value, not the peak or amplitude voltage.

A dc generator can be made by using a split-ring commutator as in a dc motor. Here, the voltage polarity in the loop changes, but the change of commutator contacts each half-cycle maintains the same output polarity. The voltage output varies or pulsates.

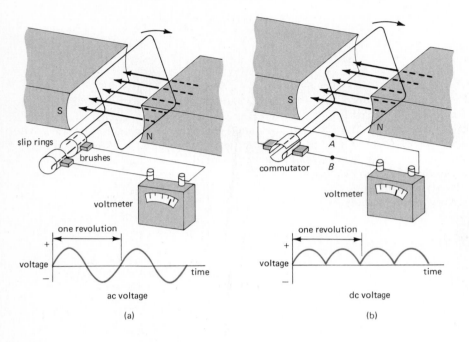

Figure 19.7
Simple generators. (a) In an ac generator, the output voltage alternates in polarity, and the current alternates. (b) In a dc generator, the voltage output varies but does not change polarity, so the current is direct.

Direct-current generators are not used much anymore. It is easier to "rectify" or convert ac to dc. For example, alternators are now used in automobiles, and the alternating current is rectified to direct current. (Why?) Alternators are more efficient than the old-style dc generators because of greater outputs at lower rotational speeds.

Of course, there is a great deal more to the generation of electricity than we have mentioned here (Fig. 19.8). Steam from the heat of combustion of fuels (or nuclear energy) or hydropower is used to turn turbines (mechanical energy) that drive the electrical generators. A generator has several sets of armature coils (not one loop). When several electromagnets are used, the magnetic field can be made to rotate, and the armature remains stationary. The induction effect is the same — relative motion.

■ TRANSFORMERS

Another important application of electromagnetic induction is the transformer. There's one outside your house, and you've probably seen them in other electrical equipment (Fig. 19.9). As shown in the figure, a **transformer** is simply two coils (primary and secondary) wrapped around an iron core. There are no moving parts.

Suppose you put a battery and a switch in the primary-coil circuit and closed the switch; what would happen? Similar to the case of the parallel loops in Figure 19.4, while the current is building up from zero, the magnetic field increases. In the transformer, this field also goes through the secondary coil and induces a voltage. If the secondary has more windings than the primary, then the voltage pulse in the secondary is increased by this factor, for example, twice the voltage if there are twice as many windings. A similar situation occurs when the switch is opened.

But for continuous operation, why not use an ac source? Here, the current turns on and off automatically 120 times a second for 60-Hz alternating current. Then if the primary input voltage were 120 V and there were twice as many windings on the secondary, the output voltage of the secondary would be 240 V! For obvious reasons, we call this a **step-up transformer.**

By the same reasoning, if the secondary has fewer windings than the primary, the output voltage will be less than the input voltage, and we have a **step-down transformer.** For example, for a 120-V input and a secondary with half the number of windings as the primary, the output voltage would be 60 V. This can be expressed by the ratios:

$$\frac{\text{Number of primary windings}}{\text{Number of secondary windings}} = \frac{\text{Primary voltage}}{\text{Secondary voltage}}$$

or

$$\frac{N_p}{N_s} = \frac{V_p}{V_s}$$

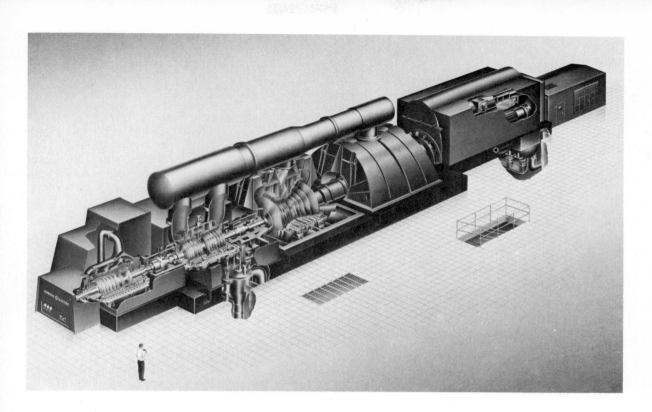

(b)

Figure 19.8
Electrical generation. (a) A cutaway view of a steam-turbine generator unit. The unit is rated at more than 500 megawatts and is capable of producing enough electricity for a city of more than half a million people. (b) A turbine for a hydroelectric generator. Moving water turns the turbine.

Solving for V_s, we have

$$V_s = \frac{N_s}{N_p} V_p$$

For example, for the case of twice as many windings on the secondary ($N_s = 2N_p$, or $N_s/N_p = 2$), the voltage is stepped up by a factor of two, $V_s = 2V_p$. Very large voltages can be obtained using step-up transformers. See Special Feature 19.2 for a practical application of this.

In stepping up the voltage, it might seem that you are getting something for nothing, but not so. Nature doesn't work that way, as you've no doubt learned by now. Remember that the voltage is energy/charge, and you can't double the voltage without giving something up. To see what this is, let's look at the conservation of energy in terms of electrical power (energy/time and $P = IV$).

Assuming that the power losses (joule heating) in the transformer are negligible, we have

Power to primary = power out of secondary

or $$I_p V_p = I_s V_s$$

Hence, if you step up the voltage, say from 120 V to 240 V, then the current is *stepped down* by the same factor.

The Auto Spark Ignition System

The common ignition system that "fires" spark plugs in an automobile engine has an ignition "coil," which is actually a transformer (figure below). The 12 V from the battery would not cause the spark discharge in the spark-plug gap to ignite the air-fuel mixture.

When a spark plug is supposed to fire, a switch (the "breaker points" or "points") breaks the circuit between the battery and the primary of the coil. (The points are located in the distributor, which distributes the secondary voltage to the spark plugs in turn.) Voltages of 12,000 V or more are generated in this manner.

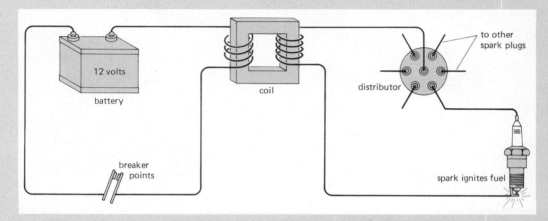
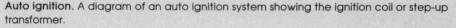

Auto ignition. A diagram of an auto ignition system showing the ignition coil or step-up transformer.

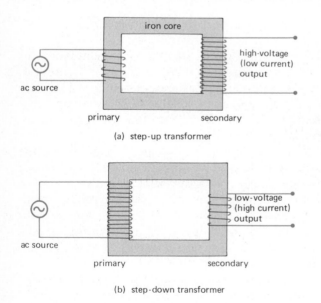

(a) step-up transformer

(b) step-down transformer

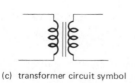

(c) transformer circuit symbol

(d)

Figure 19.9
Transformers. Transformers can be used to step up (a) or step down (b) ac voltage, depending on the relative number of turns on the transformer primary and secondary. (c) The circuit symbol for a transformer, and (d) some actual transformers.

For example, if the primary current is 10 amps, then the secondary current is 5 amps:

$$(10 \text{ amps})(120 \text{ V}) = (5 \text{ amps})(240 \text{ V})$$

So a step-up transformer steps up the voltage but *steps down* the current. Can you guess what a step-down transformer does? You've got it—it steps down the voltage but *steps up* the current.

ELECTRIC POWER TRANSMISSION

The first central electric generating and distributing facility in the world went into operation in 1882. It was built by Thomas Edison on Pearl Street in New York City and supplied direct current at 110 V through underground mains to an area roughly 3.2 km (2 mi) in diameter. For some years, dc electricity was practically the only type in use, but today almost all the electricity generated is ac. When ac systems first appeared, there were some definite opinions against alternating current. Advocates of dc branded ac as dangerous because of the high voltage used.*

The development of the transformer and of the induction motor helped lead to ac electric power systems. The transformer provided a convenient way to vary voltages and made high-voltage power transmission possible. The ac induction motor is simple and economical. In your home, the motors on the washing machine, furnace fan, and refrigerator are induction motors. In such motors, the current in the armature or rotor is induced by the changing magnetic field of alternating current in the stationary (stator) field coils. With a current in the rotor in a magnetic field, the rotor experiences a force and rotates.

A diagram of a typical ac power transmission system is shown in Figure 19.10. Notice the high voltages used in the transmission lines. Line losses are due to the I^2R (joule heat) losses of the power

---

*"My personal desire would be to prohibit entirely the use of alternating currents. They are unnecessary as they are dangerous. . . . I can therefore see no justification for the introduction of a system which has no element of permanency and every element of danger to life and property.

"I have always consistently opposed high-tension and alternating systems of electric lighting, not only on account of danger, but because of their general unreliability and unsuitability for any general system of distribution." Thomas A. Edison, 1889. (From *A Random Walk in Science,* compiled by R. L. Weber, The Institute of Physics, London, 1973.)

lines themselves. Although metal wires are used, there is still appreciable resistance, particularly over long distances. For example, heavy aluminum wire has a resistance of about 0.3Ω per kilometer. Aluminum has almost, if not completely, replaced copper in long-distance transmission lines because of its lightness, strength, and cost. An important point to notice is that the I^2R losses depend on the *square* of the current.

Let's do a little arithmetic to show why electrical power is transmitted at such high voltages. Suppose a generator has an output of 10 A at 240 V, and we want to transmit this electric power over a 50-km line with a 15Ω line resistance (50 km $\times 0.3 \Omega =$ 15Ω). If we just sent this power over the line as generated, the line I^2R losses would be

$$P_{loss} = I^2R = (10 \text{ A})^2(15 \Omega) = 1500 \text{ W}$$

Hence, the line losses would be more than 60 percent of the input power, $P = IV = (10 \text{ A})(240 \text{ V}) = 2400$ W.

But suppose we use a step-up transformer and step the voltage up by a factor of 10 to 2400 V before transmission. The current would be stepped down by a factor of 10 to 1 A. The line losses in this case would be

$$P_{loss} = I^2R = (1 \text{ A})^2(15 \Omega) = 15 \text{ W}$$

The line losses are reduced by a factor of 100 (as a result of the I^2 term) and would be less than 1 percent of the power input!

After high-voltage transmission, the power voltage is stepped down in a series of steps (Fig. 19.10) until the output voltage is the normal 240 V at your home.

Another source of power loss is through "leakage." At high transmission voltages, the air surrounding the lines may be ionized. This provides a conduction path to nearby trees, buildings, or the ground and gives rise to leakage losses. Long insulators are used on powerline towers to prevent the wires from coming too close to the tower (Fig. 19.11).

Leakage losses are generally greater during wet weather because moist air is more easily ionized than dry air. Under certain atmospheric conditions, the ionizing of air near power lines with very large voltages gives rise to corona discharge. The corona can sometimes be seen and the accompanying crackling noise heard.

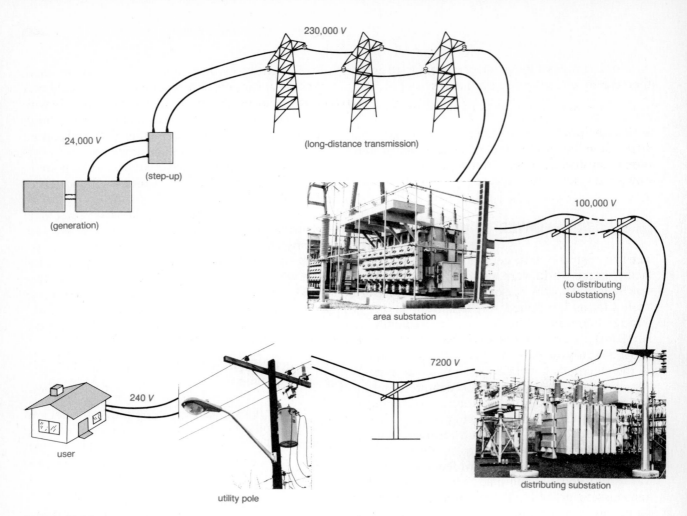

230,000 V

(long-distance transmission)

24,000 V

(step-up)

(generation)

100,000 V

(to distributing substations)

area substation

7200 V

240 V

user

utility pole

distributing substation

Figure 19.10
Electrical transmission. A diagram of a typical ac power transmission system.

Figure 19.11
Extra-high-voltage (500,000-V) transmission line. Notice
the long insulators at the towers and the insulators holding
the suspended lines apart.

■ ELECTROMAGNETIC WAVES

We have considered electromagnetic waves or radiation generally in the study of heat and energy transfer. However, we are now in a position to understand the electromagnetic nature of these waves that are all around us. James Clerk Maxwell (1831–1879), an English scientist, collectively summarized the relationships between electricity and magnetism. The relationships between electric and magnetic fields are expressed in what are known as **Maxwell's equations**. A qualitative description of the general results is as follows:

> A changing magnetic field produces a changing electric field.
> A changing electric field produces a changing magnetic field.

Here, *changing* means "varying with time."

The first statement reflects the fact that a changing magnetic field gives rise to an induced voltage, or *an electric field in space*. The second statement implies that a changing electric field in space gives rise to a changing magnetic field in space.

Electromagnetic waves are radiated by accelerating electric charges. Suppose electrons oscillate back and forth in simple harmonic motion. This could occur in a radio transmitter, where the oscillations have frequencies on the order of 1 million hertz. During the oscillations, the electrons are accelerated and radiate electromagnetic waves (Fig. 19.12).

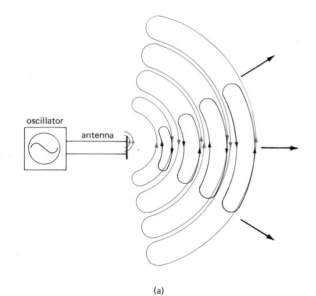

(a)

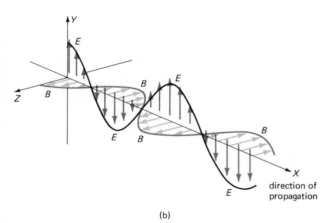

(b)

Figure 19.12
Electromagnetic waves. (a) Electromagnetic waves are generated from electron oscillations in antennas. (b) The oscillating electric and magnetic fields radiating outward are perpendicular to each other and to the direction of propagation.

The continual oscillations of the charges produce oscillating electric and magnetic fields. These propagate outward, with the wave energy contained in the fields. The oscillating electric and magnetic fields are perpendicular to each other and to the direction of propagation.

Recall that electromagnetic waves need no medium of propagation and that all travel in vacuum at the same speed—the speed of light ($c = 3 \times 10^8$ m/s $= 30,000$ km/s $\approx 186,000$ mi/s). What distinguishes one wave from another is the frequency or wavelength ($\lambda f = c$), which forms an electromagnetic (EM) spectrum (Fig. 19.13). Because of the short wavelengths involved with electromagnetic waves, the small length unit of the nanometer (nm) is used, where 1 nm $= 10^{-9}$ m $= 10^{-7}$ cm. Notice that the greater the frequency of a wave, the shorter its wavelength, and vice versa ($\lambda = c/f$).

Similar to the sound-frequency spectrum, different regions of the EM spectrum are given different names. Let's take a look at these frequency regions.

Power Frequency Waves

Electromagnetic waves of the 60-Hz power frequency result from ac currents moving back and forth in electrical power lines and circuits. With wavelengths of about 5000 km (3100 mi), these waves are of little practical importance. Occasionally you may pick them up as a 60-Hz ac hum in a stereo.

Radio and TV Waves

AM radio waves are generally in the kilohertz (kHz) range and FM radio waves in the megahertz (MHz) range. Higher frequencies are used in "short-wave" radio. Television channels are also in the megahertz range.

Microwaves

Microwaves, with a frequency range of 10^9 to 10^{11} Hz, are used in communication and radar applications. Probably the most common use today is in microwave ovens (Chapter 12).

Infrared Radiation

Just below the visible region is the invisible infrared region. This type of radiation was considered earlier

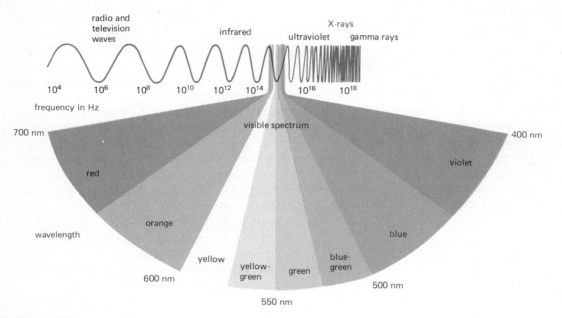

Figure 19.13
The EM spectrum. Note that the visible region is only a small part of the spectrum.

Visible Light

The visible spectrum is a very small part of the total EM spectrum, occupying a region between about 4×10^{14} Hz and 7×10^{14} Hz. As the activator of one of our major senses, visible light will be treated in greater detail in a later chapter.

Ultraviolet Radiation

Fortunately, most of the nonvisible ultraviolet (uv) radiation is absorbed in the atmosphere by the ozone layer. If this were not the case, we would be badly burned and our environment would not be as it is today. [You may recall that there was concern about depletion of the ozone layer, which has its maximum concentration at an altitude of about 30 km (20 mi), through chemical reaction with the Freon gas used as a spray-can propellant. Other gases are now used, but the depletion is still occurring.] A small portion of the solar uv radiation is

in the greenhouse effect and in radiation heat transfer (Chapter 12).

transmitted to the Earth's surface, and you go out and expose yourself to this uv radiation if you go sunbathing to get a tan.

The protective mechanism of the skin against uv radiation is pigmentation. The degree of penetration of uv radiation depends on the amount of melanin pigment in our skin, which all people have in varying amounts, except albinos. When skin is exposed to sunlight, the uv radiation induces the production of more pigment, causing the skin to "tan."

People with little pigment may initially redden and burn. We now have chemicals or "sunscreens" with molecules that absorb part of the uv radiation before it reaches the skin and prevents severe burning. Incidentally, ultraviolet radiation is also absorbed by ordinary glass, so you cannot get a tan through glass windows. You feel the "heat" coming through the window as a result of transmitted infrared radiation.

Ultraviolet radiation can be harmful to the eyes. Welders wear goggles or face masks with special uv-absorbing glass to protect their eyes from the large amounts of uv radiation produced by welding arcs. If you're a (snow) skier, you may also wear protective glasses, particularly in going cross-country. Sunlight reflected from snowy surfaces can produce "snow blindness" due to uv radiation.

X-Rays

Most of us have had an "X-ray" to check for a broken bone or some other medical condition. X-rays are highly penetrating electromagnetic radiations that can pass through many solid bodies that are opaque to other types of radiations. More dense, solid objects absorb more of the X-rays than less dense materials. In the human body, the bones absorb more X-rays than the surrounding tissue. As a result, a contrasted picture of internal body structures is obtained (Fig. 19.14).

Another type of X-ray technique, the CAT scan, used to "see" inside the body is discussed in Special Feature 19.3.

X-rays are used to check metal pieces for internal cracks, for example, stress cracks in airplane wings. At the airport, your bags get a shot of X-rays as they go through the security check so that the security officer can check out any metal objects in your bag. (When you walk through the check port and your keys set off a buzzer, this is a magnetic effect; the metal in the keys takes energy from a magnetic field, and this sets off an alarm.)

The basic principle of an X-ray tube is illustrated in Figure 19.15. Electrons emitted from the heated cathode are accelerated toward a metal anode "target" by an applied high voltage across the tube (typically 50 to 150 kV). As the "cathode rays" collide with the target, they interact electrically with the electrons of the target material. The repulsive electric forces cause the cathode electrons to slow down. In doing so, they lose energy in the form of electromagnetic radiation, X-rays. In German the name for X-rays is *bremsstrahlung*, meaning "braking rays."

Gamma Rays

The gamma rays (γ-rays) at the upper end of the electromagnetic spectrum are produced in nuclear reactions and other processes in particle accelerators. This type of radiation will be considered in later chapters, where these processes are discussed.

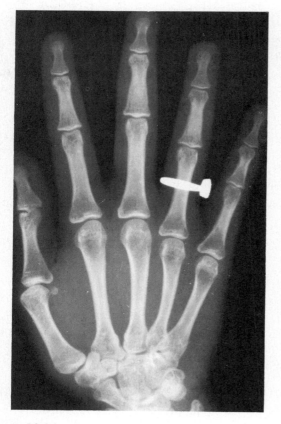

Figure 19.14
X-rays. Variations in X-ray absorption give a contrasted picture of the internal structures of a hand (and external structures — the ring).

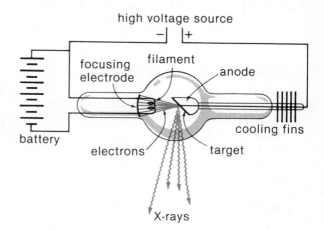

Figure 19.15
An X-ray tube. High-energy electrons from the cathode are decelerated by the electrons in the target anode. Radiation (energy) in the form of X-rays is emitted as a result of this "braking" action.

The X-ray CAT Scan

CAT scan refers to a relatively new X-ray technique (computerized axial tomography), not a feline look. For conventional medical and dental X-ray photographs, the X-rays emerging from the tube are detected on a photographic film. The X-rays themselves may expose the film or excite some fluorescent material that produces light for the film exposure. The latter method reduces the amount of X-rays needed. The difference in the absorption by different structures in the body gives rise to the image production — the less the absorption, the greater the transmission and the darker the film. In a sense, the image is a "shadow" of what the rays have passed through.

In the 1970's a new technique called computerized tomography (CT) was developed. In conventional X-ray images, the entire thickness of the body is projected on the film and structures overlap, making them difficult to distinguish in many cases. A tomographic image, on the other hand, is an image of a "slice" through the body. (Tomography comes from the Greek word *tomo*, meaning "slice," and *graph*, meaning "picture.")

Measurements are made at a larger number of points along a "slice" as the X-ray source and detector move past the body together. The apparatus is rotated slightly about the body axis and another scan is made, and so on. This would take some time to get data for a complete picture, but fan beams and multiple detectors speed up the process.

The image of a slice is perpendicular to the long axis of the body. For this reason, *CAT* stands for **computerized axial tomography.** However, the abbreviation *CAT* can also stand for **computer-assisted to-**

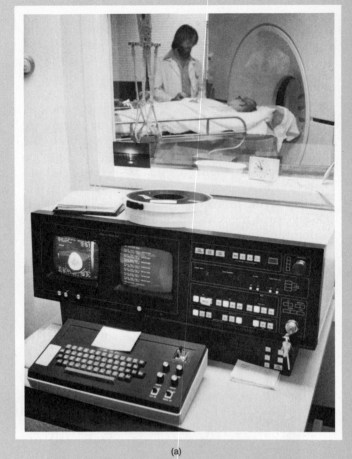

(a)

mography. The computer assists in reconstructing the angular slices to give pictures with resolution that cannot be achieved by conventional X-ray photographs (figure above).

SUMMARY OF KEY TERMS

Electromagnetic induction: the inducing of a voltage or electric field as a result of a changing (time-varying) magnetic field.

Faraday's law of induction: a voltage is induced in a conducting loop when there is a change in the number of field lines through the loop.

Lenz's law: an induced current is in such a direction that its effects oppose the change that produces it.

Generator: a device that converts mechanical energy into electrical energy.

Transformer: a device that increases or decreases the magnitude of a voltage or current through electromagnetic induction. The change depends on the primary and secondary windings:

$$\frac{\text{No. of primary turns } (N_p)}{\text{No. of secondary turns } (N_s)} = \frac{\text{Primary voltage } (V_p)}{\text{Secondary voltage } (V_s)}$$

and

$$\text{Power to primary } (I_p V_p) = \text{power out of secondary } (I_s V_s)$$

if there are no losses.

Electromagnetic waves: propagating electric and magnetic fields radiated from accelerated charged particles.

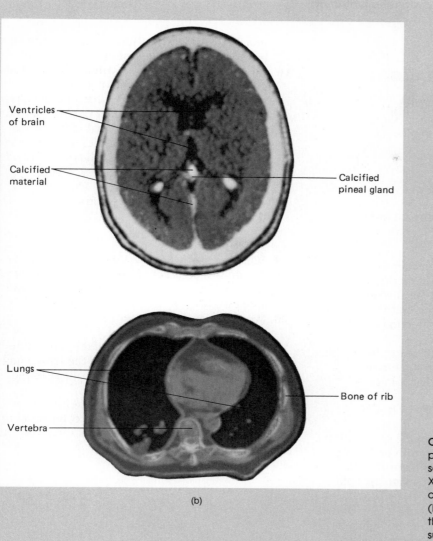

Ventricles of brain

Calcified material

Calcified pineal gland

Lungs

Bone of rib

Vertebra

(b)

CAT scan. (a) A patient being prepared for a CAT (or CT for short) scan. A computer is combined with X-ray equipment to produce images of cross sections through the body. (b) CAT scans through the brain and the chest. The densest structures, such as bone, appear white in a CAT scan.

QUESTIONS

Faraday's Law of Induction (and Lenz's Law)

1. Give three ways you could induce a voltage in a wire loop. Could you make it four or five ways?

*2. If a thin coil with many loops is rotated in the Earth's magnetic field, a voltage is induced in the coil. Since the Earth's magnetic field is "free," why isn't this method used to generate electricity?

3. Does the orientation of a loop with respect to a magnet make any difference in inducing a voltage through relative motion? Explain.

4. A voltage is induced in a stationary loop when a magnet is brought toward it or when the loop is brought toward a stationary magnet. What would be the effect if (a) both the magnet and the loop were brought toward each other? (b) if both the magnet and loop had the same velocity (both magnitude and direction)?

5. Suppose the loop in Figure 19.3 were moved parallel to the field. Would there be an induced voltage in the loop? Explain.

*6. A metal frame with a movable metal bar is in a uniform magnetic field (Fig. 19.16). (a) If the bar is moved to the right, as shown in the figure, would there be an induced voltage in the frame? Explain. If so, what is the direction of the induced magnetic field? (b) What would be the case if the bar is moved in the opposite direction?

351

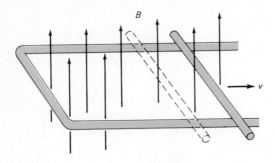

Figure 19.16
Changing the number of field lines. See Question 6.

•7. Referring to Figure 19.16, consider the effect on an electron in the bar being in motion in a magnetic field (cf. Chapter 18). In what general direction is the magnetic force on the electron? How does this relate to the direction of the induced current?

•8. Where does the induced voltage in a loop come from? Is this a case of getting something for nothing? (*Hint*: Remember that voltage is energy per charge.)

9. In inducing a current in a loop by bringing a magnet toward the loop, would it make any difference in either the magnitude or direction of the induced current if the north or south pole end of the magnet were used? Explain. (*Hint*: Think in terms of the induced magnetic field opposing the change.)

10. When a magnet is brought toward a loop or coil of wire, the magnetic field from the induced current is like that of a repelling magnet (see Fig. 19.6). As a result, work must be done in moving the magnet into the coil. Why is this?

•11. A bar magnet is dropped through a coil (Fig. 19.17). (a) What will be observed on the galvanometer? (b) As the magnet enters, will the acceleration be greater or less than *g*? How about as it leaves? Explain.

12. Distinguish between self-induction and mutual induction.

13. How will MagLev trains' levitation, guidance, and propulsion depend on electromagnetic induction?

Generators

14. Suppose the frequency of a generator is increased from 60 Hz to 120 Hz. What effect would this have on the ouput voltage?

15. A motor and a generator basically perform opposite functions. Yet someone makes a state-ment that a motor really acts as a motor and a generator at the same time. Is this really true?

•16. An inventor proposes to revolutionize electrical generation with a new type of generator, as shown in Figure 19.18. (a) What type of voltage would the generator produce? (b) The inventor states that the internal frictional losses of the spring are practically negligible, so the magnet will oscillate indefinitely and therefore provide a cheap source of electrical power. Would you financially back this invention? Explain.

17. Give three ways the voltage output of a generator could be increased.

Transformers

18. What is actually transformed in a transformer?

19. Would two coils act as a transformer without an iron core? If so, why not omit the core to save money?

20. Would a transformer operate on (a) a battery? (b) a dc generator?

21. What would be the effect(s) if the primary and secondary of a transformer had the same number of windings? What kind of "step" transformer would this be?

22. Some transformers have various terminals or "taps" on the secondary so that connecting to different taps puts different fractions of the total number of secondary windings into a circuit. What is the advantage of this?

Figure 19.17
A student generator? See Question 11.

Figure 19.18
A bright idea? See Question 16.

pull down
and release
to turn on
lamp.

*23. A fluorescent lamp needs an initial voltage of 12,000 V to ionize the internal gases. This is supplied by a transformer with a 120-V input. How many more windings are on the transformer secondary than on the primary?

24. An ac source has a 10-V output. A particular circuit requires only a 2-VAC input. How could you accomplish this? Explain.

*25. A person has a single transformer with 50 turns on one part of the core and 500 turns on the other. Is this a step-up or a step-down transformer? Explain.

26. Does a transformer operate on self-induction or mutual induction? Does it have both? If so, what are the effects?

27. In the ignition system shown in the figure in Special Feature 19.2, if 12,000 V is required to fire a spark plug, what can you say about the "coil" windings?

Electric Power Transmission

*28. If the line losses in electric power transmission are reduced by stepping up the voltage, why not transmit electric power at even higher voltages than those normally used since the step-up depends only on the relative number of windings on the transformer core?

29. What is a major disadvantage of long-distance dc power transmission?

30. Where are step-down transformers used in electrical power transmission systems?

31. Does the weather affect power line losses? Explain.

32. Figure 19.19 shows a satellite photo of the United States at night. What can you say about the relative use of electricity? Can you pick out some specific cities in the United States?

Figure 19.19
All lit up. See Question 32.

Electromagnetic Waves

•33. Are all electromagnetic waves the same? Explain.

34. How do radio waves differ from television waves?

35. Why do some people get darker and tanner when continually exposed to sunlight? Why do most people get sunburned initially with a long exposure?

•36. When you are sunbathing and the Sun "goes behind a cloud," it feels cool. It is also common to get a sunburn on a cloudy, cool day at the beginning of summer (before you tan). Explain these experiences in terms of electromagnetic absorption and transmission. (*Hint*: Think about the "greenhouse" effect, Chapter 12.)

37. How do "sunscreens" in suntan oils and lotions prevent sunburns? [*Hint*: Think in terms of molecular resonance and absorption as in microwave ovens (cf. Chapter 12).]

38. In movies and actual practice, money is sometimes "marked" and messages written with substances that can be seen only under ultraviolet light. This has to do with phosphor materials and fluorescence. Read ahead in Chapter 20 and find out and explain why.

39. Why is an X-ray tube sometimes called a cathode-ray tube?

40. Why is an X-ray CAT scan different from and so much superior to the normal X-ray plate? CAT scans haven't replaced the normal plate X-rays. Why not if CAT scans are so superior?

•41. Electromagnetic waves are generally radiated by accelerating electric charges. Yet X-rays are called "braking rays" instead of "accelerating rays." Explain why.

LIGHT

VI

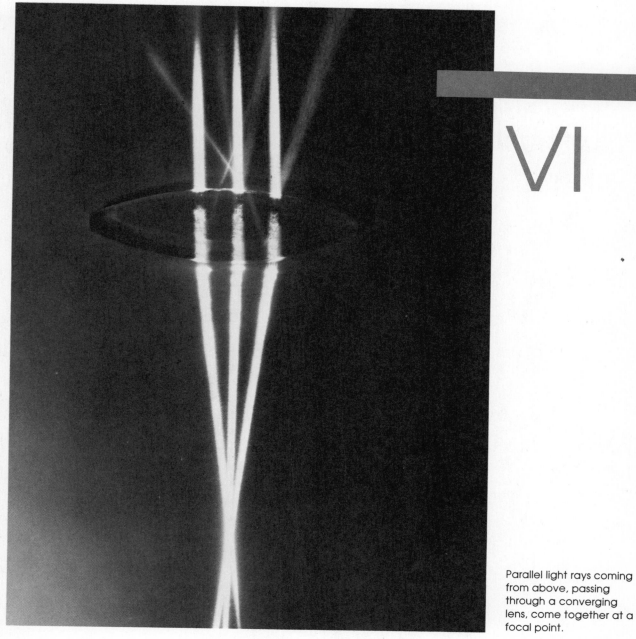

Parallel light rays coming from above, passing through a converging lens, come together at a focal point.

PART VI
The Nature
of Light

The light entering our eyes brings to us the visual information about our world—the beauty of sunsets; the features of objects, including our own and other persons' "looks"; the existence of stars; and so on. It is a small wonder that the nature of light has occupied scientific thought from the earliest times. Have you ever asked yourself the question, "What is light?"

One theory, dating back to the ancient Greeks, considered light to be composed of "corpuscles" or particles, and this theory survived relatively unquestioned for many centuries. But, evidence began to be noted that supported a wave nature of light. Leonardo da Vinci (circa 1500), noting the similarity between sound echoes and the reflection of light, speculated that light might have a wave nature. By the 17th century a definite controversy existed over the nature of light. Isaac Newton favored a theory based on light being a stream of particles.

However, by the late 19th century there appeared to be overwhelming evidence in support of the wave theory. Experiments showed that light exhibits wave properties such as interference and the Doppler effect, so it must be a wave. Then, studies in electromagnetism showed that light is a special kind of wave—an electromagnetic wave (Chapter 19).

Interesting enough, we have been forced to return to a "particle" description of light in certain cases. In some instances light behaves like a wave, whereas in other instances it must be described as a particle or packet of energy in order to explain what is observed. This has given rise to what is called the *dual nature of light*. Have you ever heard of this? For our present discussion, we will focus on the wave nature of light and look at some interesting practical applications of light. The particle nature of light will be discussed later in the text (Part Seven: Modern Physics).

Light Waves

20

Light from the Earth's main source of energy, the Sun.

LIGHT EMISSION

Electromagnetic radiation spans a wide range of frequencies, but light is generally taken to mean electromagnetic radiation in or near the visible region. For example, radio waves are electromagnetic waves but are not generally thought of as light. To produce radio waves, electrons are oscillated back and forth in metal antennas hundreds of thousands of times a second.

The oscillation frequencies are controlled by crystals, for example, quartz, that vibrate naturally at radio frequencies. These resonant frequencies depend on the elasticity of the crystal material and on the size and shape of the crystal. In general, the higher the frequency is, the smaller the crystal, in much the same way as a high sound frequency or pitch generally comes from smaller bells.

For light to radiate from an antenna, the electrons would have to oscillate many billions or trillions of times each second. This is beyond the range of crystal vibrations in both size and elasticity. The "antennas" of visible light emissions are atoms themselves. In our simplified model of the atom (Chapter 7) we speak of transitions rather than vibrations. When atoms become excited, electrons are raised to higher orbits or shells (Fig. 20.1).

These are states of higher energies. The electrons generally remain in the higher energy levels for only fractions of seconds and then spontaneously return to their normal orbits or states. When they do so, energy is emitted as light or electromagnetic radiation, with different frequencies of light corresponding to transitions between different energy levels.

The excitation in most materials is short-lived, but the electrons in some materials remain in excited states for long periods of time—minutes and hours. As a result, light is emitted from these materials long after they have been excited, and they can glow in the dark. You have probably seen novelty items and toys that have done this, as well as some luminous clock dials. This process of electrons being "hung up" in excited states is called **phosphorescence**. As you might expect, phosphorus is a good phosphorescent material.

Atoms may be excited by various means, all of which involve a transfer of energy. Phosphorescent materials are excited by exposing them to light. (Remember that waves carry energy.) In the case of a gas discharge tube, electrons boiled off the cathode excite the gas molecules in the tube through collisions. Depending on the type of gas in the tube, light of certain frequencies or colors is emitted.

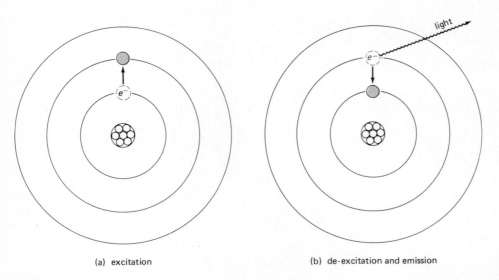

(a) excitation (b) de-excitation and emission

Figure 20.1
Light emission. (a) An atomic electron is excited to a higher orbit. (b) In de-excitation, the electron returns to a lower orbit, with energy being emitted as light or electromagnetic radiation. (It will be learned in a later chapter that the orbits are not evenly spaced, as drawn here for simplicity.)

For example, neon tubes used in "neon signs" emit light primarily in the red end of the visible spectrum. Other gases produce different colors, since different atoms and molecules have different orbit spacings and emit different, but particular, light frequencies that determine the color of the emitted light. Light from a sodium lamp, for example, is yellow, and the mercury vapor in mercury lamps commonly used for street lights has a large number of transitions, emitting light in the blue end of the visible spectrum.

The situation is different if we look at light from an incandescent light bulb. In the common light bulb, a solid tungsten filament glows as a result of electrical joule heating and emits visible light of all wavelengths or colors. This is called white light, which is a mixture of all visible wavelengths or frequencies. (Light with a single frequency or wavelength is called **monochromatic** — "one color.") In solids, the atoms are closely packed together. This influences the spacings of the atomic electron

(a)

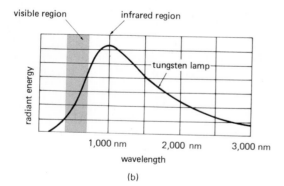

(b)

Figure 20.2
The radiant energy spectrum for an incandescent lamp. Note that only a small portion of the energy is emitted in the visible region of the spectrum, making the lamp very inefficient as a light source.

levels to the point that an electron from one atom can jump to the level of another atom. There are so many possibilities that transitions for all the frequencies in the visible spectrum exist and occur. The incandescent lamp is very inefficient. Most of the radiation is in the infrared portion of the spectrum (Fig. 20.2), and 95 percent or more of the energy is given off as heat rather than visible light. (Recall that infrared radiation is called heat waves.)

◼ FLUORESCENCE

Another common light source is the fluorescent lamp. This is basically a gas-discharge tube containing mercury vapor. When the lamp is in operation, electrons from the electrode excite the mercury atoms, which emit ultraviolet radiation in the de-excitation transitions (Fig. 20.3). The ultraviolet light is absorbed by a thin coating of phosphor material on the inner surface of the glass tube, and the phosphor emits visible light in a process called **fluorescence**.

A fluorescent material absorbs ultraviolet light and emits light at a lower frequency, either because part of the energy goes into the kinetic energy of an entire atom or because an excited electron jumps from orbit to orbit, or energy level to energy level, down the "energy ladder." Perhaps a rung is skipped now and then instead of going directly from the initial excited state to its normal state (Fig. 20.4).

In either case, many frequencies of light are emitted that combine to produce white light. Because of the fluorescent conversion and lower operating temperatures, fluorescent lamps are considerably more efficient than incandescent lamps. Notice in the graph in Figure 20.3 that most of the light is emitted in the visible region. Different types of phosphors emit light of different colors. The phosphors in the so-called "cool white" fluorescent lamps emit nearly all visible colors and give very white light. "Warm white" lamps have phosphors that emit more red light, giving a warm glow.

The fluorescent excitation by ultraviolet (uv) radiation has another common use. "Black lights" used at discos emit radiation in the violet and near-ultraviolet regions. The uv black lights cause fluorescent paints and dyes on signs, posters, and performers' clothes to fluoresce with brilliant colors. Many products, such as laundry detergents, are

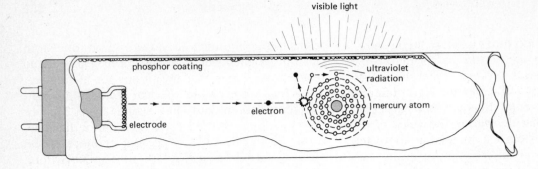

(a)

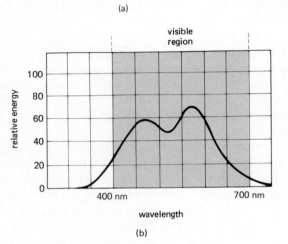

(b)

Figure 20.3
The fluorescent lamp. (a) Ultraviolet light emitted by mercury atoms is absorbed by the phosphor material on the inner surface of the tube and is re-emitted as visible light. (b) As a result, most of the emitted light is in the visible region as the spectrum shows. (Compare with Figure 20.2.)

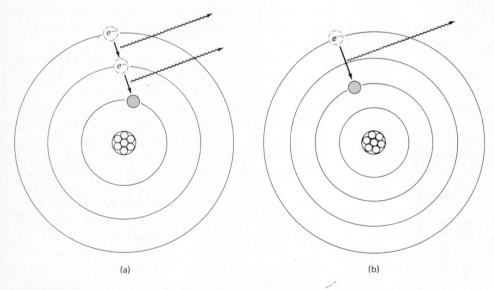

(a)   (b)

Figure 20.4
Energy levels. (a) In the de-excitation process, an electron may jump from orbit to orbit (energy level to energy level) down the "energy ladder." (b) It may also skip a "rung" or two in going back to its normal unexcited state. (It will be learned in a later chapter that the energy ladder "rungs" aren't evenly spaced as drawn here for simplicity.)

Electromagnetism and Light

All lit up. A view of the Space Needle and the Seattle skyline at night. More than half of the area's electricity is marketed by the federal government agency that maintains a network of more than 12,000 miles of transmission lines in the Pacific Northwest.

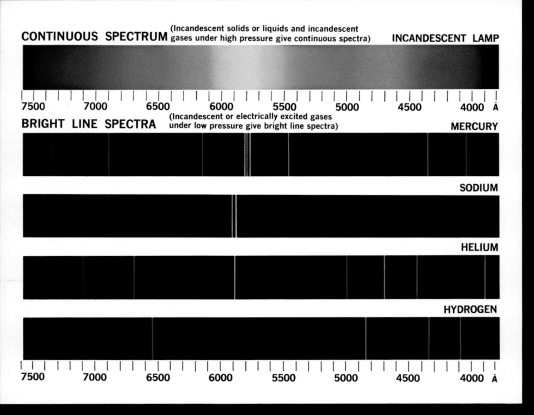

CONTINUOUS SPECTRUM

(Incandescent solids or liquids and incandescent gases under high pressure give continuous spectra)

INCANDESCENT LAMP

7500   7000   6500   6000   5500   5000   4500   4000 Å

BRIGHT LINE SPECTRA

(Incandescent or electrically excited gases under low pressure give bright line spectra)

MERCURY

SODIUM

HELIUM

HYDROGEN

7500   7000   6500   6000   5500   5000   4500   4000 Å

Continuous and line spectra.

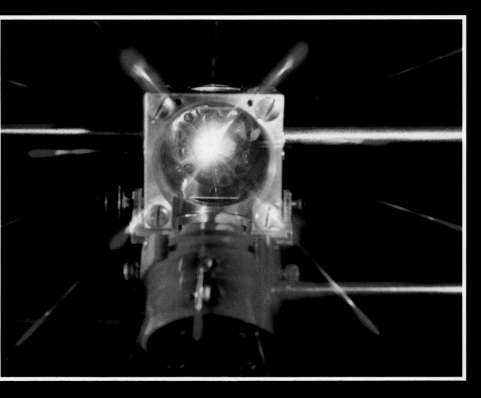

Solar cell research. To improve efficiency, a "tandem" solar cell using two photoelectric materials is tested. A spectrum-splitting filter is used to divide sunlight into two beams, one containing wavelengths best suited to produce a current in silicon and the other containing wavelengths more suited to gallium arsenide.

packaged in boxes that have bright colors containing phosphors in the ink. These fluoresce somewhat under the store's fluorescent lamps and appear brighter to get your attention. In the dark, when illuminated with black light, the boxes glow brilliantly. Some natural minerals are also fluorescent. Illumination with black lights (uv) is a method of identifying these materials.

■ INTERFERENCE

As a wave, light shows interference phenomena. Recall from Chapter 14 that waves superimpose and interfere constructively and destructively. One of the earliest demonstrations of interference of light was performed in 1801 by Thomas Young, an English scientist. This experiment helped establish the wave nature of light, as well as giving a means to measure the wavelength of light. It is often called Young's double-slit experiment because he used light from a single source passing through two small slits (Fig. 20.5). The slits act as two similar "sources" that emit the same light together.

Because of the path difference from the slits to the screen, there are bright and dark regions of constructive and destructive interference. At positions where the waves arrive in phase — for example, wave crests arriving at the same time — there is constructive interference, and a bright "fringe" is seen on the screen. At intervening positions, the waves arrive out of phase, and dark fringes occur as a result of destructive interference.

By taking a photograph of the interference pattern and measuring the distance of one of the bright fringes from the central maximum, which is on the order of millimeters, the wavelength of a monochromatic light source can be computed from the interference conditions and geometry of the experiment. This is a neat way of measuring something that is less than one millionth of a meter.

An interference effect involving another electromagnetic wave is sometimes observed on TV. You have probably noticed how a TV picture flutters when an airplane flies over. This arises when TV waves are reflected from the plane. The TV set then receives two signals, one reflected from the plane and the direct signal from the television transmitter. The waves travel different distances and may arrive at the TV set either in or out of phase. But as the plane moves at a high speed, the path length changes, and the interference at the set shifts rapidly back and forth between constructive and destructive interference, giving a fluttering in the picture.

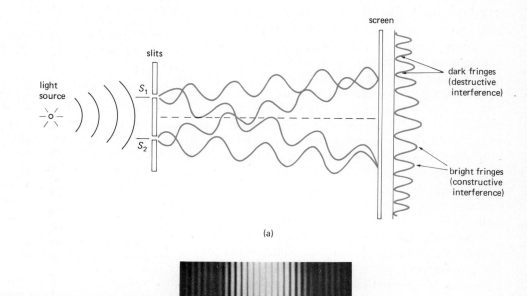

(a)

Figure 20.5
Young's double-slit experiment. (a) Light from the slits arriving in phase at some points on a screen interferes constructively and gives bright fringes. Destructive interference gives dark fringes. This produces an interference pattern (b).

(b)

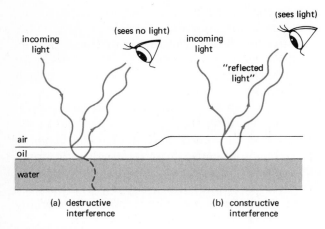

Figure 20.6
Thin film interference. Reflections from the top and bottom surfaces give rise to interference. (a) If the film thickness is such that the reflected waves are completely out of phase, destructive interference occurs, and no light is seen. (The light is transmitted, not destroyed.) (b) With the film thicknesses such that in-phase constructive interference occurs, light is seen (reflected).

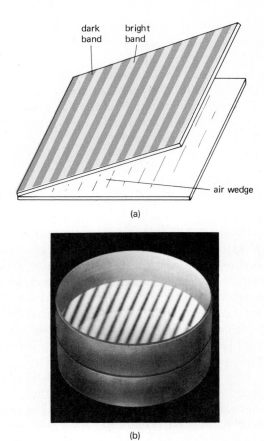

Figure 20.7
Interference in action. An air wedge (a) and an optical flat (b) can be used to check the smoothness of a reflecting surface. If the surface is smooth, a regular interference pattern occurs.

■ THIN FILM INTERFERENCE

Have you ever noticed the multicolored patterns on oil or gasoline slicks on a wet pavement or in soap bubbles and wondered what caused them? These are common examples of interference—thin film interference. Even though the films are transparent, there is some reflection at the surfaces, and interference occurs for light reflected from the top and bottom surfaces of a film (Fig. 20.6). When the film thickness is such that the reflected waves are in phase, constructive interference occurs, and the light is reflected from the film. When the film thickness is such that the reflected waves are completely out of phase, destructive interference occurs for the reflected waves, and the light is transmitted (not destroyed).

The path difference of the reflected waves also depends on the angle at which the light strikes the film. In any case, when white light falls on a thin film, some wavelengths are reflected and others are transmitted. For a film of relatively uniform thickness, the reflection may be selective for a particular wavelength region of the visible spectrum, and the film is seen to be of one particular color. However, oil and soap films generally vary in thickness, and a vivid display of different colors corresponding to different thicknesses of film is seen.

For the oil film in Figure 20.6, with normal (direct) incidence, destructive interference or transmission occurs for light when the film thickness is one fourth of a wavelength. The path difference is $\frac{1}{4}\lambda$ down and $\frac{1}{4}\lambda$ back, so the two reflected waves are out of phase by half a wavelength, or a crest meets a trough when the reflected waves interfere.*

* When light is reflected from a more "optically dense" medium, the wave is inverted or undergoes a 180° phase shift. This is analogous to a wave in a rope being reflected from a rigid support (see Chapter 14). For the case in Figure 20.6, the reflections at both surfaces undergo phase shifts, so there is no change in the phase difference of waves. On the other hand, for a soap film the top surface reflection is phase-shifted, but the bottom surface reflection is not (as in a rope with a movable support). As a result, destructive interference occurs for film thicknesses of $\lambda/2$ for normal (direct) incident light.

Nonreflecting Lenses

Thin films are used to make nonreflective glass lenses, such as those used on cameras and binoculars. Interference properties allow for the reduction of reflected light from the lenses or the greater transmission of light, which is needed for exposing camera film or for binocular viewing. A nonreflective lens has a thin film coating with a thickness of $\frac{1}{4}\lambda$ for a particular light component. This is usually chosen in the yellow-green portion of the visible spectrum (about 550 nm or 5500 Å wavelength) for which the eye is most sensitive. Other wavelengths are still reflected and give the glass a somewhat bluish appearance. You have probably noticed this bluish hue for coated lenses in cameras and other optical instruments.

The film coating serves a double purpose. Not only does it promote nonreflection from the front of the lens, but it also cuts down on back reflection (figure below). Some of the light transmitted through the lens is reflected from the back surface. This could be reflected again from the front surface of an uncoated lens and give rise to a poor image on the photographic film in a camera. However, the reflections from the two film surfaces of a coated lens interfere destructively, and there is no reflection.

Nonreflective coatings for lenses. (a) Reflections from the interior surface of a lens can give a poor focus or image on the camera film. (b) With a proper thin film coating, the reflections from the film surfaces interfere destructively, and there is no reflection.

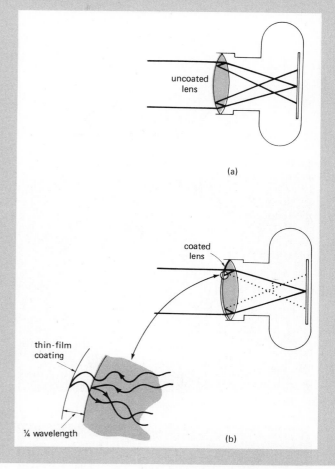

An application of this type of destructive thin film interference is given in Special Feature 20.1.

Thin film interference is used to check the smoothness and uniformity of optical components such as mirrors and lenses. Glass plates, called optical flats, are ground and polished so they are as flat or as smooth as possible. The smoothness of the flats can be checked by putting the plates together so there is a uniform air wedge between the flats (Fig. 20.7). If they are smooth, regular interference patterns occur as a result of the uniformly increasing path differences between the plates. If a surface (or surfaces) is not smooth, there would be an irregularity in the interference pattern in that region. Once you have an optical flat, it can be used to determine the flatness of a reflecting surface such as a glass mirror or metal plate.

A similar technique is used to check the smoothness and symmetry of lenses (Fig. 20.8). When a curved lens is placed on an optical flat, there is a curved air wedge between the lens and the flat. The regular interference pattern in this case is a set of concentric bright and dark circular fringes or rings. They are called **Newton's rings**, after Sir Isaac,

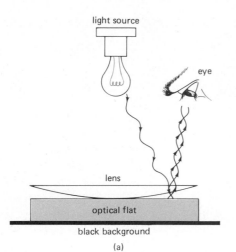

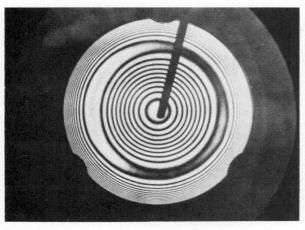

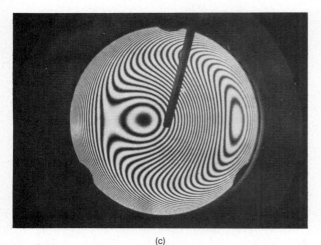

Figure 20.8
Newton's rings. (a) Lens inspection using an optical flat.
(b) Circular interference fringes or rings indicate that the
lens is of high quality and smoothly ground. (c) A distorted
interference pattern indicates irregularities.

who described this interference effect. If there are irregularities in the lens surface, a distorted ring pattern is observed.

■ DIFFRACTION

We think of light waves and waves in general as traveling in straight lines, unless they are deviated from their straight-line paths by reflection or refraction as for sound waves in Chapter 15. However, waves may be deflected or "bent" by another means called **diffraction**, which is the bending of waves around corners. The diffraction of sound waves is a common experience. For example, even though you cannot see someone around the corner of a tall outside wall, you can hear him or her speaking. If the sound waves traveled in straight lines, you would not receive the sound. But since you do, the sound waves must be diffracted or bent around the corner of the wall.

You might wonder why light waves are not similarly bent around the corner of the doorway so you could see as well as hear someone. This is because diffraction depends on the wavelength of the wave and the size of the diffracting object. Recall from Chapter 7 that for you to be able to "see" or detect an object with a wave, the object must be the right size to influence the wave. For example, long-wavelength water waves may pass by a thin reed without any indication of its presence being given by the wave. But if the wavelength of the waves approaches the size of the reed, the waves are affected.

Sound waves have wavelengths ranging from a few centimeters to a few meters, so they are diffracted or bent around common-sized objects such as trees and people. An "object" may also be an opening, such as a door or a window. Here, too, sound is diffracted or bent around the corners of the opening. The effect of the relative size of an object or opening and wavelength on diffraction is illustrated in Figure 20.9.

Diffraction effects are greater when the object or opening size is of the same order as or smaller than the wavelength. Consider radio waves: The AM band (550 to 1600 kHz) ranges in wavelength from 550 m to 190 m, and the FM band (88 to 108 MHz) ranges from 3.4 m to 2.7 m. Hence, FM radio waves may be blocked by large objects such as buildings or hills, whereas AM radio waves are diffracted around them and are received in a "shadow" area.

small pinhole may be fuzzy as a result of diffraction effects.

Light diffraction is used practically through diffraction gratings. A diffraction grating consists of many equally spaced parallel lines—thousands of lines per centimeter. A transmission grating has lines cut on a piece of glass, typically 6000 lines per cm (Fig. 20.10). The transparent spaces between the grooved lines act as a series of parallel slits. A reflection grating can be made by cutting lines on a piece of metal instead of glass. Light diffracted by the grating slits interferes, and an interference pattern similar to that of Young's double-slit experiment is formed. In effect, the grating extends the interference from a two-slit situation to a large number of slits.

When illuminated with white light, the bright fringes of the interference pattern are spectra of wavelengths or colors. Each wavelength is diffracted at a slightly different angle, which spreads out or separates the component wavelengths into a spectrum. You have probably noticed this colorful diffraction-grating effect on a phonograph record when it is viewed at a glancing angle. The spacings between the record grooves act as individual reflectors, and in effect the record is a reflection grating that separates white light into a spectrum of colors or wavelengths. Because of this property, diffraction gratings are used in instruments called spectrome-

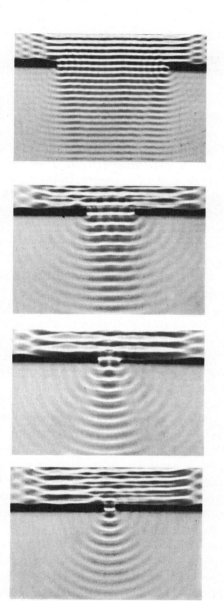
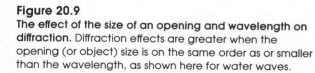

Figure 20.9
The effect of the size of an opening and wavelength on diffraction. Diffraction effects are greater when the opening (or object) size is on the same order as or smaller than the wavelength, as shown here for water waves.

With light waves you must have some mighty small objects or openings to produce diffraction. Recall that the wavelength of visible light is on the order of 0.00001 or 10^{-5} cm. Hence, diffraction is not observed for ordinary-sized objects or openings, and their corners cast sharp shadows when illuminated with light. However, the shadow around a

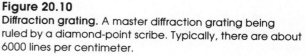

Figure 20.10
Diffraction grating. A master diffraction grating being ruled by a diamond-point scribe. Typically, there are about 6000 lines per centimeter.

ters to select specific wavelengths of light. When the grating is slightly rotated, the various wavelengths "parade" by a given point and particular wavelengths may be selected. These wavelengths are used to optically analyze substances.

POLARIZATION

Another interesting wave property of transverse light waves is polarization. This is the principle of the "polarized" sunglasses that many of us use. Recall how light is an electromagnetic wave with vibrating electric (E) and magnetic (B) field vectors perpendicular to the direction of propagation (Chapter 19). Light from an ordinary light source consists of many waves emitted by the atoms or mol-

ecules of the source. Each atom produces a wave with its own orientation of the E (and B) vibration corresponding to the direction of the atomic vibration.

However, with numerous atoms, all directions are possible. This is schematically represented in Figure 20.11 in terms of the electric field or E vectors. When the field vectors are randomly oriented, we say the light is **unpolarized**. If there is some partial preferential orientation of the field vectors, the light is **partially polarized**. If the field vectors are in a plane, the light is **linearly polarized** or plane polarized (or simply polarized).

Light may be polarized by several means. One common method is absorption. Certain crystals have the ability to selectively absorb field vector components with certain orientations, and light passing through such crystals is polarized. Around 1930, Edwin Land, an American scientist, found a way to align tiny needle-shaped crystals in sheets of transparent celluloid, thereby producing a thin, polarizing material now known commercially as Polaroid.

Improved Polaroid films have been developed using polymer materials. During the manufacturing process, the film is stretched so as to align the long chain molecules of the polymer. With proper treatment, the outer (valence) molecular electrons can move along the oriented chains. As a result, the molecules readily absorb light with E vectors parallel to the oriented chains and transmit light perpendicular to the chains.

The direction perpendicular to the oriented chain molecules is commonly called the *transmission axis* or *polarization direction*. Hence, when unpolarized light falls on a polarizing sheet, the sheet acts as a polarizer and polarized light is transmitted (Fig. 20.12).*

The human eye cannot distinguish between polarized and unpolarized light. Hence, we need an analyzer to detect polarized light, which can be another sheet of polarizing film. When the analyzer has the same orientation as the polarizer, light is transmitted. But, when rotated 90°, or when the po-

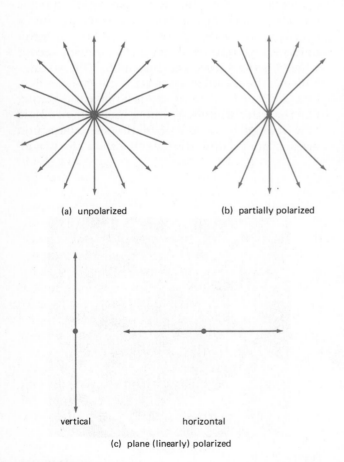

(a) unpolarized   (b) partially polarized

vertical   horizontal

(c) plane (linearly) polarized

Figure 20.11
Polarization. Polarization is graphically represented by field vectors.

---

* A common analogy is to view a polarizer as a picket fence through which a random wave in a rope will pass only parallel to the pickets and be polarized in this direction. However, in a polarizing sheet, the blocking "pickets" (absorbing molecular chains) are perpendicular to the transmission axis.

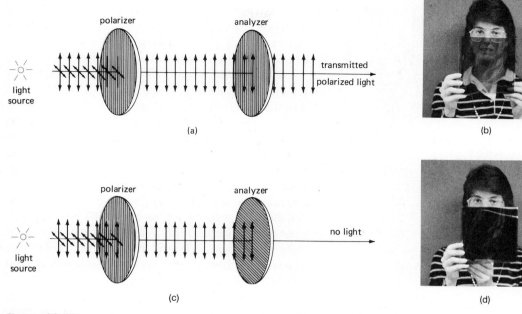

Figure 20.12
Polaroid films. When the film sheets are oriented with the same polarization direction, the transmitted light is polarized (a and b). One film acts as a polarizer and the other as an analyzer. When one of the sheets is rotated 90° so that the Polaroids are "crossed," little or, ideally, no light is transmitted (c and d).

larizing sheets are "crossed," the analyzer absorbs the polarized light, and ideally, no light is transmitted (Fig. 20.12).

You can demonstrate this yourself with a couple pairs of polarizing sunglasses. Try it next time you are at the drugstore or somewhere that sells polarizing sunglasses (if you don't personally have two pairs of polarizing sunglasses). Incidentally, polarization proves that light is a transverse wave instead of a longitudinal wave because longitudinal waves cannot be polarized. Why not?

Light is also polarized by reflection. The direction of polarization is parallel to the reflecting surface, and the degree of polarization depends on the material and the angle at which the light strikes the surface. This is why we use polarizing sunglasses (Fig. 20.13). Light reflected from roads, water, and other surfaces gives a large intensity and glare. When the polarizing lenses of the glasses are properly aligned (transmission axis vertically), some of the partially polarized reflected light is absorbed, which reduces the glare.

There is also polarization by scattering. In the scattering process, when light strikes a suspension of particles, such as the molecules of the air, elec-

trons are set into vibration by the light, and light waves are reradiated. The oscillating electrons do not radiate in the direction of their oscillation, so the scattered light is partially polarized. As a result, sunlight scattered by the atmosphere and received at the Earth's surface is partially polarized.

You can check this yourself by looking at a blue sky through polarizing sunglasses and then rotating the glasses. Light from different areas of the sky will show different degrees of transmission according to the degree of polarization, which depends on the relative direction of the Sun. Some insects, such as bees, are believed to use the polarized sky light for navigation. Scientists have put bees in a box with a Polaroid film over the top. When the film is rotated, the flying bees change direction.

In concluding this chapter, let's take a look at a couple of practical applications of polarization. One is discussed in Special Feature 20.2. Another involves movie entertainment. Recently, there has been a comeback of 3-D (three-dimensional) movies. This is the process through which objects seemingly jump out of the screen at you when you view a movie with special glasses, which give a third dimension or depth perception to a two-dimen-

(a)

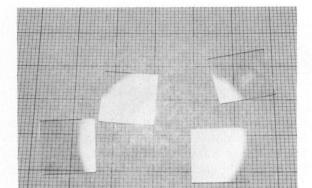

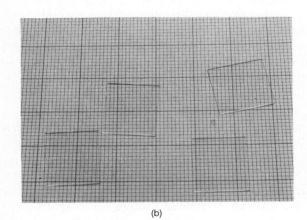

(b)

Figure 20.13
Glare reduction. (a) Light reflected from surfaces is partially polarized. Polarizing sunglasses reduce the intensity or glare by blocking out the polarized light. The top photo in (b) shows glare from glass surfaces, and the bottom photo shows the effect of polarizing glasses.

sional picture on the screen. We have visual depth perception in part because we see simultaneously with both eyes, each having a slightly different view or perception. The brain interprets the differences of the views in terms of depth perception.

Three-dimensional motion pictures are projected on the screen by two projectors, each with a slightly different image (Fig. 20.14). One image is for one eye and the other for the other eye so that the brain will interpret this as depth or a third dimension. So that you get a different image for each eye, the light from each projector is polarized but in perpendicular directions to each other. The 3-D glasses are really polarizing glasses with perpendicular polarization directions for each eye. Hence, one eye sees one image and the other eye sees the slightly different image. Viewed without the polarizing glasses, the combined pictures on the screen appear blurred.

Other methods of using two images and color to fool the brain have now been used in TV applications. With this new technology, a viewer without glasses doesn't see a blurred picture.

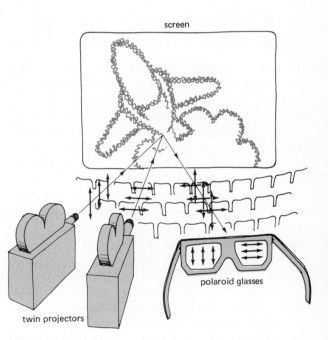

Figure 20.14
Polarized light. Three-dimensional movies use twin projectors and polarizing glasses so that slightly different images are seen by each eye, giving the perception of depth or a third dimension.

SPECIAL FEATURE 20.2

LCDs — Liquid Crystal Displays

A now very common application of polarization is in LCDs, or liquid crystal displays. These displays are found on calculators, wristwatches, and even gas pumps. One type of LCD makes use of the light-polarizing properties of some liquid crystals. (Recall that liquid crystals are liquids that show some degree of molecular order. See Chapter 8.) Such crystals have the ability to rotate or "twist" the polarization direction of polarized light by 90° (figure below). In a so-called twisted nematic display, a liquid crystal is sandwiched between crossed polarizer sheets and backed by a mirror. Light falling on the surface of the LCD is polarized, twisted, reflected, and twisted again, and then it leaves the LCD. Hence, the display appears light when illuminated.

But when a voltage is applied to the crystal, it loses its twisting property, and the crystal appears dark when illuminated because the light is blocked by the polarizing sheet and not reflected. Such voltage-induced dark regions of the crystal are used to form numbers and letters of the display. You can show that the light coming from such an LCD is polarized by a Polaroid sheet (or polarizing sunglasses), as shown in the figure below.

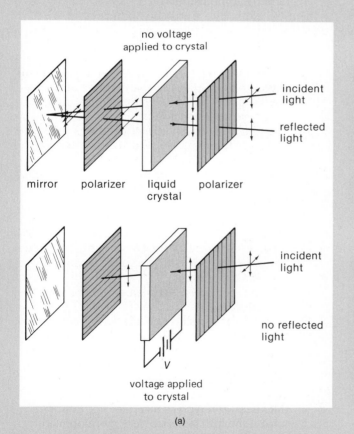

(a)

(b)

Twisted nematic LCD. (a) When no voltage is applied to the liquid crystal, it "twists" the light polarization through 90°. When voltage is applied, there is no twisting, and light does not pass through the second polarizer and is not reflected. The crystal then appears dark. (b) The bright areas of an LCD are due to reflected polarized light, as can be seen with a polarizing sheet (analyzer). (There's a bit of glare from the polarizing sheet. Should a polarizer be used to reduce this?)

SUMMARY OF KEY TERMS

Phosphorescence: the process whereby the electrons in some materials remain in excited states for long periods of time and emit light long after being excited, or "glow in the dark."

Fluorescence: the process whereby certain materials absorb ultraviolet light and emit light in the lower-frequency visible region.

Thin film interference: the interference of light reflected from the top and bottom surfaces of thin films.

Newton's rings: the interference ring pattern seen in a lens on an optical flat as a result of thin film interference.

Diffraction: the bending of waves around the corners of objects or openings that occurs when the wavelength is on the order of the size of the object or opening or smaller.

Polarization: a preferential orientation of the field vectors of light waves.

QUESTIONS

Light Emission

1. Cold is sometimes said to be the absence of heat. Is dark the absence of light? Explain.

*2. Quartz crystals are used in watches. How can a quartz crystal keep time?

3. What determines how long a phosphorescent material will glow in the dark? What determines how brightly a material will glow?

4. During the summer, "brownouts" or voltage reductions are experienced in some large cities because of electrical demand exceeding the supply. How would this affect the light emitted from an incandescent lamp? (*Hint:* The temperature of the filament would be less. See also Special Feature 11.3.)

*5. Mercury-vapor street lights have a bluish hue, and the newer sodium arc lamps have a yellow appearance. Explain this difference.

6. At a disco, the black light over the band blows out and there are no more bulbs available. The proprietor decides to use an infrared lamp instead to get the fluorescent effects. Is this a good idea? Explain.

7. Why can we still see things when only black lights are used in a dark room?

8. Some soap manufacturers add to their detergents a blue dye that fluoresces with a slight blue color. What is the purpose of this? (*Hint:* "Whiter whites" have a cleaner appearance when they are slightly bluish in color.)

Interference

9. When washed dishes aren't rinsed well, they sometimes show a display of colors while drying. What causes this?

10. What would be the effect if a thin film were illuminated with monochromatic light instead of white light? Consider films of uniform and nonuniform thicknesses.

*11. Will a thin film always be nonreflecting (i.e., will destructive interference occur) for normal incidence when the film thickness is one fourth of a wavelength of the light? Explain.

12. Explain why the film thickness on nonreflecting glass is one fourth of the wavelength of yellow-green light. What would happen if the film thickness were half a wavelength?

*13. The film on most nonreflecting lenses is just the thickness to produce destructive interference for yellow-green light. If you wanted to have destructive interference primarily for blue light, would you make the film thinner or thicker? Explain.

14. For destructive interference with thin films, is the light really "destroyed"? Explain. How about the light for the dark fringes or destructive-interference regions in the interference pattern of Young's experiment?

15. Why are camera lenses coated to make them nonreflecting?

16. Is there a phase shift or shifts for light during the back reflections from a coated camera lens? Explain. How does this affect the interference?

*17. Suppose you want to use thin film interference to make "reflecting" glass or a mirror. How could you do this? (Discuss in terms of film thickness *and* type of material. For example, what if the film material were more optically dense than glass?)

18. Explain why a uniform air wedge between two optical flats produces an interference pattern of alternate bright and dark lines when illuminated with monochromatic light. What would be seen if white light were used?

19. Some of the vivid colors of the fanned tail of a peacock are due to diffraction. Why is this? (*Hint:* Think in terms of a feather diffraction grating.)

Polarization

20. How do polarizing sunglasses reduce glare?
21. Does it make any difference how the plane or direction of polarization of the lenses of sunglasses is oriented? Explain. What would be the effect if the direction of polarization of one lens was one way and the direction of the other lens perpendicular to this?
•22. Given two pairs of sunglasses, how could you tell if they were polarizing or nonpolarizing, or could you?

•23. Suppose you had two sheets of Polaroid film and you held one in front of the other while looking through them (see Fig. 20.12). (a) If you rotated one of the sheets through one complete rotation while holding the other still, how many times would the Polaroids be crossed or darken? (b) Think about this one — rotating both Polaroids through one complete rotation at the same time but in opposite directions: How many times would the Polaroids cross or darken in this case? (c) What would be the case if the Polaroids were both rotated in the same direction at the same speed? How about different speeds?
24. Rather than wear the flimsy glasses provided at the a 3-D movie, Joe Cool decides to wear his designer polarizing sunglasses instead. Is this a cool idea?

Reflection
and Refraction

Part of a light beam is
reflected at one surface,
and the transmitted
portion is refracted. At a
second interface it
happens again.

Were it not for the reflection of light, our world would appear dark, except in the vicinity of light sources. Look around you. You see things because of light reflected from them. Without reflection, we would not see the moon. (How about the stars?) Many optical phenomena involve refraction. Without refraction, there would be no rainbow; and eyeglasses, which many of us wear, would not help correct vision problems. Reflection and refraction as they relate to sound were touched upon briefly in Chapter 15. In this chapter we'll take a look at these effects as they involve light.

LIGHT RAYS

The wave nature of light was quite important in considering interference phenomena in Chapter 20. The wiggly waveforms allow us to visualize and analyze constructive and destructive interference. However, for reflection and refraction it is convenient to think of light *rays*. This is the representation of light in what is called geometrical or ray optics. In effect, we ignore the wave nature of light and represent it as rays or straight lines.

To see how this is done, consider the diagrams in Figure 21.1. As waves spread out circularly (or spherically) from a point source, the adjacent portions of a wave that are in phase are said to form a **wave front**. For example, if a line were drawn through adjacent crests, it would define a circular wave front.

The wave front propagates outward with the same velocity as the wave and in a direction perpendicular to the wave front. A line drawn perpendicular to the wave fronts that indicates the direction of propagation is called a **ray**. There are also plane wave fronts, for example, at distances far from a point source, and the rays are parallel. We often speak of a beam of light, by which we mean a group of parallel rays. Unless reflected, refracted, or diffracted, light travels in a straight line, as represented by a light ray.

REFLECTION

Reflection might be thought of as light "bouncing" off a surface. Actually, it is much more complicated than this. It involves absorption and emission associated with complex electron vibrations in the atoms of the reflecting medium. But with ray optics we don't have to worry about this. We can describe what happens very simply in terms of rays. It is found that light rays are reflected from a smooth surface in a particular way (Fig. 21.2). Measuring the angles of the incident ray and reflected ray from the "normal," that is, a line perpendicular to the reflecting surface, gives the **law of reflection**:

> The angle of incidence is equal to the angle of reflection, $\theta_i = \theta_r$

(where the incident and reflected rays and the normal to the surface all lie in the same plane).

Question: If you have a mirror positioned in a doorway so you can see someone coming down the hall, can they see you?

Answer: Yes. As children we probably all played this game, thinking that the person could not see us around the corner. But the law of reflection works both ways or in both directions; the rays are just reversed.

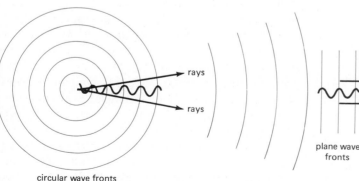

circular wave fronts

plane wave fronts

rays

Figure 21.1
Rays. From a point disturbance, waves spread out with circular wave fronts. Lines perpendicular to the wave fronts in the direction of the propagation are called rays. At great distances, the wave fronts are in planes and the rays are parallel. A group of parallel rays forms a beam.

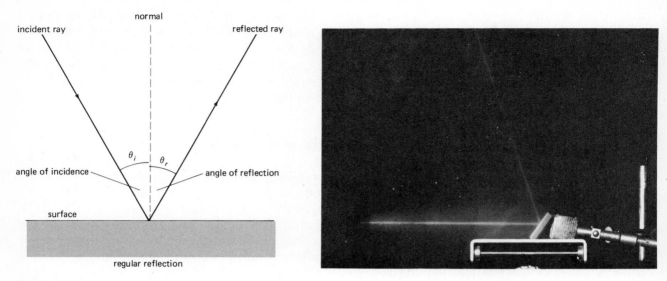

Figure 21.2
Law of reflection. The angle of incidence θ_i is equal to the angle of reflection θ_r, where the angles are measured from a normal (line) perpendicular to the reflecting surface.

You see the person because light rays are reflected to your eye. Light rays from your face at the same angle are reflected back along the same path. This is called reverse-ray tracing. If you can see their eyes, they can see you.

The reflection from very smooth or mirror surfaces such as in Figure 21.2 is called **regular** or **specular reflection**. Of course, not all surfaces are smooth enough to act as mirrors in which we can see images. The pages of this book are reflecting light to your eye; otherwise, you could not see them.

Other than light sources, in general we see objects because of the light they reflect.* A page of this book is relatively rough or irregular, and the reflection from such surfaces is called **irregular** or **diffuse reflection** (Fig. 21.3).

The law of reflection still applies in diffuse reflection. It's just that the small reflecting surface irregularities are at different angles to each other, and so are the reflected rays. It is diffuse reflection

*The black print on this page absorbs most of the incident light, and we "see" it by contrast with the reflected white light from the white portion of the page. More about this in the next chapter in the discussion on color.

from particles in the air that allows us to see light beams such as those from a flashlight or searchlight. Light is reflected from the beam by dust particles and so on; otherwise, we wouldn't see it (no light to our eye).

This is why we can see "shafts" of light when sunlight shines through the leaves of trees in a forest or through clouds. The latter is referred to as "the Sun drawing water" in weather folklore and is said to be indicative of rain. (Particles or "nuclei" are needed in condensation and formation of raindrops:)

Mirrors

When you look at your "reflections" in a mirror, these are regular reflections. In general, a **mirror** is any smooth surface that regularly reflects light. This may be a smooth water surface. When the water surface is rough, the reflection is diffuse, and the mirror quality is lost. A surface is "mirror"-smooth when the surface irregularities are on the same order as or smaller than the wavelength of light (10^{-5} cm for visible light).

Commercially, flat or **plane mirrors** are available as polished metal surfaces, but most commonly, mirrors are made of glass with surfaces that

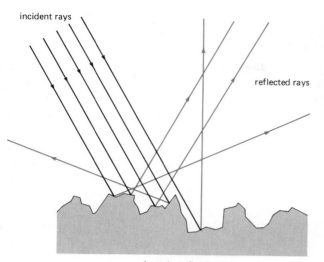

incident rays

reflected rays

irregular reflection

Figure 21.3
Diffuse or irregular reflection. The law of reflection still applies, but the surface irregularities cause the reflections to be reflected in various directions and to be diffuse.

have been coated or "silvered" with compounds of tin and mercury. (Silver compounds can be used, but they are expensive.) A glass mirror may be front-coated or back-coated, depending on the application. Common household mirrors are usually back-coated, but mirrors for telescopes are front-coated (Chapter 22).

Question: At night, a glass windowpane acts as a mirror. Why isn't it a mirror during the day?

Answer: When light strikes a transparent medium, most of the light is transmitted, but some is reflected. During the day, the light

reflected from the inside of the windowpane is overwhelmed by the light coming through the window from the outside. However, in the evening or at night when the light transmitted from the outside is reduced, the inside reflections can be discerned, and the windowpane acts like a mirror.

Because of regular reflections, mirrors produce images. We look at our images in mirrors everyday, usually starting off with the bathroom mirror in the morning. Light reflected by such plane mirrors causes us to see images "inside" the mirror. An image appears to be located the same distance behind the mirror as the object is in front of the mirror (Fig. 21.4).

There is also a right-left reversal of the object and image in a plane mirror. As we have all observed when looking into a mirror, if you raise your right hand, your image will raise the left hand. Your image's hair will appear to be parted on the opposite side as yours (unless you part your hair in the middle, have no part, or are bald). A practical example of the right-left reversal of a plane mirror is shown in Figure 21.5.

Another type of mirror is a spherical mirror, which is just a reflecting surface that has a spherical contour. There are two kinds of spherical mirrors, depending on which surface of the spherical section is reflecting (Fig. 21.6). If the concave (inside) surface is reflecting, this is a **converging spherical mirror**. Notice in the figure how rays parallel to the mirror axis converge or come to a focus at a point, which is called the **focal point**.

The converging properties of these spherical mirrors allow light to be converged to form an upside-down image on a screen (when object is beyond focal point). The image may also be magnified. Converging spherical mirrors are used as cosmetic mirrors. When you bring your face close to such a curved mirror, you see a right-side-up magnified image of yourself (when you, the object, are closer to the mirror than the focal point).

If the convex (outside) surface is reflecting, this is a **diverging spherical mirror** [Fig. 21.6(b)]. Here, light rays parallel to the mirror axis are reflected as though they come from an inside focal point. A reflecting, spherical Christmas-tree ornament or a spherical car hubcap would act as a diverging mirror.

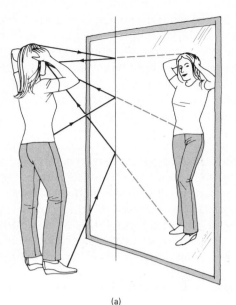

(a)

(b)

Figure 21.4
Plane mirrors. (a) The image in a plane mirror appears to be located the same distance behind the mirror as the object is in front of the mirror. (b) It's not hot. The finger is being held over the image of the candle behind a glass window in a laboratory cabinet that acts as a mirror.

Some practical applications of spherical mirrors are shown in Figure 21.7. The first two are similar to the diagrams in Figure 21.6, but with reverse-ray tracing. If a light source is at the focal point of a converging mirror, the rays are reflected in a parallel beam. This is the purpose of the spherical reflector in a flashlight or searchlight. Large diverging mirrors are commonly used in stores for monitoring. By reverse-ray tracing, light coming in at wide angles is reflected into a parallel beam by a diverging mirror. As a result, we have a wide field of view when looking into a diverging mirror.

Another example of a converging "mirror" is the satellite TV "dish" antenna. This is not an optical mirror. Instead of light, radio waves from distant satellites, which form nearly parallel rays, are focused or converged at the focal point, where a sensor or detector is located.

Question: The mesh TV dish in Figure 21.7(c) has holes in it. Won't the TV waves pass through and not be reflected?

Answer: No. Recall from previous chapters that a particle, or a hole in this case, must be about the same size as the wavelength of the wave to influence it.

The frequency of TV waves is on the order of megahertz (MHz). Let's say that a TV wave with a frequency of 100 MHz is incident on the antenna. This is 10^8 Hz and taking the order of magnitude of the speed of light as 10^8 m/s (actually 3×10^8 m/s). Then, the order of magnitude of the wavelength is ($\lambda f = c$):

$$\lambda = c/f = 10^8/10^8 = 1 \text{ m}$$

So, the holes in the mesh are much, much smaller than the TV wavelengths, and the TV waves don't even notice the holes and "see" a smooth reflecting surface.

■ REFRACTION

Back in Chapter 15, the refraction of sound waves was discussed briefly. Recall that refraction is the change or deviation of a wave's direction of propagation when it passes obliquely from one medium into another or into a different-density region of the

(a)   (b)

Figure 21.5
Right-left reversal. (a) The spelling of *AMBULANCE* is reversed so that it will be seen
spelled properly in a car's rear-view mirror as shown in (b). Notice that the sign painter
had a problem with reversed lettering of the *N*.

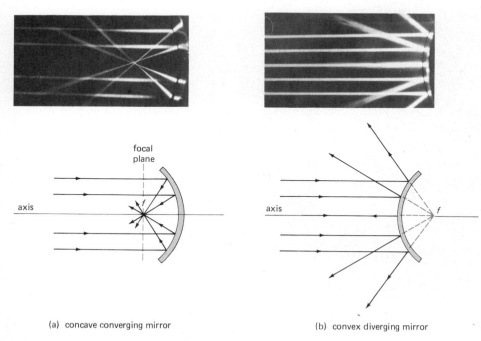

(a) concave converging mirror   (b) convex diverging mirror

Figure 21.6
Spherical mirrors. (a) Concave. Rays parallel to the mirror axis converge at the focal
point *f*. (b) Convex. Rays parallel to the mirror axis diverge on reflection, as though they
come from a focal point on the opposite side of the mirror.

(a)

(b)

(c)

Figure 21.7
Spherical mirror applications. (a) A bulb at the focal point of a spherical
flashlight reflector gives a parallel beam. (b) Convex spherical mirrors
provide a wide-angle view for store monitoring. (c) A spherical "mirror"
for converging radio and television waves from satellites. The detector is
located at the focal point of the TV "dish" antenna.

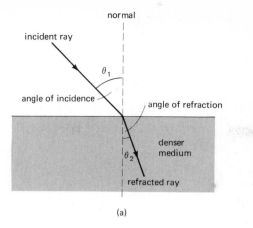

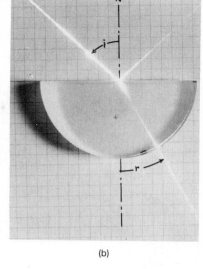

Figure 21.8
Refraction. A light ray is deviated or "bent" when it enters another medium. When entering a more optically dense medium, the ray is bent toward the normal ($\theta_2 < \theta_1$) as illustrated in (a). Notice in the photo in (b) that some of the incident light is reflected as well as refracted.

(a)

(b)

same medium.* Similarly, light is bent when it passes from one medium into another (Fig. 21.8).

The directions of the incident and refracted rays are expressed in terms of incident and refracted angles (θ_1 and θ_2, respectively), which are measured from a normal or perpendicular line to the surface boundary of the medium. When light passes obliquely into a denser medium, for example, from air into glass or water, light rays are bent toward the normal ($\theta_2 < \theta_1$).

It is the slowing down of the light as it enters the denser medium that gives rise to refraction. This involves complex processes, but in a simple manner, one might expect the passage of light by atomic absorption and emission through the denser medium to take longer. For example, the speed of light in water is about 75 percent of that in air or vacuum, and in glass it is about 67 percent or less.

To help you understand how light is bent or refracted when it passes into a medium, consider a group such as your school band marching across a field. The marching column then obliquely (at an angle) enters a muddy, wet region (Fig. 21.9). As the marchers in the rows hit the wet, slippery region, they keep marching at the same cadence (frequency). But slipping in the muddy earth, the marchers in that portion of the row are slowed down.

The marchers in the other part of the same row on solid ground continue on with the same stride, and as a result, the direction of the marching col-

Obliquely means slanting or inclined at an angle, as opposed to direct or normal incidence (incident angle of 0°).

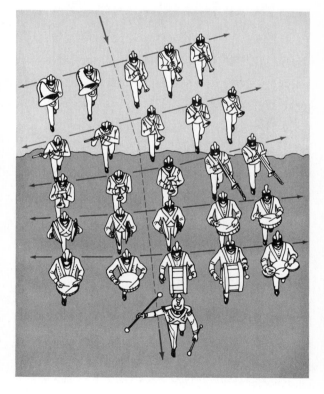

Figure 21.9
Marching "refraction." When the rows of the marching band enter a slippery (muddy) region, the marchers in a particular row in the mud slip and change speed while keeping the same cadence or frequency. As a result, the direction of the marching column is changed.

(a)

(b)

Figure 21.10
Refraction effects. (a) When refraction is not taken into account, the fish appears to be at a different location than it actually is, which makes catching it difficult. (b) The bending of light rays makes the pencil appear to be almost severed.

umn is changed when it is in the muddy region. (This change in direction with changes in marching speeds is also seen when a marching column turns a corner.) We might think of the marching rows as wave fronts. In this case, the wave frequency (cadence) remains the same, but the wavelength and wave velocity change in the medium. Notice in Figures 21.8 and 21.9 that by reverse-ray tracing, when light enters a less dense medium, it is refracted or bent away from the normal.

We have all observed refraction effects. Have you ever tried to catch a fish in a bowl or tank, and the fish is seemingly never where it appears to be? This is because it isn't where it appears to be when we think of light traveling in straight lines, as we usually do (Fig. 21.10). Owing to refraction, the fish is actually at a lower depth than we think it to be. Our minds don't usually take refraction into account. An object in a glass of water appears to be bent or almost severed. Refraction in the atmosphere also produces some interesting effects. See Special Feature 21.1.

Total Internal Reflection

As we have learned, light entering a less dense medium is bent away from the normal. How far can the light be bent? The answer is "all the way," so to speak. If the incident angle is increased, the refracted light rays will just graze the surface at some critical angle of incidence (Fig. 21.11). Beyond this critical angle, the incident light is totally reflected back into the medium.*

This total internal reflection allows specially cut pieces of glass (prisms) to be used as mirrors that reflect light at angles of 90° or even 180° (backward), as illustrated in Figure 21.12. The critical angle of glass is about 42°, so light internally incident on a surface at a greater angle, for example, 45°, is totally internally reflected. By the law of reflection, the angle of reflection is equal to the angle of incidence, so the light ray is changed in direction by 90°. Two 90° internal reflections give a 180° change in direction.

Internal reflection also enhances the brilliance of cut diamonds. A diamond is cut with many faces or facets—58 in the so-called brilliant cut. Light

*Although called *total* internal reflection, a small part of the light is always transmitted into the second medium.

Atmospheric Refractions: Mirages, Hot Air, and Twinkling Stars

Atmospheric refraction effects are common. One of these effects is a mirage. The term *mirage* generally brings to mind a thirsty person in a desert "seeing" a pool of water that really isn't there. This is not the mind playing tricks but is an optical illusion. A more common mirage is the "wet spot" we often see on a hot highway in the summer that we can never seem to reach. What we are actually seeing is a reflection of the sky [Fig. 1(a)].

The warm air layer near the road is less dense than the cooler air above it. This density difference causes sky light to be refracted or bent to the eye. In the desert, a mirage may be due to the refraction of light from the leaves of a tall, distant palm tree.

It is also refraction that allows us to "see hot air" rising from hot surfaces such as a road or a hot plate. If you stop and think for a moment, you'll know that you can't see air. What you are seeing is the refraction or bending of light waves passing through the different-density regions of the rising convection currents. Similarly, atmospheric motions and density changes refract incoming starlight, causing stars to twinkle.

When watching a sunset (or moon set), you may have noticed that the Sun near the horizon appears to be flattened [Fig. 1(b)]. This is also due to atmospheric refraction. The density of the atmosphere decreases with altitude, so the rays from the top and bottom portions of the Sun on the horizon are refracted by different degrees. This produces the observed flattening. Rays from the sides of the Sun on a horizontal plane are generally refracted by the same amount, so the Sun still appears circular along its sides. Also, because of refraction the sun can be seen for several minutes after it actually sinks below the horizon.

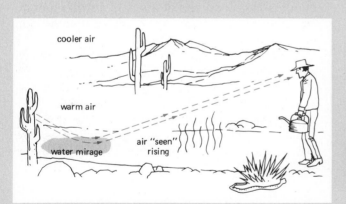

(a)

(b)

Figure 1
Atmospheric refraction. (a) Layers of air at different temperatures and densities act like different media and refract or bend sunlight so that we see a water mirage on hot road surfaces. Refraction also accounts for the phenomenon of "seeing" hot air rising. (b) The Sun appears flattened because of different refractions of the light rays from the top and bottom portions.

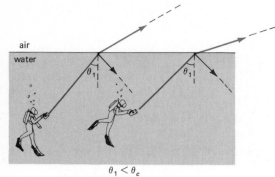

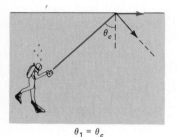

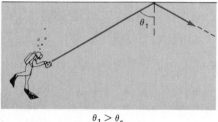

$\theta_1 < \theta_c$

(a)

$\theta_1 = \theta_c$

(b)

$\theta_1 > \theta_c$
total internal reflection

(c)

Figure 21.11
Total internal reflection. (a) When light is directed at a less optically dense medium with an incident angle less than a certain critical angle (θ_c), the light is transmitted above the surface (and a small amount reflected). (b) At an angle of incidence equal to θ_c, light is transmitted along the surface. (c) At an angle of incidence greater than θ_c, the light is totally reflected and the surface acts like a mirror.

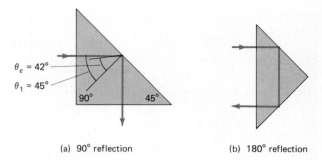

(a) 90° reflection   (b) 180° reflection

Figure 21.12
More internal reflection. Glass prisms become mirrors using internal reflection. The light can be reflected (a) 90° and (b) 180°.

entering the diamond is internally reflected so it emerges from the upper facets, giving rise to the diamond's sparkling brilliance. Check this out on a diamond ring.

Because of total internal reflection, light can be "piped" from one location to another in glass or plastic rods (Fig. 21.13). On entering the "light pipe" at an angle greater than the critical angle of the pipe material, the light undergoes repeated internal reflections and follows the contour of the pipe.

Flexible light pipes can be made by fusing bundles of transparent fibers together. This has given rise to the relatively new field of **fiber optics** (Fig. 21.14). Images can be transmitted through the fiber bundles. With two sets of fibers in a well-randomized fiber bundle, light can be transmitted through one set and returned by the other.

This arrangement is used in a fiberscope, which allows visual inspection of places that cannot be viewed directly. Such techniques have made it possible for physicians to view internal parts of the human body, such as the stomach and heart valves. In addition, fiber-optics light pipes or "wires" are now being used to optically transmit telephone conversations.

■ DISPERSION

When white light is passed through a prism, a spectrum of colors is observed to emerge (Fig. 21.15). This separation of light into its component frequencies or colors is called **dispersion**. To get an idea of how this occurs, consider the electron oscillators of the atoms making up a transparent medium. When light is transmitted through the medium, it is transferred by atoms through absorption and emission by the electron oscillators. Since an oscillator has a

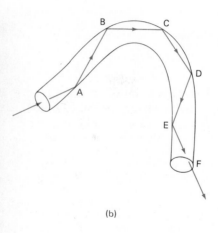

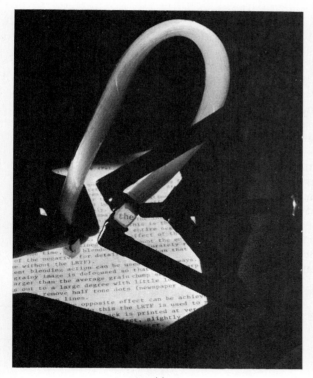

(a)

(b)

Figure 21.13
Internal reflection is the principle of light ''piping'' in fiber optics. (a) The word *the* can be seen at the opposite end of a light pipe with the other end on a printed page. (b) The angle of incidence is greater than the critical angle of the material for each reflection, and the light is internally reflected along the pipe.

Figure 21.14
Fiber-optics applications.
Light and images can be transmitted through light pipes and fibers. They can be used in making decorative lamps (a) and make it possible to view otherwise inaccessible places (b).

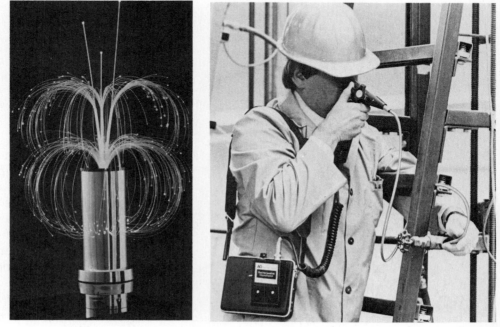

(a)   (b)

The Rainbow

What causes a rainbow? Rain, of course, or more correctly, water droplets in the air after a rain, and the Sun
must be there too. But we do not see a rainbow after
every rain, so there must be some special conditions.

Dispersion, refraction, internal reflection, and rainbows.
(a) Single and (b) double internal reflections can occur in
water droplets, giving rise to primary and secondary
rainbows (c). Note the faint secondary rainbow above the
lower primary rainbow in the photo. Dispersion separates
the component colors of sunlight at different angles, and
an observer sees different colors coming from different
altitudes, or a rainbow of colors. The colors of the primary
and secondary rainbows are reversed.

Basically, whether or not we see a rainbow depends on
the relative positions of the Sun and the observer. You
may have noticed that the Sun is behind you when a
rainbow is seen.

The rainbow is produced by refraction, dispersion,
and internal reflection in water droplets. Occasionally
more than one rainbow can be seen. The main or primary rainbow is sometimes accompanied by a fainter
and higher secondary rainbow arc. Have you ever seen a
secondary rainbow? For the primary rainbow, light is
reflected once inside a water droplet (figure below).
Being refracted and dispersed, the sunlight is spread
out into a spectrum of colors.

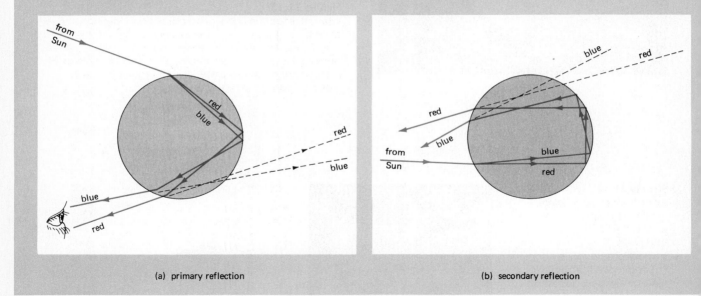

(a) primary reflection   (b) secondary reflection

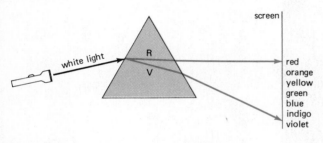

Figure 21.15
Dispersion. The dispersive properties of glass cause a prism
to separate white light into a spectrum of colors.

resonant frequency, there is more absorption of the
components of the light with frequencies closer to
electron-oscillator resonant frequency.

The natural or resonant frequency of atoms in
most transparent materials is in the ultraviolet region. Hence, light in the higher-frequency blue end
of the visible spectrum is absorbed more by the
electron oscillators and takes a longer time to get
through the material or travels more slowly than
light at the lower-frequency red end of the visible
spectrum. Blue light is then bent or refracted more
than red light. As a result of the different velocities

Because of the conditions on the refraction and internal reflection, the angles between the incoming and outgoing visible-light colors lie in the narrow range of 40° for violet to 42° for red. This means that you see a rainbow only when the Sun is positioned so that the dispersed light is reflected to you through these angles. With millions of water droplets in the air, a colorful arc is seen, running vertically in color from violet to red. (Below the rainbow arc the light from the droplets combines to form a bright, illuminated region.)

The less frequently seen secondary rainbow is caused by a double reflection in the water droplets. This gives rise to an inversion of the sequence of colors in the secondary rainbow from that in the primary rainbow.

As the Sun rises, less of the rainbow arc is seen. In fact, an observer on the ground cannot see a primary rainbow when the Sun's altitude (angle above the horizon) is greater than 42°. The primary rainbow is then below the horizon and will not be seen unless viewed from a height. As an observer is elevated, more of a rainbow arc is seen. It is common to see a completely circular rainbow from an airplane (no rainbow's end, no pot of gold). You have probably seen a circular "rainbow" in the fine spray of a garden hose.

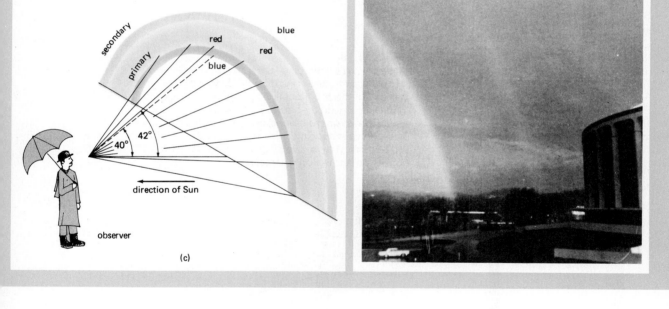

(c)

for light with different frequencies, the components of white light are refracted at different angles and spread out into a spectrum of colors or frequencies.

Because of dispersion, prisms are used in spectrometers to analyze the components of light. In addition to a diamond's brilliance due to internal reflection, it is said to have "fire" as a result of colorful dispersion. Dispersion can cause problems with lenses, as will be learned later in this chapter, but if not for dispersion, we would not have one of the beautiful sights of nature — the rainbow. See Special Feature 21.2.

■ LENSES

Refraction is to lenses as reflection is to mirrors. Many of us wear lenses in frames on our noses or in contact with our corneas to correct vision problems. The term *lens* comes from the Latin word for lentil, a plant with a seed that resembles a common lens shape. An optical lens is made from a transparent material, most commonly glass but also plastic and crystal, and usually has a spherical contour on one or both surfaces. There are a variety of lens shapes, but we will limit our discussion to two simple types.

If two prisms are placed base to base, they tend to refractively converge incident light rays parallel to their common baseline (Fig. 21.16). As such, the prisms approximate a **spherical converging lens**. The converging light energy will form an image of an object on a screen. We use lenses in projectors to project images on screens. A single lens inverts the image, but another lens can be used to invert it again so that it is right side up on the screen.

Similarly, if two prisms are placed point to point, they approximate a **spherical diverging lens**. The light energy diverges, so an image cannot be formed on a screen with a diverging lens. However, diverging lenses have various uses, as we will see in the next chapter when both types are discussed.

Aberrations

Lenses are used to form images, magnify, and so on in many optical instruments. Examples of some of these will be considered in the next chapter. Lenses are introduced here to give you an idea of how they incorporate refraction. Spherical lenses may not form perfect images, owing to material defects and to various natural and inherent effects called **aberrations**. Aberrations cause blurred or out-of-focus images. Let's consider a couple of the more common ones.

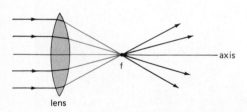

prisms

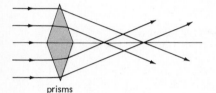

lens

(a) converging lens

Figure 21.16
Lenses. (a) A converging lens is similar to two prisms placed base to base and converges light parallel to the lens axis at the focal point. A converging lens can be used to form an image on a screen. (b) A diverging lens is similar to two prisms placed tip to tip, and a light beam is diverged as though it comes from a focal point. The image formed by a diverging lens is said to be virtual and cannot be formed on a screen.

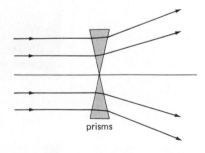

prisms

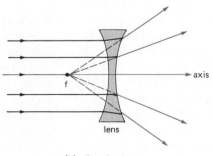

lens

(b) diverging lens

Spherical aberration results from rays nearer the edges of a converging spherical lens being brought to focus in front of the focal point (Fig. 21.17). This can be remedied by covering the outer part of the lens, as with a diaphragm in a camera. Spherical aberration also results for spherical converging mirrors. This can be eliminated by making parabolic-shaped converging mirrors. All the reflected rays from a parabolic mirror converge at a focal point or in the focal plane.

Chromatic aberration arises from dispersion. The dispersion of light passing through a lens results in rays of different colors coming together at different points. This aberration effect can be eliminated by using a lens combination called an achromatic doublet (achromatic means "without color"). The converging and diverging lenses are made of two different materials, so one lens refractively counteracts the dispersion of the other.

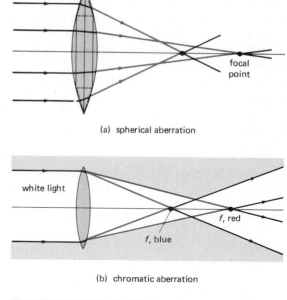

(a) spherical aberration

(b) chromatic aberration

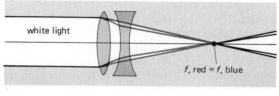

(c) achromatic doublet

Figure 21.17
Lens aberrations cause distorted images. (a) Spherical aberration. Rays near the edges of a converging lens are brought to focus in front of the focal point. (b) Chromatic aberration results from dispersion, causing different colors to have different focal points. This can be corrected with a achromatic doublet (c).

SUMMARY OF KEY TERMS

Wave front: the front or form defined by adjacent portions of a wave that are in phase, for example, a line drawn through adjacent crests, which may form a circular, spherical, or plane wave front.

Ray: a line drawn perpendicular to wave fronts that indicates the direction of wave propagation and represents the wave.

Reflection: the return of incident light from a surface such that the angle of incidence is equal to the angle of reflection (law of reflection). Reflection from a smooth, mirror surface in which images can be seen is called regular reflection. Reflection from a rough surface is called irregular reflection and is in all directions.

Converging spherical mirror: a mirror surface on the inside of a spherical section that, on reflection, converges or focuses rays parallel to the mirror axis at a focal point.

Diverging spherical mirror: a mirror surface on the outside of a spherical section that, on reflection, diverges rays parallel to the mirror axis so that

they appear to come from a focal point inside the mirror.

Refraction: the bending or deviation of a wave's direction of propagation when it passes at an angle from one medium into another or into a different-density region of the same medium. When light passes into a denser medium, it is bent or refracted toward the normal.

Total internal reflection: the reflection of light at a transparent-medium surface when the angle of incidence is greater than some critical angle. At incident angles above the critical angle, the light is internally reflected instead of being transmitted.

Dispersion: the refractive separation of light into component colors as a result of light of different frequencies having different velocities in a medium.

Lenses: glass, plastic, or crystal forms that use the refractive properties of light to form images, magnify, and so on.

Converging lens: a lens that converges or focuses rays parallel to the lens axis at a focal point.

Diverging lens: a lens that diverges rays parallel to the lens axis as though the rays come from a focal point on the incident side of the lens.

Aberration: the distortion of an image due to the failure of the light to converge at a point or in a plane. Aberration can result from light rays passing through the edge of a lens or reflected from the edge of a spherical mirror that do not converge with the rays transmitted near the axis (spherical aberration) and from dispersion in a lens (chromatic aberration).

QUESTIONS

Light Rays

*1. How do wave optics and geometrical optics differ?
2. What is a wave front?
3. Is a light ray part of a light wave? Explain.
4. What does a beam of light mean in terms of rays?

Reflection

*5. A supplementary statement to the law of reflection as given in the chapter is that the incident and reflected rays and the normal to the surface lie in a plane. Discuss what this means.
*6. For regular reflection, what is the situation with an angle of incidence of (a) 0° and (b) 90°?
7. If the moon had a smooth surface, how would a full moon appear?
8. Why are only parts of the laser beam visible in Figure 21.18? What kind of reflection is this? (Blackboard erasers are being clapped together in the background.)
9. When you look at a window-glass "mirror" at night, two similar images, one behind the other, are often seen. Why are there two images? (*Hint:* Remember the glass has two surfaces.)
10. If you walk toward a plane, full-length mirror, what does your image do? How fast does the image move? Is the image in step with you (like marching in a band)?

11. In detective movies and in studies of children's activities, one-way mirrors are used for secret observations. (A one-way mirror is seen as a mirror from one side, but can be seen through from the other side.) Reflecting sunglasses are another example. How do one-way mirrors work? (*Hint:* At night a glass windowpane is a one-way mirror.)
*12. A good reflecting surface such as a plane mirror has a reflectivity of about 95 percent, or 5 percent of the incident light is absorbed on reflection. If four mirrors are set up so that an incident beam is reflected from one to the other, what percentage of the incident light energy will be reflected from the fourth mirror? (*Hint:* The answer is *not* 80 percent.)
13. Explain the purpose of dual truck mirrors, as shown in Figure 21.19.

Figure 21.18
A laser beam made visible. See Question 8.

Figure 21.19
Two-in-one mirror. See Question 13.

14. Operating room and dentists' lamps have large spherical reflectors. What is their purpose?

15. A dentist sometimes uses a small converging spherical mirror to examine cavities when filling them. What is an advantage of a converging mirror over a plane mirror? (Although not discussed in the chapter, another advantage is that converging mirrors magnify.)

Refraction

16. When light enters a denser medium, how is the wavelength of light affected? (*Hint:* Think about the marching analogy given in the text.)

•17. Is there refraction for the incident angles of (a) 0° and (b) 90°? Explain.

18. Explain how refraction causes the pencil in Figure 21.10 to appear almost severed.

19. A design or picture on the bottom of a swimming pool has varying distortions when viewed through the water in a full pool. What causes these distortions?

20. While sitting by a swimming pool on a fixed bench before the pool is filled in the spring, a person notices that he can see only a limited portion of the design on the bottom of the pool (Fig. 21.20). However, when the pool is filled, he notices that he can see more of the design previously hidden by the pool wall. How is this possible?

•21. Does atmospheric refraction have any effect on the length of the day? That is, would the daylight hours be longer or shorter if we had no atmosphere?

22. A thick beer or root-beer mug looks as though it holds more than it actually does because the liquid appears to be closer to the sides than it actually is (Fig. 21.21). Explain this "optical illusion" that makes you think that you're getting more for your money.

Total Internal Reflection

•23. Is it possible to have total internal reflection with light in air incident on water? Explain.

24. Consider light in water incident on an air surface. What is the limiting angle of refraction that can be achieved? Explain. (*Hint:* Think of reverse ray tracing.)

25. We sometimes say that total internal reflection can be used to make an internal "mirror." Would it be possible to see an image in such an internal mirror? Explain.

26. Design a periscope to be used to see over and around objects using (a) plane mirrors and (b) prisms.

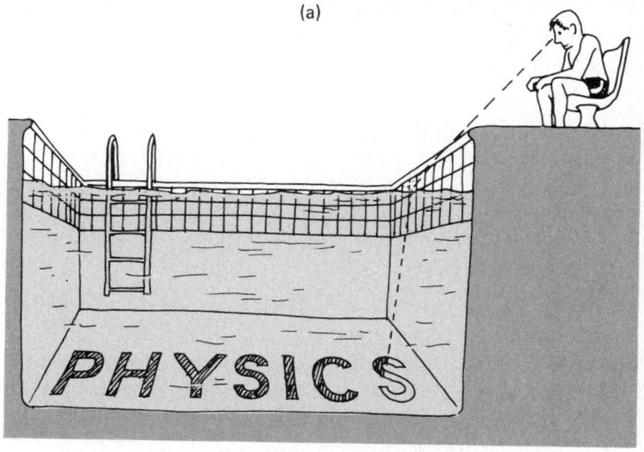

Figure 21.20
Water makes things clear. See Question 20.

Figure 21.21
More for your money? See Question 22.

•27. (a) Could you see underwater in all directions when looking into water at various angles of incidence? (b) How would a fish see the "above-water" world when looking up at various angles? *(Hint:* Think in terms of critical angle and what this means in terms of the "cone" of light coming in from above the water in reverse ray tracing.)

Dispersion

28. Why is a spectrum of colors not seen when light passes through a glass windowpane since it is dispersed?

•29. A prism can be used to separate white light into a spectrum of colors. Could another prism be used to recombine the spectrum colors into white light? (*Hint:* Make a sketch and remember that light is refracted away from the normal when entering a denser medium.)

30. Is there an absence of rainbows after some rains, or is there always a rainbow and an observer may not be in the proper location to see it? Explain. (Rains after the Sun goes down are excluded.)

•31. Why are rainbows seen in the form of "bows" or circular arcs? Why is the secondary rainbow fainter than the primary rainbow?

32. Do two observers at different locations see the exact same rainbow? Explain your answer.

33. With dispersion and a single internal reflection in a water droplet, it might appear that the red arc of the primary rainbow would be below the violet or blue arc (see figure in Special Feature 21.2). However, the colors of the primary rainbow are seen to run vertically from violet to red. Explain why this is the case.

34. A halo or ring is sometimes seen around the moon or Sun as a result of high, thin cirrus clouds that are composed of ice crystals. What causes these halos? (*Hint:* Keep in mind that the clouds are between the light source and the observer and that there are circular rainbows.)

Figure 21.22
Brought to focus. See Question 36.

Lenses

35. Distinguish between converging and diverging lenses. What is the spherical shape of each?

•36. A magnifying glass is a converging lens. Sunlight can be focused to a small spot using such a lens (Fig. 21.22). What is the small spot an image of? Why are holes burnt in paper or leaves when the small bright spot is focused on them? (If you've ever focused the spot on your skin, you know it gets hot.)

37. What would be observed coming from a converging lens if a small (point) source were placed at the focal point on the opposite side of the lens? (*Hint:* Think in terms of reverse ray tracing.)

•38. Discuss the types of aberration that may affect the images formed by (a) a front-coated spherical converging mirror (i.e., coating on the concave surface) and (b) a back-coated spherical mirror (i.e., coating on the back of the glass).

Vision and Optical Instruments

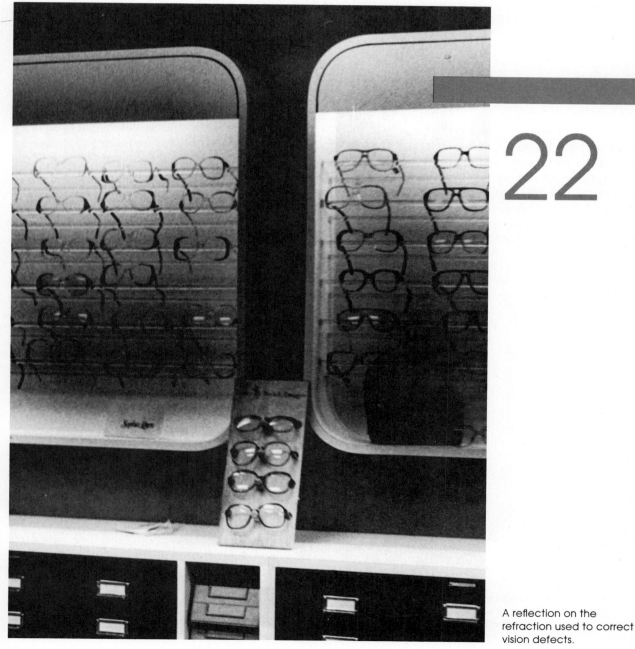

22

A reflection on the refraction used to correct vision defects.

Vision or sight is perhaps our most important sense. We tend to take vision for granted—until things start to appear blurred. Then we can often correct the problem with lenses (glasses). Even when we don't have vision defects, we use optical instruments containing mirrors and lenses to improve or extend our visual observations. For example, with microscopes and telescopes we can see very small and very distant objects that are not visible to the unaided eye.

In this chapter we'll take a look at these things so you can understand how you "see" the world.

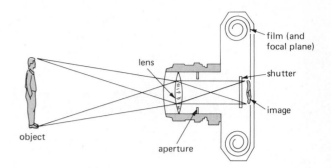

Figure 22.2
The operation of the eye is similar in several ways to that of a simple camera. Compare with Figure 22.1.

THE HUMAN EYE

The eye is a unique optical instrument and one of our most valuable possessions. Without it, we would not see the beauty of the sunset or be able to read this and other books to gain knowledge. The anatomy of the eye is shown in Figure 22.1.

The eyeball is nearly spherical, with a white outer covering called the sclera (the "white of the eye"). Light enters the eye through a curved, transparent tissue called the cornea. Behind the cornea is

a circular diaphragm, the iris, which has a central hole called the pupil. Our knowledge of everything we see depends on the information conveyed through the tiny pupils of our eyes. The size of the pupil aperture or opening is adjusted by muscle action and controls the amount of light entering the eye. For example, in very bright light the iris diaphragm closes, and the pupil becomes very small. How about in dim light?

A converging lens composed of glassy fibers is situated behind the iris. The shape or curvature of the lens is controlled by the ciliary muscles. By adjustments and changes in the curvature of the lens, which is called **accommodation**, the images of objects at different distances from the eye can be focused on the **retina** on the back wall of the eyeball. The eyeball contains a fluid in front of the lens and a gelatinous material in the space behind the lens.

The operation of the eye is similar in several ways to that of a simple camera (Fig. 22.2). Both have a lens, but the curvature of the camera lens cannot be changed. Instead, the camera lens is moved back and forth so as to focus images on the film. Both have variable diaphragms and both have a shutter. The shutter of the eye is the eyelid, which, unlike the shutter of a camera, is open for continuous exposure. Because of the refractive properties of the converging lenses, both form inverted or upside-down images on a light-sensitive surface (the film of the camera and the retina of the eye).

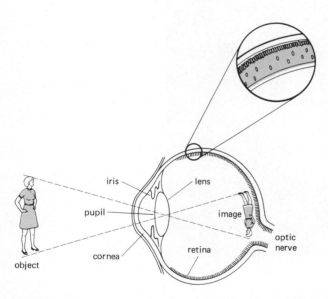

Figure 22.1
The anatomy of the human eye. Basically, the crystalline lens focuses the image on the retina. The blowup of the retina illustrates the rods and cones that are used in twilight and color vision. See text for detailed description.

Question: If the images in our eyes are upside-down, why don't we see the world that way?

Answer: By some means, the brain learns early in life to interpret the inverted image of the world right side up. Experiments have been done with persons wearing special glasses that give them an inverted view of the world. After some initial run-ins, they become accustomed to and function quite well in their "upside-down" world.

The retina "film" of the eye is composed of two types of photosensitive cells called **rods** and **cones** (names descriptive of shapes). The more numerous rods have a greater sensitivity to light and can distinguish between low light intensities for twilight (black-and-white) vision. The cones respond selectively to certain colors of light, some to one color and others to other colors. Cones are considerably less sensitive to light than are rods. This is why we cannot see color in very dim light. The rods and cones of the retina are connected to optic nerve fibers, which relay the light-stimulated signals to the brain.

In the region where the optic nerve enters the eyeball (Fig. 22.1), there are no rods or cones. As a result, your eye has a "blind spot" for which there is no optical response. Figure 22.3 can be used to demonstrate this blind spot. Hold the book at arm's length and, with your right eye closed, look intently at the black cross. Then slowly bring the book toward your face. At certain points, you will see the square and dot alternately disappear and reappear. Where did they go? Nowhere. Their images were just crossing your eye's blind spot. The blind spot is not noticed in ordinary vision because of eye and object movements and binocular vision.

Binocular (two-eyed) vision also accounts for some of our depth perception, which depends on slight differences between the shapes and positions of the images on the retinas of the two eyes. This occurs because the eyes are set several centimeters apart and get slightly different views of objects. Some other visual effects are discussed in Special Feature 22.1.

VISUAL DEFECTS

Optical illusions as discussed in Special Feature 22.1 aren't visual defects. Defects generally occur because the eye is abnormal for some reason. This can be due to a variety of reasons, including disease and injury. Let's take a look at several common vision defects that can be corrected with glasses.

The points between which the eye can see distinctly are called the far point and the near point. The far point is normally without limit (infinity), and the near point depends on the accommodation of the crystalline lens. Bring your finger slowly toward your nose. At some point (your near point), you will see the finger blur.

The normal eye produces sharp images on the retina for objects between the near and far points. However, a person may be nearsighted or farsighted and see blurred images (Fig. 22.4). This results from images not being focused on the retina because the size and shape of the eyeball and the crystalline lens are not properly matched. Recall that the crystalline lens has different shapes for near and far visions.

Nearsightedness, or myopia, arises when the image is formed in front of the retina. A nearsighted person can see close or near objects clearly, but not distant objects. This defect is corrected by wearing glasses with a diverging lens for the nearsighted eye. The lens diverges the incoming rays so that the image of a far object formed by the crystalline lens is moved backward to the retina [Fig. 22.4(a)].

Farsighted persons can see far objects clearly, but near objects are blurred or out of focus. **Farsightedness**, or hyperopia, is due to the image of a nearby object being formed behind the retina. A converging lens will correct this by converging the incoming rays so that the image is moved forward to the retina [Fig. 22.4(b)].

Farsightedness occurs naturally with age. You may have noticed an older person without glasses holding reading material away from himself or herself, even at arm's length. Children can see objects

Figure 22.3
Demonstration of the blind spot. Hold the book at arm's length and, with your right eye closed, look intently at the black cross. Slowly bring the book toward your face. What happens to the square and the dot on the left?

SPECIAL FEATURE 22.1

Optical Illusions

"Seeing is believing," goes the old saying. But this is not always the case, or at least you shouldn't believe everything you see. We can be visually fooled by several means. For example, fooling our depth perception causes objects to seem to "jump" out of the screen in 3-D movies (Chapter 20), and atmospheric refraction produces mirages (Chapter 21).

Some optical illusions have to do with the eye itself and interpretations by the brain. For example, in Figure 1 you may see fleeting patches of gray on the white areas between the black diamonds (spots before your eyes!). This is because the stimulation of one area of the retina can affect the sensations in an adjacent region, producing an illusion.

Also in Figure 1 is a design that can be used to illustrate an "after-image." When a particular area of the retina is stimulated, it may "remember" the stimulation after you have shifted your gaze to another object. Look intently at the white dot on the black Y for 15 to 20 s (quite a long time), then transfer your gaze to the white dot between the E's. You'll see a ghostly white after-image Y.

Other optical illusions are shown in Figure 2.

(a)   (b)

Figure 1
Optical illusions. (a) Fleeting gray patches are usually seen on the white areas between the black diamonds (spots before your eyes!). (b) This one calls for concentration. See text for description.

Figure 2(a) through (e)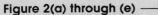
Some geometrical illusions. Answer the questions under the figures.

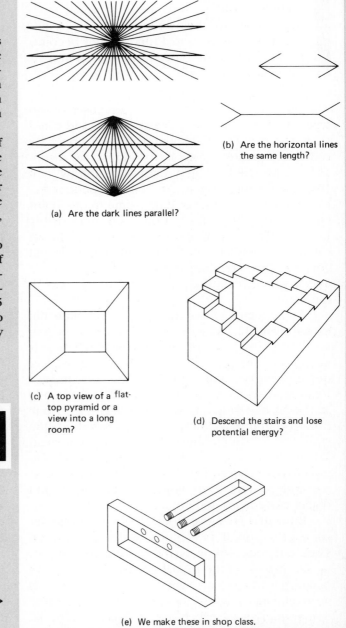

(a) Are the dark lines parallel?

(b) Are the horizontal lines the same length?

(c) A top view of a flat-top pyramid or a view into a long room?

(d) Descend the stairs and lose potential energy?

(e) We make these in shop class.

clearly as close as 10 cm from the eye (Table 22.1). However, as a person grows older the ciliary muscles weaken, and the crystalline lens loses its elasticity or hardens, limiting the eye's accommo-dation. As a result, the near point recedes with age. When a person's "arms get too short," he or she has to get "reading glasses." The converging lens forms an image of a close object outside the near point,

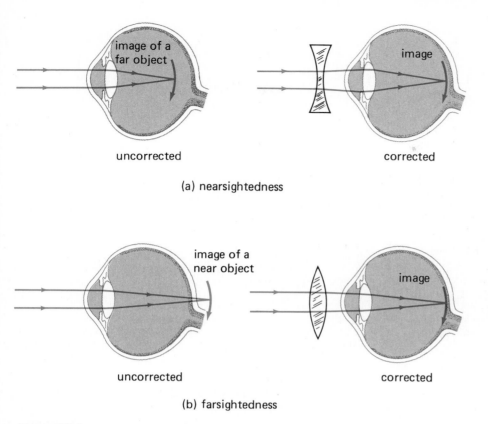

(a) nearsightedness

(b) farsightedness

Figure 22.4
Common visual defects. (a) Nearsightedness. A nearsighted person can see nearby objects clearly, but the images of far objects are formed in front of the retina. This is corrected with a diverging lens. (b) Farsightedness. A farsighted person can see far objects clearly, but the images of nearby objects are formed behind the retina. This is corrected with a converging lens.

which the eye sees clearly (Fig. 22.5). Notice that the image is larger than the object. Reading glasses magnify things slightly.

This type of farsightedness may be called a *normal* defect, since it occurs naturally with age. So the next time you see someone holding something at arm's length to read, don't laugh. You'll be there someday.

Another common defect, **astigmatism**, occurs when the cornea and/or the crystalline lens of the eye are not perfectly spherical. As a result, the light rays have different focuses in different planes or directions. That is, a viewed object may be distinct in one direction and blurred in another. Astigmatism may be corrected with a lens having a greater curvature in the plane in which the cornea or crystalline lens has deficient curvature.

Bifocal glasses are sometimes used to correct a combination of defects. The bifocal lens, invented by Ben Franklin, consists of two lenses on the same piece of glass. (Ben glued two lenses together.) For example, a small lower lens can correct for farsightedness (for reading), and the upper lens can be used to correct for nearsightedness or astigmatism. In some cases, "trifocals" are used to correct three conditions, or the center portion doesn't correct at all when a correction for intermediate "seeing" is unnecessary.

Table 22.1
Approximate Near Points of the Normal Eye

Age (yr)	Near Point (cm)
10	10
20	12
30	15
40	25
50	40
60	100

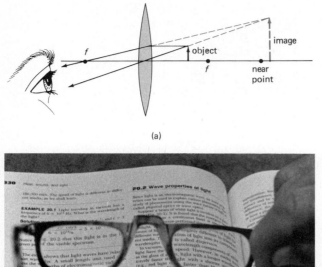

(a)

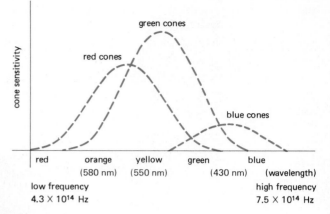

Figure 22.6
Cone sensitivity. Different cones of the eye are believed to respond to different frequencies of light to give three basic color responses.

(b)

Figure 22.5
Near-point correction. The near point of the eye recedes with age, and closer objects appear blurred (farsightedness). (a) A converging lens "projects" the image of a close object beyond the near point, and it can be seen clearly. (b) People with this normal farsightedness commonly wear glasses for reading.

■ COLOR VISION

Color is actually a physiological sensation of the brain in response to the light excitation of cone receptors in the retina. Many animals have no cone cells and are color-blind or live in a black-and-white world. In the human eye, the cones are sensitive to light with frequencies between about 7.5×10^{14} and 4.3×10^{14} Hz (400 nm to 700 nm wavelength). Different frequencies of light are perceived by the brain as having different colors. The association of colors with particular light frequencies is subjective. Monochromatic light has a particular physical frequency, but as pitch is to sound, color is to vision — it may vary from person to person.

The concept of color vision is not well understood. One of the most popular theories is that three types of cones are contained in the retina, each of which responds to light from different parts of the visible spectrum (Fig. 22.6). The "blue" cones have

maximum response for light with a wavelength around 430 nm, the "green" cones for wavelengths around 550 nm, and the "red" cones for wavelengths around 580 nm.*

Combinations and different degrees of cone stimulation give rise to intermediate colors. For example, when red and green cones are equally stimulated by light of a particular frequency, the brain interprets this as yellow. But when the red cones are stimulated more strongly than the green cones, the brain "sees" orange.

Color-blindness results when one type of the primary cones is lacking. Occasionally this occurs because of failure to inherit the appropriate gene for the cone formation. The color genes are found in the female sex chromosome, so almost all color-blind people are male (about 4 percent of the male population).

For example, if a person completely lacks red cones, he can see green through orange-red colors by use of his green cones. However, he is not able to distinguish among these colors satisfactorily because he has no red cones to contrast with the green ones. A similar condition exists if the green cones are missing. In either case, it is difficult or impossible to distinguish colors of the larger wavelengths, and the condition is called red-green color blindness.

* The cone "color" indicates only a general response region of the spectrum. For example, light with a wavelength of 580 nm is orange, but the major cone response in this end of the spectrum is from a "red" cone.

COLOR MIXING

White light is a mixture of all visible frequencies or colors, although perhaps not of equal intensities. For example, sunlight has a predominant yellow-green component. Even so, the composite colors of white light can be easily demonstrated by dispersion with a prism.

When colors of light are mixed together, it is found that one doesn't need all colors to make white light — just red, green, and blue. When light beams of these colors are projected on a screen so the beams overlap, additive mixtures of colors and white are produced (Fig. 22.7). Evidently the excitation of all types of cones in the eye causes the combination of signals to be interpreted as white and other combinations as different colors. This is referred to as the **additive method of color production.**

By adding varying amounts of red, green, and blue light, we find that a wide variety of colors and hues of the visible spectrum can be generated. The triad dots on a color TV screen have red, green, and blue phosphors.

As a result, **red, green,** and **blue** are called the **additive primaries** or **primary colors.** Not only does a mixture of the additive primaries appear white to the eye, but many pairs of color combinations do also. The colors of such a pair are said to be **complementary colors.** For example, the complement of blue is yellow, of red is cyan (turquoise), and of green is magenta (purplish red). This is not surprising, since from Figure 22.7 we can see that a combination of red and green is interpreted by the brain as yellow. Presumably, yellow light stimulates the red and green cones, which along with blue cone stimulation is a "white" combination.

Thus we see that an object has color because of the light coming from it. Other than light sources, objects have color when illuminated with white light because they reflect the wavelengths of the color they appear to be. The light coming from an object can be looked at in terms of selective reflection or selective absorption.

When white light strikes a colored surface, certain light frequencies cause resonance electronic oscillations in the surface atoms, somewhat as sound causes a tuning fork to vibrate resonantly. The electrons of a particular atom have a narrow range of vibrational frequencies. The light components in this range are absorbed and re-emitted (reflected). The other light frequencies are absorbed by the material but go into heat (internal energy) rather than being re-emitted.

The paper of this page has atoms with a great range of resonance frequencies. When white light strikes the page, enough frequencies are reflected to your eye so that it appears white. The black print, on the other hand, has very few atoms with vibrations in the visible range of frequencies, so the light is almost totally absorbed, and the print appears black. (The light goes into the internal energy of the print ink. For the same reason a thermometer with its bulb painted black registers a higher temperature than one with an unpainted bulb.) When white light strikes transparent red glass or a red rose, only red light is transmitted through the glass or reflected from the rose. All of the other colors are absorbed.

Selective absorption is important in the mixing of pigments for color production, such as in making paints and dyes for clothing. The additive method discussed previously for light does not apply here. You probably know or can guess, or any artist will tell you, that if you mix red, green, and blue pigments or paints you won't get white as in the case of light. You'd end up with paint with some sort of dark brown color.

But how about mixing blue and yellow paints? You might know that this produces green. The same effect is obtained with light by passing white light through blue and yellow filters (Fig. 22.8). This is because the blue pigment in the paint or filter absorbs the wavelengths or colors, except those in the blue region of the spectrum. Likewise, the yellow

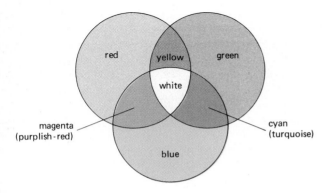

Figure 22.7
The additive method of color production. Light mixing of the primary colors (additive primaries) produces the colors as illustrated in the regions of overlapping light beams.

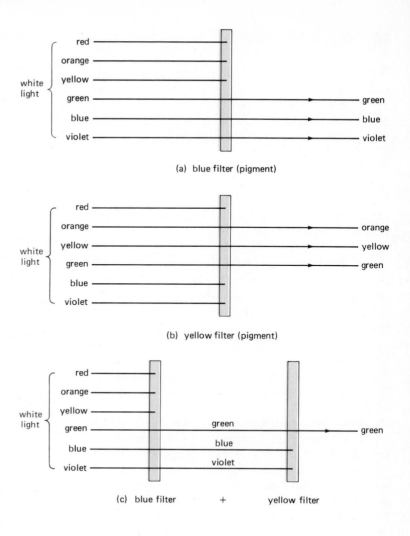

Figure 22.8
Selective absorption. A particular filter (or paint pigment) absorbs wavelengths or colors in a particular region. By the subtraction (i.e., absorption) of component colors of white light, a particular color is produced.

pigment in the paint or filter selectively absorbs the wavelengths of light, except those in the yellow region. The wavelengths in the intermediate green region are not strongly absorbed by either pigment, and hence, green light is transmitted through the filters or is reflected from the paint mixture.

This is an example of the **subtractive method of color production or mixing**. A mixture of absorbing pigments results in the subtraction of colors, and the eye sees the color that is not subtracted or absorbed. Three particular pigments—**cyan**, **magenta**, and **yellow**—are called the **subtractive primaries** or **primary pigments** (colors). Various combinations of two of the subtractive primaries produce the three additive primary colors (red, blue, and green), as illustrated in Figure 22.9.

Notice that the magenta pigment "subtracts" the green color from the cyan-yellow overlap. As a result, magenta is sometimes referred to as "minus

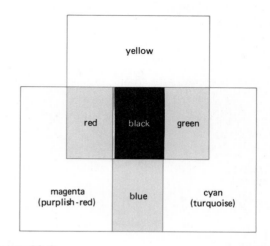

Figure 22.9
Subtractive primaries. When the subtractive primaries or pigments—cyan, magenta, and yellow—are mixed, the colors (or lack of color) are obtained, as illustrated in the overlapping regions.

green." (Think about adding a magenta filter to Figure 22.8.) Similarly, cyan is "minus red," and yellow is "minus blue."

When all of the primary pigments are mixed in the proper proportions, almost all of the wavelengths of the visible spectrum are absorbed or subtracted, and the mixture appears black. People in the paint and dye businesses commonly refer to the subtractive primaries as being red, yellow, and blue, because magenta is a shade of red and cyan is a shade of blue. In paint stores, paint of any color can be produced by mixing proper combinations of subtractive primaries.

Question: Suppose a yellow banana and a red apple are illuminated with green light. How would they look?

Answer: The banana would appear green. Since green is close to yellow, some green light would be reflected by the "yellow" oscillators in the banana peel. The apple would appear dark because most of the green light would be absorbed. If there were some leaves on the apple stem, you'd see green leaves and a dark apple.

See Special Feature 22.2 (*next page*) for another color phenomenon:

Question: If the scattering by atmospheric gases is greater for visible light of greater frequency, why isn't the sky violet? It should be scattered more than blue light.

Answer: Violet light is scattered more than blue light, but the sky is blue for a couple of reasons. First, and more important, the eye is more sensitive to blue light than to violet light. (See Figure 22.6.) Second, sunlight contains more blue light than violet light. The greatest color component is yellow-green, and the distribution generally decreases toward the ends of the spectrum.

OPTICAL INSTRUMENTS

There are a wide variety of optical instruments that use mirrors, lenses, prisms, diffraction gratings, fiber optics — any optical application to achieve the desired purpose of the instrument. Before leaving the subject of optics, let's consider some of the optical instruments that you use or are probably familiar with.

Projectors

The projection of the images of slides and films on a screen is commonplace. We all like to show slides of our vacation or of the party last week or to show a home movie. This is done by means of a projector. The basic components of a slide projector are shown in Figure 22.10, along with some common types of projectors.

A pair of "condensing" lenses concentrates light from a source on a slide. The slide is placed just beyond the focal point of the converging lens, and a magnified image of the slide is formed on the screen. Because of the refractive properties of the

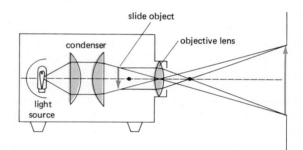

(a)

(b)

Figure 22.10
Optical instruments. (a) Diagram of a simple projector and (b) various kinds of projectors.

Why the Sky Is Blue (and Sunsets Are Red)

If we lived on the moon, which has no atmosphere, the sky would appear black, except in the vicinity of the Sun. This is what astronauts on the moon see. On Earth, however, our sky is blue as a result of the scattering of sunlight in the atmosphere. As sunlight passes through the atmosphere, the nitrogen and oxygen molecules of the air absorb some of the light and re-emit it. The light is scattered in all directions from the free gas molecules.

This scattering is selective, with the resonant frequencies of the small molecular oscillators in the ultraviolet region. The frequencies of light in the visible region are below the resonant frequencies but are close enough to be absorbed and scattered somewhat, particularly in the blue end of the spectrum. Thus, the light in the visible spectrum is preferentially scattered, with the light at the blue end of the spectrum being scattered about ten times more than light at the red end. This is called **Rayleigh scattering** after Lord Rayleigh, the British scientist who explained it. Some of the scattered light reaches the Earth, which we see as blue sky light (see figure, next page).

It should be kept in mind that *all* colors are present in sky light, but the dominant wavelength or color lies in the blue. You may have noticed that the sky light is more blue directly overhead or high in the sky and less blue toward the horizon, becoming white just above the horizon. This is because there are fewer scatters along a path through the atmosphere directly overhead (your zenith position) than toward the horizon, and multiple scattering along the horizon path gives rise to the white appearance.

By analogy, if you add a drop of milk to a glass of water and illuminate the suspension with intense white light, the scattered light has a bluish hue. And yet, a glass of milk is white (owing to multiple scatterings). Atmospheric pollution may enhance the milky white appearance of the sky.

The scattering of sunlight by the atmospheric gases *and* small particles gives rise to red sunsets. Generally it might be thought that since the distance sunlight travels through the atmosphere is greater to an observer at sunset, then most of the higher-frequency colors of the visible spectrum are scattered from the sunlight and only light in the red end of the spectrum reaches the observer.

However, it has been shown that the dominant color of this light, due solely to molecular scattering, is orange.* Hence, there must be scattering by small particles in the atmosphere that shifts the light from the setting (or rising) sun toward the red. Foreign particles in the atmosphere are not necessary to give a blue sky;

* Bohren, C. F., and A. B. Fraser, Colors of the Sky, *The Physics Teacher,* 23, 5, p. 267, 1985.

---

lens, which are indicated by the light rays in the diagram, the image of an upright object is inverted. Therefore, slides are placed in the projector upside-down so the image on the screen is seen right side up.

Motion picture projectors have basically the same components, along with the machinery to advance the film at a given speed. By quick advancement of one slightly different picture frame after another, the projected image is given the illusion of motion.

A motion picture with sound also involves optical methods. When a motion picture is filmed, a recording amplifier drives a special lamp that exposes a narrow "sound track" along the side of the film. This track is a variation in optical density (light and dark regions) that follows the variations in the lamp voltage induced by sounds.

To reproduce the sound when the film runs through the projector, an optical system sends a beam of light through the sound track to a photocell that detects the light. The variations in the light and dark portions of the track cause fluctuations in the light beam, which are monitored by the phototube, and the electrical impulses are fed into an amplifier-speaker system that reproduces the original sounds.

Microscopes

When we want to see small objects better, we use a magnifying glass or a microscope. The single, conthey even detract from it. Yet, such particles are necessary for deep red sunsets and sunrises, which occur most often when there is a high-pressure air mass to the west (for sunsets) and to the east (for sunrises), since the particle concentration is generally higher in a high-pressure air mass than in a low-pressure air mass.

Red sunrises and sunsets are often made more spectacular by layers of pink clouds. The cloud color is due to the reflection of red light. Since the water droplets of clouds scatter visible light of all wavelengths about equally, the clouds do not affect the color of the light but merely diffusely reflect the incident red light.

Atmospheric scattering. Scattering in the atmosphere accounts for (a) the blueness of the sky and (b) red sunsets.

(a)   (b)

verging-lens magnifying glass is called a **simple microscope**. To see how a lens magnifies, think of how we judge the size of an object. A car in the distance looks quite small, as compared with when it is close by (Fig. 22.11). The difference is the angle of view. When the car is close by, it looks much larger because it is viewed through a much wider angle.

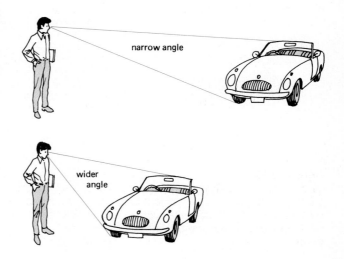

Figure 22.11
How big? The angle of view determines how large an object appears.

A lens can make an object appear larger than it is because the refraction or bending of light widens the angle of view (Fig. 22.12). But the eye sees the light as if it had been traveling in a straight line, and the object appears to be much larger than it really is, or magnified.

To obtain greater magnification than that given by a single lens or simple microscope, a **compound microscope** may be used. Here, more than one lens is used, as shown for a basic compound microscope in Figure 22.13. The objective (lens) forms a magnified image of the object, in much the same way as the lens of a projector does. The eyepiece or ocular (lens) acts as a magnifying glass and further magnifies the image as seen by an observer looking into the microscope. Actual microscopes are much more involved; only the basic principles are presented here.

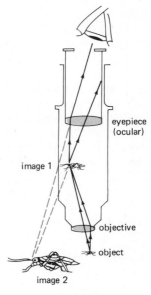

(a)

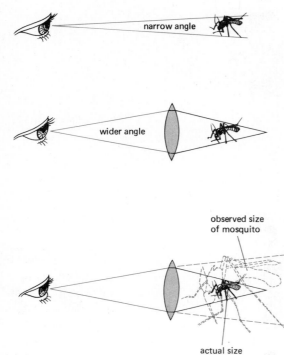

Figure 22.12
Magnification. A single lens or simple microscope expands the angle of view so that an object appears larger. (Note: the image is not formed at the same place as the object, but farther away. They are drawn together here for general size comparison.)

(b)

Figure 22.13
The compound microscope. (a) A basic compound microscope uses two lenses, an objective and an eyepiece, for greater magnification. (b) A modern binocular compound microscope. Two eyepieces make for easier viewing, and three objectives mounted on the central rotatable turret allow a selection of magnifications.

Telescopes

Telescopes are used to see distant objects or to make them appear closer. Telescopes collect and concentrate light energy to form images. This is particularly true for astronomical telescopes used to view distant stars and galaxies. Smaller telescopes are used to view things on Earth, for example, a transit telescope used by a surveyor. The two general types of telescopes are refracting telescopes, based on lens refraction, and reflecting telescopes, based on mirror reflection.

The Refracting Telescope

The optics of an astronomical refracting telescope are similar to those of a microscope, but the objective lens is made very large so as to collect a large quantity of light from a distant object so that an image may be seen (Fig. 22.14). The final image of an **astronomical telescope** is inverted or upside-down, but this poses no problem in astronomical work.

However, it is not convenient to have an inverted image when viewing objects on Earth with a telescope. In a **terrestrial telescope**, the image of an object is viewed right side up, which is important, for example, in a surveyor's telescope (Fig. 22.15). This is accomplished by using a diverging lens as an eyepiece, which is called a **Galilean telescope**, after Galileo, who built one in 1609.

A third, "erecting" converging lens may be placed between the objective and the eyepiece to invert the image so that the final image is right side up. This is the principle of the "spyglass" telescope, which must be extended to a great length (like the type used in pirate movies). The inconvenient length increase can be avoided by using internally reflecting prisms. This is the principle of the common prism binoculars [Fig. 22.15(c)].

The Reflecting Telescope

To form images of distant stars and galaxies, enough light energy must be collected or received by the telescope. The amount of light entering a refracting telescope can be increased by enlarging the objective lens. However, there are physical limitations to this approach in grinding the lens, as well as material defects and aberration effects. The largest refracting telescope is the 40-in. (102-cm; objective diameter) telescope at the Yerkes Observatory in Williams Bay, Wisconsin, shown in Figure 22.14.

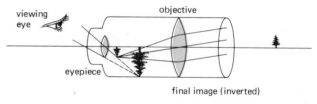

(b)

Figure 22.14
Refracting astronomical telescope. (a) The image is inverted in an astronomical telescope, but this poses no problem in such work. (b) The 40-in. (102-cm; objective diameter) refracting telescope at Yerkes Observatory.

Another approach is to use a reflecting telescope. The basic components of this type of telescope are shown in Figure 22.16. In this case, the objective is a concave mirror, which collects and focuses the light. To make the image conveniently accessible to the eye or a camera, the rays may be deflected by a plane mirror to the side of the tube and observed, with the image magnified by the eyepiece.

However, if the telescope is large enough, the observations may be made inside the telescope tube. This is the case for the 200-in. (5.1-m; mirror

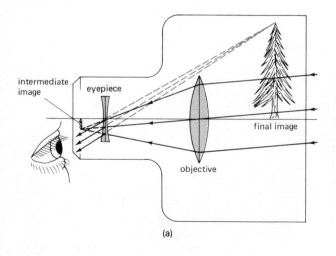

(a)

(b)

(c)

Figure 22.15
Terrestrial telescopes. (a) With use of a diverging lens for an eyepiece, the final image is upright. This is called a Galilean telescope. (b) A surveyor's transit is a type of terrestrial telescope. (b) Binoculars shorten the length of a telescope by using prisms.

diameter) Hale Observatories telescope on Palomar Mountain in California, which at the time of this writing is the United States' largest and the world's second largest reflecting telescope (Fig. 22.16).* There are plans calling for reflecting telescopes using multiple mirrors and computer-combined images that give a large effective area.

Large mirrors can be constructed more easily than large lenses, since only one surface need be ground and silvered. Reflecting telescopes are also free of chromatic aberration. Parabolic mirrors are usually used, which makes them free of spherical aberrations as well.

* The largest reflecting telescope is in the Soviet Union (6-m or 236-in. reflector).

Plans are also being made for orbiting space telescopes that will be free of atmospheric effects and background light from cities that hamper Earth-bound observation. An infrared telescope has been put into orbit about the Earth. It surveyed the "infrared sky" and sent back a great deal of data before it stopped transmitting.

Among the initial findings are what is believed to be very small solid particles orbiting the star Vega, the first direct evidence of solid material around a star other than the Sun, and the discovery of five new comets in the solar system. Another type of telescope, the radio telescope, which gives another "view" of the sky, is considered in Special Feature 22.3.

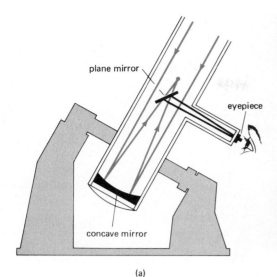

plane mirror

eyepiece

concave mirror

(a)

(b)

(c)

Figure 22.16
Reflecting telescope. (a) A concave mirror collects and focuses the light. A mirror may be used to bring the rays to a side eyepiece. (b) The Hale 200-in. (5.1-m; mirror diameter) telescope. The glass mirror weighs 15 tons and is 24 in. thick at the edges and 20 in. thick at the center. A thin coating of aluminum provides the reflecting surface. (c) Observations are made at the prime focus position at the center of the telescope. See also Figure 27.1.

SUMMARY OF KEY TERMS

Retina: the "film" on the back surface of the eyeball on which images are formed and transmitted to the brain.

Accommodation: the changing of the curvature of the crystalline lens so that the images of objects at various distances are formed on the retina.

Rods: photosensitive cells of the retina that are responsible for twilight (black-and-white) vision.

Cones: photosensitive cells of the retina that are responsible for color vision.

Nearsightedness: a vision defect in which a person can see near objects clearly but cannot see distant objects clearly.

Farsightedness: a vision defect in which a person can see far objects clearly but not near or close objects.

Astigmatism: a vision defect caused by the cornea and/or the crystalline lens not being spherical, so the image on the retina is out of focus in particular planes or directions.

Additive primaries or primary colors: the colors red, green, and blue. Light of these colors can be mixed or "added" to give any color of the visible spectrum or white.

Complementary colors: pairs of colored light combinations that appear white to the eye, for example, blue and yellow.

SPECIAL FEATURE 22.3

Radio Telescopes

Although not an optical instrument, the radio telescope is an important astronomical tool. Stars and galaxies emit radio waves as well as light waves. This fact was discovered accidentally by an electrical engineer named Carl Jansky in 1931 while working on a static problem in intercontinental radio communications. He found that an annoying static hiss was coming from a fixed direction in space, which later was found out to be the center of our galaxy (the Milky Way).

Thus, Jansky discovered radio waves coming from space. This was quickly recognized as another source of astronomical information, and radio telescopes were built (Fig. 1). A radio telescope operates similarly to reflecting light telescopes, inasmuch as radiation is collected by a large-area reflector and focused to form an "image." However, the parabolic collector of a radio telescope does not look like a mirror surface, since it is covered with wire mesh. To avoid confusion, the collector of a radio telescope is referred to as a "dish" instead of a mirror. You now commonly see satellite TV antenna dishes.

The radio telescope dish is not a mirror for light waves, but it is for radio waves. This is because electromagnetic waves cannot detect any hole or surface irregularity that is smaller than the wavelength of the radiation. Since radio waves range from about 1 cm to several meters in wavelength, the wire-mesh surface of the metal dish acts as a good reflecting surface for such waves.

Another noticeable difference between optical and radio telescopes is that there is no film or eyepiece for a radio telescope. Instead, the radio waves are detected by an antenna positioned at the focal point of the dish. The signals are amplified, and the information received by the telescope is displayed on a recorder so it can be "seen."

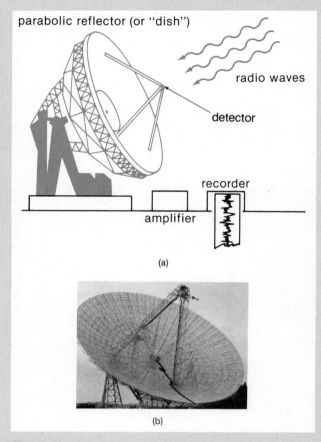

(a)

(b)

Figure 1
Radio telescope. (a) The telescope "dish" reflects radio waves to a focus at the antenna or detector, which feeds the signal into an amplifier and recorder. (b) The radio telescope at Green Banks, West Virginia.

Subtractive primaries or primary pigments: the pigments cyan, magenta, and yellow, which can be mixed to give paints or dyes that appear as any color or black through selective absorption of light.

Rayleigh scattering: the preferential scattering by molecules of the air of high-frequency visible light that gives rise to the blueness of the sky.

Refracting telescope: a telescope that uses lenses to view distant objects.

Reflecting telescope: a telescope that uses a converging mirror to view distant objects.

Figure 2
The 305-m (1000-ft)–diameter radio telescope at Arecibo, Puerto Rico. The dish was made by placing wire mesh over a natural bowl in the mountains. The antenna is mounted on a trolley suspended by cables 500 ft above the valley floor. Although the dish is fixed, signals are received from many directions as a result of the Earth's rotation and revolution.

Radio telescopes have added a new dimension to astronomy. They supplement optical telescopes and offer some definite advantages. Radio waves pass freely through the huge clouds of dust that exist in our galaxy and hide a large part of it from optical view. Radio waves easily penetrate the Earth's atmosphere, whereas a large part of the incoming light is reflected and scattered.

Radio astronomy has extended the dimensions of the known universe almost twofold. With radio telescopes astronomers can detect galaxies that are two times farther away than those detected by optical telescopes. Although the emission of radio waves may be less intense, the penetration of the radio waves and construction of large dishes permit their detection (Fig. 2). Visual light from distant galaxies is blanketed by background illumination of the night sky. Radio telescopes may be operated around the clock, and observations can be made day or night.

QUESTIONS

The Human Eye and Visual Defects

1. Many people wear "contacts." On what part of the eye are contact lenses worn?
2. How do the operations of a camera shutter and of the eyelid generally differ? Can you think of an instance when a camera shutter is operated like an eyelid?
3. If upside-down images are formed on the retina, how do we know that the world isn't really upside-down when we see it?
*4. Does the eye "camera" have black-and-white or color film? Is there an analogy with the type of film used in a camera?

5. What is meant by "twilight vision"?

6. During eye examinations, the ophthalmologist usually dilates your eyes. This is done by adding fluid to your eyes that causes the pupils to widen (dilate) and remain open so that the inside of the eyeball can be examined. The fluid also temporarily paralyzes the ciliary muscles. How does this affect the person whose eyes are dilated?

7. Answer the questions in Figure 22.17.

*8. A person wears bifocal glasses to correct for nearsightedness and astigmatism. How would the bifocal lens be ground for these vision defects?

9. Why are "reading glasses" that are used to correct normal farsightedness also "magnifying glasses?" (See Fig. 22.5.)

Color Vision and Mixing

*10. Moonlight is reflected sunlight. Why don't we see color when viewing objects on the moon?

11. Is white a color? Is black? Explain.

12. In very rare instances of color-blindness, a person lacks blue cones (a condition called blue weakness). What colors would this person have difficulty in distinguishing?

13. Suppose that the beams of red and blue spotlights overlapped each other on a wall. (a) What color would the wall appear to be if it were white? (b) How about a blue wall? (c) A green wall?

14. What do light filters filter out?

*15. In a department-store window display, white light passing through a red filter falls on cyan and yellow objects. What colors are seen by someone looking at the display through the store window?

16. How do different-colored spotlights influence the appearance of a performer's clothes, for example, a red spotlight on a performer in a dark blue suit? What color of light would you use if you wanted to make the performer really stand out?

*17. (a) What color would a green chalkboard appear to be when viewed through yellow sunglasses? (b) How about when viewed through rose-colored (red) glasses? (c) If the chalkboard had writing on it in white chalk, what color would the writing appear in these cases?

18. How would the American flag appear when illuminated with (a) red light, (b) blue light, and (c) green light?

19. Could you ignite a piece of paper with a magnifying glass and sunlight more quickly if the paper were white or black? Explain.

20. Why is the inside of a camera black?

21. Why do fluorescent lamps tend to enhance blue clothing more than incandescent lamps do?

*22. Two complementary colors produce white, but if colored light from a spotlight falls on an object with its complementary color, it appears black. Why?

23. Why does a white piece of paper appear the color of whatever type of light that illuminates it?

(a) Which cube does not have all of its sides?

(b) Are the diagonal lines parallel?

(c) Are all the men the same size?

(d) Is the lower line on the right an extension of the line on the left?

(e) Is line AB equal in length to line AC?

Figure 22.17
More optical illusions. See Question 7.

24. By which process of color production could red and cyan (purplish red) colors be combined to get white?

25. Can white be obtained by the subtractive method of color production? Explain.

26. Is color TV based on the additive or subtractive method of color production? Explain.

27. A painter mixes paints with primary pigments of (a) yellow and blue and (b) blue and red. What colors are the resulting paint mixtures?

•28. Clouds generally appear white, but nimbus clouds, which threaten or produce rain, appear dark. Why is this?

29. Stars are in the sky during the daytime as well as at night. Why can't we see them? Why can we see the moon during the day and not the stars? Do astronauts on the moon see stars during the day?

30. Why does the Sun generally appear yellowish orange at sunset? Why aren't there red sunsets every evening?

•31. There is an old weather saying, "Red sky at night, sailors' delight. Red sky in the morning, sailors take warning."* Explain the possible validity of this prediction, considering that fair weather is generally associated with high-pressure air masses and that the conterminous (being adjacent or in contact) United States lies in the Westerlies wind zone (air mass movement generally from west to east).

* A similar quote is made in the Bible. See Matthew 16:1–4.

Optical Instruments

32. Could slides be put in a projector right side up and have a right-side-up image on a screen if other lenses were added to the system? Explain.

33. How are animated cartoons made?

34. Why are upside-down images no problem in astronomical telescopes?

•35. What determines the distance at which telescopes can be used to see objects? Do reflecting or refracting telescopes offer any advantage in this respect?

36. A detective uses a small astronomical telescope to view a crime committed in a distant apartment building. How would he describe what he saw? If you were a defense attorney at the accused person's trial, how would you try to discredit the detective's testimony? (Would it have helped if the detective had stood on his head while observing the crime?)

37. Why do astronomers want to put telescopes in high-altitude, Earth-orbiting satellites? Would being that much closer to stars and galaxies make a difference?

•38. The reflecting telescope mirror at Palomar Mountain Hale Observatories is 200 in. in diameter. What is the diameter in meters? How does the circular area of the mirror surface compare with the floor area of your room? (The area of a circle is $\pi d^2/4$, where d is the diameter.)

39. Spherical and parabolic mirrors can be used for solar heating, for example, to cook foods directly or to light a fire. Discuss the advantages and disadvantages of this source of heating.

MODERN
PHYSICS

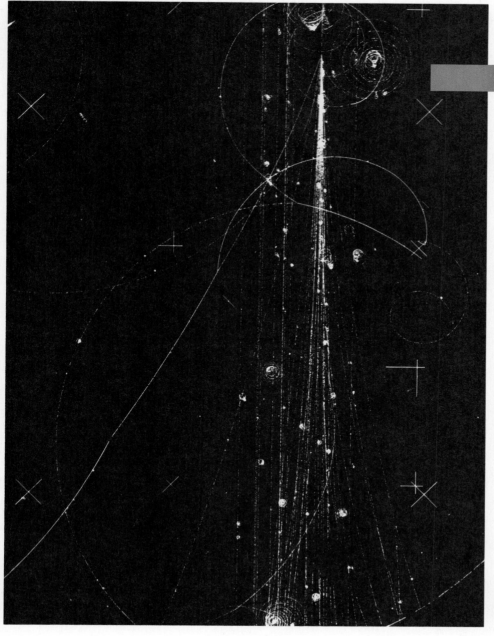

VII

Electrically charged
particles in a bubble
chamber with a magnetic
field move in spiral paths.

PART VII

The term *modern physics* generally means the developments in physics since around 1900. But there's more to it than that. Modern physics also means a revolution in scientific thought. There were many developments in "classical" physics during the 1700's and 1800's. The principles of mechanics, gravitation, heat and thermodynamics, electricity and magnetism, optics — all the things we have studied — were reasonably well understood. Scientists felt secure in their descriptions of nature. The theories perhaps needed a little polishing to make them just right.

Then, in the late 1800's, things began to happen. The prevailing classical theories failed to explain various experimental results — sometimes rather badly. There were problems with explaining electromagnetic wave propagation, line spectra, and photoelectricity (photocells). In the 1900's new areas of physics opened up, for example, nuclear physics. Here, too, there were some problems with classical theories. By the scientific method, something had to change — theories needed to be modified or scrapped.

It would have been foolish to throw out the classical theories that explained so many phenomena, particularly when there were no replacements. What evolved was the modification of classical theories and principles with new ideas *and,* in some instances, completely new theories to explain the modern physics observations. The remaining chapters of this text are about the "new" or modern physics. We'll kick it off with a "classic" example of how new thoughts and ideas had to be added to physics in the case of relativity.

Relativity

23

Albert Einstein, although famous for his theory of relativity, won a Nobel prize for the explanation of the photoelectric effect.

413

THE BACKGROUND

The background of the theory of relativity is full of discovery and debate. For most people, the term *relativity* or the theory of relativity immediately brings to mind Albert Einstein (and vice versa). Indeed, Einstein did formulate a theory of relativity, which consists of two main parts: the special theory (circa 1905) and the general theory (circa 1915). But before looking at this theory, let's set the stage for its development.

The theory of relativity involves light, in particular, the speed of light. From common experience we see that light appears to travel instantaneously from one place to another. Measuring its speed is no easy task. Galileo tried to do this by measuring the time it took light signals to travel several kilometers. Galileo's signals came from a lantern. When a covering bucket was removed, a light beam was sent to an assistant, who removed a bucket from another lantern and sent a signal back to Galileo.

As you might imagine, the experimental results were not very good (in fact, they were useless), since the assistant's and Galileo's reaction times were involved. (For typical people, this reaction time would allow light to go all the way around the Earth.) The speed of light was subsequently measured using astronomical methods.

The first successful terrestrial method of measuring the speed of light was carried out in 1849 by the French physicist H. L. Fizeau. The idea of Fizeau's cogwheel method is illustrated in Figure 23.1. If the cogwheel is not moving, then the light from a candle will go through the opening between cogs 1 and 2, travel to the mirror, be reflected back, and pass again through the opening between the cogs to an observer behind the candle. However, if the wheel is rotated, the beam is "chopped" up.

When the wheel is rotating slowly, a segment of the light beam—for example, that chopped off between cogs 1 and 2—will be reflected back and will arrive while cog 2 is in front of the observer. But at a faster rotational speed, the cog will be out of the way and the light-beam segment will pass between cogs 2 and 3 and will be seen by the observer. The time it takes for the next cog space to rotate in front of the observer can be calculated from the rotational speed, and if we know the distance traveled by the light, the speed of light can also be calculated.

Fizeau's results were not extremely accurate, but they were pretty good considering the equip-

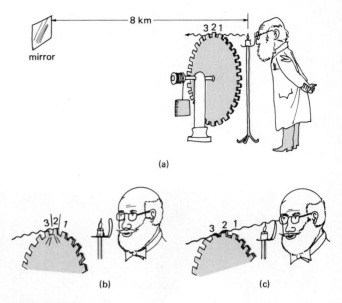

Figure 23.1
The speed of light. Fizeau's cogwheel method for measuring the speed of light. See text for description.

ment. His experimental result for the speed of light was about 5 percent off the present-day accepted value:

(speed of light)* $c = 3 \times 10^8$ m/s ($\approx 186{,}000$ mi/s)

At this rate, it takes about 8 min for light to reach us from the Sun. The closest star to our solar system, Alpha Centauri, is about 4.3 light years away. A light year is the *distance* light can travel in a year (about 10 trillion kilometers or 5.8 trillion miles). Hence, the light we see from Alpha Centauri is more than four years "old," since it takes 4.3 years for the light to reach us. How about the light from a star 1000 light years away?

THE ETHER

With the wave nature of light being demonstrated through interference experiments (Chapter 20) and the speed of light being reasonably well known by astronomical methods prior to the 1800's, scientists turned their attention to the consideration of the medium that carried or propagated light waves. From general experience, it was thought that a me-

* This is the speed of light in vacuum (or approximately in air). Recall from Chapter 21 that the speed of light in a transparent material is less than c.

dium was necessary for wave propagation. For example, sound waves propagate in air (and other media). It could be shown that sound waves could not travel through a vacuum—some material substance was needed for their propagation. Similarly, you couldn't have water waves without water. By such reasoning, it was also believed that light waves had to have a medium of propagation.

How then did light propagate through the void of space, for example, from the Sun to Earth? Not being able to see how light could travel through nothing, scientists created a hypothetical medium that was called luminiferous **ether**, or just plain ether. This substance was presumed to occupy interstellar space and to be present in all materials through which light travels. The idea of the existence of the ether seemed so logical that it quickly gained widespread acceptance.

It is easy to *postulate* the existence of something, but by the scientific method, a theory must be substantiated by experiment. So scientists set out to detect the ether experimentally. This set the stage for Einstein's theory of relativity.

■ THE MICHELSON-MORLEY EXPERIMENT

Since the ether permeated all space, it was reasoned that it was the one thing that remained fixed in the universe. This belief led to an idea about how to detect the ether and prove its existence. The Earth's average speed through the stationary ether in its revolution about the Sun is about 30 km/s. Moving through the ether would cause an ether "wind" to blow over the Earth (Fig. 23.2). This is similar to the "wind" you experience when traveling in a car or on a motorcycle through still air.

From experience with real winds and sound waves, it was known that the measured speed of sound in air depends on the wind speed and direc-

tion. For example, suppose the speed of sound in still air were 335 m/s. If there were a 25-m/s wind blowing in the direction in which the sound wave was propagating, then an observer would measure the sound speed to be $v = v_s + v_w = 335$ m/s $+ 25$ m/s $= 360$ m/s (Fig. 23.3). Similarly, if the wind were blowing in the opposite direction, the observer would measure the sound speed to be $v = v_s - v_w = 335$ m/s $- 25$ m/s $= 310$ m/s.

Thus, the motion of the air, or wind, changes the observed speed of sound. It was reasoned that the ether wind should do the same with the speed of light and thereby would prove that the ether existed.

This was the purpose of the famous Michelson and Morley experiment, which was performed in 1887 by two American scientists, A. A. Michelson and E. W. Morley.* An atmospheric wind analogy of the experiment is shown in Figure 23.4.

Two equally powerful planes fly the same distance in a race. The plane flying perpendicular to the wind must steer slightly into the wind on both legs of the race, which slows the plane somewhat. The other plane, flying directly into the wind on the first leg of the race, is slowed much more, but the wind adds to the speed on the return leg. Even so, it can be shown by analysis or by an actual race that this plane always loses the race or takes a longer time to cover the distance. Your instructor can prove this to you.

If the atmospheric wind were replaced by the ether wind and the airplanes by light beams, the light beams would then arrive back at the finish line at different times, as the airplanes do. Michelson used an ingenious method to develop an instrument to measure the expected time difference (Fig. 23.5). The method was based on the wave interference of light. The light beams interfering at a detector (finish line) produce a pattern of alternate bright and dark fringes. This pattern is noted for one position of the apparatus, which is called an interferometer, and then the apparatus is turned through an angle, say 90°. If this change in direction changes the travel times of the light beams, as it should in an ether wind, then there would be a shift in the interference pattern.

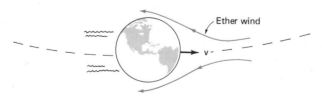

Figure 23.2
The ether wind. The orbital motion of the Earth through the ether should give rise to the "ether wind."

* Because of his work in the development of optical instruments and light investigations, Albert A. Michelson was the first American to receive the Nobel Prize in physics (1907). Edward W. Morley, Michelson's associate, was a chemist by training.

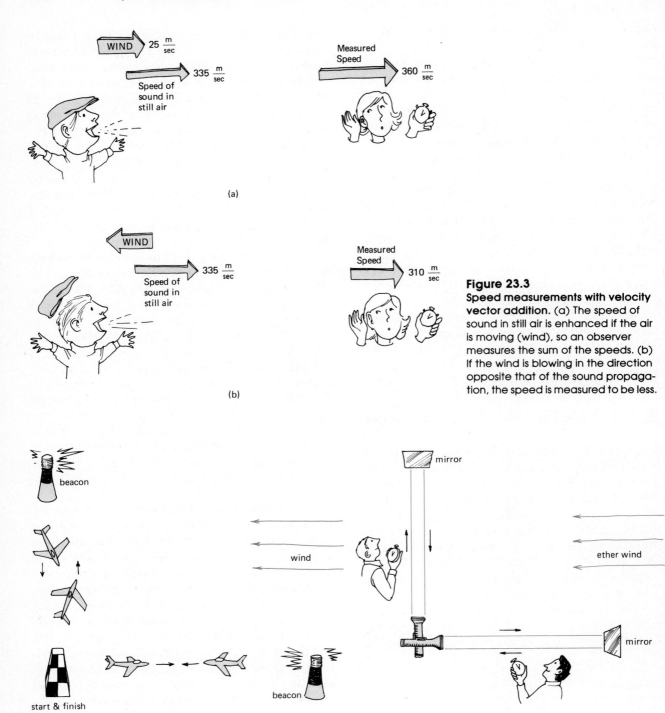

Figure 23.3
Speed measurements with velocity vector addition. (a) The speed of sound in still air is enhanced if the air is moving (wind), so an observer measures the sum of the speeds. (b) If the wind is blowing in the direction opposite that of the sound propagation, the speed is measured to be less.

Figure 23.4
Principle of the Michelson-Morley experiment. (a) Airplane analogy. When equal round-trip distances are flown as shown, the trip parallel to the wind always takes longer. (b) Similarly, light beams would classically be expected to arrive back at the starting point at different times.

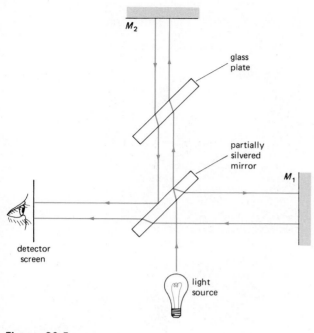

Figure 23.5
The Michelson interferometer. Light beams arriving at the detector show an interference pattern. If the apparatus is rotated 90°, any differences in the travel times of the light beams would cause a shift in the interference pattern.

The interferometer was sensitive enough to detect a time difference resulting from adding or subtracting the Earth's orbital speed of 30 km/s to or from the 300,000 km/s speed of light, but nothing was observed! Measurements were made at different times (day and night, different seasons) and at different locations (America and Europe), but always with a null result. There was no fringe shift!

Where was that ether? Several explanations were suggested to make the null result of the Michelson-Morley experiment consistent with classical wave theory. Perhaps the ether in the immediate vicinity of the Earth was "dragged" along with it so the interferometer was at rest with respect to the ether. This would explain the lack of an interference fringe shift, since there would be no wind. But if this were the case, light waves coming to the Earth would also be dragged along, since they travel in the ether. As a result, we would always see the light from a distant star coming from the same direction —which is not the case.

Another popular explanation at the time was offered by G. F. Fitzgerald, an Irish physicist. He suggested that all objects contracted or shrank in the direction of their motion through the ether. This

so-called **Fitzgerald contraction** would adequately explain the results of the Michelson-Morley experiment if the arm of the interferometer moving in the direction of the ether wind was shortened just enough so that no shift in the interference pattern would occur. It was a clever idea, but not without problems, since similar contractions weren't observed in other situations.

This was the unsettling state of affairs at the turn of the century that paved the way for a major reassessment of physical theories. In 1905 Albert Einstein published his special theory of relativity, which set forth the currently accepted explanation for the Michelson-Morley experiment. (See Special Feature 23.1.) At the time, Einstein's theory was rejected by many scientists. It required some thinking that went against "classical common sense."

THE SPECIAL THEORY OF RELATIVITY

The special theory is based on two postulates:

I. The Principle of Relativity. All the laws of physics are the same for all observers moving at a constant velocity with respect to one another.

II. The speed of light in free space is the same ($c = 3 \times 10^8$ m/s) for all observers regardless of the motion of the source or the motion of the observer.

Let's take a look at each of these postulates to get a grasp of their meanings. The first postulate, the Principle of Relativity, has subtle implications. It is really an extension of Newtonian relativity for the laws of mechanics to *all* the laws of physics. The fact that physical laws or experimental results of these must be the same for all observers moving uniformly with respect to each other was recognized by Newton, as well as Galileo.

Another way of stating the first postulate is that the laws of nature are the same in a laboratory at rest as they are in any *uniformly* moving laboratory. Everything would apear the same. For example, if you were in a ship or an airplane moving with a constant velocity, the making and pouring of a cup of coffee or throwing something up and catching it would be exactly the same as when the ship or plane were at rest. Or, looking at it another way, you cannot tell, by any experiment, whether you are at rest or moving uniformly.

Albert Einstein

Albert Einstein (1879–1955) was one of the greatest figures in physics. Born on March 14, 1879, in Ulm, Germany, he received his early education in Germany. School instruction seemed to bore him, and he showed no particular intellectual promise.

In 1896 at the age of 15, he entered the Swiss Federal Polytechnic school in Zurich to be trained as a teacher of physics and mathematics. In 1900 he received his diploma and acquired Swiss citizenship. After graduation, he found work in the Swiss Patent Office in Berne, where his main duty was the preliminary examination of patent applications. This job provided a livelihood and left him with ample time to work on fundamental problems in physics.

In 1905 he received a Ph.D. degree from the University of Zurich, and in the same year he published three papers of immense importance. Each contained a great discovery in physics. One dealt with quantum theory and an explanation of the photoelectric effect (Chapter 24). This paper formed the basis for his receipt of the 1921 Nobel Prize in physics. Another addressed molecular motions and sizes and an analysis of Brownian motion (Chapter 7). The third paper provided the special theory of relativity, which revolutionized modern ideas about space and time.

Following this, Einstein had no difficulty in becoming a professor at various universities. In 1915, while at the University of Berlin, he published his paper on the general theory of relativity, which provided a new theory of gravitation that included Newton's theory as a special case. During the late 1920's the political situation in Germany deteriorated severely. In 1933, when Hitler and the Nazis came to power, Einstein immigrated to the United States and joined the Institute for Advanced Study in Princeton, New Jersey, where he settled permanently. He became a U.S. citizen in 1940. The rest of his life was spent working on a unified theory that would include both gravitation and electromagnetism.

Near the beginning of World War II and following the discovery of nuclear fission, Einstein was asked to write a letter to President Roosevelt by other emigrant scientists, who along with him recognized the tremen-

Albert Einstein.

dous military potential of nuclear fission, particularly if Germany should develop it first. Einstein's famous letter was instrumental in starting the Manhattan Project and in developing the atomic bomb. After World War II, Einstein devoted much of his time to organizations advocating agreements to end the threat of nuclear war.

Suppose a person is watching uniformly moving automobiles, as illustrated in Figure 23.6. The velocities of the cars are shown referenced to the ground or the "stationary" observer. But if car A is taken as a *reference system,* car B is not moving and car C is moving with a speed of 10 km/h. As a matter of fact, with respect to car A, the "stationary" observer is moving with a speed of 40 km/h in the direction opposite to that of car C. Hence, the motion is *relative.*

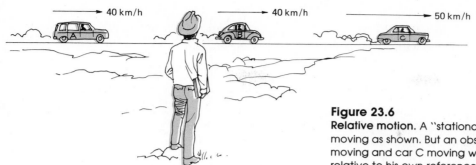

Figure 23.6
Relative motion. A "stationary" observer sees the cars moving as shown. But an observer in car A sees car B not moving and car C moving with a speed of 10 km/h relative to his own reference frame (car A).

But who is at rest? No one and everyone, in a sense. Anyone moving uniformly with respect to someone at "rest" is entitled to consider himself or herself to be at rest and the other person to be moving uniformly. A "state of rest," then, is one for which Newton's first law of motion (law of inertia) holds. We call this an inertial system — one at rest or in uniform motion.*

What all this means is that there is no "absolute" reference frame with the unique property of being at rest with respect to everything else — like the ether. So, the ether is rejected in Einstein's theory.

▪ THE CONSTANCY OF THE VELOCITY OF LIGHT

The second postulate is a bit more difficult to understand, if not believe. It says that the speed of light (in vacuum) has the same value c with respect to any inertial observer, or that everyone measures the speed of light to be the same no matter how fast he or she is moving. This postulate is a revolutionary statement, which seems to defy common sense and experience.

Classically, we expect and observe velocities to add (as vectors). For example, in Figure 23.7 the two inertial observers measure the ball to have different horizontal speeds. Relative to the observer on the train, which travels at 30 m/s, if the speed of the

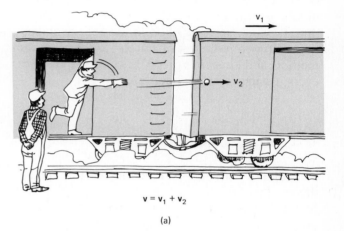

$$\mathbf{v} = \mathbf{v}_1 + \mathbf{v}_2$$

(a)

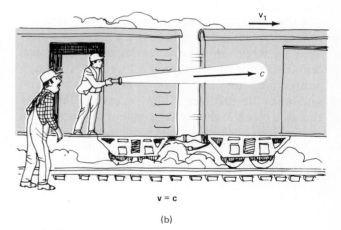

$$\mathbf{v} = c$$

(b)

Figure 23.7
Velocity addition. (a) Classically, the thrown ball has a horizontal velocity v_2 relative to the moving thrower, and the horizontal velocity is different for an observer in another system or frame of reference. (b) Relativistically, light has the same speed for all observers.

---

* The special theory of relativity is limited to inertial systems by the first postulate. Accelerated systems are treated in the general theory of relativity, discussed later in the chapter. In accelerating systems, we must either modify the laws of physics or introduce (pseudo) forces not found in inertial systems (see Chapter 4).

ball is 10 m/s, then relative to the observer on the ground, it has a speed of $30 + 10 = 40$ m/s.

Suppose that the ball is replaced with a beam or pulse of light. Would the speed of light then be 30 m/s + 300,000,000 m/s for the stationary observer? Not according to the second postulate. Both observers would measure the speed of light to be $c = 3 \times 10^8$ m/s. This would be the case even if both the light source and the observer were moving toward (or away) from each other.

The constancy of the velocity of light explains the null result of the Michelson-Morley experiment, but as you can imagine, the idea didn't receive much acceptance at the time. The postulates of the special theory go on to predict other rather "strange" things, which are described in the following sections. However, the test of a theory is the scientific method, and Einstein's theory has passed the test in every instance that has been tried.

TIME DILATION AND LENGTH CONTRACTION

Two of the "strange" predictions of the special theory involve the measurements of time and length. The strangeness comes about because we do not commonly observe the predictions at our slow pace. Noticeable relativistic effects require speeds that are apreciable fractions of the speed of light.

When this occurs, with one system moving at a constant velocity with respect to another, the special theory predicts that an observer will measure *different* times and lengths in the *different* systems. That is, when an observer compares times and lengths that he measures in the "moving" system to similar measurements made in his own system, he finds they are different. There is a **time dilation**— that is, a clock in the moving system appears to run more slowly—and a **length contraction**—i.e., a length, such as that of a meterstick, in the direction of the motion is measured to be shorter.

These ideas are illustrated in Figure 23.8. When both observers are at rest relative to each other, both of their clocks run together, and the lengths of objects are the same. But when things get moving, an observer finds (measures) that the moving clock runs more slowly and that lengths in the moving system are shorter. Of course, the situation is *relative*. An observer in the spaceship considers himself to be at rest and the "stationary" observer to be

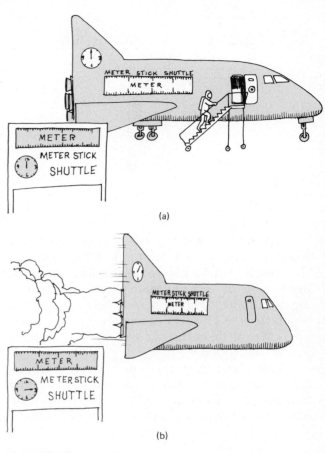

Figure 23.8
Time dilation and length contraction. When one system moves at a constant velocity relative to another, the special theory predicts that an observer will measure different times and lengths in different systems. The moving clock will appear to run more slowly (time dilation), and lengths in the moving system will appear to be shorter (length contraction).

moving relative to him. The spaceship observer would measure the same time dilation and length contraction when studying (measuring) a clock and lengths in the other system.

If we set $\gamma = 1/\sqrt{1 - v^2/c^2}$, then the magnitudes of the measured time dilation and length contraction are given by

$$t = \gamma t_o$$

and

$$L = L_o/\gamma$$

where v is the relative speed of the systems, c is the speed of light, and t_o and L_o are the *proper* time and *proper* length intervals. **Proper time** and **proper**

Table 23.1
Some Values of $\gamma = 1/\sqrt{1 - (v/c)^2}$

v	γ
$0.1c$	1.01
$0.2c$	1.02
$0.3c$	1.05
$0.4c$	1.09
$0.5c$	1.15
$0.6c$	1.25
$0.7c$	1.40
$0.8c$	1.67
$0.9c$	2.29
$0.99c$	7.09
$0.995c$	10.0
$0.999c$	22.4

length are the time and length measurements made by an observer moving along with the clock or the measured object, i.e., in the system in which the clock and the length, say a meterstick, are actually located. The t and L are the time and length intervals measured by an observer in another (relative moving) system who also measured the events or length in the proper system with his own measurement instruments.

Notice that $\gamma = \dfrac{1}{\sqrt{1 - v^2/c^2}}$ is always greater than one, since v is always less than c. Some values of γ for speed that are appreciable fractions of the speed of light are given in Table 23.1.

For example, suppose a spaceship moved with a constant velocity relative to an observer with a speed of $0.6c$ (six-tenths the speed of light). Then, with $\gamma = 1.25$, this gives $t = 1.25t_o$ and $L = L_o/1.25 = 0.8L_o$. This means that when the hour hand on the clock in the moving system goes around one time ($t_o = 1$ h), the observer in the "stationary" system observes this to take $t = 1.25t_o = 1.25(1) = 1.25$ h, or 1 h and 15 min, according to his clock. Or when the observer sees 0.8 h [48 min] pass on the moving clock (proper time), 1 h has elapsed on his own clock: $t = 1.25t_o = 1.25(0.8) = 1$ h. Hence, he observes things to be happening at a slower rate in the moving system.

Also, according to the preceding length equation, a meterstick on the spaceship [$L_o = 1$ m] would be measured to be contracted or shorter in the direction of motion. The meterstick would appear to be 80 cm long as compared with the observer's own meterstick [$L = L_o/\gamma = 1/1.25 = 0.8$ m $= 80$ cm].

Notice that if the spaceship were moving with a speed of $0.999c$, events on the spaceship would appear to be in really slow motion and "crunched up." For every hour that elapsed on the observer's clock, less than 3 min would be measured to pass on the moving clock, and the moving meterstick would be measured to be less than 5 cm long!

Question: All of this sounds nice, but is there any experimental evidence for time dilation or length contraction?

Answer: Yes. To observe these effects, we need to have a situation in which something is moving at a very high speed. This is provided by subatomic particles. One such particle, called a muon, has the same charge as an electron but is about 200 times more massive. Muons are created in the atmosphere as a result of cosmic ray collisions with the nuclei of the gas molecules of the air. The muons then approach the Earth with speeds near that of light ($\sim 0.998c$).

However, the muon is unstable and quickly decays into other particles. The average lifetime of a muon at rest in the laboratory is about 2 μs (2×10^{-6} s). During this time, even traveling at the speed of light, a muon would travel only 600 m ($d = ct$). But muons are created at altitudes of several kilometers, so one would expect the vast majority of the muons to decay before reaching the Earth. Yet an appreciable number of muons are detected at the Earth's surface.

The disagreement arises because time dilation was not taken into account. The muon decays by its own clock. To an observer on Earth, the fast-moving muon "clock" runs more slowly (Fig. 23.9). A speed on the order of $0.998c$ gives a $\gamma = 15$, and for a $t_o = 2$ μs, the time on an Earth clock is

$$t = \gamma t_o = (15)(2 \ \mu s) = 30 \ \mu s$$

During this time, according to an Earth observer, the muon would travel a distance

$$d = vt = (0.998c)\gamma t_o$$

$$= (0.998c)(30 \ \mu s) = 8900 \ m$$

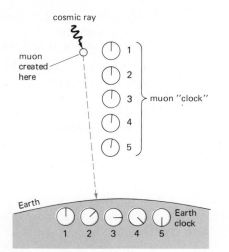

Figure 23.9
Experimental evidence of time dilation in muon decay.
See Question and Answer.

and many muons would reach the Earth before decaying.

You can also look at this from a muon's point of view. Suppose it were created at an altitude of 8000 m (Earth system). Because of length contraction, the muon traveling at $0.998c$ measures this distance to be

$$L = L_o/\gamma = 8000 \text{ m}/15 = 530 \text{ m}$$

which can be traveled in its 2-μs lifetime.

THE TWIN PARADOX

Time dilation gives rise to another relativistic topic — the twin paradox. According to the result of the special theory, a clock in a moving system runs more slowly than a clock in an observer's system. For example, with a $\gamma = 4$, for every hour that elapses on an observer's clock, only one fourth of an hour ticks off on a clock in a moving system by his observations.

Similarly, observing one year of events in the moving system takes four years in the observer's time frame. Keeping in mind that we "age by the clock" (heartbeat and age measured by proper time), a good question quickly arises: Does an observer in one system age more quickly than another in a relatively moving system?

The "twin paradox" states the problem in terms of a set of twins (Fig. 23.10). Suppose one of the twins takes a high-speed space journey that takes 40 years according to the twin who stays on Earth. Would the Earth twin be older than the space twin on return? With $\gamma = 4$, the Earth twin would spend 40 years observing the 10 years (t_o) that elapsed on the spaceship clock. If the twins were 25 years old at blast-off, then does the 35-year-old returning space traveler find his twin getting ready for retirement at age 65? The answer is yes.

(a)

(b)

Figure 23.10
The twin paradox. The theory of relativity predicts that identical twins (a) would age differently if one went on a space trip (b).

But what happens from the point of view of the space traveler? From the *relative* part of relativity, you know that the space twin looking back at the Earth "sees" his twin's clock running slowly. So from this point of view, the Earth twin ages more slowly, and he is the one who is younger. (Do not get the false idea that relativity allows one to go back in time. This is *not* the case.)

The difficulty with the analysis is that the space traveler does not remain in an inertial system. What happens when he is accelerating in starting and stopping? To investigate the paradox in detail, we should apply the general theory, which treats accelerating reference frames. However, by special applications of the special theory, it too predicts that the twin who has been accelerated will be younger than the one who stayed at home. You may be wondering how we know the "correct" answer, since this requires experimental testing.

The twin paradox *has* been tested. Not with real twins, of course, but with atomic-clock "twins." Four cesium atomic clocks were flown around the world in *opposite* directions on commercial aircraft.* The clocks had been previously synchronized with stationary cesium-clock "twins" on Earth. Afterward, the moving clocks were found to be "out of sync" (showed a different time) in accord with the relativistic predictions. Ultra-accurate cesium clocks had to be used instead of common wristwatches because the time differences were on the order of billionths of a second. Even so, the flying clocks came back "younger."

In a more recently reported experiment (1985), rather than physically moving clocks, researchers used Earth-based clocks located in different countries. Pairs of these Earth stations simultaneously viewed signals from global positioning satellites, which, depending on the sequence observed, gives an east-west effect similar to that of the flying clocks. Time differences were noted as predicted by the theory of relativity. This research will help in synchronizing clocks around the world to subnanosecond accuracy.

Experiments with unstable particles accelerated to high speeds in particle accelerators also provide experimental testing that involves the twin paradox. These particles are stable for only a certain time, and it is observed that accelerated particles

* Hafele and Keating, *Science,* 117, 4044, July 14, 1972, pp. 166–170.

live longer, on the average, than those at rest or unaccelerated.

The twin paradox gives rise to such limericks as

A precocious student quite bright
Could travel much faster than light,
He departed one day
In an Einsteinian way
And arrived on the previous night.

MASS IN MOTION

If relative motion affects the measurements of two fundamental properties, length and time, can mass be far behind? The answer is no. The theory of relativity predicts that an object's inertia increases as its speed increases. Similar to time dilation or increase, there is an inertia or apparent mass increase or "dilation" according to the equation

$$m = \gamma m_o$$

where m_o is the rest mass of the object, or the mass of the object when measured in its rest frame. The observed mass m is commonly called the apparent relativistic mass.

There are experimental observations of the relativistic mass increase. Particle accelerators accelerate beams of charged particles, usually electrons or protons, to very high speeds using electric fields. The directions of the particle beams are controlled by magnetic fields. In linear accelerators, beams are deflected or turned by "deflecting" magnets (Fig. 23.11).

It is found that a greater magnetic field is required to deflect the particles than would be needed as predicted by using their rest mass, indicating a mass increase. In circular accelerators such as cyclotrons, the particles are given a timed "kick" of energy in certain parts of their orbits. As the particles gain speed, it is found that they arrive increasingly late. The "lateness" is explained by a relativistic increase in the masses of the particles, which follows the previous equation.

The mass equation can be used to illustrate the basis of the phrase, "Nothing can travel faster than the speed of light." From the equation, we see that as v approaches c, the term v^2/c^2 approaches a value of one

$$v \sim c \quad \text{and} \quad v^2/c^2 \sim c^2/c^2 = 1$$

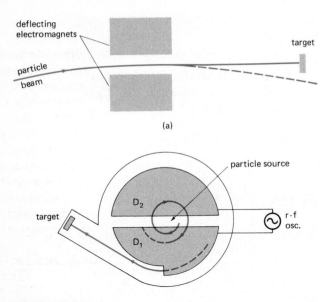

(a)

Figure 23.11
Accelerators and relativity. (a) When a particle is accelerated to high speeds, a greater magnetic field is needed to deflect it than would be predicted using the particle's rest mass. (b) In a cyclotron, a particle travels more slowly than expected. Both of these situations indicate a mass increase.

This makes the term $\sqrt{1 - v^2/c^2}$ approach zero, and the mass m is infinitely large. Since we don't have enough energy to accelerate a mass that grows infinitely large (which is difficult to imagine), the theory of relativity sets an upper limit for the speeds of objects. We could modify the foregoing phrase somewhat: "Nothing (no material object) can travel *as fast as the speed of light.*"

■ CLASSICAL OR RELATIVISTIC?

The ideas of length contractions, time dilations, and mass increases may seem "strange," inasmuch as we do not ordinarily observe examples of these effects. This is because our everyday observations are "classical" observations. To observe relativistic effects, we need speeds that are appreciable fractions of the speed of light. Our fastest race cars, and even our fastest rockets, do not even come close.

But notice that the theory of relativity takes our relatively slow-paced world into account. With v very much less than c, the ratio v/c is very small or

practically zero. Since $v^2/c^2 = (v/c)^2$, the relativistic equations become

$$t = \frac{t_o}{\sqrt{1 - (v/c)^2}} \approx \frac{t_o}{\sqrt{1 - 0}} = t_o$$

$$L = L_o \sqrt{1 - (v/c)^2} \approx L_o \sqrt{1 - 0} = L_o$$

$$m = \frac{m_o}{\sqrt{1 - (v/c)^2}} \approx \frac{m_o}{\sqrt{1 - 0}} = m_o$$

Hence, the relativistic equations predict that at relatively slow (classical or nonrelativistic) speeds, we will observe classical results.

■ MASS AND ENERGY

One of the most remarkable results of Einstein's special theory of relativity is the relationship between mass and energy. It had previously been believed that both energy and mass could be neither created nor destroyed. Energy could be changed in form, but the total energy was constant, just as was the total mass. What Einstein showed was that mass is a form of energy, and the relationship for this is given by the famous equation

$$E = mc^2$$

where c is a constant, the speed of light.*

The validity of this relationship has been shown in nuclear processes such as fission and fusion (Chapter 26). Here, the total mass of the products after a reaction is less than that before the reaction. The mass loss is just equal to the energy gained in the reaction according to $E = mc^2$. In other reactions, energetic particles can cause nuclear reactions in which there is more total mass after the reaction but less energy, by an amount $E = mc^2$. Hence, we now talk about the conservation of mass-energy instead of separate conservation laws.

The mass-energy equivalence does not mean that we can convert any amount of mass into energy at will. If we could, our energy problems would be solved. For example, if only 1 gram of mass could be completely converted into energy, the amount

---

* Technically we should write $E_o = m_o c^2$, where E_o is the rest energy and m_o is the rest mass, but it is common to write and say $E = mc^2$. The total mass referred to in this discussion is the total rest mass. The total relativistic energy of a moving particle (with kinetic energy K) is $E = K + E_o$ or $mc^2 = K + m_o c^2$, so for a particle at rest, $E = E_o = mc^2$.

would be equivalent to the electrical energy used on the average in 20,000 homes in a month. The c^2 in $E = mc^2$ is a big number, so a little mass goes a long way in creating energy.

Mass-energy conversion takes place on a limited scale in nuclear reactions, as will be considered in more detail in Chapter 26. It also takes place in ordinary chemical reactions, such as in the burning of a match. Mass is changed into energy that is lost as heat and light. However, the mass loss is so slight that it is difficult to detect.

On the other hand, the amount of mass lost by the Sun in mass-energy conversion is enormous. If the energy radiated by the Sun is converted into units of mass using Einstein's $E = mc^2$ equation, it is found that the Sun loses about 4.5 billion kilograms of mass per second (about 5 million tons per second in customary weight units). You might wonder if the Sun will waste or radiate away to nothing. Don't worry. At this rate, in a billion (10^9) years the Sun will lose only about 1/100 of 1 percent of its mass. This is still a good bit of matter—equal to about 25 Earth masses.

■ THE GENERAL THEORY OF RELATIVITY

The special theory, after being put forth by Einstein in 1905, gave scientists a great deal of food for thought. Then, ten years later in 1915, he did it again and brought forth his general theory of relativity. This complex mathematical theory extends to relativistic applications to accelerated systems. As you might imagine, the general theory was also met with a bit of skepticism. We have already seen some of its predictions—that a twin in an accelerated system grows older (twin paradox) and that gravity is a "warping" of a space-time continuum (Chapter 5).

The general theory is a result of Einstein's belief that the laws of physics should be the same in *all* reference frames, both nonaccelerated (inertial) *and* accelerated. It is essentially a gravitational theory directed toward this end. To illustrate the equivalence in accelerated and nonaccelerated systems, consider the astronaut in the two situations shown in Figure 23.12. If the spaceship were at rest in free space (negligible gravity), then objects, including the astronaut, would float around in the ship in a state of "zero g." Then suppose the spaceship were in free fall near the Earth. As discussed in Chapter 5, the astronaut would observe the same

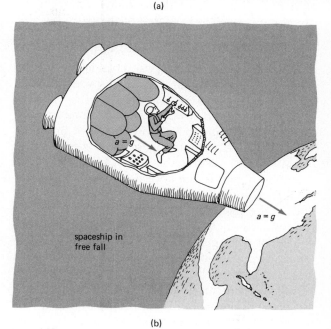

(a)

(b)

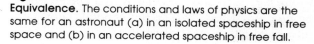

Figure 23.12
Equivalence. The conditions and laws of physics are the same for an astronaut (a) in an isolated spaceship in free space and (b) in an accelerated spaceship in free fall.

conditions in a state of *apparent* weightlessness in the accelerated system. In fact, no experiment could be performed in the closed spaceship system that could distinguish between the nonaccelerated and free-fall conditions.

Further, let the astronaut in free space turn on the rocket engines and accelerate with an acceleration equal to g (Fig. 23.13). The astronaut would then feel the reaction force of the floor or "weight," and objects would fall "down" to the floor. Everything would be exactly the same as in a stationary ship in a gravitational field with a force of 1 *g*. These results are summed up in the **principle of equivalence:**

> No experiment performed inside an accelerated, closed system can distinguish between the effects of a gravitational field and the effects of an acceleration.

In other words, observations made in an accelerated system are indistinguishable from observations made in a gravitational field.

According to Einstein's general theory, the principle of equivalence holds not only for mechanical phenomena, such as in dropping an object, but also for *all* phenomena, including electromagnetic phenomena. This gives rise to some new ideas and predictions. For example, suppose a ball is thrown parallel to the floor of a stationary spaceship in a gravity-free region (Fig. 23.14).

The ball would be observed to follow a straight-line path according to Newton's first law. But if the spaceship were accelerating, say with *a* = *g*, the astronaut would observe the ball to follow a curved (parabolic) path to the floor. (An outside observer would see the ball moving in a straight line, and the floor of the spaceship accelerates up to the ball.)

By the principle of equivalence, the astronaut could not distinguish between an accelerated system and a gravitational field. The ball follows a path just as if it were thrown in a stationary spaceship sitting on a launching pad on Earth. Since the principle of equivalence holds for *all* phenomena, let's replace the ball with a light beam (Fig. 23.15). In an accelerated system, *or equivalently in a gravitational field,* it is predicted that the light beam would be bent or curved. That is, the general theory predicts that light should be bent in a gravitational field!

No such bending of light is commonly observed in the Earth's gravitational field. This is because the effect is too small. For the prediction to be tested experimentally, large distances and a very strong gravitational field are needed, such as that of the Sun—and this is what has been used. Distant stars appear to be motionless and are measured to be a constant angular distance apart (at night, Fig. 23.16).

The same stars are "out in the daytime" at other times of the year (as a result of the Earth's revolution about the Sun). Light from one of the stars may pass

near the Sun, but any bending would not be observed because the starlight is masked by sunlight scattered in the atmosphere (which is why we can't see stars in the daytime). But when a total eclipse occurs, the moon coming between the Earth and the Sun blocks the sunlight, and an observer in the

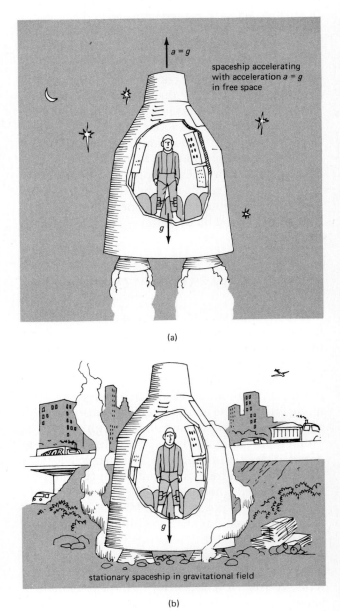

(a)

(b)

Figure 23.13
Equivalence. The conditions and laws of physics are the same for an astronaut (a) in an accelerating spaceship (*a* = *g*) in free space and (b) in a stationary spaceship on the Earth's surface.

moon's shadow can see stars during the darkened midday.

During a total solar eclipse, a distant star appearing near the surface of the Sun will have an apparent location that is different from its actual location, owing to the bending of light by the Sun's gravitational field (Fig. 23.16). The apparent angle between the two stars on opposite sides of the Sun should be slightly larger during the eclipse. Such measurements have been made during total

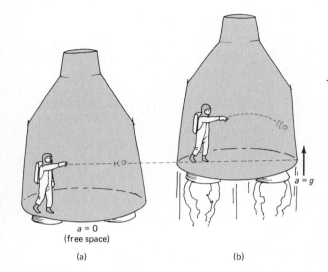

Figure 23.14
Effects of acceleration. (a) In a nonaccelerated system in free space, a thrown ball follows a straight path. (b) In an accelerated system, the astronaut sees the thrown ball deflected downward as though in a gravitational field.

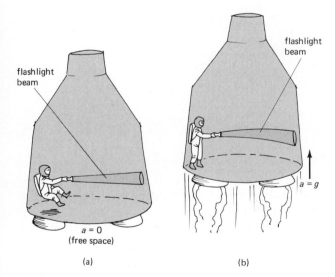

Figure 23.15
Effects of acceleration. (a) In a nonaccelerated system, a light beam follows a straight-line path. (b) In an accelerated system, the light beam appears to be deflected downward. By the principle of equivalence, a similar deflection should occur in a gravitational field.

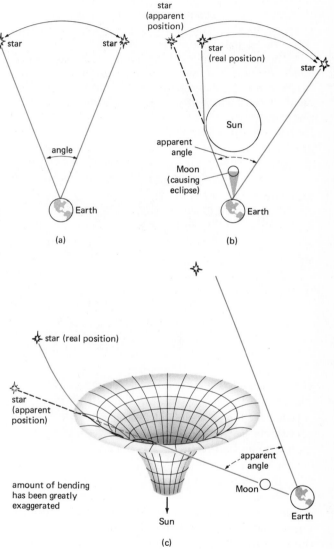

Figure 23.16
Experimental evidence of the gravitational attraction of light. (a) Normally, two stars are viewed to be a certain angular distance apart. (b) During a solar eclipse, with one star behind the Sun, the star can still be seen because of solar gravitational effects on light, as evidenced by a larger apparent angular separation. (c) This effect is illustrated in Einstein's space-time gravity. Light is deflected in the vicinity of the Sun, as a mass would be.

Space-Age General Relativity

In addition to light sources appearing to be displaced slightly when seen near the Sun, the radiation from them is delayed slightly in reaching the Earth. We have no way of measuring the delay in receiving light from stars, but we can detect it in radio broadcasts from space probes because we know where they are and when the signals should arrive at Earth.

The experiment has been performed with several space probes, but most precisely with the *Viking* landers on Mars. When Mars is on the far side of the Sun, signals from a *Viking* lander must pass through a gravi-

tational field or a space-time warp or "sink" caused by the Sun (figure below). Signals from *Viking* have been observed to be delayed by about 100 μs. This is as though Mars had jumped about 30 km out of its orbit. The time delay is very close to what is predicted by the theory of relativity.

More evidence of the general theory. Signals from the *Viking* lander on Mars were delayed when Mars was on the far side of the Sun because they had to pass through the gravitational field (space-time warp) of the Sun.

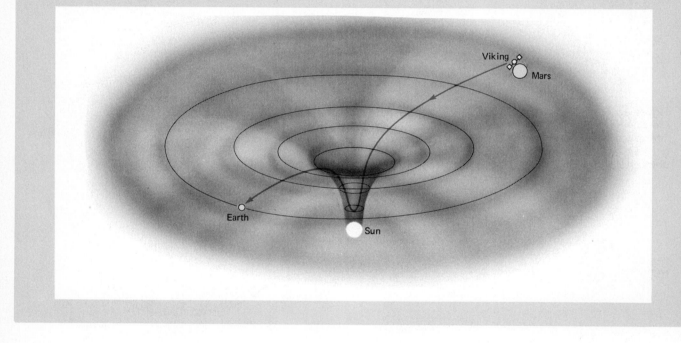

eclipses, and the stars were measured to be slightly farther apart, as predicted by Einstein's general theory. Other experimental evidence is also available, as discussed in Special Feature 23.2.

Actually, this result of general relativity was presented way back in Chapter 5. Recall that a black hole is a burned-out, gravitationally collapsed star that is so incredibly dense that light coming very close to it would be "bent" into the black hole and could not escape. Of course, a distant star and a less distant black hole would still be relatively motion-

less, and we would never see the star if its light were captured by the black hole.

If and when the existence and location of a black hole are established by the light of a binary star and X-ray sources (see Chapters 5 and 27), then we might test the general theory with some *big* light bending, if we could get a spaceship in an orbit around the black hole (but not *too* close an orbit!).

Another result of the general theory is that gravitation affects time — gravity causes time to slow down, and the greater the gravitational field, the greater the slowing of time. There is evidence of this effect in the gravitational "red" shift of light from the Sun. The gravitational field of the Sun is quite large, and it would be expected to cause the electronic vibrations in atoms to slow down. As a result, light emitted from these atoms would have a lower frequency or would be shifted toward the lower-frequency red end of the visible spectrum. When light from the Sun is compared with similar light from a source on Earth where the gravity is less, a gravitational red shift of the solar light is observed.

This gravitation-time effect of the general theory means that clocks near the surface of the Earth should run more slowly than those at heights above the Earth, which would be in a weaker gravitational field. Although extremely small, this effect has been measured between the bottom and top floors of a building using "nuclear clocks." So if you work or live on the bottom floor of a building, your clock runs more slowly, or you age more slowly than a person on the top floor — a little general-relativity youth "tonic."

SUMMARY OF KEY TERMS

Ether: a hypothetical medium for propagation of light waves.

Michelson-Morley experiment: an experiment that was designed to detect the ether through velocity (vector) addition and interference. No effects were observed.

Fitzgerald contraction: a proposed contraction or shortening of the length of the arm of the Michelson-Morley interferometer in the direction of the ether wind that would explain the null result of the experiment.

Special theory of relativity: Einstein's theory of relativity that deals with nonaccelerating or inertial systems.

Principle of relativity: all the laws of physics are the same for all observers moving at a constant velocity with respect to one another.

Constancy of light: the speed of light in free space is the same for all observers regardless of the motion of the source or the motion of the observer.

Time dilation: the observation (t) of a clock in a moving system (t_o) running more slowly, according to the equation $t = \gamma t_o$.

Length contraction: the observation (L) of a shortening of a length (L_o) in the direction of motion in a moving system, according to the equation $L = L_o/\gamma$.

Proper time and length: the time (t_o) and length (L_o) measurements made by an observer moving with the clock or object, i.e., in the system in which the clock and length, say a meterstick, are actually located.

Twin paradox: the paradox of a space-traveling twin returning to Earth younger than his Earth-bound twin, which is predicted by the general theory of relativity.

Relativistic mass: the apparent mass (m) of an object in a moving system owing to an observed increase in inertia: $m = \gamma m_o$, where m_o is the rest mass or the mass of the object measured in its rest frame.

Mass-energy conversion: the changing of mass into energy and vice versa, according to the equation $E = mc^2$. As a result, mass is considered to be a form of energy.

General theory of relativity: Einstein's theory of relativity that deals with accelerated systems.

Principle of equivalence: observations made in an accelerated system are indistinguishable from observations made in a gravitational field.

QUESTIONS

Background and the Ether

*1. How long does it take light to travel to Earth from (a) the nearest star (93 million miles) and (b) the next-nearest star (4.2 light years)?

*2. Galileo could not get a consistent result for the speed of light because of the reaction times of his assistant and himself. Assuming the human reaction time to be 0.20 s, how far could light travel during the experimental time error due to the reaction times? (*Hint: d = vt* or *d = ct.*)

3. How might a supporter of the ether theory explain (a) a transparent object, (b) a translucent object, and (c) an opaque object?

4. When you look at a clear night sky, it might be said that you are looking into the past. Explain why.

Time Dilation and Length Contraction

5. If there really were a Fitzgerald contraction in the Michelson-Morley experiment, why not just

measure the lengths of the interferometer arms and see if there was a difference?

6. When the driver of a constantly moving car blows the horn, he hears a frequency (pitch) of 1000 Hz, but a stationary observer in front of the oncoming car hears a frequency of 1010 Hz (Doppler effect, see Chapter 15). Do the laws of physics vary in different systems? Explain.

7. Do the relativistic effects of mass, length, and time apply to ordinary observations? Explain.

8. Describe what is observed by the drivers of cars B and C in Figure 23.6.

•9. In Table 23.1, why aren't the units of γ given?

10. Two observers moving at a large relative constant velocity watch the same half-hour programs on their TV sets (from stations in their own reference frames). What is the proper time of the TV program? What would one observer note if she compared the program in the other system with her own?

•11. A meterstick moving a constant velocity is observed to be only one half as long as the observer's meterstick. How would the mass of the meterstick compare with that of the observer's?

12. What would happen to the length and mass of the meterstick in Question 11 if it were turned perpendicular to the direction of motion?

13. An observer sees a friend in a spaceship traveling with a large constant velocity. He knows his friend should be the same height as he, 1.8 m, but somewhat bigger around (fatter). How does his friend appear to him in the moving spaceship?

14. A distant star system is 100 light years away. Would it be possible for an astronaut to go in a spaceship to this star system and come back if the average life span is 70 years? Explain in terms of (a) an Earth observer and (b) the astronaut.

•15. Suppose you are an astronaut and traveling away from Earth at a large constant velocity. What effects would you observe on your (a) pulse rate, (b) mass, and (c) volume? How about an Earth observer?

16. Would it be possible to use the twin paradox and have a son or daughter older than the parents? (*Hint:* Either a child *or* a parent(s) could be the space travelers.)

Figure 23.17
What time is it? See Question 21.

17. If mass is a form of energy, why isn't the price of energy cheaper since we have all kinds of mass available?

The General Theory of Relativity

18. What would you as an outside observer see for the case shown in Figure 23.14? Show by drawing several sequential sketches.

•19. When a ball is thrown horizontally, it falls a distance of 4.9 m in the first second. By the principle of equivalence, a horizontal light beam also drops 4.9 m in 1 s. Why isn't this drop observed?

20. Laser beams are used for surveying, for example, in building or digging tunnels. Won't the gravitational bending of a beam give rise to errors? Explain.

•21. A "gedanken" or thought experiment on the effect of gravity on time is illustrated in Figure 23.17. Clocks A and B are at rest relative to each other, so they read the same time. Clock C is moving relative to B and so runs more slowly. Since A and B are equal, C also runs more slowly relative to A, even though they are in the same system. What causes this?

Quantum Physics

An argon laser used in scattering experiments to investigate atomic structure.

The question of how light propagates through the void of space or vacuum led to the theory of relativity (Chapter 23) and the nonclassical idea that light or electromagnetic waves simply propagate through empty space without requiring a medium. However, there were also problems on a more "local" level.

The wave theory of light did not accurately predict other observed features of electromagnetic phenomena. It did adequately explain some things, such as interference and polarization, but it failed in other instances. In some cases, light seemed to behave as a quantum or "particle" instead of as a wave. This quantum behavior of light led to what is known as quantum theory and the "dual nature of light." In this chapter we shall review the development of quantum physics and some of its results. Let's see what was happening in the early 1900's.

■ THE ULTRAVIOLET CATASTROPHE AND QUANTIZATION

Even before the turn of the century, it was well known that light of all frequencies was emitted from an incandescent solid. The relative brightness or intensity of the different frequencies, which we see as colors, depends on the temperature of the solid (Fig. 24.1).

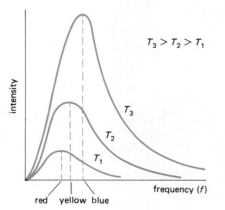

Figure 24.1
Radiation and temperature. For an incandescent solid, the relative brightness or intensity of the different frequencies of radiation depends on the temperature of the solid. As a result, hot objects glow with different colors as the temperature increases.

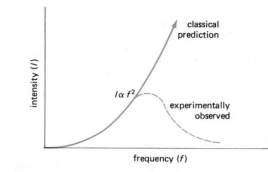

Figure 24.2
The ultraviolet catastrophe. Classical theory predicts that $I \propto f^2$ for a thermal radiator, which requires the emitted energy at higher frequencies (ultraviolet) to be infinitely large (catastrophe).

As the temperature is increased, more radiation is emitted at every frequency, and the radiation component of maximum intensity is shifted to a higher frequency. As a result, a very hot solid appears to go from a dull red to a blue-white as the temperature is increased (see Chapter 11). This would be expected, since the hotter the solid, the greater the electron vibrations and the higher the frequency of the emitted radiation.

When the emission of thermal radiation was analyzed using classical wave theory, it was predicted that the intensity [energy/(area/time)] should be proportional to the square of the frequency, $I \propto f^2$, which would be as shown in Figure 24.2. This gives rise to what was called the **ultraviolet catastrophe**: "ultraviolet" because the difficulty occurred at high frequencies and "catastrophe" because it predicted that the intensity of emitted energy should be overly large. Thus, the classical theory "bombed out" again, disagreeing not only with experiment but also with common sense.

The dilemma was resolved by Max Planck (pronounced "Plonk"), a German physicist, who in 1900 formulated a theory that correctly predicted the frequency distribution of thermal radiation — but only with the introduction of a radically new idea. Planck considered the oscillators of the body to have only discrete energies. Classically, an oscillator is like a spring or pendulum, and it vibrates at a certain frequency. Presumably, an electron in a solid would vibrate at a certain frequency and can have any amount of energy from zero up to some maximum value. This is a continuous energy distribution (Fig. 24.3).

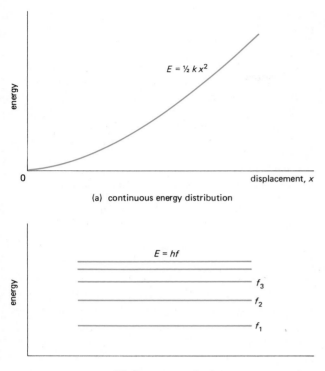

$E = \frac{1}{2}kx^2$

energy

0   displacement, x

(a) continuous energy distribution

$E = hf$

energy

f_3

f_2

f_1

(b) discrete energy levels

Figure 24.3
Energy distribution and quantization. (a) For a classical oscillator, such as a spring, there is a continuous energy distribution, and an oscillator can have any amount of energy from zero to some maximum value. (b) According to Planck's quantum hypothesis, a thermal oscillator could have only discrete amounts of energy or be in a particular "energy level."

Planck's hypothesis was that the energy was quantized or that the oscillators in a thermal radiator could have only *certain* or discrete energies. Moreover, the energy of an oscillator depended on its frequency, according to the relation

$$E = hf$$

where h is a constant, called **Planck's constant** for obvious reasons. The discrete amount of energy of an oscillator was called a quantum of energy (plural, quanta; Latin *quantus,* meaning "how much").

Although the theory fit the experimental data, there was a great deal of skepticism. Even Planck himself was not convinced about why the energy should be quantized. However, the quantum hypothesis set the stage for the development of modern quantum physics, which is needed to explain other phenomena. For his initial work, Planck was awarded a Nobel Prize in 1918.

THE PHOTOELECTRIC EFFECT

There were some other problems with the wave theory of light in predicting observed results, and it wasn't long before the quantum theory scored some other successes. One of these was the explanation of the photoelectric effect. In the latter part of the 19th century, it was observed that certain metallic materials, such as selenium oxide and cesium oxide, were "photosensitive." That is, electrons were emitted from the surfaces of these materials when they were exposed to light. The emission of electrons from a material when exposed to light is called the **photoelectric effect**.

Light then, as a form of energy, must be doing work in freeing the electrons from the surface. You might think of the light waves as causing electrons to vibrate with greater and greater amplitudes until they break free of the material — something like the resonance vibration of a pushed swing and a person not being able to hang on at a large amplitude.

Of course, electrons cannot be observed "jumping out" of a material. But if a photoelectric material is used as the cathode of a vacuum tube, a phototube is formed (Fig. 24.4). When the phototube is exposed to light, a current flows in the phototube circuit. The photoelectric effect also occurs in semiconducting diodes or photocells. Such cells are commonly used in photographic light meters and in solar panels.

In investigating the characteristics of a phototube or photomaterial, it is found that the "photo-

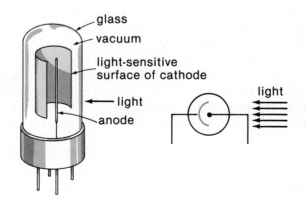

Figure 24.4
Phototube. The direct conversion of light energy to electric energy. When the phototube is exposed to light, current flows in the circuit.

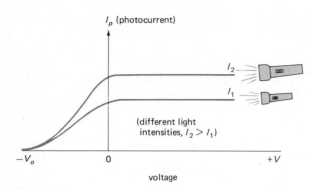

Figure 24.5
Photocurrent and light intensity. The current of a photo-tube or cell is proportional to the intensity of light—the greater the intensity, the greater the photocurrent. This effect is used in photographers' light meters.

Table 24.1
Photoelectric Characteristics

1. The photocurrent is proportional to the intensity of light. Predicted by classical theory?	YES
2. The kinetic energy of the emitted electrons is dependent on the light frequency but not on the light intensity. Predicted by classical theory?	NO
3. No photoemission occurs for light having a frequency below a certain cutoff frequency f_o, regardless of the intensity. Predicted by classical theory?	NO
4. A photocurrent is observed immediately on illumination of a photomaterial when the light frequency is greater than f_o, even if the intensity is very low. Predicted by classical theory?	NO

current" is proportional to the intensity of light (Fig. 24.5). That is, the greater the intensity or energy received, the greater the photocurrent. (This is what is used in photographers' light meters to measure the brightness or intensity of light.) Classically, one would expect this to be the case, since the greater the intensity or amplitude of the light wave vibration, the more energy there is available to free electrons from the material.

However, there are other effects that cannot be explained by classical wave theory. It is found that the kinetic energy of the emitted electrons depends on the frequency of the light but not on its intensity.* But no photoemission occurs for light having a frequency below a certain **cutoff frequency** f_o. For example, photoemission is observed for violet light but not for red light (lower f and below f_o). Also, when a photocurrent is observed, it is observed *immediately* on illumination without any delay, even though the intensity is low.

These characteristics are summarized in a "photoelectric scoreboard" in Table 24.1. Notice that classical wave theory is not having a good season in this case. Let's take a closer look at the classical losses.

Classically, the electric field of a light wave would interact with the electrons at the surface of a material. It would be thought that the stronger electric field of more intense light would interact more

with the electrons so that they would be emitted with higher speeds or greater kinetic energies. But this is not the case (number 2 in Table 24.1). Yet, a low-intensity or dim light of a higher frequency produces a smaller number of photoelectrons, but they have much greater kinetic energies. This is rather strange and doesn't conform to classical ideas.

Also, according to wave theory, photoemission should occur for any frequency of light, provided that the intensity of the light is sufficient. Yet below a certain frequency there is no photoemission, regardless of the intensity (number 3 in Table 24.1).

The last item in Table 24.1 was very perplexing. According to wave theory calculations, the time for free electrons to acquire enough energy from a continuous light wave to have the observed kinetic energies of photoelectrons is on the order of minutes. Add to this the energy needed to do the work in freeing an electron from the surface of a material and you have quite a long time!

In 1905 Albert Einstein put forth an explanation of the photoelectric effect, for which he received a Nobel Prize. He applied Planck's hypothesis to light. In his theory, light itself was quantized, being made up of discrete packets or quanta called **photons**. The energy of a photon of light depended on the frequency f, with a single photon having an energy of $E = hf$.* Planck's constant h is a very small number (on the order of 10^{-34}), so most beams of light have many photons.

---

* The kinetic energy can be measured by applying a reverse bias or stopping potential $-V_o$ to a phototube such that no current flows (Fig. 24.5).

* It is interesting to note that Newton first thought that light was made up of "corpuscles" or particles. (See the introduction to Part VI.)

Einstein applied the conservation of energy to the situation. If an electron in a photomaterial did absorb a photon, then

$$\text{Energy of photon} = \text{energy needed to free electron} + \text{kinetic energy of emitted electron}$$

or $\qquad E = hf = \phi + K$

where ϕ is called the work function of the material and is the energy needed to free an electron. Quantizing this, $\phi = hf_o$, where f_o is called the threshold frequency and is the same as the cutoff frequency, we have

$$hf = hf_o + K$$

Hence, light with a frequency f less than f_o will not have enough energy in its photons to free electrons. When f is greater than f_o, electrons will be freed, with the leftover photon energy going into the kinetic energy of an electron. Increasing the light intensity only increases the number of photons and emitted electrons, not their kinetic energy.

Also, no time delay for the emission of photoelectrons is expected in the quantum theory, since a bundle or packet of energy is "dumped on" an electron. That is, in absorbing a photon or quantum of energy, the electron receives a definite amount of energy all at once, instead of receiving it continuously as in wave theory. This is somewhat like filling a bucket by dumping another bucket (quantum) of water in it rather than filling it with a continuous stream of water from a hose, as would be analogous to a continuous wave.

THE DUAL NATURE OF LIGHT

Such developments as the photoelectric effect led to what is known as the **dual nature of light**. That is, light or electromagnetic radiation sometimes behaves as a wave and sometimes as a quantum or "particle." In order to explain classical phenomena, such as interference and polarization, light is treated as a wave. Yet for other physics phenomena, light is viewed as a packet or particle.

It is important to note that light has a single consistent nature that is described without ambiguity by modern quantum theory. This, of course, is beyond the scope of this text. The "dual nature" arises when we try to force it into a classical model

as a "wave" or "particle." Neither fits entirely, but one *or* the other generally seems to work.

A good analogy used by one of the reviewers of this text involves an airplane. Suppose a strange society knew about cars and birds but not airplanes. Upon discovery or observation of an airplane, they might attribute to it a dual nature. Under some circumstances a plane runs along the ground on wheels and carries people. At other times it soars through the sky like a bird—different from other birds perhaps, but certainly more like a bird than a car. With more knowledge and a more modern outlook, this society would describe an airplane as a single consistent entity without ambiguity.

You may wonder why we don't observe the particle or quantum nature of light in everyday observations. For example, we do not see discrete packets of light or see intermittently. This is because the energy of a single photon is very small, and many quanta must be absorbed each second to give enough energy for ordinary observations. As a result, we are macroscopically unaware of the discrete nature of the energy absorption, just as we are not aware of the individual molecules in liquid flow because they are so numerous.

Other examples of the quantum nature of light follow, but first some practical applications of the photoelectric effect are illustrated in Figure 24.6.

THE BOHR THEORY OF THE HYDROGEN ATOM

The quantum theory played an important role in the development of our simplified model of the atom. Recall that we visualize a planetary model of the atom, with the electrons being in orbits about the nucleus (protons and neutrons), similar to the planets in orbits about the Sun. In 1911 the British physicist Lord Ernest Rutherford showed that the protons of the atom were in a "nucleus" in the atom (Chapter 25). The idea of orbiting electrons came from another source and another consideration.

Around the turn of the century, a great deal of experimental work was being done with gas discharge tubes, for example, mercury (vapor), neon, and hydrogen. When the light from these tubes was analyzed, instead of continuous spectra as are observed from incandescent sources, discrete or line spectra were observed. That is, only light or spectral

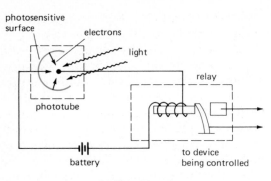

(a) Electric eye

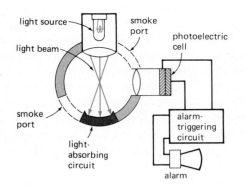

No Alarm Condition

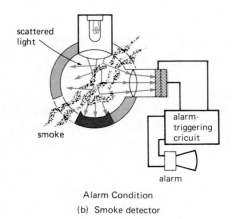

Alarm Condition
(b) Smoke detector

Figure 24.6
Photoelectric applications. (a) Electric eye. When light strikes a phototube or cell, current flows in the circuit. Interrupting the light beam opens the circuit, and the relay (magnetic switch, Chapter 18) controls some device. (b) Smoke detector. When smoke enters the detector, the scattered light is detected by the photocell and the current triggers an alarm.

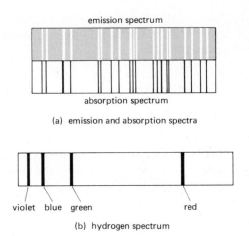

(a) emission and absorption spectra

(b) hydrogen spectrum

Figure 24.7
Line spectra. (a) Emission spectrum *(top)*. Light is emitted from gases only at certain frequencies, giving a discrete line emission spectrum. Absorption spectrum *(bottom)*. If white light is passed through a relatively cool gas, the same frequencies are absorbed, and these lines are missing in an absorption specturm. (b) The visible line spectrum of hydrogen.

lines of certain frequencies or wavelengths were found (Fig. 24.7).

Light coming directly from a tube gives an emission or bright-line spectrum. If white light is passed through a relatively cool gas, it is found that certain frequencies or wavelengths are missing in the resulting absorption spectrum, which shows dark lines. Atoms absorb light as well as emit it, and when we compare the emission and absorption spectral lines of a particular gas, we see that they occur at the same frequencies. These puzzling atomic line spectra gave the clue to the electronic structure of the atom.

The spectrum of hydrogen, the lightest element and simplest atom, received much attention because of its relative simplicity (Fig. 24.7). In the late 1800's a Swiss mathematician and physicist, J. J. Balmer, found an empirical formula that gave the positions of the spectral lines. Some new lines were found as a result, but Balmer gave no reason why the formula worked or the physics of the spectral lines.

An explanation of the spectral lines in a theory of the hydrogen atom was put forth in 1913 by a Danish physicist, Niels Bohr (Fig. 24.8). Using the results of Planck's and Einstein's quantum theories along with the Balmer formula, Bohr was able to calculate the energy states of the electron in a hydrogen atom (one electron and one proton), which

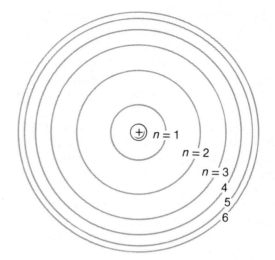

Figure 24.9
Bohr orbits. In the Bohr theory of the hydrogen atom, the hydrogen electron could only be in quantized or discrete orbits.

Figure 24.8
Niels Bohr (1885–1962). A photograph of Bohr at the time he developed his theory of the hydrogen atom.

corresponded to the energies of certain *discrete* circular orbits about the nuclear proton (Fig. 24.9).

The possible electron orbits were characterized by a "quantum number" n, called the principal **quantum number**, that could have any whole-number value, $n = 1, 2, 3, 4, \ldots$. The *quantized* circular orbits in Bohr's theory were the result of the quantization of the orbital angular momentum of the electron, which occurred in whole-number multiples of $\hbar = h/2\pi$, or $n\hbar$, where $n = 1, 2, 3, 4, \ldots$ are the principal quantum numbers. (Notice that Planck's constant comes into the picture again.)

However, there was a classical problem with Bohr's planetary theory. According to classical theory, an accelerating electron radiates electromagnetic energy, and in a circular orbit an electron is centripetally accelerating. Thus, an orbiting electron should radiate energy and quickly spiral into the nucleus, similar to an Earth satellite in a decaying orbit.

This doesn't happen in an atom, and so Bohr had to make some bold, nonclassical assumptions in

his theory. To account for this and the observed discrete-line spectrum, Bohr postulated that the hydrogen electron *does not* radiate energy when in a bound, discrete orbit but does so only when it makes a "jump" or transition to another orbit or energy state.

The allowed orbits of the hydrogen electron are commonly expressed in terms of energy states or levels, with each state corresponding to a particular orbit (Fig. 24.10). With the electron being bound to

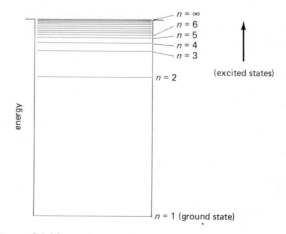

Figure 24.10
Energy levels. Each orbit corresponds to a particular energy level. The electron makes transitions between energy levels when changing orbits.

the nuclear proton, it is in a potential energy "well," similar to a gravitational potential well (Chapter 5). The electron is normally at the bottom of the well, or in the **ground state** ($n = 1$), and must be given energy or be "excited" to raise it up in the well to an **excited state**. Thus, the hydrogen electron can be excited only by discrete amounts.

It is sometimes convenient to consider the energy levels to be like steps on an energy "ladder." As for a real ladder, a person going up or down a ladder must do so in discrete steps on the ladder rungs. Notice that the energy "rungs" of the hydrogen atom are not evenly spaced. Given enough energy to excite the electron to the top of the well, the electron is no longer bound to the nucleus, and the atom is ionized.

An electron does not usually remain in an excited state for long and quickly decays or makes a transition to a lower energy level. In doing so, the electron can radiate energy by the Bohr theory and does so by emitting a photon or a discrete amount of light energy (Fig. 24.11). The energy of the photon is equal to the energy difference between the two energy levels, or $\Delta E = hf$. Hence, the hydrogen atom emits only discrete frequencies of light.

The electron does not always "jump" down the energy ladder one rung at a time. It can skip rungs, just as a person coming down a ladder can. In an excited hydrogen gas, as in a discharge tube, with many, many atoms, various transitions are made, and these correspond to the frequencies of the observed spectral lines (Fig. 24.12).*

* In an incandescent solid, such as a lamp filament, the atoms interact strongly, and the electrons are not associated with a single atom. In this case, the electron energies have a continuous range and hence give rise to a continuous spectrum.

Thus, quantum physics and the quantum nature of light scored another success. The electron energy levels for other, more complex atoms are more difficult to calculate, but the line spectra of various atoms are indicative of their energy spacings and provide a characteristic "fingerprint" by which an atom may be spectrographically identified.

■ THE LASER

We hear more and more about lasers. Laser beams have been sent to the moon and have been used in eye surgery, in surveying, and in drilling holes. We now hear talk of laser weapons capable of destroying satellites and missiles. Because of such publicity, the laser is sometimes thought of as an incredible source of energy. This is not the case. The laser is a light source, and like other light sources, it converts one form of energy to another—always at a loss. The uniqueness of the laser as a light source comes about because of the special properties of the light it produces.

The development of the laser was one of the crowning successes of the modern approach to science. In many instances, scientific discoveries have been made accidentally, and even though these discoveries were applied, no one fully understood how or why they worked. The discovery and use of X-rays are an example. Often, in inventing something, a trial-and-error or Edisonian approach was used. (In developing the incandescent lamp, Edison tried many filament materials until he found one that worked—a carbonized thread from his wife's sewing basket.)

However, the laser was developed first "on paper" and then built as predicted. Atomic theory

Figure 24.11
Photo emission. (a) When an electron is excited to a higher orbit or energy level, it eventually "decays" (b), with the emission of a photon or a discrete amount of energy at a certain frequency, as determined by the difference in the values of the energy levels (c).

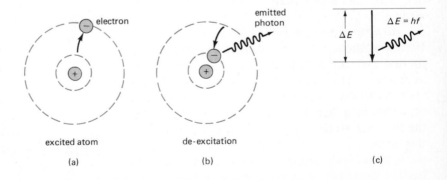

excited atom

(a)

de-excitation

(b)

(c)

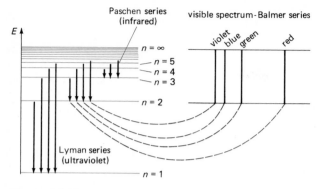

Figure 24.12
Transitions and line spectra. The visible spectrum or Balmer series of the hydrogen atom is given by transitions to $n = 2$. Other "series" are transitions to other energy levels. Bohr's theory correctly predicted the wavelengths of the observed spectral lines of hydrogen and aided in the discovery of the Lyman series.

allowed scientists to "look" into the atom for previously unobserved and unapplied phenomena. As a result of quantum theory, we now have the laser. The theory behind the laser was developed during the middle and late 1950's by the American scientists Arthur Schalow and Charles Townes, and the first working device was constructed in 1960.

The term **laser** is an acronym for *l*ight *a*mplification by *s*timulated *e*mission of *r*adiation (*l-a-s-e-r*). The laser was the first device capable of amplifying light waves themselves. You may think that the amplification of electromagnetic radiation is not new. For example, radio waves are amplified in a radio. But here it is the amplification of the current signals *after* the electromagnetic energy has been converted to the vibrational energy of electrons in the antenna (Chapter 19). Electrical circuits are incapable of handling the frequencies of light (a million times those of radio waves), because the circuit electrons cannot oscillate at these frequencies. This limitation is overcome in the laser, in which light is amplified using energy stored in atoms.

The principle of the laser was first developed for *m*icrowave frequencies, and the first device was called a *m*aser. The laser is an optical maser. The same principles are involved, but the laser involves higher frequencies. The first maser used ammonia as an amplifying medium. Other substances, such as ruby crystals and various gases, have been found to be suitable for higher light frequency "lasing" action or amplification. To understand this action,

let's take a look at the "aser" part of a laser— *a*mplification by *s*timulated *e*mission of *r*adiation.

When an electron is excited to a higher energy level in an atom, it generally "decays" after a short time (on the order of 10^{-8} second) and returns to a lower energy level with the emission of a photon. This is called **spontaneous emission** (Fig. 24.13).

However, an excited atom can be *stimulated* to emit a photon. In a **stimulated emission** process, an excited atom is struck by a photon of the same energy of the allowed transition, and the atom emits an identical photon. After emission, the two photons have the same frequency, are emitted in the same direction, and are in phase. Since one photon goes in and two come out, this is an amplification process (which was predicted by Einstein in 1917).

The monochromatic (single frequency), phase, and directional properties of the light of the lasing action give laser beams some unique properties. Light from common sources, such as an incandescent lamp, is incoherent. That is, the waves have no particular relationship to each other. For example, if you throw a handful of gravel into a pond, the resulting waves would be incoherent. In light sources, excitation occurs randomly, and atoms emit randomly and at different frequencies (different transi-

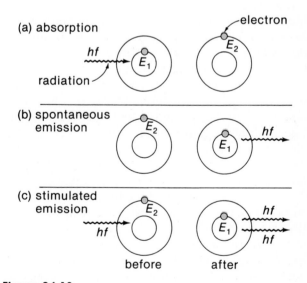

Figure 24.13
Radiation emission. (a) An atom absorbs energy and becomes excited. (b) It generally decays spontaneously after a short time, with the emission of a photon. (c) However, if another photon strikes an excited atom, it is stimulated into emission.

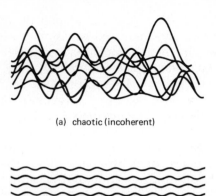

(a) chaotic (incoherent)

(b) coherent

Figure 24.14
Laser light is coherent. (a) Light from common sources is chaotic or incoherent, with the waves having no particular relationship to each other. (b) Coherent laser light waves have the same frequency, phase, and direction.

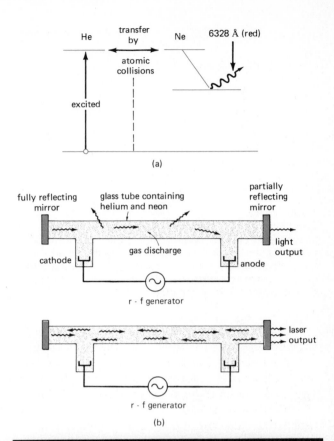

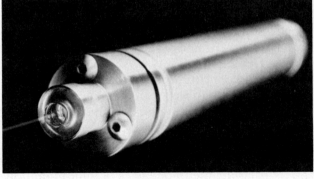

(c)

Figure 24.15
He-Ne laser. (a) The laser light comes from a neon atom transition. (b) A schematic diagram of a gas laser. (c) A laser. Steel-ceramic tubes, which are much more reliable than glass tubes, are commonly used.

tions). As a result, an incoherent light beam is "chaotic" (Fig. 24.14).

Such a beam spreads out and becomes less intense. To get monochromatic light, all but one frequency must be filtered out of the beam, which leaves it weak and still incoherent. When the waves (or photons) of a beam of light have the same frequency, phase, and direction, the light beam is **coherent**. It is possible to make a coherent beam that does not spread out appreciably and that, if amplified, can be very intense.

There are pulsed lasers, which produce pulses of coherent light, and continuous lasers. Let's focus on the continuous-beam He-Ne (helium-neon) gas laser, which you are most likely to see in the classroom and in some common applications. The gas mixture is subjected to a high voltage in a laser tube, and an electrical discharge is induced (Fig. 24.15). Electron collisions efficiently excite He atoms to an excited state that has a relatively long lifetime ($\sim 10^{-4}$ s). The Ne atom also has an energy state at the same energy. There is a good chance that before an excited He atom emits a photon spontaneously, it will collide with an unexcited Ne atom and transfer energy to it. (There are about 10 Ne atoms to one He atom in the gas mixture.) The excited Ne atoms decay after a short time, and the photon of the transition has a frequency that is in the visible red region.

The amplification of the light emitted by the Ne atoms is accomplished by reflections from mirrors placed at each end of the laser tube [Fig. 24.15(b)]. Reflecting back and forth in the tube, the photons cause stimulated emissions, and an intense beam of photons builds up along the direction of the tube axis. Part of the beam emerges through one of the

Lasers in the Supermarket

You may have noticed the pink glow of He-Ne laser light in a supermarket checkout that uses a scanner for the optical reading of the Universal Product Code (UPC) or "bar code" on prepackaged items (figure below). The checkout person moves an item across an opening through which a laser beam is directed toward the bar code label. The dark areas of the code absorb light, and the light areas reflect. A light-sensitive detector reads the pattern of bars from the reflection pattern. This information goes into a central computer, which identifies the product and sends the programmed price for that item to the electronic cash register.

(a)

Code-bar scanning system. He-Ne lasers are used in the optical scanning systems (a) that read product codes (b) on grocery and other items. The zero to the left of the code bar indicates a grocery item. The next five digits identify the manufacturer (13,000, H. J. Heinz Co.), and the next five digits identify the product (45090, 16-oz can of pork and beans).

0

13000 45090

(b)

end mirrors, which is only partially silvered. The outgoing beam is highly intense, coherent, and directional.

A common He-Ne laser application is given in Special Feature 24.1. The laser light seen in this case is scattered or indirect. An intense laser beam can be dangerous to the eye, and you should never look directly into a laser because eye damage can occur. In the medical field, such *slight* damage is used to "weld" detached retinas in eyes (Fig. 24.16).

The list of laser uses is quite long and continues to grow. The distance of the moon from the Earth has been accurately measured using a laser pulse reflected from a mirror placed on the moon by astronauts. Surveyors use laser beams to "line up" mea-surements, as in drilling tunnels. Long-distance communications with laser beams in space and in optical fibers are being developed and are already in use.

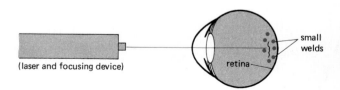

(laser and focusing device)

small welds

retina

Figure 24.16
Medical application. Lasers are used to "weld" detached retinas in eyes.

High-intensity lasers can cut metals (Fig. 24.17), as well as cloth (laser knife or scissors) in commercial garment-making. Metal welding is also done with lasers. Computer printouts are now done with laser printers, and laser videodiscs are available. The videodisc is about half the size of an LP album. More than 50,000 separate images can be stored on each side of a disc, along with two-channel audio. Information is coded in tracks of microscopic indentations. As the disc spins, a laser "needle" from a small semiconductor laser reads the tracks for playback on a conventional TV receiver. The same principle is used in the increasingly popular audio-compact discs.

The coherent property of laser light has made the production of three-dimensional images possible through **holography** (Greek *holos,* meaning "whole"). A hologram is made by using reference and interference beams (Fig. 24.17). Their interference patterns on a photographic film record the information from "in-depth" parts of the object. When developed, the interference pattern appears as a meaningless pattern of light and dark areas. However, when illuminated with a laser beam (or another light source, for some holograms), the recorded information is reproduced and perceived as a three-dimensional image. Someday, holographic, 3-D TV sets may be common.

There are a variety of new types of lasers — glass, chemical, and semiconductor. The light produced ranges from infrared through ultraviolet, and some types are "tunable" and can be adjusted to different frequencies. X-ray lasers are also being developed. This holds the promise of in-depth views, possibly three-dimensional (holographic), of microscopic structures. Also, as will be learned in the next chapter, experimental work is being done in the use of lasers to induce nuclear fusion.

(a)

Figure 24.17
Other laser applications. (a) A laser being used to cut metal. (b) An arrangement for making and viewing a hologram.

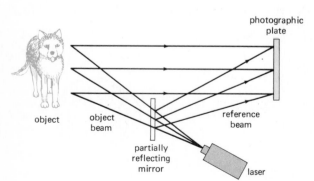

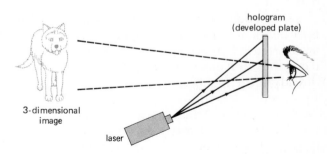

(b)

MATTER WAVES AND QUANTUM MECHANICS

Light has a dual nature: What had been thought to be a wave — with evidence to that effect — sometimes behaves as a "particle," also with evidence. A logical question follows: Do mass particles have a wave nature? Since nature exhibits a great deal of symmetry, the French physicist Louis deBroglie reasoned that matter, as well as light, had properties of both particles and waves. In 1924 he postulated that any moving particle or object has a wave associated with it, with a wavelength (λ) given by

$$\lambda = h/p = h/mv$$

where m and v are the mass and velocity of the particle (recall that $p = mv$ is momentum) and h is Planck's constant again.

The waves associated with moving particles are called **matter waves** or **deBroglie waves**. If a photon is taken to be a "massless" particle, then an electromagnetic wave is the deBroglie wave for a photon. But the deBroglie waves associated with mass particles, such as electrons and protons, are *not* electromagnetic waves.

As you might expect, deBroglie's hypothesis met with a great deal of skepticism. The idea that photons were electromagnetic waves did not seem unreasonable, but to think that a mass in motion somehow had wave properties was another matter.

In support of his hypothesis, deBroglie showed how it could give an interpretation of the quantization of the electron orbits in Bohr's theory of the hydrogen atoms. As the electron travels around in one of its circular orbits, the associated matter wave would be expected to be a standing or stationary wave. That is, only a certain number of wavelengths would fit into a given orbit, as in the case of a standing wave in a string (Chapter 14).

DeBroglie's theory predicted this (Fig. 24.18). A wave must match or meet itself after going around one full circumference. If it didn't match or wasn't in phase with itself, the wave would interfere destructively, and the associated particle could not be in that orbit. Thus, there were only discrete, or allowed, orbits.

Question: If moving masses have wave properties, why aren't these observed normally?

Answer: According to deBroglie's equation for the wavelength of a wave associated with a moving "particle," for normal objects and speeds the wavelengths are so small that the wave properties go unnoticed. For example, a 1000-kg car traveling at 100 km/h would have a wavelength on the order of 10^{-38} meter.

Recall that wave properties, like diffraction, are observed when the wavelength is on the same order as an object or opening. A wavelength of 10^{-38} m is just too small for wave effects to be observed.

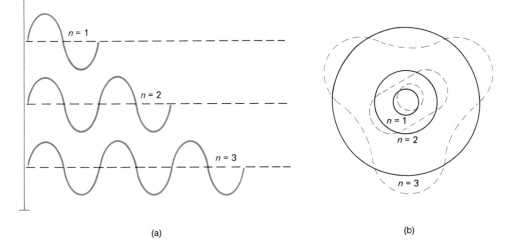

(a)   (b)

Figure 24.18
DeBroglie waves and the hydrogen atom. (a) Only a certain number of (deBroglie) wavelengths would fit in the electron orbits, and (b) discrete standing waves (characteristic frequencies or wavelengths) correspond to the discrete orbits in the Bohr theory.

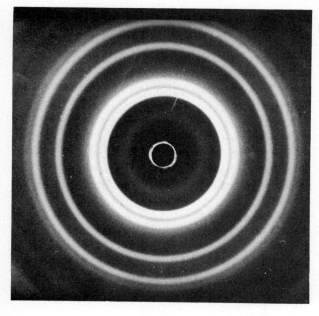

(a)

(b)

Figure 24.19
Diffraction patterns. (a) A diffraction pattern made by
X-rays passing through aluminum foil and (b) a diffraction
pattern made by a beam of electrons passing through the
same foil. (The center of the pattern of the film is blocked
off to prevent overexposure of the film.)

Although deBroglie's hypothesis provided an interpretation of discrete electron orbits, this was hardly proof. The motions of particles had been studied extensively, and no wave-like behavior had ever been observed. The preceding Question and Answer tells you why. What would be needed is a particle of very small mass so that the deBroglie wavelength would be appreciable.

In 1927 two physicists in the United States, C. J. Davisson and L. H. Germer, took advantage of this and showed that a beam of electrons could be diffracted in a nickel crystal, thereby demonstrating the wave-like properties of particles! (Recall that crystals had also been used to study the wavelengths of X-rays, which also have very short wavelengths.)

Another experiment carried out by G. P. Thomson in Great Britain in the same year supported this result. The diffraction pattern for an energetic electron beam passing through a thin metal film was obtained. This was compared to a transmission X-ray diffraction pattern (Fig. 24.19). The comparison of the patterns leaves little doubt about the wave-like properties of the electron particles. For an important application of the wave nature of particles, see Special Feature 24.2.

As a result of the dual natures of waves and particles, a new kind of physics based on the synthesis of wave and quantum ideas was born in the 1920's and 1930's. This new physics, called quantum mechanics, replaced the old mechanistic view that all things moved according to *exact* natural laws with a new concept of probability. For example, a quantum-mechanical analysis of the hydrogen atom predicts that the electron will *most probably* be in the discrete orbits of the Bohr theory. The probability of finding the electron is very low in the intervening regions. This prediction gives rise to the idea of an "electron cloud" around the nucleus, where the cloud density reflects the probability that the electron is in that region. In quantum mechanics, the matter waves "pilot" or "guide" particles, and the description or wave function gives an indication of the probability where the particle is.

▇ THE HEISENBERG UNCERTAINTY PRINCIPLE

Let's look at one other aspect of quantum mechanics. According to classical mechanics, there is no limit as to the accuracy of a measurement, but it only depends on the refinement of the measurement instrument and/or procedure. This gave rise to a "deterministic" view of the world. For example, if you know (measure) the position and velocity of a particle at a particular time, then you could calculate where it will be in the future (with any future influences also known), and similarly where it was in the past.

So, according to classical theory, it is possible, in principle, to make measurements with no uncertainty, i.e., with exact accuracy. However, the quantum theory predicts otherwise and sets limits on measurement accuracies. This idea was developed by Werner Heisenberg (1901–1976), a German physicist, and a statement of **Heisenberg's uncertainty principle** is as follows:

> It is physically impossible to simultaneously know (measure) a particle's exact position and velocity (or momentum).

To help understand this principle, consider the process of measuring the position and momentum of an electron (Fig. 24.20). We'll do this looking through a very powerful microscope. In order to see the electron, and thus determine its location, you must have at least one photon of light bounce off the electron and come into your eye. When the photon strikes the electron, it transfers some of its energy and momentum to it (a collision process, see Chapter 3). The electron recoils and in trying to locate the position very accurately (making the uncertainty in position, Δx, very small), a rather large uncertainty was caused in the electron's velocity or momentum ($m\,\Delta v = \Delta p$), since the electron went flying off.

If we use light, the position of an electron can be measured at best to an accuracy of about the wavelength (λ) of the light, i.e.,

$$\Delta x \approx \lambda$$

The photon "particle" has a momentum $p = h/\lambda$ (deBroglie hypothesis), and hence the momentum of the electron will be uncertain by the amount

$$\Delta p \approx h/\lambda$$

The total uncertainty is the product of the individual uncertainties, and we have

$$(\Delta x)(\Delta p) \approx h$$

Thus, the uncertainty principle says that the product $\Delta x\,\Delta p$ is on the order of Planck's constant. So, the more accurately we know the position of an object, the greater uncertainty there is in the value of its momentum, and vice versa. If we could measure the exact location of a particle ($\Delta x \rightarrow 0$), then we would have no idea of the value of its momentum ($\Delta p \rightarrow \infty$).

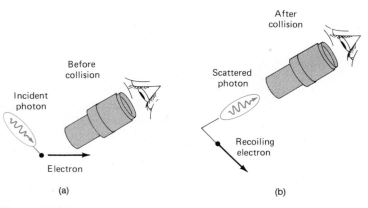

(a)   (b)

Figure 24.20
A thought or "gedanken" experiment of the uncertainty principle.
(a) For someone to be able to view or determine the position of an electron with a powerful microscope, a photon must collide with the electron and be scattered toward the microscope (b). The electron recoils (is disturbed) as a result of the collision, and its position, as well as its momentum or velocity, is uncertain.

The Electron Microscope

In some instances, electromagnetic waves have a particle-like nature — the dual nature of light. Conversely, it has been shown that particles exhibit a wave-like nature. For example, a beam of electrons can be diffracted, and the behavior is described by a wave relationship.

The "dual nature of particles" is applied in the electron microscope. A magnetic coil can focus an electron beam just as a glass lens can focus a light beam. Such magnetic "lenses" are used to focus the electron beam in an electron microscope in an arrangement similar to that of glass lenses in a light microscope (Fig. 1).

The electron beam is directed on a very thin specimen. Different numbers of electrons pass through different parts of the specimen, depending on its structure. The transmitted beam is then brought to focus, forming an image on a film or a fluorescent screen.

Specimens that cannot be thin-sectioned can be viewed by other means. A specimen is coated with a thin layer of metal. The electron beam is then made to scan across the specimen. The surface irregularities cause variations in the intensity of the reflected electrons, and the reflected beam is focused to form an image. The metal coating makes the specimen electrically conducting. Otherwise, a nonuniform charge would build up on the specimen and cause the image to be distorted.

Both the transmission electron microscope and the scanning electron microscope are enclosed in a high-vacuum chamber so that the electrons are not deflected by air molecules. As a result, an electron microscope looks nothing like a light microscope (Fig. 2). However, magnifications up to 100,000 times can be achieved with an electron microscope, whereas a light microscope is limited to a magnification of about 2000 times.

A new microscope technique is being used to enhance the study of such things as living cells (Fig. 3). This method uses X-rays to form a replica of the cell in a photosensitive material called resist. The replica is then observed in an electron microscope. Resist, which is used in fabricating electronic circuits, can record much smaller features than can the photographic films used for conventional X-ray pictures.

This method makes it possible to see cell features only 5 nm (50×10^{-10} m) wide, which is about 1/100 the wavelength of visible light. The blood platelet shown in Figure 3 is about 3 μm ($\approx 1/10,000$ in.) in diameter. The dark filaments are believed to be a kind of skeleton in the cell, probably formed of muscle filaments. This structure is of great interest to biologists and cannot be seen in conventional electron micrographs.

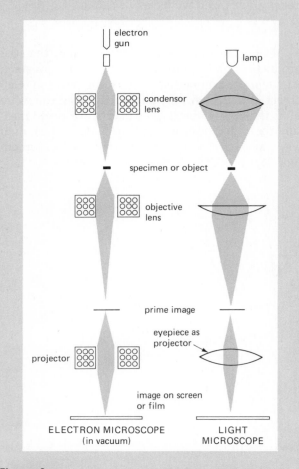

Figure 1
A comparison of **electron** and light microscopes. The electron microscope uses magnetic-coil "lenses," and the light microscope uses glass lenses.

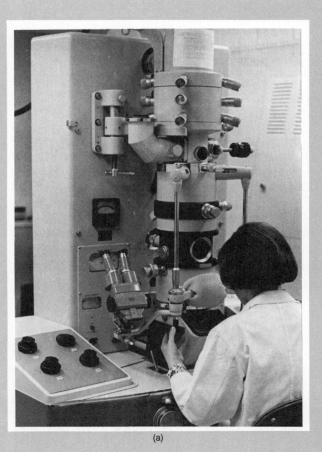

(a)

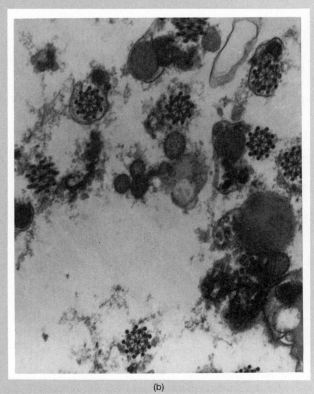

(b)

Figure 2
Electron microscopy. (a) An electron microscope. (b) An electron micrograph of biological tissue at 33,000× magnification.

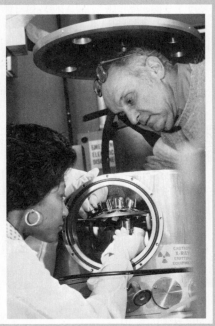

(a)

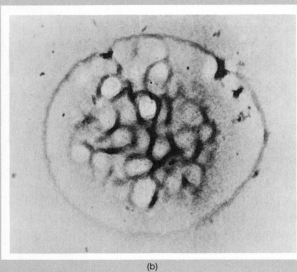

(b)

Figure 3
A new microscope technique. (a) X-rays are used to form a replica of an object in photosensitive material, and the replica is then observed with an electron microscope. (b) Electron micrograph of an x-ray replica of a blood platelet, which is only about 3 μm (3×10^{-6} m) in diameter.

SUMMARY OF KEY TERMS

Ultraviolet catastrophe: the problem with the classical wave theory prediction for thermal radiation that the radiation intensity or energy at high frequencies (ultraviolet) should be infinitely large (catastrophe).

Planck's quantum hypothesis: the oscillators of a thermal radiator could have only discrete energies given by $E = hf$, where h is Planck's constant and f is the frequency of the oscillator.

Photoelectric effect: the emission of electrons from a material that is exposed to light.

Photon: a packet or quantum of light with energy $E = hf$, where f is the frequency of the light.

Dual nature of light: light or electromagnetic radiation sometimes behaves as a wave and sometimes as a quantum or "particle."

Bohr theory of the hydrogen atom: the theory that quantizes the hydrogen electron in discrete orbits or energy levels. Photons are emitted when the electron makes transitions to lower energy levels.

Laser: a light source that amplifies light through stimulated emission (*l-a-s-e-r*, *l*ight *a*mplification by *s*timulated *e*mission of *r*adiation).

deBroglie hypothesis: particles in motion have a wave nature, with their "matter waves" having a wavelength given by $\lambda = h/mv$.

Quantum mechanics: the physics that uses deBroglie or "pilot" waves to predict the probability of a physical event.

Heisenberg's uncertainty principle: it is physically impossible to simultaneously know (measure) a particle's exact position and velocity (or momentum).

QUESTIONS

Quantization

1. When a solid, such as an iron ball, is heated, why does it first glow with a dull red color?
2. Why did Planck "invent" quanta?
•3. Which is more energetic, a quantum or photon of red light or a quantum of blue light? How about photons of infrared radiation and visible light?
4. If electromagnetic radiation is made up of quanta, why don't we hear a radio intermittently with the arrival of discrete packets of energy?

The Photoelectric Effect

5. If a photoelectric material has a cutoff frequency, why does white light always give a photocurrent in a light meter?
6. A photographer is using a light meter to check out a modeling set that is illuminated with red light and gets a zero reading. What is wrong? Is the meter broken? (It works out in the hall.)
•7. In terms of the water bucket analogy of quantum absorption given in the chapter, explain why there is no photoemission for $f < f_o$.
8. Can you have a photon of white ligh? Explain.
9. A sunburn is caused by ultraviolet light. Why doesn't visible light cause a burn?
10. Explain how the photoelectric effect is used in the application of an electric "eye" counter or a burglar alarm. (See Fig. 24.21, and note the light source and receiver on each side of the door.)

Figure 24.21
The eye is watching. See Question 10.

The Bohr Theory of the Hydrogen Atom

11. We know that the Sun contains certain elements. How do we know this when we chiefly receive light from the Sun?

•12. Incandescent lamp filaments are made of tungsten. Does the light from these lamps exhibit a discrete tungsten spectrum? Explain.

13. Does the light from neon signs have continuous spectra? Explain.

14. An absorption spectrum is obtained by passing light through a cool gas. The gas atoms absorb light, but they also re-emit it, so why are there dark absorption lines?

15. Why is the electron in a Bohr orbit accelerated if it is traveling with a constant speed?

16. Classically, why should an electron in circular orbit emit radiation? Why was it necessary for Bohr to assume that a bound electron in orbit did not emit radiation?

17. In which transition is the photon of greater energy emitted, $n = 3$ to $n = 1$ or $n = 2$ to $n = 1$? (*Hint:* See Fig. 24.10.)

•18. A hydrogen electron is in the excited state $n = 4$. How many photons of different frequencies could possibly be emitted in returning to the ground state?

19. Is a satellite in circular orbit about the Earth like the hydrogen electron in circular orbit about the nuclear proton? Give differences and/or similarities.

The Laser

20. Incoherent light is sometimes likened to a crowd of people. How would the people be moving in this case? Could a group of people be analogous to coherent light? How about a marching band? Explain.

•21. In addition to stimulated emission in a laser, another condition for laser operation is "population inversion." That is, more atoms must be in an excited state than in the ground state. Why is this necessary?

22. Would a He-Ne laser operate if one end of the tube were covered by a transparent piece of glass? Explain.

23. Could holograms be made with an incandescent lamp? Explain.

Matter Waves and Quantum Mechanics

24. What is a basic difference between classical mechanics and quantum mechanics?

25. Waves act like particles. Do particles act like waves?

•26. Estimate your deBroglie wavelength when you are running. (Recall that $h \approx 10^{-34}$ in SI units and that 1 lb is equivalent to about 0.5 kg — actually 0.45 kg.) For the computation, estimate how fast you can run in meters per second.

27. Is there any limit to the accuracy of measurement? Explain.

28. If you increased the accuracy of locating a particle by 50 percent, how would a simultaneous measurement of the particle's velocity be affected?

The Nucleus and Radioactivity

25

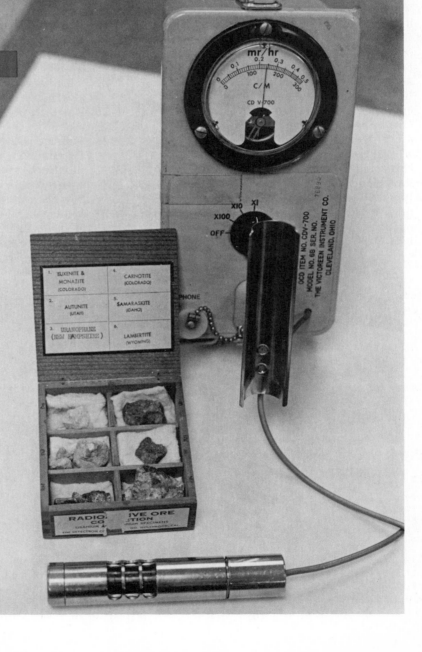

Radioactivity from uranium ore samples is detected by a Geiger counter. It is sometimes forgotten that radioactivity is a natural part of our environment.

THE ATOMIC NUCLEUS

Having looked at the orbiting "planet" electrons of the simplified planetary model of the atom, let's now look at the atomic "sun"—the nucleus of the atom. Another name for our model of the atom is the Rutherford-Bohr model. The Bohr contribution was discussed in the last chapter. The Rutherford part of the model is concerned with the central core or nucleus of the atom.

In the early 1900's the atom was thought to have a "plum-pudding" structure. This plum-pudding model viewed the electrons of the atom as being spread out like raisins in a sphere of positively charged "pudding." The pudding atom was thought to be about 10^{-8} cm across. But around 1910 there were experimental results that didn't agree with this model. The experiment was suggested by the British physicist Ernest Rutherford and was performed in his laboratory. Positively charged particles, called alpha particles, were directed toward a thin gold foil "target" (Fig. 25.1).

These alpha particles are thousands of times more massive than electrons, and the plum-pudding model predicted that they would be deflected only slightly as a result of collisions with the electrons. However, the results were completely different. Some of the alpha particles were scattered through large angles, even scattered backward. As Rutherford put it, this was "almost as if you fired a 15-inch shell at a piece of tissue paper, and it came back and hit you."

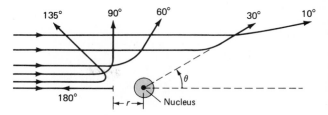

Figure 25.2
The nucleus. The scattering of the Rutherford experiment could be explained if the positive charges in the atom were concentrated in a core or nucleus.

In 1911 Rutherford offered an explanation of the alpha-particle scattering that gave a different view of the atom. If all the positive charges of an atom were concentrated in a central massive core or nucleus, then an alpha particle coming close to this region would experience a large deflecting force and would even be back-scattered in head-on collisions (Fig. 25.2). Theoretical calculations showed that the theory fit the data, and the atomic nucleus was "born." For his efforts, Rutherford was given a title (he became Lord Rutherford) and received a Nobel Prize.

The alpha scattering is an example of a "black box" experiment that is very common in nuclear physics. In general, you know what goes in, observe what comes out, and infer what happened in between. The atomic nucleus was Rutherford's "black box." Notice in Figure 25.2 that the 180° back-scat-

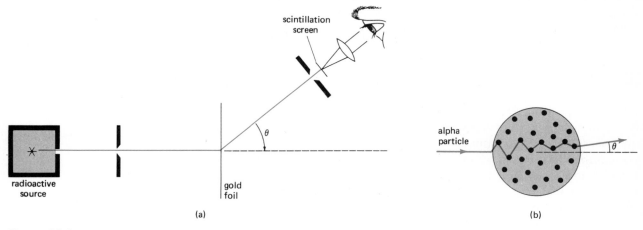

(a)   (b)

Figure 25.1
Rutherford's scattering experiment. (a) The experimental setup for observing alpha particle scattering from a gold foil. (b) The plum-pudding model of the atom predicted that the alpha particles would be only slightly deflected as a result of collisions with the electron of an atom. The experimental results were different.

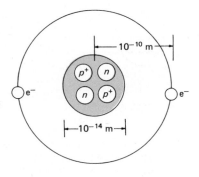

Figure 25.3
A schematic diagram of a helium atom showing atomic dimensions.

tering gives an upper limit on the nuclear size or radius. (The nuclear size must be less than a volume with a radius equal to the distance of closest approach.) This is on the order of 10^{-14} m for a typical atom. The atomic electrons in the Rutherford-Bohr model are much farther out, on the order of 10^{-10} m (Fig. 25.3).

Because of the small size of the atom, most people do not realize the extent of the "void of space" within the atom. The relative dimensions of the atom's structure have been likened to a large major-league football stadium with the nucleus being a small marble (on the 50-yd line). If the nucleus of a typical atom were expanded to the size of the Sun, then on a proportionate basis, the "inner planets" (Mercury and Venus) in Figure 25.3 would be well beyond the outer reaches of our solar system— many, many times beyond.

◼ NUCLEAR NOTATION AND ISOTOPES

Recall that the number of nuclear protons in an atom determines what kind of atom it is (Chapter 7). The nucleus of a hydrogen atom is a single proton, helium atoms have two protons, lithium atoms have three protons, and so on. However, it was found that all of the nuclear mass could not be accounted for by protons. Rutherford suggested in 1920 that there may be another electrically neutral particle in the nucleus, which he called a **neutron**. The existence of the neutron was confirmed experimentally in 1932. We now know that all nuclei, with the exception of the common hydrogen nucleus (a proton),

contain neutrons, which are electrically neutral particles with a mass about the same as that of a proton.

Protons and neutrons are about 2000 times more massive than electrons, so the vast majority (more than 99.9 percent) of the atomic mass lies in the nucleus. Nuclei are generally thought of as being spherical, but they may deviate from this shape and resemble a watermelon or a doorknob. Nuclear protons and neutrons are collectively referred to as **nucleons**, but we speak specifically of the proton number (sometimes called the atomic number) and the neutron number of a nucleus, which of course are just the numbers of protons and neutrons in a particular type of nucleus.

To designate a particular nucleus or nuclear species, which is called a **nuclide**, a special notation, shown in Figure 25.4 for a general case and a carbon nuclide, is used. Notice that the **proton number** is on the lower left of the chemical symbol of the element or atom. The proton number is also commonly called the atomic number and is designated by the letter Z. The fact that a nucleus has six protons makes it a carbon nucleus. The **neutron number** is on the lower right.

Notice that if you add the number of protons and the number of neutrons $(6 + 6)$, you get the number (12) to the upper left of the symbol. This is called the **mass number** $(p + n)$. It is customary to leave off the neutron number and simply write $^{12}_{6}C$. This still gives you the same information. (Why?) In referring to this nucleus, we say that it is a carbon-12 nucleus.

If you examined a bunch of carbon nuclei (all with six protons), you would occasionally find a nucleus with more than six neutrons—sometimes

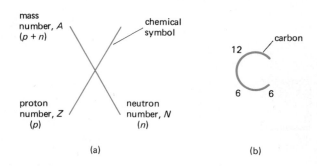

Figure 25.4
Nuclear notation. (a) The general notation with the meaning of the superscript and subscripts on a chemical symbol. (b) Notation for a carbon nucleus. The notation conveys that this carbon nuclide has six protons (as do all carbon nuclei) and six neutrons.

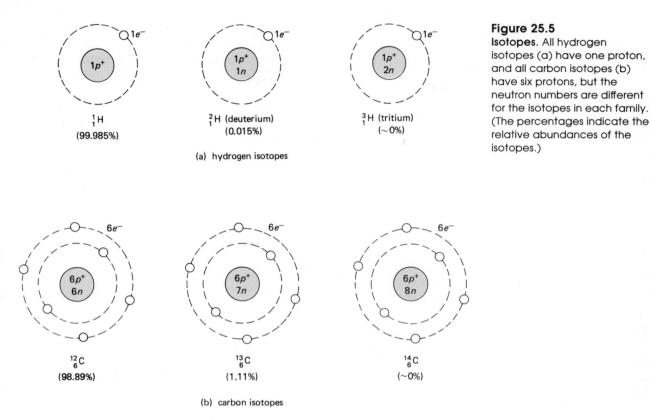

$^1_1$H
(99.985%)

$^2_1$H (deuterium)
(0.015%)

$^3_1$H (tritium)
(~0%)

(a) hydrogen isotopes

$^{12}_6$C
(98.89%)

$^{13}_6$C
(1.11%)

$^{14}_6$C
(~0%)

(b) carbon isotopes

seven neutrons and, on rarer occasions, eight neutrons. Nuclei or nuclides of the same element having different numbers of neutrons are called **isotopes**. Isotopes of hydrogen and carbon are illustrated in Figure 25.5. A group of isotopes is somewhat like a family—they are all Joneses or Smiths (for example, hydrogen or carbon), but the individual family members are different. In the nuclear case, the "family" members differ by the number of neutrons they contain. Some elements have large nuclear isotope "families" with a dozen or more members.

Only the isotopes of hydrogen are given specific names. $^1_1$H, the most common isotope, has just a nuclear proton and is referred to as **ordinary hydrogen** or simply **hydrogen**. The other isotopes, $^2_1$H and $^3_1$H, are called deuterons and tritons, respectively, or deuterium and tritium in atomic form. When the more massive deuterium atom replaces the common hydrogen atom in H_2O, we have what is called "heavy water." For every 6500 or so atoms of ordinary (light) hydrogen in water, there is one atom of deuterium. The oceans contain millions of tons of

deuterium. Tritium is radioactive. You will learn more about both these isotopes later in the discussion on nuclear energy. Other isotopes are referred to by their element-mass number designation, for example, carbon-12, carbon-13, and carbon-14.

■ THE NUCLEAR FORCE

Within the small confines of the nucleus, the nucleons are clustered together, with each nucleon presumably taking up about the same amount of space (Fig. 25.6). The nucleons are on the order of 10^{-15} m apart. This means that there are large repulsive electrical forces between the positively charged protons. What then holds the nucleus together? Since the nucleons are mass particles, the ever-present attractive gravitational force is there. However, calculations show that the gravitational force between two nucleons is a factor of 10^{-40} smaller than the electrical force between two protons. On a relative basis, with a factor of 10^{-40}, the gravitational force is so small that it is negligible, so that's out.

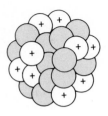

Figure 25.6
Nuclear nucleons. Nucleons are clustered in the nucleus with each nucleon taking up about the same amount of space as represented here by close-packed spheres.

Obviously, there must be some other strongly attractive force acting in the nucleus that overcomes the repulsive electrical force. We call this attractive force acting between nucleons the nuclear force, or the strong interaction.* The **nuclear force** is a fundamental force like the electrical and gravitational forces, but it is more complicated and not completely understood. From experiments, we believe that this force acts between any pair of nucleons — proton-proton, neutron-neutron, and proton-neutron.

The strong nuclear force is a short-range force. That is, it falls off very quickly with nucleon separation distance. Recall that the electrical force falls off as the inverse square of the distance ($1/r^2$). The strong nuclear interaction weakens much more quickly. For example, the repulsive electrical force between two protons on opposite sides of a sizable nucleus may be appreciable, but the attractive nuclear force is very small. An electrical force also exists between the nuclear protons and the orbiting atomic electrons (centripetal force), but the nuclear force does not extend outside the nucleus.

If the attractive nuclear and repulsive electrical forces acted only between protons, then there would be only small nuclei or atoms because of the short range of the nuclear force. However, there are very massive stable nuclei, for example, the lead isotope $^{208}_{82}$Pb, with 82 protons and 126 neutrons. The key is the number of neutrons. Remember that the nuclear force acts between all nucleons, so the neutrons act as a nuclear "glue" to hold the nucleus together. As the number of protons increases for massive *stable* nuclei, the number of neutrons increases too, so that the short-range nuclear forces

* There is also a weak nuclear force of weak interaction, which will be discussed later.

are greater than the long-range repulsive electrical forces. Neutrons provide attractive nuclear forces between *both* protons and other neutrons.

This effect is shown graphically in Figure 25.7. For lighter or less massive nuclides up to about calcium (proton number of 20), the stable nuclei have an equal or an approximately equal number of protons and neutrons. But for heavier nuclides, there are more neutrons than protons.

If you look at the magnitudes of the attractive nuclear forces between nucleons and the attractive electrical forces between nuclear protons and orbiting atomic electrons, the latter are about one billionth as great. This gives a hint about why the nuclear energy of atoms is so much greater than chemical energy.

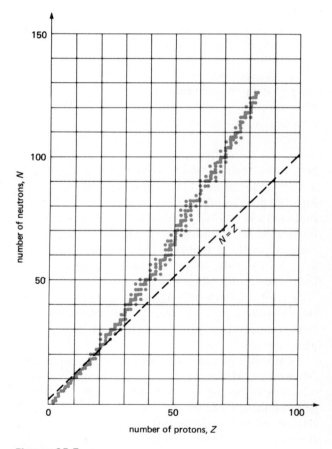

Figure 25.7
Neutron number versus proton number for nuclides. Nuclides with proton numbers greater than 20 have more neutrons than protons, as indicated by the nuclides above the $N = Z$ line. The excess neutrons act a a nuclear "glue" that holds the nucleus together.

In a chemical reaction, such as the burning or combustion of gasoline, there is a rearrangement of the electrons in the atoms or molecules. The nuclei of the atoms do not change. In a nuclear reaction, however, there is a "rearrangement" or change of nuclear particles. This involves strong forces and the release of large amounts of energy, as will be learned in the next chapter in the discussion on nuclear energy.

■ RADIOACTIVITY

There are more than 250 stable nuclei making up the graph in Figure 25.7. However, there are approximately 1400 known nuclides. The big difference reflects the known *unstable* nuclides. For some reason, the previous argument about neutron "glue" doesn't apply, and unstable nuclei spontaneously "decay" with the emission of energetic particles. Such nuclei are said to be radioactive. The "radio-" part refers to the emitted radiation, which in the modern context may be a particle or a wave. So a radioactive nuclide or isotope is "active" in emitting radiation. Notice that the graph in Figure 25.7 terminates at proton number 83. There are no stable nuclei with proton numbers greater than 83.

Of the nearly 1200 known unstable nuclides, only a small number occur naturally. The other radioactive nuclei are made artificially. The unstable nuclides found in nature decay with the emission of alpha particles or beta particles, which are sometimes accompanied by gamma rays. Using a magnetic field, which deflects electrically charged particles, it was found that alpha "rays" were positively charged particles and beta "rays" were negatively charged particles (Fig. 25.8). Because they are deflected more, beta particles must be less massive than alpha particles. Gamma rays were not deflected at all, so they must be uncharged. Actually, a gamma ray or "particle" is a photon or quantum of energy.

Radioactivity was discovered accidentally by the French physicist Henri Becquerel in 1896. The circumstances were not unlike Roentgen's discovery of X-rays. Becquerel noticed that a photographic plate in a light-tight wrapper that he had left in a drawer with a uranium compound showed signs of exposure when developed. Evidently, some type of radiation or "rays" from the uranium was able to penetrate the wrapper and expose the film.

A couple of years later, the husband-and-wife team of Pierre and Marie Curie announced the dis-

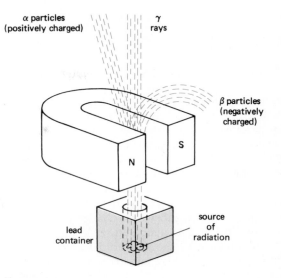

Figure 25.8
Radioactive decay particles. Passing radiation through a magnetic field shows there are positively charged (alpha) particles, negatively charged (beta) particles, and neutral (gamma) particles or rays.

covery of two new radioactive elements, radium and polonium, which they had isolated from uranium pitchblende ore. (See Special Feature 25.1.) They had painstakingly isolated 10 mg of radium and a smaller amount of polonium from 8 tons of ore! The Curies and Becquerel shared the 1903 Nobel Prize in physics for their work in radioactivity. Madame Curie, as she is commonly known, also received the Nobel Prize in chemistry in 1911 for her contributions in chemically isolating radioactive materials. Carrying on the family tradition, the Curies' daughter, Irene Joliot-Curie, and her husband Frederic Joliot won the 1935 Nobel Prize in chemistry.

Let's take a closer look at the ABCs of radioactivity: alpha (α), beta (β), and gamma (γ) decays and their respective particles.

Alpha Decay

An alpha particle consists of two protons and two neutrons. It is the same as a helium nucleus (^4_2He, sometimes written $^4_2\alpha$), with a 2+ electrical charge. When a nucleus undergoes alpha decay, an alpha particle is spontaneously emitted, and the nucleus is *transmuted* (changed) into the nucleus of another element, since two protons are lost. An example of alpha decay is given in Figure 25.9. The origi-

Marie and Pierre Curie

Marie Sklodowska (1867–1934) was born in Warsaw, Poland, and received her early scientific training from her father. She became involved in a students' revolutionary organization and found it advisable to leave Warsaw. In Paris she earned a science degree, and in 1895 she married Pierre Curie, who was a physicist well known for his work on crystals and magnetism. One of his most important discoveries was that the magnetic properties of substances change at a certain temperature, the "Curie temperature."

After their marriage, Marie did doctoral research on radioactivity, and Pierre joined his wife in this work. In 1898 they announced their discovery of two new radioactive elements that they had isolated, radium and polonium (named after Marie Curie's native country). From 8 tons of uranium pitchblende ore, they painstakingly isolated 10 mg of radium and a smaller amount of polonium.

In 1903 they were awarded the Nobel Prize in physics for their discovery in radioactivity. The prize was shared with Henri Becquerel, who discovered the radioactive properties of uranium in 1896. Pierre Curie was killed in a horse-drawn vehicle accident in 1906. Marie Curie (commonly known as Madame Curie) was appointed to his professorship at the Sorbonne, the first woman to have this post.

In 1911 Mme. Curie was awarded the Nobel Prize in chemistry for her work on radium and the study of its properties. She was the first person to win two Nobel Prizes in science. The rest of her career was spent in establishing and supervising laboratories for research on radioactivity and the use of radium in the treatment of cancer. In 1921 President Harding, on behalf of the women of the United States, presented her with a gram of radium in recognition of her services to science.

Mme. Curie died in 1934 of leukemia, a form of blood cancer, which most probably was caused by overexposure to radioactive substances. This occurred

Marie and Pierre Curie.

shortly before an event that would no doubt have made her very proud. In 1935 the Curies' daughter, Irene Joliot-Curie, and her husband, Frederic Joliot, were awarded the Nobel Prize in chemistry.

nal nucleus in a decay process is commonly called the parent nucleus (radium-226 in this case), and the resulting nucleus is called the daughter nucleus (radon-222).

The decay process is written in a nuclear reaction equation (similar to a chemical reaction equa-

tion). Notice in this reaction, *as in all nuclear reactions,* that

a. the total number of nucleons remains constant (conservation of nucleons) and

b. the total charge remains constant (conservation of charge).

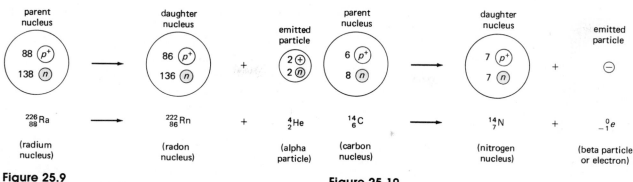

Figure 25.9
An illustration of an example of alpha decay.

Figure 25.10
An illustration of an example of beta decay.

Condition (a) means that the sums of the superscript mass numbers on each side of the equation are equal ($226 = 222 + 4$). Condition (b) requires that the sums of the subscript proton numbers on each side of the equation are equal ($88 = 86 + 2$).

Beta Decay

A beta particle is simply an electron. An example of beta decay is given in Figure 25.10. Notice in the nuclear reaction equation that the -1 subscript of the electron (its *charge* number) allows for the conservation of charge ($6 = 7 - 1$).

But where does an electron come from within the nucleus? Our nuclear model has only protons and neutrons in the nucleus. Looking at the proton and neutron numbers in the beta-decay reaction, we see that the daughter nucleus has one less neutron and one more proton than the parent nucleus. This indicates that a neutron must have been converted into a proton and an electron in the beta-decay process; i.e.,

$$_0^1 n \rightarrow _1^1 p + _{-1}^0 e$$

Notice how a neutron is written in nuclear notation. ($_1^1 p$, the proton, is equivalent to $_1^1 H$. Why?)

A nuclear force distinct from the strong interaction is associated with beta decay. This is the weak nuclear force or weak interaction. All the particles we will study interact through the weak interaction, while some also interact electromagnetically and still fewer act through the strong interaction. For example, two nuclear protons interact by all four fundamental forces—strong, electromagnetic, weak, and gravitational interaction. The relative strengths of the fundamental interactions are given in Table 25.1.

As in classical physics, we generally consider only the important forces and ignore the others or consider them negligible in a particular situation; for example, the strong and electromagnetic interactions dominate between nucleons. In the case of beta decay involving an electron, there is no strong force between the nucleons and an electron, so the weak interaction becomes important. The weak interaction is complex and not well understood.

Gamma Decay

A gamma "particle" is a quantum of electromagnetic energy. The nucleus has energy levels similar to the atomic electron energy levels. In the de-excitation of a nucleus, a gamma ray is emitted (Fig. 25.11). The asterisk in the equation indicates that the nucleus is in an excited state. It decays with the emission of a gamma ray or particle. Gamma rays are photons, just as visible light and X-rays are photons or quanta of lower energies or frequencies ($E = hf$; see Fig. 24.11). Since it is a de-excitation process, there is no nuclear transmutation with gamma decay; that is, there is no change in the type of species of nucleus, only in its state.

Table 25.1
Relative Strengths of Fundamental Interactions

Interaction	Relative Strength	Range
Strong	1	Short range, about 10^{-14} m
Electromagnetic	10^{-2}	Infinite ($1/r^2$)
Weak	10^{-13}	Extremely short range
Gravitational	10^{-40}	Infinite ($1/r^2$)

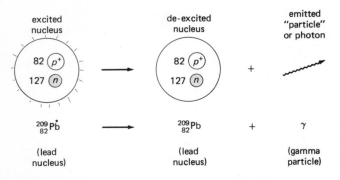

Figure 25.11
An illustration of an example of gamma decay.

DECAY RATE AND HALF-LIFE

Radioactive isotopes decay at vastly different rates. However, the decay rates of the various isotopes are remarkably constant and are unaffected by environmental conditions such as temperature and pressure. In fact, we know of no way to change the rate of decay or "activity." Radioactive nuclei just "do their own thing" and decay at their own fixed rates.

The decay rate of a radioactive isotope is measured in terms of a characteristic time called the half-life.

One half-life is the time it takes for one half of the nuclei of a sample of a given radioactive isotope to decay.

The radioactive decay curve has the form shown in Figure 25.12. Suppose we started with 1 g of strontium-90 ($^{90}_{38}$Sr), which beta-decays. After 28 years, or one half-life ($t_{1/2}$), only half of the nuclei in the sample would be ^{90}Sr, or there would only be ½ g of ^{90}Sr. The other half of the sample or the other half of the nuclei would have decayed (into the daughter nucleus, $^{90}_{39}$Y). After another half-life or another 28 years (56 years total), half of the remaining half of the sample would have decayed and only ¼ g of ^{90}S would be left, and so on. The half-lives of some radioactive nuclides are listed in Table 25.2.

Of course, in practice the half-life of a particular isotope is not determined by counting nuclei or weighing the sample. This is done by monitoring the activity of the sample. The activity of a radioactive sample is the number of nuclear decays per unit time. The result of radioactivity or nuclear decay is emitted particles, and these are what are counted. When the number of radioactive nuclei decreases by one half (in a time $t_{1/2}$), so does the number of emitted decay particles.

The common unit of radioactivity is named in honor of the Curies. One **curie** (Ci) is defined as follows:

$$1 \text{ Ci} = 3.7 \times 10^{10} \text{ decays/second}$$

The curie unit was originally defined as the activity of 1 g of radium-226. It is rather a large unit, so millicuries (mCi) and microcuries (μCi) are commonly used. The SI unit of radioactivity is the **bec-**

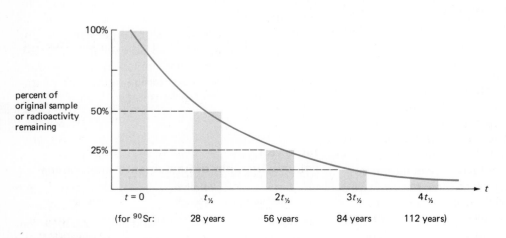

Figure 25.12
Radioactive decay curve. The percentage of original sample of radioactive material versus time. The percentage decreases by one half during each half-life. The half-life ($t_{1/2}$) of strontium-90, used here as an example, is 28 years.

Table 25.2
The Half-lives of Some Radioactive Nuclides

Nuclide		Decay mode	Half-life
Beryllium-8	($^8_4$Be)	α	1×10^{-16} s
Polonium-213	($^{213}_{84}$Po)	α	4×10^{-6} s
Carbon-16	($^{16}_8$C)	β	0.75 s
Aluminum-28	($^{28}_{13}$Al)	β	2.24 min
Magnesium-28	($^{28}_{12}$Mg)	β	21 h
Gold-194	($^{194}_{79}$Au)	β	40 h
Iodine-131	($^{131}_{53}$I)	β	8 days
Cobalt-60	($^{60}_{27}$Co)	β	5.3 years
Strontium-90	($^{90}_{38}$Sr)	β	28 years
Radium-226	($^{226}_{88}$Ra)	α	1600 years
Carbon-14	($^{14}_6$C)	β	5730 years
Uranium-238	($^{238}_{92}$U)	α	4.5×10^9 years
Rubidium-87	($^{87}_{37}$Rb)	β	4.7×10^{10} years

querel (Bq), which is simply defined as the number of decays per second, or

1 Bq = 1 decay/second

Although it is more convenient, the becquerel unit has not gained widespread use.

So a radioactive material continually decays. The activity of a sample may become so small that it is considered to be negligible; the radioisotope has "decayed away" (into stable nuclei). For a radioactive material to be found in nature, it must have a half-life that is not much smaller than the age of the Earth (~ 4.5 billion years), or it must be continually produced by the decay of another radioactive isotope — for example, as in the decay series shown in Figure 25.13 — or it must be continually produced by some other means, such as carbon-14 decay, as discussed in Special Feature 25.2.

Uranium-238 has a half-life about equal to the age of the Earth, so it is still around or found in nature, as are the products of the decay series. There are only 25 unstable naturally occurring nuclides that are not products of the decay chains of heavy nuclides. Notice how some nuclides decay by two modes. The series terminates with a stable lead isotope.

It is believed that when the universe was formed, all nuclides, both stable and unstable, were formed in varying amounts. The unstable nuclei with half-lives much less than a billion years have long since decayed away into stable nuclei. Recall from Chapter 7 that the elements technetium and promethium are not found in nature at all because these elements have short half-lives and no stable isotopes. We will discuss how artificial isotopes are made in the next chapter.

■ RADIATION DETECTORS

Since we cannot generally sense the particles of nuclear radiations when they are emitted from decaying nuclei, they must "do" something before we know they are there. For example, the detection of nuclear radiations by photographic films has been mentioned. Nuclear radiations and X-rays expose various film materials, and persons working with radiation often wear "film badges." The amount of exposure is indicated by the degree of darkening of the film after it has been developed. This is fine, but you would like to be able to detect radiations immediately, along with getting a count of the emissions or decays.

Energetic alpha, beta, and gamma rays from nuclear decay processes have the ability to ionize neutral atoms. As a result, they are commonly referred to as ionizing radiations. For example, when an energetic charged alpha or beta particle passes through a gas, it interacts with gas molecules and rips some of the loosely bound outer electrons from the gas atoms. Although some of the freed electrons might recombine or attach themselves to other atoms or molecules, most will remain free for a while.

This ionizing property of nuclear radiations forms the basis of a commonly used detection instrument, the **Geiger counter**. Your instructor probably has one of these detectors and will demonstrate it for you. It was developed chiefly by Hans Geiger, a student and colleague of Rutherford.

The basic principle of the Geiger counter is illustrated in Figure 25.14. A voltage is applied across the Geiger tube, which contains argon gas at a low pressure (about 5 percent of normal atmospheric pressure). When an ionizing particle enters the tube through a thin window and "strikes" one or more argon atoms, electrons are ejected. These electrons are accelerated by the applied voltage and produce

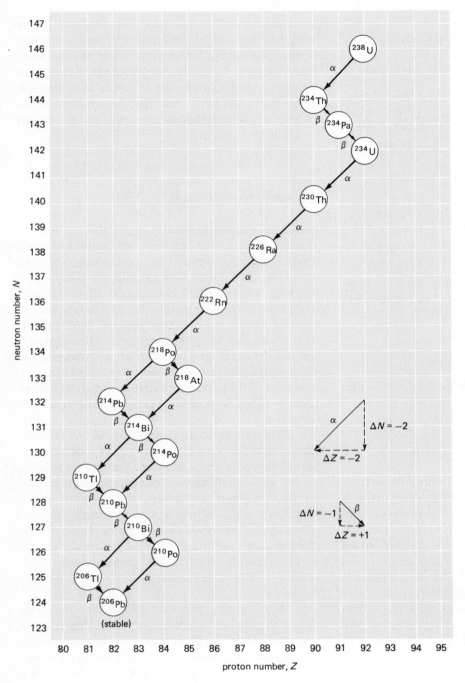

Figure 25.13
Radioactive decay series (uranium). A series of alpha and beta decays leads to a stable nuclide. The series is shown on a neutron number (N) versus proton number (Z) plot.

further ionizations. As a result, an "avalanche" current is set up momentarily in the tube. This current pulse is amplified and counted by the counter circuitry. Hence, radiation particles are detected and counted by counting the current pulses.

One of the main disadvantages of the Geiger counter is its "dead time." **Dead time** refers to the time it takes for the tube to recover or clear itself between counts. If a lot of particles enter the Geiger tube rapidly, the tube will not have time to recover,

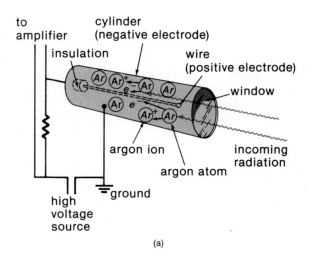

Figure 25.14
Geiger counter. (a) Radiation ionizes the gas molecules in the tube, giving rise to an electrical pulse, which is counted. (b) A portable (battery-operated) Geiger counter.

and some particles will not be counted. Another method of counting gives a shorter recovery time. This is used in the scintillation counter. Certain phosphor materials have the ability to convert the energy of a particle into visible light. For example, if an alpha particle falls on a screen coated with zinc sulfide, it produces a tiny flash of light that is bright enough to be seen in the dark with the aid of a magnifying lens.

This was the detection device used in the Rutherford scattering experiment. In this experiment, the counting was done visually, which was slow and tiring. It has been said that whenever Rutherford himself did the actual counting, he would "damn" vigorously for a few minutes and then have one of his assistants take over. Overall, more than a million scintillations or flashes were counted in the experiment.

This laborious task was eliminated with the development and application of the **photomultiplier tube** (Fig. 25.15). Photons from the scintillator strike a photoelectric material, and photoelectrons are emitted. The photocurrent is amplified in a photomultiplier tube, which consists of a series of electrodes, called "dynodes," at successively higher potentials.

The accelerated photoelectrons cause secondary electrons to be emitted when they strike a dynode, and so the current is multiplied. In this manner, weak scintillations are converted into sizable electrical pulses that can be counted electronically. Also, since the energy of the incident particle and the photon is proportionately amplified in the photomultiplier tube, the resulting output pulse gives a measure of the energy of the incident particle.

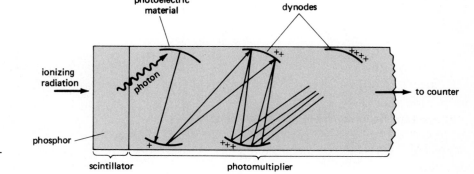

Figure 25.15
The photomultiplier tube. Radiation produces photons in the scintillator, which strike a photoelectric material. The photocurrent is amplified or multiplied at successive dynodes.

Radioactive Dating

Because of their fixed decay rates, radioactive isotopes can be used as nuclear "clocks." For example, suppose you had half a gram of ^{90}Sr (Fig. 25.12). Then you know that 28 years (one half-life) have "ticked off" on the clock since there was 1 g of ^{90}Sr. Similarly, if you had ¼ g, you would know that it was $2t_{1/2}$ (two half-lives) or $2 \times 28 = 56$ years past the 1-g mark on your nuclear "clock." It is this radioactive "ticking" and counting backward that form the basis of radioactive dating, which allows scientists to determine how old various objects are.

An important radioactive-dating process uses the radioisotope carbon-14. Carbon-14 is a natural part of the environment (figure below). It is formed in the upper atmosphere as a result of cosmic ray interactions. (Cosmic rays are particles from outer space, mostly protons.) The carbon-14 enters the biosphere, and all living things — plants and animals — obtain a certain amount of carbon-14.

A radioactive isotope behaves chemically just like a stable isotope, and part of the carbon in your body is radioactive carbon-14. The activity of carbon-14 in a living organism remains fairly constant because of life processes. In each gram of carbon in a body, there is enough carbon-14 so that there are about 16 beta emissions per minute (per gram of carbon). Recall from Figure 25.10 that carbon-14 beta-decays. However, when the organism dies and the carbon-14 is not replenished by the intake in life processes, the activity decreases.

Suppose an old bone found in an archaeological "dig" has a carbon-14 activity of 4 beta emissions per minute per gram of carbon. When the bone was a part of a living organism, there were 16 beta emissions per minute. So, counting backward in time ($4 \rightarrow 8 \rightarrow 16$),

we know that the carbon-14 in the bone has gone through two half-lives. Carbon-14 has a half-life ($t_{1/2}$) of 5730 years (Table 25.2), and $2t_{1/2} = 2(5730) = 11,460$ years. Hence, the bone is about 11,500 years old.

Carbon-14 dating can be used to date objects up to about 30,000 years old. Other radioactive isotopes with longer half-lives can be used to date older objects. For example, ^{48}K and ^{87}Rb are useful in determining the age of geological rock formations.

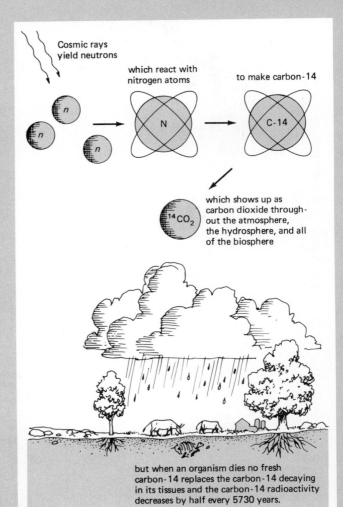

Cosmic rays yield neutrons

which react with nitrogen atoms

to make carbon-14

N

C-14

$^{14}CO_2$ which shows up as carbon dioxide throughout the atmosphere, the hydrosphere, and all of the biosphere

but when an organism dies no fresh carbon-14 replaces the carbon-14 decaying in its tissues and the carbon-14 radioactivity decreases by half every 5730 years.

Carbon-14 in the environment. Formed in the upper atmosphere, the radioactive carbon-14 eventually finds its way into the biosphere.

With the development of semiconducting materials, "**solid-state**" **detectors** are now available that can measure the energy of radiation very precisely and are capable of very high counting rates—very "fast" detectors (very small dead times). Charged particles passing through a semiconductor produce electron-hole pairs (solid-state "ionization"). With an applied voltage, these electron-hole pairs give rise to electrical pulses that can be amplified and counted.

Other types of detectors include cloud and bubble chambers. These are used primarily in research applications. Vapors and liquids are used in the respective chambers. When the temperature and pressure are properly adjusted, vapor condenses on the ionized molecules along the path of an energetic particle, leaving a trail of liquid droplets in a cloud chamber, and boiling or bubble formation is enhanced in the region of charged particles, leaving a trail of bubbles in a bubble chamber.

Photographs may be taken of the trails in these chambers, which provide a permanent record of the particle's path and the "events" that take place along the way (Fig. 25.16). When a magnetic field is applied across a chamber, electrically charged particles are deflected, and their energies can be calculated from the radius of curvature of the particle's path, as measured from the photograph. Nuclear research has found a variety of particles in the nucleus, called elementary particles. More about these shortly.

■ APPLICATIONS OF RADIOACTIVITY

Since the discovery of radioactivity less than 100 years ago, nuclear decay has been used in a variety of applications. Let's consider a few medical and industrial applications here and the application to the production of nuclear energy in the next chapter.

The fact that radioactive isotopes behave chemically just like nonradioactive isotopes of the same element makes them useful as radioactive tracers. For example, an atom with a carbon-14 nucleus will react to form a molecule of carbon dioxide (CO_2) the same way as a nonradioactive atom of carbon-12 would.

Radioactively "tagged" molecules can be traced in various processes by detecting the nuclear radiations from these molecules. For example, the location of an underground pipe or a clog can be determined by using a liquid radioactive tracer, and research in the wearing of machine parts can be carried out by tracing processes (Fig. 25.17).

Similarly, in medicine radioactive tracers allow the monitoring of body functions. For example, the iodine uptake of the thyroid gland can be measured by using radioactive iodine-131. The minimum daily requirement of iodine is 150 μg. If a patient is given some radioactive iodine, the thyroid gland doesn't know the difference, and a scan of the gland shows how the radioactive iodine is distributed and concentrated in the gland (Fig. 25.17).*

The distribution is altered if there is some abnormality in the gland function. Various other internal parts of the body can be "observed" using radioactive tracers. A detector is used to scan the area in question. The information is stored in a computer and later put together as a scan "picture."

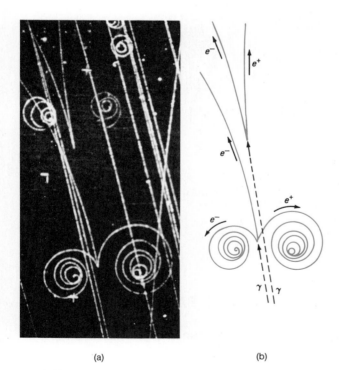

(a)   (b)

Figure 25.16
Bubble chamber photograph and diagram. (a) Particles leave trails of bubbles in their wake. (b) A magnetic field causes the opposite deflection of oppositely charged particles.

* Iodine-131 beta-decays into xenon-131, which emits gamma rays. It is the gamma radiation that is monitored in a thyroid scan.

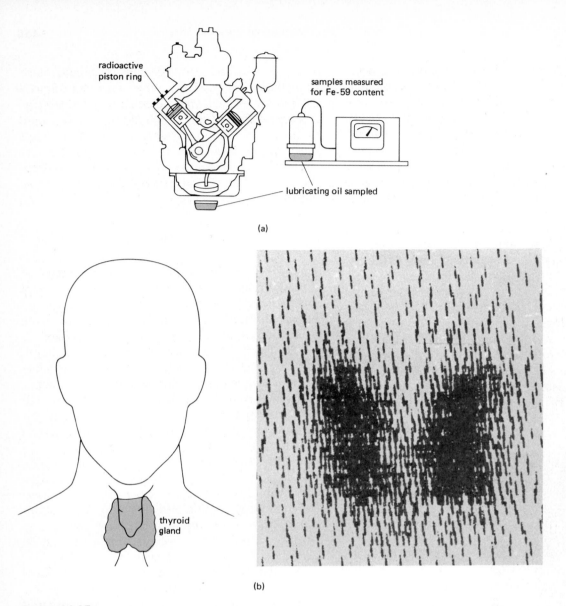

Figure 26.17
Radioactive tracers. (a) Used to test mechanical wear. (b) Thyroid scan showing distribution of radioactive iodine in a thyroid gland 24 hours after being administered. The distribution is normal, indicating that the gland is "picking up" iodine normally.

The penetration of radiation depends on the density of a particular material and is inversely proportional to the thickness of the material—the greater the thickness, the less penetration. This follows a given relationship, so radiation absorption can be used in thickness measurement and monitoring, for example, in metal sheet and foil manufacturing (Fig. 25.18). Another common application is discussed in Special Feature 25.3.

ELEMENTARY PARTICLES

The search for the ultimate or *fundamental* "building blocks" of nature is as old as physics itself. The simple picture of an indivisible atom gave way to an atom with subatomic particles. By the 1930's scientists had identified four of these: the electron, the proton, the neutron, and the photon. Since then, more than 200 various particles have been discov-

(a)

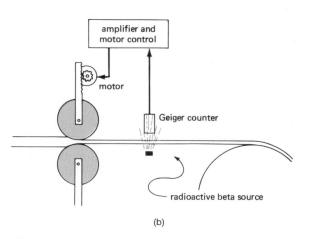

(b)

Figure 25.18
Thickness measurement. The penetration of radiation depends on the density and thickness of a material. Radioactive particles can be used in thickness monitoring and control, as shown here for a metal milling process.

ered coming from the nucleus. They are called "elementary" particles, elementary because we are not certain which of them are really fundamental. You may have heard of some of these — neutrinos, positrons, muons, mesons, pions, lambda particles, and others.

It is believed that most elementary particles do not exist outside the nucleus but come into being when the nucleus is disrupted. Some elementary particles are thought to be the "exchange" particles responsible for the four fundamental forces or interactions. For example, in one theory the "exchange" particle for the strong nuclear interaction is the pi meson or pion. Nuclear particles with a strong interaction are viewed as interacting by exchanging

pions back and forth (the way you interact with someone by exchanging a beach ball back and forth). All strongly interacting particles, which include the nuclear protons and neutrons, are called **hadrons**.

There have been efforts to simplify matters. In 1964 Murray Gell-Mann at the California Institute of Technology and George Zweig in Switzerland independently suggested that all hadrons, the elementary particles with the strong interaction, were made up of no more than three subparticles with fractional electronic charges, called **quarks.*** For example, a proton is thought to be made up of three quarks (two with a $+\frac{2}{3}$ electronic charge and one with a $-\frac{1}{3}$ electronic charge).

Quarks apparently do not exist as free particles, but are permanently bound inside hadrons, so we never see fractional electronic charges in nature. There is, however, indirect experimental evidence for their existence. The theoretical exchange particle between quarks is the **gluon** — the "glue" that holds the world together.

At first there were only two types, or "flavors," of quarks, the u and d (for up and down, associated with spins), and their antiparticles or antiquarks. Then a new quark flavor, the strange quark, denoted by the letter s, joined the club. The discovery of a new elementary particle in 1974 added the necessity of a fourth flavor, a "charmed" or c quark.

Then, to make the physics agree with experimental observations, it was assumed that each quark comes in three different "editions," which are called colors, for example, red, green, and blue u and d quarks. In the exchange of gluons between quarks, it seems that a gluon can change the color of a quark, or is "color-carrying."

The story does not stop here. The discovery of still another elementary particle in 1978 gave evidence for a new flavor, the b quark. To round out the quark symmetry (u-d, s-c), the b quark is associated with a t quark. There was some effort to call the b-t quark pair *b*eauty and *t*ruth, but this seems to have lost out to *b*ottom and *t*op. So we might say that quarks come in six flavors, each of which can have three colors.

---

* From a line in James Joyce's novel *Finnegans Wake,* which reads, "Three quarks for Muster Mark!" The "three quarks" denote the children of Mr. Finn, who sometimes appears as Mister (Muster) Mark. A bit strange, but there are also "strange" quarks. Read on.

The Smoke Alarm

Smoke alarms are common in many homes. The early warning of smoke is important in personal and fire protection. Smoke may be detected by several means. A photoelectric type of smoke detector is described in Chapter 24. However, more common is the ionization type of smoke detector, which uses radioactivity (figure below).

A weak radioactive source ionizes the air, setting up a small current in a detector circuit. If smoke is present, the ions become attached to the smoke particles. The slower movement of the heavier, charged smoke particles causes a reduction in the detector current. This is electronically sensed and causes an alarm to sound.

Such smoke detectors are quite sensitive. Smoke vapors from cooking (from burnt foods) often set them off.

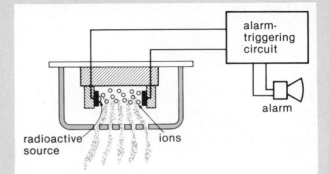

Smoke detection. In an ionization-type smoke detector, a weak radioactive source ionizes the air and sets up a small current. Smoke affects the current and causes an alarm to sound.

Thus, according to this current popular theory, quarks are "glued" together by the strong force. But, they are also subject to the weak force, which changes the "flavor" or kind of quark and makes it unstable so the nucleus decays radioactively as in beta decay. The exchange particles for the weak force are the **W and Z particles**. The latter was postulated to be related to the **photon**, which is the exchange particle for the electromagnetic force. In 1983 the W and Z particles were discovered, which gave evidence that the electromagnetic and weak forces are merely two components of a single **electroweak force**. [Recall how Maxwell effectively combined the electric and magnetic forces into a single electromagnetic force (Chapter 19).] To round out the picture, the exchange particle for the gravitational force is called the **graviton**. However, this elusive particle has not been detected experimentally to date.

There is now a theory that merges the electroweak force with the strong gluon force in a "grand unified force." This is called the **grand unified theory (GUT)**. Of course, scientists would like to believe that all forces are manifestations of a single **superforce**. This idea is called a unified field theory. Einstein suggested such a theory, but at the time only gravity and the electromagnetic force were definitely identified. A single superforce would be nice, but there is a lot of work to be done to find out if this is the case.

The preceding discussion, particularly of quarks, was not meant to be facetious, but rather to simply describe the enormous difficulties in trying to understand the intricacies of nature and the enormous strides scientists have made. Scientists apply the scientific method, and nature will be the final judge of their theories and understanding.

SUMMARY OF KEY TERMS

Atomic nucleus: the central core of the atom in which the protons and neutrons of an atom are located.

Nucleon: a nuclear proton or neutron.

Nuclide: a particular nucleus or nuclear species.

Proton number: the number of protons in a nucleus, which defines its atoms as being a particular element.

Mass number: the sum of the protons and neutrons in a nucleus.

Isotopes: nuclei or nuclides of the same element with different numbers of neutrons.

Nuclear force: the strong attractive interaction that acts between nucleons.

Radioactivity: the spontaneous decay of certain isotopes with the emission of energetic particles.

Alpha particle: a particle consisting of two protons and two neutrons, which is the same as a helium nucleus.

Beta particle: an electron.

Gamma particle: a quantum or photon of energy.

Half-life: the time it takes for one half of the nuclei of a sample of a given radioactive isotope to decay.

Geiger counter: a common radiation detector based on the ionizing nature of radiation.

Dead time: the time required for a detector to recover for another detection or count.

Scintillation counter: a radiation detector based on the ability of a phosphor material to convert the energy of a particle into visible light.

Elementary particles: the *fundamental* building blocks of nature, i.e., the basic or "elementary" particles that make up matter and account for the interactions thereof.

Exchange particles: the elementary particles responsible for fundamental forces or interactions. These are as follows: for the strong force, the **gluon**; for the weak force, the **W and Z particles**; for the electromagnetic force, the **photon**; and for gravity, the **graviton**.

Hadron: any particle that interacts by the strong force.

Quarks: subparticles with fractional electronic charges thought to make up hadrons.

Electroweak force: the single force believed to incorporate both the electromagnetic and weak forces.

Grand unified theory (GUT): a theory that combines the strong and electroweak forces into a single force.

Superforce: a single force that would describe all fundamental interactions in a unified field theory.

QUESTIONS

Nuclear Notation and Isotopes

1. Given only the proton numbers or neutron numbers or mass numbers of several nuclei, would you *always* be able to distinguish between different nuclides? Explain.

•2. (a) Write the nuclear notation for the isotopes of hydrogen using the symbols H, D, and T. (b) Using H, D, and T symbols, it is possible to write six chemical formulas for water. Write these, and identify which three would be radioactive forms of water.

3. Tin ($_{50}$Sn) has 23 isotopes. Write the complete nuclear symbol (with neutron number) for each of the isotopes of tin with mass numbers from 114 to 120. (The complete family of isotopes has mass numbers from 108 to 130.)

4. Discuss how the gravitational and electrical forces might contribute to the stability or instability of an atom, taking into account nucleons *and* orbiting electrons.

5. The diameter of a uranium atom is the same as that of a hydrogen atom. Discuss what this might imply about the structure of these atoms.

6. Considering the following nuclides, show that the neutron number generally exceeds the proton number for nuclei with proton numbers greater than 20: lithium-6, bromine-80, silicon-28, titanium-48, fluorine-19, and platinum-179. (*Hint:* See Tables 7.1 or 7.2.)

7. Write the complete nuclear symbol for the heaviest (most massive) stable nucleus. Specu-late why heavier nuclei are not stable in terms of the forces acting in the nucleus. (*Hint:* Think about the range of forces.)

Radioactivity

8. Speculate about why the Curies named one of the new elements they discovered in 1898 "radium." (If you're wondering about polonium, it was named after Madame Curie's native country. See Special Feature 25.1.)

•9. Suppose an electric field were used in Figure 25.8 instead of a magnetic field to distinguish between nuclear radiations. What would be observed?

10. Write the nuclear equations for (a) the alpha decay of $^{226}_{88}$Ra, (b) the beta decay of $^{60}_{27}$Co, and (c) the gamma decay of $^{210}_{84}$Po.

11. Write the nuclear decay equations that show how gamma rays are used to monitor the thyroid gland in iodine-131 uptake. (*Note:* Iodine-131 beta-decays.)

12. A radioactive decay series, called the neptunium series, is shown in Figure 25.19. None of its members has a half-life comparable to the age of the Earth, so the nuclides in this series do not occur naturally. However, they can be made artificially. Identify the members of this decay series.

13. Protactinium-233 ($^{233}_{91}$Pa) undergoes beta decay. (a) What is the daughter nucleus in this process?

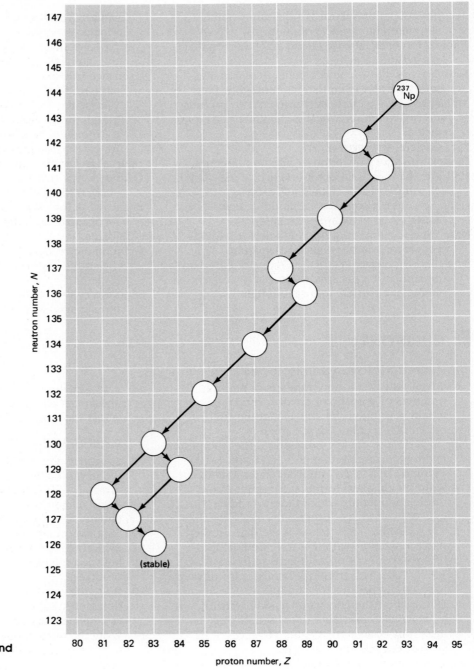

Figure 25.19
Where does neptunium-237 end up? See Question 12.

Write the equation for it. (b) The daughter nucleus then undergoes alpha decay. What is the "granddaughter" nucleus? Write the equation for this decay process.

*14. Radon-222 ($^{222}_{86}$Rn), a radioactive inert gas, undergoes alpha decay. (a) What is the daughter nucleus of this process? Write the equation for the decay process. (b) The daughter nucleus undergoes both beta decay and alpha decay.

What is the "granddaughter" nucleus in each case? Write the equations for the decay processes.

15. Complete the following decay reactions:

a. $^{64}_{29}$Cu → $^{64}_{30}$Zn + ___

b. $^{190}_{78}$Pt → $^{186}_{76}$Os + ___

c. $^{123}_{52}$Te → $^{123}_{52}$Te + ___

d. $^{3}_{1}$H → $^{3}_{2}$He + ___

Decay Rate and Half-Life

16. Why can't the half-life of a radioactive sample be determined by weighing the sample after periods of time?

17. Bismuth-209 ($^{209}_{83}$Bi) is said to be the heaviest stable nucleus. Yet, this nuclide alpha-decays with a half-life greater than or equal to 2×10^{18} years. Why is it justified to say that this bismuth isotope is "stable"?

18. Some short-lived radioisotopes are found in nature. Why? Shouldn't these have decayed a long time ago?

•19. How many becquerels are there in 1 Ci?

20. Why can't carbon dating be used accurately for ages of 25,000 to 30,000 years and older?

21. The technique of carbon dating relies on the assumption that the cosmic ray intensity has been constant for the last 25,000 to 30,000 years. Suppose we found out that the cosmic ray intensity was much greater 10,000 years ago. How would this affect the ages of samples that have been dated by carbon-14 dating? Explain.

Elementary Particles

22. What are the names of the particles believed to "carry" the fundamental interactions (forces) between objects?

23. What are the elementary particles believed to make up larger particles such as protons and neutrons? Are these particles electrically neutral?

24. How many quark flavors and colors are there? What is the total number of quarks accounted for by this scheme? (Don't forget antiquarks.) Can you name all of them?

EXERCISES (See Table 25.2 for needed half-lives.)

1. A radioactive isotope has a half-life of 1 h. If a sample contains 100 μg of this isotope, how much is left after (a) 1 h, (b) 2 h, and (c) 3 h?

   Answers: (a) 50 μg, (b) 25 μg, and (c) 12.5 μg

2. A sample of cobalt-60 has an activity of 120 μCi. How long will it take for the activity of the sample to decrease to 15 μCi?

   Answer: 15.9 y

3. If the bubbles in Figure 25.20 "decay" like radioactive nuclei, (a) what is the half-life of this soap-bubble "isotope"? (b) How many bubbles would be present when the clock reads 6 s?

   Answers: (a) 2.0 s, (b) 3

4. A Geiger counter near a radioactive sample registers 1600 cps (counts per second). Twelve hours later, the count rate is 200 cps. What is the half-life of the radioactive nuclide?

   Answer: 4 h

5. A particular radioactive nuclide has a half-life of 10 min. By what percentage would the activity of a sample of this isotope decrease in 1 h?

   Answer: 98.4 percent

6. A patient in a hospital is given 100 μg of iodine-131 for a thyroid test. Assuming that none is lost through body functions, how much of the iodine remains after one month?

   Answer: about 6.3 μg

7. A petrified wood carving is found in an archaeological dig. It is determined that there is an average of two beta emissions per gram of carbon per minute coming from a sample of the carving. Approximately how old is the wood carving?

   Answer: 17,000 years

8. If a large animal died today, in how many years would there be an average of four beta emissions/g of C/min from its bones?

   Answer: in about 11,500 years

Figure 25.20
Bubble half-life. See Exercise 3.

Nuclear Energy: Fission and Fusion

Three Mile Island near
Middletown, Pennsylvania.

THE BACKGROUND OF THE "NUCLEAR AGE"

The dropping of the "atomic" bomb on Hiroshima, Japan, in 1945 was the devastating beginning of what is sometimes called the atomic or, more correctly, the nuclear age. It was a public announcement of the awesome power that could be released from the nucleus. Prior to this, the secret of "splitting the atom" was known to only a handful of scientists. The scenario began in the first part of this century, not long after the discovery of radioactivity and the introduction of the nuclear model of the atom. If a radioactive nucleus can decay into the nucleus of another element with the emission of a particle, how about the reverse reaction — putting a particle into a nucleus and getting a nucleus of another element? "Reverse radioactivity," so to speak.

Scientists quickly thought of this and showed that it could be done. Attempts were then made to make new, heavy elements. Uranium, the heaviest element known at the time, was bombarded with particles — in particular, neutrons — to do this. One of the unexpected findings was that the uranium nucleus sometimes ended up in fragments or broken ("split") apart.

In 1938 the German chemists Otto Hahn and Fritz Strassman identified some uranium nucleus fragments as the nuclei of lighter elements. Hahn related this discovery to Lise Meitner, a longtime coworker, who was forced to flee Nazi Germany

Figure 26.1
Lise Meitner and Otto Hahn. The pair collaborated in scientific research for 30 years. In 1938 Meitner was forced to leave Nazi Germany. She was in Sweden when she published the first report on nuclear fission with her nephew Otto Frisch.

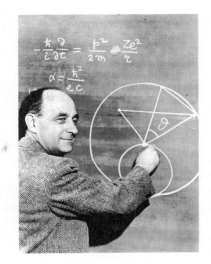
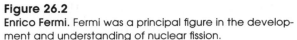

Figure 26.2
Enrico Fermi. Fermi was a principal figure in the development and understanding of nuclear fission.

because she was "non-Aryan" (Fig. 26.1). Meitner and her nephew, Otto Frisch, worked out the theory of this reaction in Sweden, calling it "fission" in analogy with biological fission or division of a living cell into two parts. In addition to the emission of neutrons from the fission reaction, there was a release of energy.

They sent their results to a scientific journal in Denmark, which you may recall was the native country of Niels Bohr (Chapter 24). Bohr carried the news of the discovery of fission with him to a scientific meeting in Washington, D.C. There, scientists realized the potential of nuclear fission in the production of energy. Albert Einstein, who now lived in America (having also fled Germany), wrote a now-famous letter in 1939 to President Roosevelt to make him aware of the potential of nuclear fission. Subsequently, the Manhattan Project was initiated, which resulted in the development of the "atomic" bomb.

The first fission reactions had actually been produced in 1934 by Enrico Fermi and his coworkers in Italy when they bombarded many elements with neutrons (Fig. 26.2). However, the reaction was unrecognized at the time. Even so, Fermi won the 1938 Nobel Prize in physics for his work in creating artificial isotopes by neutron bombardment. The Italian dictator Mussolini allowed Fermi and his family to go to Stockholm for the presentation ceremony, but Fermi didn't return. He came to the United States, where he was a principal figure in the Manhattan Project.

Although first used for military applications, nuclear energy has peaceful uses. One of the major ones is the production of nuclear energy in reactors for the generation of electricity. This has been going on since the late 1950's. However, some 20 years later in 1979, an "incident" at Three Mile Island (see opening photograph) dramatically brought to the public attention the potential dangers associated with nuclear reactors. In more recent memory is the nuclear "accident" that occurred at Chernobyl (U.S.S.R) in 1986. In this case, there was a major release of radioactive material into the environment, and deaths resulted. (See Special Feature 26.1.)

Now, well into the "nuclear age," we hear such terms as *meltdown, radioactive wastes, nuclear warheads,* and *nuclear proliferation* and debates between "nuke" and "antinuke" groups. One of the main purposes of this chapter is to make you aware of the facts about nuclear energy so that you can better understand the issues in the nuclear debate. You may be called on to make a voting decision on one or more of these issues someday.

NUCLEAR REACTIONS

Thus far, we have considered only nuclear decay reactions in which a radioactive nucleus spontaneously changes or transmutes into the nucleus of another element with the emission of an alpha or beta particle. However, it is possible to induce a nuclear reaction by "adding" particles to the nucleus, which creates different nuclides.

Lord Rutherford produced the first induced nuclear reaction in 1919 by bombarding nitrogen (N-14) gas with alpha particles from a natural bismuth-214 source (Fig. 26.3). The particles coming from the reactions were identified as protons. Rutherford reasoned that an alpha particle colliding with a nitrogen nucleus can occasionally knock out a proton. This results in the *artificial* transmutation or disintegration of a nitrogen nucleus to an oxygen nucleus:.

$$\underset{\text{Alpha particle}}{{}^{4}_{2}\text{He}} + \underset{\text{Nitrogen}}{{}^{14}_{7}\text{N}} \longrightarrow \underset{\text{Oxygen}}{{}^{17}_{8}\text{O}} + \underset{\text{Proton}}{{}^{1}_{1}\text{H}}$$

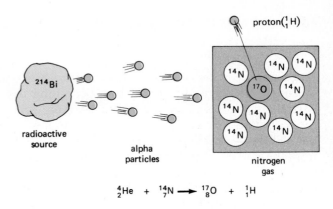

$${}^{4}_{2}\text{He} + {}^{14}_{7}\text{N} \longrightarrow {}^{17}_{8}\text{O} + {}^{1}_{1}\text{H}$$

Figure 26.3 .
Rutherford's reaction. Alpha particles from a radioactive source cause nuclear reactions that change nitrogen to oxygen.

This reaction was discovered almost by accident because it took place so infrequently in Rutherford's experiment. One proton is produced for about every 1 million alpha particles that pass through the gas. Notice that the conservation of nucleons and the conservation of charge hold in this and all nuclear reactions. That is, the sums of the mass numbers and proton numbers (symbol superscripts and subscripts) on the opposite sides of the equation are equal.

But think of the implications of this reaction. By use of such bombardment processes, one element can be changed into another. In a sense, it is the old dream of the alchemists come true. Of course, the alchemists' main concern was the changing of common elements into gold. This can be done too with nuclear reactions, for example,

$${}^{2}_{1}\text{H} + {}^{196}_{78}\text{Pt} \longrightarrow {}^{197}_{79}\text{Au} + {}^{1}_{0}n$$

$$\gamma + {}^{198}_{80}\text{Hg} \longrightarrow {}^{197}_{79}\text{Au} + {}^{1}_{1}\text{H}$$

Gold (Au)-197 is the only stable isotope of gold. However, making gold by this process is not too economical. It costs more to produce the gold than the gold is worth.

The neutrons produced in nuclear reactions with charged particles (protons, deuterons, or alpha particles) can in turn be used to induce nuclear reactions to make artificial radioactive isotopes for industrial and medical applications, for example,

$$\underset{\text{Neutron}}{{}^{1}_{0}n} + \underset{\text{Aluminum}}{{}^{27}_{13}\text{Al}} \longrightarrow \underset{\text{Magnesium}}{{}^{27}_{12}\text{Mg}} + \underset{\text{Proton}}{{}^{1}_{1}\text{H}}$$

Table 26.1
Artificial Transuranic Elements

Symbol	Name	Named for	Year
$_{93}$Np	Neptunium	The planet Neptune	1940
$_{94}$Pu	Plutonium	The planet Pluto	1940
$_{95}$Am	Americium	America	1944
$_{96}$Cm	Curium	Marie and Pierre Curie	1944
$_{97}$Bk	Berkelium	Berkeley, California	1949
$_{98}$Cf	Californium	California	1950
$_{99}$Es	Einsteinium	Albert Einstein	1952
$_{100}$Fm	Fermium	Enrico Fermi	1953
$_{101}$Md	Mendelevium	Dmitri Mendeleev	1955
$_{102}$No	Nobelium	Alfred Nobel	1957
$_{103}$Lw	Lawrencium	Ernest Lawrence	1961
$_{104}$Rf	Rutherfordium	Ernest Rutherford*	1964
$_{105}$Ha	Hahnium	Otto Hahn	1970

* Called Kurchatovium (Ku) by Russian scientists and named after Igor Kurchatov, who independently created the element in Russia at about the same time as it was created in America.

Magnesium-27 is radioactive, and it beta-decays with a half-life of 9.5 minutes.

Neutrons are especially effective as nuclear "bullets" in initiating reactions because they have no electric charge. They are not subject to the repulsive electrostatic forces experienced between positively charged "bullets" and the nucleus and are more likely to penetrate nuclei than protons or alpha particles.

Examples of other important neutron-induced reactions are

$$_0^1n + {}_{92}^{238}U \longrightarrow ({}_{92}^{239}U)* \longrightarrow {}_{93}^{239}Np + {}_{-1}^{0}e$$

$$_0^1n + {}_{92}^{235}U \longrightarrow ({}_{92}^{236}U)* \longrightarrow {}_{56}^{141}Ba + {}_{36}^{92}Kr + 3({}_0^1n)$$

In the first reaction, a *transuranic* (above uranium) element, neptunium, is made. There are no naturally occurring elements above uranium. Neptunium beta-decays into plutonium. Some artificial transuranic elements are listed in Table 26.1. The second reaction is an example of the dividing ("splitting") or fissioning of a large nucleus. More about this shortly.

* These intermediate, unstable nuclei are called "compound" nuclei and quickly decay. They are commonly omitted from the nuclear reaction equations.

The early study of nuclear reactions was limited by the kind of particle "bullet" that could be used to bombard nuclei. Only alpha particles from natural radioactive sources could bring about nuclear reactions, and these particles were low in energy and intensity. The nucleus is a "hard nut to crack," and a particle must have enough energy to penetrate it. When heavier elements are bombarded with alpha particles, the greater repulsive electric force from the greater charge in the heavier nuclei makes it difficult for alpha particles to penetrate the nuclei. There is an advantage to using protons or deuterons, since the repulsive force on them is not as great as that on an alpha particle.

Since the 1930's many devices for accelerating charged particles have been invented. In all of these, charged particles are accelerated by electric fields, and the energetic beams can be guided or steered by magnetic fields. Particle accelerators are used to produce radioactive isotopes for military, medical, and industrial purposes, but very large accelerators are used mainly to probe the nucleus in elementary particle research (Fig. 26.4).

The use of electric fields or voltage differences to accelerate charged particles in accelerators has brought about a commonly used unit of energy in nuclear physics—the electron volt (eV). An **elec-**

(a)

(b)

Figure 26.4
Accelerators. (a) An early cyclotron with inventors Ernest
O. Lawrence *(left)* and M. S. Livingston. (b) Fermi National
Accelerator Laboratory near Batavia, Illinois.

tron volt is simply the amount of energy received by an electron when accelerated through a potential of 1 V. (Recall that voltage is the work or energy per charge, so eV is a unit of energy.)

The electron volt is a small unit (1 eV = 1.6 × 10^{-19} J). Accelerator energies are expressed in MeV (M, mega-, 10^6 or million electron volts) and GeV (G, giga-, 10^9 or billion electron volts).

■ ENDOERGIC AND EXOERGIC REACTIONS

If you compare the sum of the masses of the reactants of a nuclear reaction with the sum of the masses of the reaction products, you might be surprised to find that they are not the same. For example, in Rutherford's original experiment of bombarding nitrogen nuclei with alpha particles, there is more mass at the end than was started with (Fig. 26.5). This might seem strange, but remember that according to Einstein, mass is a form of energy ($E = mc^2$, Chapter 23). So if the total mass increases, the increase must come from energy—the energetic alpha particle has energy, some of which is converted to mass in the reaction.

This type of reaction is called an **endoergic reaction.** That is, energy (−*ergic*) must be put into (*endo*−) the reaction for it to occur or proceed. This is analogous to an endothermic chemical reaction in which heat (energy) must be added for the reaction to proceed, or more energy is put in than is obtained. (A chemical reaction involves the atomic electrons, whereas a nuclear reaction involves the nucleus of an atom.)

However, some chemical reactions are *exo*thermic. Here, you get more (heat) energy out than you put in, for example, in combustion such as burning logs in a fireplace. Similarly, for nuclear reactions there are **exoergic reactions.** An example is the fission reaction shown previously (Fig. 26.6). Here, there is less total mass after the reaction than before, on the order of 200 MeV mass equivalents less. Hence, we have a *source* of energy—nuclear energy.

Scientists were quick to realize this, as evidenced by Einstein's letter to President Roosevelt in 1939 (mentioned previously). A result was the awesome releases of energy over Hiroshima and Nagasaki, Japan, on August 6 and 9, 1945, and the world has never been the same. We had entered the "nuclear age."

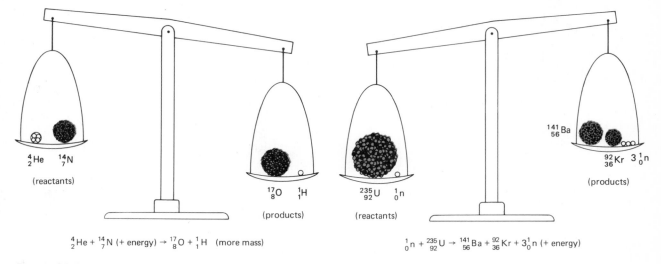

$$^{4}_{2}\text{He} + ^{14}_{7}\text{N} \;(+ \text{energy}) \rightarrow ^{17}_{8}\text{O} + ^{1}_{1}\text{H} \quad (\text{more mass})$$

Figure 26.5
Endoergic reaction. The total mass of the products of the reaction is greater than that of the reactants. Energy is converted to mass.

$$^{1}_{0}\text{n} + ^{235}_{92}\text{U} \rightarrow ^{141}_{56}\text{Ba} + ^{92}_{36}\text{Kr} + 3^{1}_{0}\text{n} \;(+ \text{energy})$$

Figure 26.6
Exoergic reaction. The total mass of the products is less than that of the reactants. Mass is converted to energy.

BINDING ENERGY

An indication of two methods for the release of nuclear energy may be obtained by looking at the binding energy of nuclei. The binding energy is the energy that would be needed to separate a nucleus into free protons and neutrons, i.e., how much work or energy you'd have to do to tear a nucleus completely apart, so to speak. The greater the binding energy of a nucleus, the more tightly bound are its nucleons, and, in general, the more stable the nucleus.

The binding energies of nuclides can be computed, and to get an average value, we divide the (total) binding energy (BE) of the nucleus by the number of nucleons it has so as to get the binding energy per nucleon. For example, a helium nucleus (or alpha particle) has a total binding energy of 28.3 MeV. In an illustrative equation we might write

$$\overset{pp}{\underset{nn}{\bigcirc}} + 28.3 \text{ MeV} \longrightarrow \textcircled{p} + \textcircled{p} + \textcircled{n} + \textcircled{n}$$
(mass + energy) (greater total mass)

By the conservation of mass-energy, the total mass of the individual nucleons must be greater than the mass of the nucleus itself. Then, the binding energy per nucleon is 28.3 MeV/4 = 7.08 MeV, since there are four nucleons. When the BE/nucleon for the various nuclei are plotted versus mass number, a curve as shown in Figure 26.7 is obtained.

Notice that the curve peaks or has a maximum in the vicinity of iron (Fe). Having a very large BE/nucleon, iron is one of the most stable nuclei. The curve also tells us something else. If it takes energy to separate nuclei into smaller particles or nucleons, how about going the opposite way? That is, putting lighter (less massive) particles together would release energy. This is in effect like turning the arrow around in the preceding equation and converting mass into energy. The same would be true if we combined two light nuclei into a single nucleus (with a resulting release of energy). This process is called fusion, and the resulting nucleus is higher on the BE curve (Fig. 26.7).

Notice that we can also get higher on the BE curve on the other side of the maximum by taking a large nucleus and splitting or breaking it apart so as to have two or more lighter nuclei. This process is called fission, and it too results in a conversion of mass to energy. Let's take a closer look at these two major methods for the release of nuclear energy.

FISSION

The exoergic fission reaction forms the basis for the current production of nuclear energy. In a **fission reaction**, a heavy nucleus divides into two intermediate lighter nuclei with the emission of two or more energetic neutrons (Fig. 26.8). Energy is

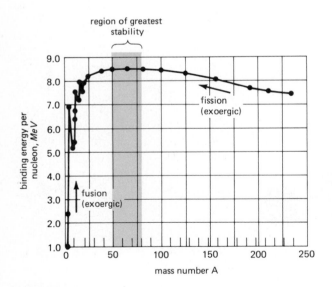

Figure 26.7
Binding energy per nucleon versus mass number. The curve has a maximum in the vicinity of iron, which is one of the most stable nuclei. Energy is released in fission and fusion reactions, which result in nuclei with greater binding energies.

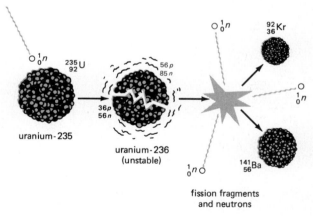

Figure 26.8
Fission. An incident neutron causes a uranium-235 nucleus to divide into two intermediate lighter nuclei with the emission of two or more energetic neutrons.

emitted in the process, being carried off by the fission fragments and neutrons.

Only certain nuclides undergo fission, and the probability of the reaction occurring for a particular nuclide depends on the energy of the incident neutrons. For example, neutrons of any energy will cause U-235 and Pu-239 to undergo fission, but a reaction is more probable for neutrons with low energies of less than 1 eV. We say that these nuclei have a greater cross-section for "slow" neutrons. (Cross-section is a measure of the probability for a reaction to occur.) Thorium-232, on the other hand, prefers "fast" neutrons with energies of 1 MeV or more.

A fission reaction can proceed in a variety of ways. For example, some fission modes other than those given previously are

$${}^{1}_{0}n + {}^{235}_{92}U \longrightarrow {}^{140}_{54}Xe + {}^{94}_{38}Sr + 2({}^{1}_{0}n) + \text{Energy}$$

$${}^{1}_{0}n + {}^{235}_{92}U \longrightarrow {}^{132}_{50}Sn + {}^{101}_{42}Mo + 3({}^{1}_{0}n) + \text{Energy}$$

Each fission reaction produces about 200 MeV of energy. MeV (*million* electron volts) may sound like a large amount of energy, but it is not really. Recall that $1\ \text{eV} = 1.6 \times 10^{-19}$ J, so 200 MeV is only 3.2×10^{-11} J of energy, which isn't much. When you pick your physics book up from a table, you do about 5 J of work or expend 5 J of energy. However, keep in mind that there are billions of nuclei in a small sample of fissionable material. The trick is to get enough fission reactions to produce practical amounts of energy.

This is accomplished by a **chain reaction** (Fig. 26.9). The neutrons released in a fission reaction induce other fission reactions, and so on, so that a growing "chain" reaction takes place. Energy is released with each fission reaction. The chain reaction proceeds very quickly, and there is a tremendous release of energy from the millions of nuclei undergoing fission.

For a chain reaction to occur, a sufficient amount of the fissionable material must be present. Otherwise, neutrons would escape from the sample before reacting with a nucleus, and the chain reaction would not proceed. The chain would be "broken," so to speak. Scientists call the amount of fissionable material necessary to sustain a chain reaction a **critical mass**.

Natural uranium is composed of 99.3 percent U-238 and only 0.7 percent of the fissionable U-235. To have more fissionable U-235 nuclei present in a

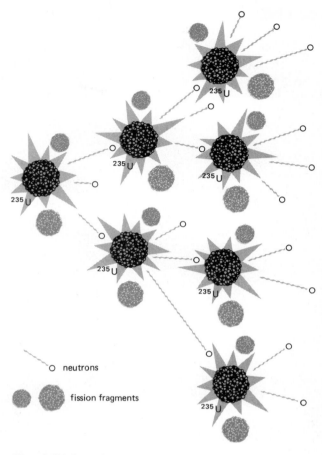

Figure 26.9
A fission chain reaction. Neutrons from one nucleus having undergone fission produce other fissions in a growing chain reaction.

sample, the U-235 is concentrated or "enriched." Weapons-grade material may be enriched to 99 percent or more, whereas the enriched uranium used in nuclear reactors for the production of electricity in only 3 to 5 percent U-235.

In a bomb, a critical mass of highly enriched material must be held together for a short time to get an explosive release of energy (Fig. 26.10). Segments of the fissionable material in an "atomic" bomb or "nuclear device" (modern terminology) are separated before explosion, so there is not a critical mass for the chain reaction. A chemical explosive is used to bring the segments together, and the explosive force on interlocking configurations holds them together long enough for a large fraction of the material to undergo fission. The result is an

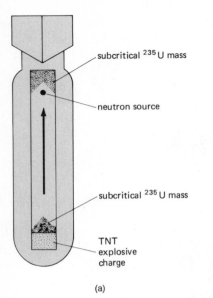

subcritical ^{235}U mass

neutron source

subcritical ^{235}U mass

TNT
explosive
charge

(a)

(b)

(c)

Figure 26.10
Uncontrolled nuclear fission. (a) In an "atomic" bomb, a critical mass is obtained by bringing two masses together by a small explosive charge. (b) An atomic bomb of the "Little Boy" type detonated over Hiroshima, Japan, during World War II. The bomb is 28 in. in diameter and 120 in. long, and has an equivalent yield of 20,000 tons of TNT. (c) An underwater nuclear detonation. Notice the ships in the foreground.

explosive release of energy. A bomb is an example of *uncontrolled* fission, which we hope will never be used again. A nuclear reactor is an example of *controlled* fission, where we control the growth of the chain reaction and the release of energy.

Fission Reactors

The basic design of a fission nuclear reactor is shown in Figure 26.11. Fuel pellets in tubes form the fuel rods in the reactor core. The chain reaction

and energy output are controlled by means of boron or cadmium **control rods**. These materials absorb neutrons. Hence, by insertion of the control rods between the fuel rod assemblies, the number of available neutrons for inducing fission and the chain reaction can be controlled.

The control rods can be adjusted so that energy is released at a relatively steady rate. This requires that, on the average, one neutron from each fission event initiates only one other event. If more energy is needed, the rods are withdrawn farther. When fully inserted, the control rods can curtail the chain reaction and shut the reactor down. A nuclear reactor can run for about four years before having to be "refueled."

> *Question:* If the control rod mechanism in a nuclear reactor malfunctioned and the chain reaction proceeded uncontrolled, would the reactor explode like a nuclear bomb?
>
> *Answer:* No. A nuclear reactor cannot expode like an atomic or nuclear bomb. Reactor-grade uranium contains only about 3 percent U-235, whereas weapons-grade material is more than 90 percent. The high-grade material must be held together briefly to achieve an explosive release of energy.
>
> If control is lost in a reactor and energy is not removed, the energy release is not concentrated enough to create an explosion. Confined gases caused by chemical reactions or the conversion of water coolant to steam might give rise to an explosion, but without a "mushroom" cloud. However, there are important environmental concerns in the event of an uncontrolled release of energy and a "meltdown" (a term you've probably heard). See the discussion of the "China syndrome" that follows and Special Feature 26.1.

Water flowing through the fuel assemblies and the reactor vessel acts not only as a coolant–heat transfer agent, but also as a **moderator**. The U-235 nuclei react best with "slow" neutrons (kinetic energies of only fractions of an electron volt). The "fast" neutrons emitted from the fission reactions (energies up to 20 MeV with an average of about 2 MeV) are slowed down or "moderated" by energy losses through collisions. Hydrogen atoms in water are very effective in slowing down neutrons because the mass of the hydrogen nucleus is about the same as that of a neutron. Recall from Chapter 3 that when particles of nearly equal masses collide, there is a large transfer of energy. It takes only about 200 collisions on the average to slow fast neutrons down to energies less than 1 eV.

Other materials, such as beryllium and graphite (carbon), have been used as moderators. Because these have heavier nuclei, not as much energy is transferred per collision. About 120 collisions with carbon atoms are needed to slow fast neutrons down to an optimal "slow" speed. Although it is not the best moderator, carbon in the form of graphite permits a chain reaction to occur in natural (unenriched) uranium fuel arranged in a large mass of graphite.

The first self-sustaining fission chain reaction was accomplished in such a reactor in 1942 by a team working with Enrico Fermi (Fig. 26.12). It was called a "pile" because it essentially consisted of a pile of graphite blocks. The graphite carbon atoms acted as a moderator for the neutrons from the fission processes.

This was the first "proof" that the fission chain reaction would work. The Manhattan Project involved many scientists and engineers, including refugees from Europe. They worked under the pressure of World War II to develop a nuclear weapon before the Germans, who were also working on one. The first nuclear bomb was detonated in a test explosion over the desert in New Mexico on July 16, 1945, only 2½ years after the first reactor was built.

In the United States, regular or "light" water is used as a moderator-coolant in nuclear reactors. The hydrogen nuclei of regular water have a high tendency to capture neutrons when colliding with them.* As a result, it is impossible to achieve a chain reaction with natural uranium and ordinary water. Ordinary water can be used only with enriched uranium that has a greater neutron output. After a uranium enrichment plant was built in Oak Ridge, Tennessee, during World War II and enriched uranium became available, reactors were built using this. The first nuclear reactor for the purpose of electrical generation was a "light"-water reactor of this type. It went into operation in 1957 at Shippingsport, Pennsylvania.

The deuterium nuclei of "heavy" water, on the other hand, do not absorb neutrons very readily, and so heavy water can be used as a moderator-coolant in reactors with natural uranium fuel. The Canadians designed a reactor using heavy water as a moderator-coolant, calling it CANDU, for *Can*adian *Deu*terium *U*ranium, after its coolant and (natural) fuel.

One of the major concerns in reactor safety is that a reactor may get out of control or in a "runaway condition" because of a malfunction of the control rods or a **loss-of-coolant accident** (LOCA). With an uncontrolled release of energy, the coolant may not be able to handle the excess heat generated, or if

---

* $_0^1 n + {_1^1}H \rightarrow {_1^2}H + \gamma$

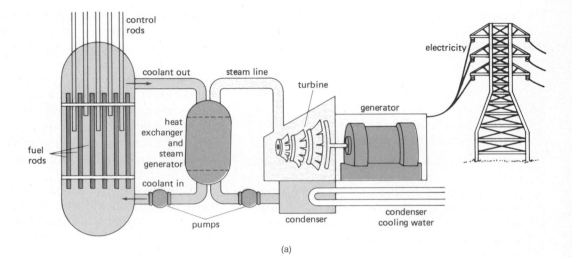

(a)

(b)

Figure 26.11

Nuclear reactors. (a) A schematic diagram of a nuclear reactor and electrical generating plant.
(b) An actual nuclear power plant with three reactors (in cylindrical containment buildings).
Opposite: (c) A fuel rod assembly and a partially dissembled reactor.

Figure 26.12

Birthplace of the atomic age. (a) The west stand of Stagg
Field at the University of Chicago (being demolished in
1957). It was under this stand in a squash court that the first
nuclear chain reaction was initiated on December 2, 1942.
Opposite: (b) and (c) The plaque installed on the wall of
the stand in 1952 remains at the location. Enrico Fermi
stands under the plaque *(second from the right)* at the
unveiling.

(a)

(c)

(b)

(c)

there were a loss of coolant, heat could build up in the core. In either case (or both) the reactor core might get so hot that the fuel rods would melt and the fused mass would burn through the reactor vessel and the containment floor. Such a "**meltdown**" would cause leakage of dangerous radioactive materials into the environment.

This is the so-called China syndrome, in which the fissioning mass might melt its way through the Earth to China. Calculations show that this cannot really happen. The introductory photo of this chapter shows the Three Mile Island nuclear facility, where a LOCA occurred and where a meltdown almost happened in 1979. (See Special Feature 26.1 at the end of the chapter for more details.)

As pointed out earlier, a reactor cannot explode like a nuclear bomb. The uranium enrichment is too low for the chain reaction to proceed fast enough for the required energy release in a short time. Moreover, to explode, a system must be prevented from flying apart before this time elapses. Should the uncontrolled pressure buildup in a reactor core cause it to burst apart, the resulting separation of the fuel rods would give neutrons more opportunity to escape from the system, and the chain reaction would stop.

■ THE BREEDER REACTOR

We have about a 50-year known domestic reserve of uranium. As with oil and gas, our domestic supply of uranium will run out in the not-too-distant future. However, it is possible to extend the supply of fissionable material through breeder reactors. A breeder reactor converts nonfissionable material, such as U-238, into fissionable material, and the reactor produces or "breeds" more fuel than it consumes.

This is not a case of getting something for nothing. Rather, it is a process of converting a nonfuel into a fuel. For example, in an ordinary nuclear reactor there is a great deal of U-238 that "goes along for the ride"; the 3 percent or so of U-235 is the fissionable fuel that does the job. A U-238 nucleus can absorb a fast-moving neutron and undergo a change to plutonium-239 (Pu-239) by the reaction

$$^{238}U + {}^1n \longrightarrow {}^{238}U \longrightarrow {}^{239}Np \longrightarrow {}^{239}Pu$$

Fertile material   Unstable products   Fissionable material

Pu-239 is a fissionable material that can do the same job as U-235. The original material is called a fertile material. Other fertile materials that can be used in a

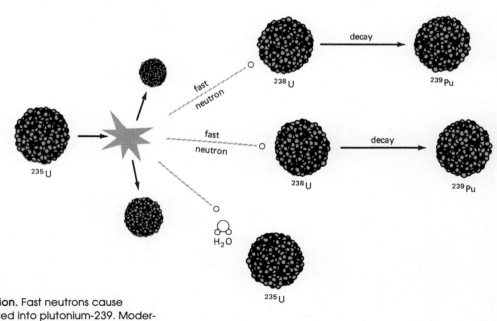

Figure 26.13
Plutonium-breeding reaction. Fast neutrons cause uranium-238 to be converted into plutonium-239. Moderated neutrons cause U-235 nuclei to undergo fission and sustain the chain reaction. In a breeding reaction, more fissionable nuclei are produced than consumed.

Figure 26.14
Nuclear power plants in the United States. (Dots are representative, not at actual geographic locations.)

● in operation or startup
○ with construction permits

breeding reaction are Th-232 and Pu-240. These breed fissionable U-233 and Pu-241 isotopes, respectively.

In a U-238 breeder reactor, the moderation or slowing down of fast neutrons is minimized, and with the presence of an appreciable number of fast neutrons, Pu-239 is bred from U-238 while energy is obtained from U-235 fission. The breeding reaction is illustrated in Figure 26.13. France, which gets about 60 percent of its electrical generation from nuclear energy, and several other countries have breeder reactors, but only experimental breeder reactors have been operational in the United States because of limited funding. One of the reasons for the curtailment of the development of the breeder reactor follows in the next discussion.

■ NUCLEAR WASTES AND PROLIFERATION

The building and planning of nuclear reactors were greatly curtailed after the Three Mile Island incident (see Special Feature 26.1). This was not only because of public concern for safety, but also because there was not the demand for electricity that had been projected, and new reactors were not needed. People were conserving energy. Another factor was falling fossil-fuel costs. Overall, it was cheaper to build and operate a fossil-fuel plant than a nuclear one, so economics played a role.

Even so, there are about 70 nuclear reactors operating in the United States (Fig. 26.14). (There are about 130 or more nuclear reactor generating facilities operating in other countries.) They are creating large amounts of **nuclear wastes**. These are the "ashes" from the fission process. A large variety of fission products are formed in nuclear reactors, most of which are radioactive. Some of these products have very long half-lives, up to tens of thousands of years or more. This means that these "high-level" wastes will be around for hundreds of years, and the concentrations make them highly dangerous.

There are still useful fissionable materials in the spent fuel assemblies, and it would be desirable to recover them. Currently, however, nuclear wastes are being stored in double-walled storage tanks that are covered with earth (Fig. 26.15). Plans call for the "ultimate" disposal of these wastes to be in underground storage in stable geological repositories, such as salt or rock formations. A bill passed by Congress calls for these repositories to be available by 1996. "Permanent" underground burial requires the development of long-lasting containers or the conversion of wastes into solid, insoluble form, for example, embedded in glass (Fig. 26.15). Stainless-steel containers are currently being considered. Appropriate geological formations must be found where the wastes could cause no contamination of underground water should they leak.

In addition to the radioactive fragments of the fission process, the transuranic element plutonium-239 is formed in the normal operation of a reactor.

Some U-238 is converted to Pu-239 through "fast" neutron capture and two beta decays. This is one of the reactions promoted in a breeder reactor, discussed previously. Pu-239 is a fissionable material and does the same job as U-235. In the reactor operation, the freshly produced Pu-239 also undergoes fission and produces energy. As a result, it is possible to extend the normal "lifetime" of the reactor fuel elements, and they do not have to be replaced as quickly.

However, it is possible to remove the fuel elements before the Pu-239 is used up and to separate it through difficult and complicated procedures. The Pu-239 can then be processed into nuclear weapons. (The second "atomic bomb" that destroyed the city of Nagasaki, Japan, used Pu-239 in-

HIGH LEVEL WASTE STORAGE TANK
CAPACITY 1,300,000 GALLONS

TOP & INSIDE	
1. CONCRETE	10. RECIRCULATING COOLING WATER
2. PRIMARY TANK	11. INSTRUMENT PROBE
3. SECONDARY TANK	12. PUMP OUT JET
4. INSULATING REFRACTORY	
5. ANNULUS AIR SUPPLY	OUTSIDE
6. ANNULUS EXHAUST	13. INLET PIPE
7. TANK EXHAUST	14. INLET-OUTLET PIPE
8. REMOVABLE PLUGS	15. STEAM PIPE
9. PUMP OUT JET	16. EARTH

(a)

(b)

Figure 26.15
Radioactive waste. (a) Nuclear waste is stored in double-walled tanks that are covered with earth. (b) Embedding radioactive waste in glass is a possible method of storage.

stead of U-235 as in the first one dropped on Hiroshima. The Pu-239 had been made in an experimental reactor.) A nuclear power plant reactor can generate up to 200 to 300 kg of Pu-239 a year. A great deal of this is used up in the fission process, but only 5 to 10 kg of Pu-239 is needed to make a nuclear bomb capable of destroying a medium-size city.

Thus, a nuclear reactor can be used to produce electricity and/or to produce Pu-239, which could be used as material for weapons. Such capability gives rise to the problem of **nuclear proliferation**.

Fissionable materials can also be produced in research reactors. This was made particularly evident in 1974 when India exploded a "nuclear device" as a test for "peaceful applications," according to that government. The explosion showed that countries other than the five "nuclear powers" (those that had tested and demonstrated nuclear weapons) could produce fissionable material and use it in nuclear weapons if they so desired.

In the early days of the "nuclear age" only three countries had the bomb. The United States, Great Britain, and the Soviet Union test-exploded several bombs in the atmosphere after World War II. The large amounts of radioactive fission products from atmospheric explosions that come to Earth are called radioactive fallout. These products can be carried around the world by winds and eventually settle out or are carried to the Earth by rain and snow.

As you may imagine, radioactive fallout can be dangerous. The greatest source of radioactive fallout in the absence of atmospheric nuclear explosions is from coal-fired plants. Radioactive materials in the coal are vented to the atmosphere in combustion gases and particles. (The biological effects of radiation will be discussed shortly.)

As a result of public and scientific concern, the three nuclear powers and many other countries signed the Nuclear Test Ban Treaty in 1963, by which they agreed to a moratorium on further tests in the atmosphere. Underground testing was still permitted. The most notable exceptions that did not sign the treaty were France and China. Both countries have since exploded bombs, now more politely called "nuclear devices," in the atmosphere and have joined the ever-growing nuclear weapons club.

Not only are the immediate results of a nuclear war a concern, but also the climatic aftereffects. The burning from the maelstrom would send tons of smoke particles into the atmosphere. These would be carried around the globe by the prevailing winds, and the layer of particles would block out a large part of the incoming solar radiation. As a result, the surface of the Earth would become cold, and we (those who survived the nuclear explosions) would experience a "**nuclear winter**."

■ FUSION AND FUSION REACTORS

Another class of nuclear reactions that release energy is called **fusion reactions**. In these exoergic reactions, light nuclei are "fused" to form heavier nuclei with the release of energy, for example, hydrogen fusion, as illustrated in Figure 26.16.

Fusion is the source of energy of stars, which includes our Sun. In the Sun, the fusion process results in the production of a helium nucleus from four protons (hydrogen nuclei). The process does not take place in a single reaction but goes through different sets of reactions with the net result of

$$4 {}^{1}_{1}\text{H} \longrightarrow {}^{4}_{2}\text{He} + 2({}^{0}_{+1}e)$$

(The symbol ${}^{0}_{+1}e$ represents a positron, which is the antiparticle of the electron.)

Many scientists look upon fusion as an ultimate and ideal source of energy for several reasons.

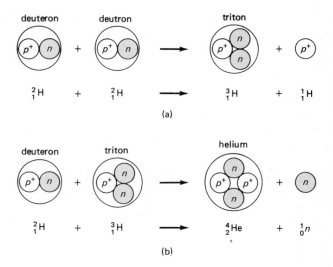

Figure 26.16
Illustrations of fusion reactions. (a) D-D reaction and (b) D-T reaction.

Among these are the availability or supply of raw materials and advantages in waste disposal. "Heavy hydrogen" or deuterium ($_1^2H$) is present in all water (heavy-water molecules). For about every 6500 atoms of ordinary (light) hydrogen in water, there is one atom of deuterium. This may not seem like much, but taking into account the vast amounts of water in the oceans and other surface waters, it is estimated that there is enough deuterium to supply the world's energy needs for more than a million years!

The other heavy isotope of hydrogen, tritium ($_1^3H$), is very rare in nature. It is radioactive, with a half-life of about 12 years. As "nuclear waste" of the first or "D-D" reaction in Figure 26.16, it would be a relatively short-term disposal problem, as compared with fission wastes. Notice that the tritium formed by deuterium fusion could immediately react by the second "D-T" reaction with available deuterium. For use as a fusion "fuel," tritium is produced by the nuclear reaction of neutron capture by lithium (Li).* As a raw material, lithium is widely distributed over the Earth, being estimated to be more abundant than tin. Should this supply of lithium be inadequate, it could be extracted from the oceans, which contain large amounts (but in small concentrations).

The energy releases from the D-D and D-T reactions are about 4.0 MeV and 17.6 MeV, respectively. This is significantly less than the 200-MeV energy release from a typical fission reaction. But keep in mind that, "kilogram for kilogram," a given mass of hydrogen isotopes has many, many more nuclei than an equivalent mass of a "heavy," fissionable isotope.

In the case of fusion, there is no critical mass or size, since there is no chain reaction to maintain. However, there are problems in practical fusion energy production. A major problem is in obtaining a self-sustaining reaction. The repulsive electric force between positively charged nuclei opposes their coming together and fusing. For hydrogen (one nuclear proton), this force is not excessive, which is why hydrogen is the fusion candidate. Fusion reactions can be produced in particle accelerators. For example, deuterons (deuterium nuclei) can be accelerated and "slammed" into a solid deuterium target to produce D-D fusion reactions. This shows fusion to be possible, but much more energy is

* $_0^1n + _3^6Li \rightarrow _1^3H + _2^4He$

spent in accelerating the deuteron than is produced by the small number of fusion reactions that occur.

To get an appreciable energy output, one might use a confined gas of hydrogen isotopes that are to undergo fusion. If the temperature of the gas is raised, the kinetic energy of the molecules would be increased, and hence it is only a matter of attaining a sufficiently high temperature for fusion to occur. In the confined space of the gas, the moving nuclei would collide repeatedly until fusion reactions take place. However, it is found that this requires temperatures on the order of millions of degrees (Fahrenheit, Celsius, or kelvins, take your choice).

At such high temperatures, almost all of the hydrogen-isotope atoms will be stripped of their electrons. Such a gas, consisting almost entirely of positively charged ions and free, negatively charged electrons, is called a **plasma**. Plasmas have several special properties that have caused them to be referred to as a *fourth state (phase) of matter,* a term used in 1879 by William Crookes, an English chemist who generated plasmas in gas-discharge or Crookes tubes. Some of these properties can be used to advantage in potential fusion-reactor techniques, but others create problems.

Large amounts of fusion energy have been released, but this occurred in uncontrolled fusion in the form of the hydrogen (H) bomb. In the H-bomb, a nuclear fission bomb is used to "trigger" the fusion reaction or to supply the energy to get it started (Fig. 26.17). Because of the high temperatures required to bring about fusion reactions in plasmas, they are called thermonuclear reactions, and fusion weapons are called thermonuclear weapons or "devices."

The fusionable material in the bomb is in the form of solid lithium hydride, LiH, where the H is deuterium, an isotope of hydrogen (^2H or D). With the explosion of the fission bomb, a hot plasma develops in which there are D-D reactions. Neutrons from the fission bomb react with the Li to produce tritium (^3H or T) for D-T reactions. The neutrons emitted in the D-T reactions come off with high energy (most of the 17 MeV released in the reaction). These neutrons are speedy enough to induce fission in ^{238}U, so the bomb is surrounded by a layer of uranium, which caps off the explosion with another fission energy release. This fission-fusion-fission process takes place in about a millionth of a second.

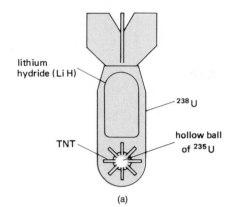

(a)

(b)

Figure 26.17
Uncontrolled fusion. (a) In a hydrogen bomb, a small fission bomb is used to "trigger" the fusion reaction. (b) A thermonuclear explosion in the Pacific in 1954.

The controlled release of fusion energy might be accomplished by adding fuel in small amounts to a fusion reactor, similar to the way we control the amount of fuel in conventional energy (heating) sources, such as a gas furnace. However, fusion reactors are not yet practical. There are major problems with confinement—confinement of an ultra-hot plasma and confinement of sufficient energy in the plasma to initiate and sustain fusion.

No material could be used to "hold" a plasma. Tungsten, the material with the highest melting point, melts at around 3370°C. Also, heat would be readily transferred from the confinement region. Considering all of the various factors, it is calculated that to achieve a net energy production from D-T fusion would require a reactor operating tempera-

ture of at least 100 million degrees. Even higher temperatures would be required for D-D fusion.

Two types of confinement, magnetic and inertial, are now under development.

Magnetic Confinement. Since the plasma is a gas of charged particles, it can be controlled and manipulated with electric and magnetic fields. In magnetic confinement, magnetic fields are used to hold the plasma in a confined space (a magnetic "bottle" or "ring"). Electric fields produce electric currents that heat and raise the temperature of the plasma.

The energy break-even point (getting as much out as put in) depends on two things: (1) the temperature of the plasma and (2) its density and confinement time. Temperatures of 100 million kelvins have been achieved in tokamaks (see Fig. 26.18), and in other experiments the density–confinement time requirement seems to be satisfied. The trick scientists must now perform is to achieve these requirements at the same time, which may prove more difficult than reaching them separately.

Inertial Confinement. In this technique, it is hoped that laser, electron, or ion beams can be used to initiate fusion. It calls for hydrogen fuel pellets to be dropped into a reactor chamber (Fig. 26.19). Energetic beam pulses would cause a pellet to "implode," causing compression and high temperatures. Fusion could occur if the pellet stays together for a sufficient time. This depends on the inertia of the pellet (hence the name inertial confinement). At the present time, lasers and particle beams are not powerful enough to induce fusion by this means, but research is under way (Fig. 26.20).

If you think about it, the technological problems of fusion reactors are enormous, but so are the benefits. Practical energy production from fusion is not expected until well into the 21st century.

BIOLOGICAL RADIATION HAZARDS

You may have seen signs like the one shown in Figure 26.21 if you have been in the vicinity of a radioactive source or a general radiation area. It is a notification or a *warning* (caution) sign. But why the warning? What is the potential danger from the improper use of radiation or radioactive materials?

The main hazard of radiation is to living cells. When "radiation," such as X-rays, gamma rays, beta

(a)

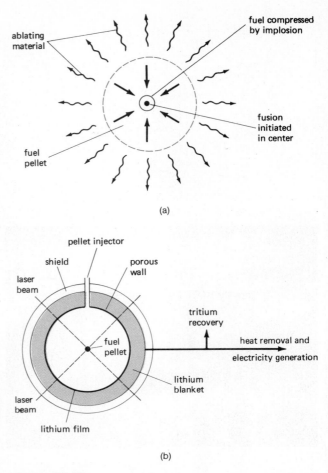

(a)

(b)

Figure 26.19
Inertial confinement and implosion. (a) The compression of a fuel pellet is enhanced by use of an outer shell of material—for example, a plastic—that is ablated or vaporized. This drives the inner shell inward and compresses the fusion fuel. (b) A design of a laser fusion reactor cavity. The laser beams compress and heat the fuel pellet. Neutrons released from the fusion are captured in the lithium blanket, creating tritium, which can be recovered and used as a fuel.

(b)

Figure 26.18
Magnetic confinement. (a) ORMAK fusion research device. The magnetic system used is known as a tokamak (a Russian word for toroidal magnetic chamber) and is a circular device shaped much like a donut. (b) The Doublet III magnetic fusion research machine. It uses a unique confinement concept invented by General Atomic Company for use in fusion power reactors.

particles, and alpha particles, passes through matter, it can knock electrons out of the atoms, forming electron-ion pairs. (Recall from Chapter 25 that this allows the detection of radiation, such as by Geiger counters.) Hence, we refer to such radiations as ionizing radiations. The amount of ionization and the penetration of radiation in a material, including living tissue, depend on the radiation energy, the electrical charge, and the density of the material.

A major injury can occur to the reproductive mechanism of cells. Because of radiation ionization, living cells may receive slight radiation damage, heal, and resume their normal activities and ability to divide and reproduce. In extreme cases,

Modern Physics

Fission and fusion. Irradiated reactor assemblies exhibit a characteristic glow when immersed in water; this is known as the Chrenkov effect. Light is emitted because the speed of charged radioactive decay particles is greater than the speed of light in water.

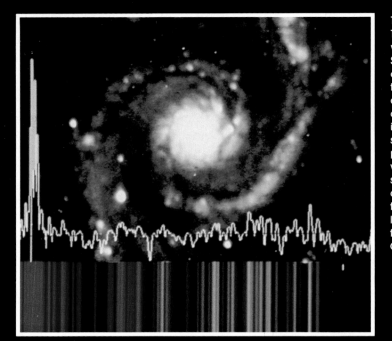

The spectrum of spiral galaxy M51. This splendid spiral galaxy shines with the light of billions of stars. Spectrographs break the light down into its component colors to render a pattern of bright and dark lines, shown here as a brightly colored spectrum and as an intensity trace at the center. This spectrum reveals the average properties of the stars and gas within the galaxy, and when this spectrum is compared with that of the Sun, it can be inferred that the average star in M51 is quite like our own star. (Courtesy of the Smithsonian Astrophysical Observatory)

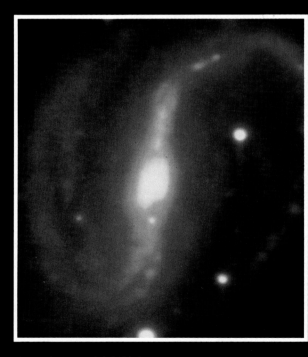

Brightness profile of Comet IRAS-Araki-Alcock. In May, 1983, this comet (1983D) came unusually close to Earth, providing astronomers with an exceptional view of the gas and dust emitted by a comet nucleus. As the nucleus spins, the apparent direction of gas emission also rotates, causing subtle brightness changes. False color contours are used here to emphasize these changes. (Courtesy of the Smithsonian Astrophysical Observatory)

Spiral Galaxy NGC 7479. To emphasize the fact that the spiral structure (red) in this typical spiral galaxy is superposed upon a uniform disk-like background of old stars (blue), this image is processed in false colors. The four white stars are foreground stars in our own galaxy. (Courtesy of the Smithsonian Astrophysical Observatory)

(a)

(b)

Figure 26.20
Fusion research. (a) Model of the Shiva laser system for laser fusion research at the Lawrence Livermore Laboratory. (b) The 20 laser beams are pointed at and focused onto a tiny fuel target about the size of a grain of sand located in the center of the target chamber. (c) This huge laser system will deliver more than 30 trillion watts of optical power in less than one billionth of a second.

(c)

the cells may be mutilated so that they cannot reproduce. If this happens to enough cells, the irradiated tissue may eventually die.

In other instances, damage to the chromosomes in a cell nucleus (genetic damage) can cause the cells to lose their identity and/or to reproduce at an abnormal rate. The condition of an abnormal reproduction rate of cells with "amnesia" is called cancer.

Cancer cells may grow slowly with little effect on the surrounding normal tissue, forming a benign tumor. When cancer cells grow at the expense of the surrounding tissue, we call it a malignant tumor. Skin cancer due to overexposure to ultraviolet radiation (too much sun) and leukemia ("cancer of the blood") are examples of the cancerous effects of

(Continued on page 494 after Special Feature)

Figure 26.21
Radiation. The symbol used to indicate the presence of a radioactive source or a general radiation area.

Case Studies: Three Mile Island and Chernobyl — the "Incident" and the "Accident"

The question of nuclear reactor safety immediately brings to mind for most people the name *Chernobyl,* which we learned to pronounce "Cher-noble" from hearing it so much on newscasts in 1986. Chernobyl, a Russian city, is now synonomous with the most serious nuclear "accident" (at the time of this writing).

Some memories don't stop there but go back to 1979 and the name of Three Mile Island. It was there in Pennsylvania that a nuclear "incident" took place. This is very close to home and doesn't sound so distant.

Are nuclear reactors safe? In general, the answer is yes — provided they are *properly constructed and operated.* There are various examples of construction specifications not being followed, and any operation has the possibility of human error since humans are involved. One can only hope (and insist) that there are checks on both to ensure that there are no "incidents" or "accidents" in any case.

Let's look at the recorded cases of a major nuclear incident (no major release of radiation to the environment or loss of life) and a major nuclear accident (radiation release, deaths, and contamination) and see how and why they happened. Hopefully we can learn from these mistakes.

Three Mile Island*

The Three Mile Island nuclear plant in Middletown, Pennsylvania, is a subsidiary of the General Utilities Corporation (see introductory photo in this chapter). Starting at 4 A.M. on March 28, 1979, a series of mishaps occurred. The sequence is enormously complicated; some 40 events have been identified during the subsequent investigations.

The operation that seems to have started the trouble was a routine one — the changing of a batch of

---

* Adapted from Turk, J., and A. Turk, *Physical Science,* Second Edition, Saunders College Publishing, Philadelphia, 1981. The purpose here is to give the general sequence of events that occurred at Three Mile Island (TMI), which came dangerously close to a "meltdown" situation. The TMI incident has been examined many times to determine the underlying causes of the accident and who was at fault. If you are interested in this aspect, there is a public report of the President's Commission on the Accident at Three Mile Island, issued on October 30, 1979. This report is known informally as the "Kemeny Report" because the Commission was chaired by John G. Kemeny, President of Dartmouth College. The Commission found enough fault to go around.

water purifier in a piping system. For a nuclear plant, this job is considered to be as routine as, say, changing the oil filter in an automobile. But a problem developed; some air got into the pipe, interrupting the flow of water. Of course, the backup systems in a nuclear plant are designed to respond automatically to such an event, but in this instance, several other things went wrong.

- Two spare feedwater pumps were supposed to be ready to operate at all times. However, the valves that control the water from these pumps were out of service for routine maintenance; therefore, the spare pumps could not deliver water. The controls for these valves were provided with tags to indicate that they were being repaired. The tags hung down over red indicator lights that go on when the spare pumps are not feeding water. Since the lights were obscured by the tags, the operators did not realize that no water was flowing. The control of a nuclear power plant is an array of indicator lights and switches (Fig. 1).

- As a result, pressure built up in the reactor core. A relief valve in the primary coolant loop then opened automatically (as it should have) to let out superheated steam. But *the relief valve failed to close,* causing a dangerous drop in pressure. This malfunction is considered to be the crucial failure of equipment in the entire sequence.

- When the emergency core-cooling system came on automatically, the pressure gauges in the control room gave a false reading, leading operators to think that the water level was still above the fuel rods. It wasn't. Instead, bubbles of gas from below were pushing the water up, leaving part of the core exposed.

- The primary and the emergency cooling pumps, which should have been left on, were turned off twice by operators misled by the faulty pressure gauges.

The net result of this confusing sequence of mishaps was that the nuclear core overheated. The indicators of the temperatures inside the reactor vessel climbed off the recorder charts. For 13½ hours the situation was very unclear. It seemed that the reactor core was partially exposed above the cooling water and that there were voids or perhaps bubbles in the system. In fact, subsequent investigations have shown that the entire core was exposed for some time. This means that only steam, not liquid water, was circulating through the core to remove excess heat.

The "void" or "bubble" that caused the problem was something entirely unexpected—it was a 1000-cubic-foot (28,000-L) volume of steam and hydrogen gas. Where did the hydrogen come from? There are two possibilities, both of which probably played a part. One was radiolysis, which is chemical change produced by radiation. In other words, the radioactivity in the core chemically decomposed the water and produced hydrogen (as well as oxygen).

The other possibility was the chemical reaction of water with metals, which also produces hydrogen (but not oxygen). The metal tubes holding the uranium fuel are made of zirconium, which reacts with water when the temperature gets hot enough. Pure hydrogen gas is not explosive. Mixtures of hydrogen with air or oxygen are explosive, but only in certain concentrations. Therefore, radiolysis (which produces both hydrogen and oxygen) differs in explosive potential from the reaction of water with metals (which produces hydrogen but not oxygen).

The superheated steam released to the atmosphere early in the emergency was radioactive, and for this reason Pennsylvania's Governor Thornburgh ordered the evacuation of children and pregnant women from the area near the plant. Others left of their own accord. As the problem subsided, the evacuees returned, with misgivings one can only imagine.

(Continued)

An array of lights and switches. The control room of a nuclear power plant showing the array of indicator lights and control switches.

Chernobyl

Although this disaster occurred in the Soviet Union, the first warning came in Sweden. At 9 A.M. on April 28, 1986, a technician at a nuclear power plant 60 miles north of Stockholm noted some unusually high radiation levels on a detector. Thinking there was a problem in their own plant, engineers checked for a leak, but could find none. Abnormal levels of radiation were also found outside the plant, which compounded the problem and explanation. Reports then began coming in from neighboring and nearby countries that they too were recording much higher than normal radiation levels (four to five times higher). Over such a wide area, the radiation must have come via the atmosphere.

The prevailing winds were from the south—from Russia. Queries to the Soviet government were met with denials. Finally, at 9 P.M. it was announced that "an accident has taken place at the Chernobyl power station and one of the reactors was damaged." Thus began the unfolding of the story of the world's worst nuclear accident to date.

There was a great deal of speculation as to what actually happened. The Soviet Union released little information. An official report was issued on the April reactor accident at a conference in Vienna in August. The bottom line for the cause of the accident was a flawed reactor design, a lack of safety procedures, and a lack of competence in the reactor's operation, i.e., "human error."

Chernobyl is a town of 25,000 people about 80 miles north of the city of Kiev, which has a population of 2.5 million. The nuclear power plant at Chernobyl supplied electricity to Kiev. The reactors at this plant were part of 28 graphite reactors that made up a total of 50 Soviet reactors. A graphite reactor in Great Britain experienced a similar crisis in 1957. Since then, graphite reactors have been almost entirely abandoned outside the Soviet Union. The most notable exception is the production of weapons-grade plutonium, for which the United States has several such reactors.

The graphite reactor is totally different from the light-water U.S. commercial reactors. In light-water reactors water is used as a moderator and a heat-transfer agent, and the water flows around the fuel rods in the reactor core (see Fig. 26.11). In graphite reactors, the fuel rod–cased assemblies are cooled by water, which is converted directly to steam at the high operating temperatures (Fig. 2). The fuel assemblies, some 1661 of them, are encased in 1700 tons of graphite blocks. The graphite (carbon) acts as a moderator to slow down the neutrons from the fission processes (similar to the Manhattan Project atomic "pile").

Chernobyl Unit 4 had been scheduled for shutdown for annual maintenance. An experiment had been planned to see how long after the steam supply was shut off the spinning turbogenerator could continue to supply electricity to run necessary equipment. (According to the official Soviet report, this experiment was not

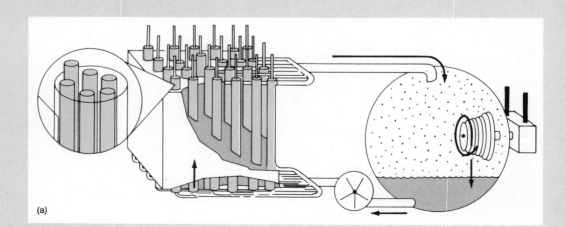

(a)

properly planned, nor had it received approval.) In any case, the automatic emergency core cooling system was apparently disconnected to keep it from interfering with the experiment.

As the reactor power output was reduced without the automatic control system, the power dropped below the required experiment level. The operator withdrew more of the 211 control rods to compensate for the power reduction until only 6 to 8 rods were still in place — well below the operating instructions not to go below the equivalent of 30 rods.

What happened in the next few critical seconds remains unclear. Scientists postulate the following:

- A sudden power surge due to lack of control shattered the fuel rods into fragments, and heat transferred to the water coolant caused it to "flash" boil into steam. The steam caused a pressure shock violent enough to blast a hole through the top of the reactor.
- This "steam explosion" exposed the red-hot fuel rods, and air rushed in and mixed with the explosive hydrogen gas formed when the zirconium fuel jackets reacted with the steam. Within a few seconds a "hydrogen explosion" occurred that blew off the top of the reactor building (Fig. 2).

- The fuel rods melted down, and the graphite blocks burned like a massive coal pile. Radioactive material was spewn into the atmosphere and carried by the prevailing winds.

Some 135,000 people were evacuated from a 30-km (about a 19-mi) zone around Chernobyl. There were 31 reported fatalities, and it has been estimated that there could be 24,000 fatal cancer deaths and hundreds of thousands of nonfatal cancer cases from the fallout. (No one knows for sure.) A large area of farmland around Chernobyl was contaminated and cannot be used without proper treatment.

There is a containment building of steel-reinforced concrete around U.S. reactors that is designed to contain any release of water or steam as well as radioactive materials in case of an accident. Chernobyl did not have such a containment building. Would such a building around one of our reactors withstand the Chernobyl-type explosions? It is difficult to say; however, more important, such explosions should not occur in our light-water reactors. Increases in temperature and steam would cause the reactor reactivity to go down, which is a built-in safety factor.

Chernobyl. (a) A diagram of a graphite reactor. See text for description. (b) Chernobyl Unit 4.

radiation. (Leukemia is the unregulated production of white cells—leukocytes—in the blood.)

One of the most abundant fission products is strontium-90 ($^{90}_{38}Sr$). It is chemically similar to calcium-40 ($^{40}_{20}Ca$)—check to see in which chemical groups of the periodic table these elements are located (Chapter 7). From radioactive fallout, the strontium-90 can get into out bodies and into our bones—for example, it falls on grass, the grass is eaten by cows, it appears in milk, the milk is drunk by humans. The radiation from the strontium-90 may cause cancer, such as bone tumors and leukemia, particularly in children.

On the other hand, cell damage may serve a useful purpose. Radiations, such as gamma rays from cobalt-60, are used in medical treatment to control the growth of tumors by producing ionization in the tumor-cell nuclei so that some of the tumor cells are "killed," and the growth of the tumor is slowed.

SUMMARY OF KEY TERMS

Artificial transmutation: the changing of the nuclei of one element into the nuclei of another element through an induced nuclear reaction.

Electron volt (eV): a unit of energy, the energy an electron receives when accelerated through a potential of 1 V, with $1 eV = 1.6 \times 10^{-19}$ J.

Endoergic reaction: a nuclear reaction in which energy is converted to mass, or in which there is a net energy input.

Exoergic reaction: a nuclear reaction in which mass is converted to energy, or in which there is a net energy output.

Binding energy: the energy needed to separate a nucleus into free protons and neutrons.

Fission: the dividing or "splitting" of a heavy nucleus into two intermediate lighter nuclei with the emission of two or more energetic neutrons.

Chain reaction: a growing series of induced fission reactions due to the emitted neutrons of previous fission reactions.

Critical mass: the amount of mass or concentration of fissionable material needed for a sustained chain reaction.

Breeder reactor: a reactor that produces more fuel than it consumes by converting nonfissionable nuclei into fissionable nuclei.

Nuclear waste: the unwanted radioactive by-products from the production of nuclear energy.

Nuclear proliferation: the increased availability of radioactive materials that could be used to produce nuclear weapons.

Fusion: a nuclear reaction in which two light nuclei are "fused" or reacted together to form a heavier nucleus, with the release of energy.

QUESTIONS

The Background of the "Nuclear Age"

1. Discuss the pros and cons of the free exchange of scientific information among scientists of different countries.

2. Identify each of the following and the role he or she played in the development of the atomic bomb: (a) Hahn, (b) Meitner, (c) Fermi, (d) Bohr, (e) Einstein.

*3. Was the Manhattan Project carried out in Manhattan (New York City)? Explain.

4. Many people think that it was not necessary to drop atomic bombs on Japan in 1945, since World War II could have been brought to an end by an invasion of the Japanese Islands. Discuss the pros and cons of dropping the bombs.

Nuclear Reactions

5. A "mad" scientist wants to sell you a large quantity of gold-194 that he made by artificial nuclear reactions for only one fourth the current price of gold. Should you buy it? (*Hint:* Check Table 25.2 before you do.)

*6. In trying to initiate nuclear reactions, would there be any advantage to using protons or hydrogen nuclei instead of alpha particles as Rutherford did? Explain.

7. Explain what is meant by the conservation of nucleons and the conservation of charge in nuclear reactions.

8. Complete the following reactions:

   (a) $^{45}_{21}Sc + ^{4}_{2}He \rightarrow \underline{\quad} + ^{1}_{1}H$

   (b) $^{27}_{13}Al + ^{1}_{0}n \rightarrow ^{28}_{13}Al + \underline{\quad}$

   (c) $\underline{\quad} + ^{4}_{2}He \rightarrow ^{35}_{17}Cl + ^{1}_{1}H$

   (d) $^{1}_{0}n + ^{235}_{92}U \rightarrow \underline{\quad} + ^{92}_{36}K + 3(^{1}_{0}n)$

   (e) $^{2}_{1}H + ^{2}_{1}H \rightarrow \underline{\quad} + ^{1}_{0}n$

*9. Write the fission reaction for uranium-235 "splitting" into rubidium-94 and cesium-139.

10. Why are only two transuranic elements named after planets (Table 26.1)?

11. Identify people and places in Table 26.1. (Use an encyclopedia if necessary.)

12. Why is the symbol for the transuranic element Rutherfordium Rf instead of Ru?

•13. How big is an electron volt? Give some comparisons with other units.

Fission and Fission Reactors

14. Give an example of a *chemical* chain reaction. (*Hint:* Think of how matches might be used.)

15. Why is it impossible to have a sustained fission chain reaction in a very small piece of fissionable material?

16. Why can't a nuclear reactor blow up like a nuclear bomb?

•17. Could the moderator (or lack of it) be used to control the chain reaction in a reactor? Explain.

18. Is it always necessary to use enriched uranium in a reactor? Explain.

19. What is meant by a "meltdown"? How is this related to the "China syndrome"?

20. A reactor can be shut down in a few seconds using the control rods, but a melting of the fuel rods is still possible. Why? Discuss the importance of a backup cooling system in a reactor.

21. Discuss the advantages and disadvantages of nuclear electrical generation as compared with fossil-fuel electrical generation.

Fusion and Fusion Reactors

22. Why is an A-bomb needed to start an H-bomb?

•23. Tritium can be made by neutron bombardment of lithium-6. Write the nuclear equation for this reaction.

24. The energy release for each fusion reaction is less than 20 MeV, whereas about 200 MeV of energy is released from a fission reaction. Even so, discuss the advantages that fusion reactors would have over fission reactors.

25. Why would a LOCA be less serious or less likely in a fusion reactor than in a fission reactor?

26. Distinguish between magnetic and inertial confinements. Why is confinement such a problem for fusion?

27. The Sun is a big fusion reactor. How are the fusion reactions confined in this case?

•28. How does fusion get started in stars?

Biological Radiation Hazards

29. Why is strontium-90 so readily absorbed in bones?

30. Why is iodine-131 such a hazard when released into the environment?

31. On the average, people normally receive a certain amount of radiation exposure or dose of radiation each year, for example, from diagnostic X-rays. About half of the radiation comes from what is called natural background radiation. What gives rise to "natural" radiation? How about "unnatural" radiation?

Astrophysics

A 10-hour exposure taken in Australia shows star trails around the celestial south pole — apparent motion is due to the Earth's rotation.

27

When taking a long look at the sky on a clear, starry night, have you not been awed by the vastness and the beauty? The distant stars and galaxies appear to be fixed and unchanging, but actually you are observing a small portion of a dynamic and evolving universe.

This intrigue of the heavens prompted the development of one of the earliest sciences— astronomy. At first, astronomy was the study and observation of "objects of the sky" that could be seen with the unaided eye. Our "known" universe was relatively very small, since the universe is all the matter and energy in existence anywhere, observable or not (known or unknown). The invention of the telescope (Chapter 22) expanded the known universe tremendously. The most distant observable object that now denotes the boundary of the known and unknown universe is believed to be some 16 billion light years away!

Into all of this is woven astrophysics, which is physics applied to astronomical situations. One of the major applications of astrophysics is in **cosmology**, the study of the nature and structure of the universe. This term is often broadened to include the origin and evolution of the universe—its past, present, and future.* The "cosmos" is the world or universe regarded as an orderly system. In this chapter we'll take a look at how astrophysics is applied in various theories of cosmology, and then we will focus on a particular evolution process—the life and death of a star.

■ THE EXPANDING UNIVERSE

The known universe has "expanded" as our observations and knowledge have increased. However, there is evidence that our universe is physically expanding—flying apart. During the early 1900's V. M. Slipher, an astronomer at the Lowell Observatory in Arizona, studied the spectra of light from nearby galaxies and noted that there were Doppler shifts (Chapter 15) in the spectra. By 1914 a dozen or more of the brightest galaxies had been studied. Strangely enough, most of the spectra showed a red shift rather than a random mixture of red and blue shifts, as might be expected. Slipher's measurements showed that most of these galaxies were

* Technically, the study of the origin or creation of the universe is called cosmogony.

Figure 27.1
The 200-in. (5.1-m) Hale telescope. Moonlight view of the dome with shutters open, showing the instrument. (See also Fig. 22.16.)

moving away from the Earth at high speeds. The most notable exception was the Andromeda galaxy, which showed a blue shift and hence is approaching.*

In the 1920's Edwin Hubble studied the spectra and measured the speeds of many galaxies at greater distances, using the largest telescope in the world at the time, the 100-in. (2.5-m) reflecting telescope on Mt. Wilson in California. Without exception, all of the distant galaxies were moving away from the Earth at high speeds. After World War II, the new 200-in. (5.1-m) telescope on Mt. Palomar was focused on even more distant galaxies and galactic clusters in the cosmos (Fig. 27.1), and all were found to be receding from the Earth at even greater speeds.

One result of Hubble's observations of a number of distant galaxies was that the more distant a galaxy, the faster it moved away. This was shown

* The Andromeda galaxy, which is relatively close to the Earth, is approaching our planet at a speed of 300 km/s. Galaxies have a certain amount of random motion, and for nearby neighboring galaxies, a small receding motion may be masked by random motions.

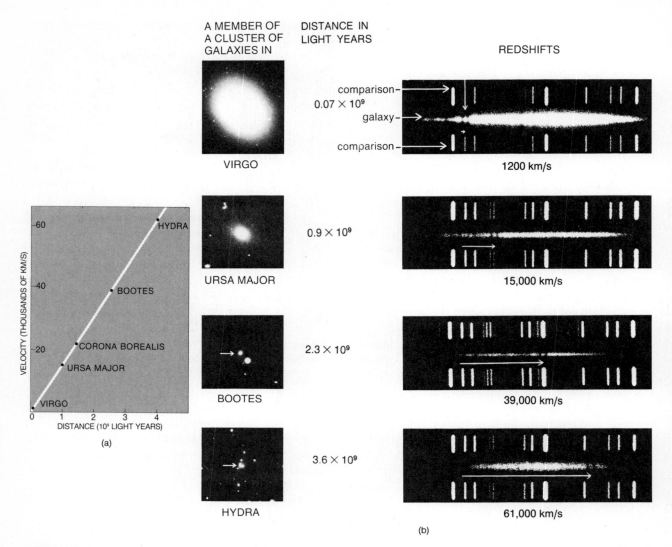

A MEMBER OF A CLUSTER OF GALAXIES IN

DISTANCE IN LIGHT YEARS

REDSHIFTS

VIRGO — 0.07×10^9

comparison —
galaxy —
comparison —

1200 km/s

URSA MAJOR — 0.9×10^9 — 15,000 km/s

BOOTES — 2.3×10^9 — 39,000 km/s

HYDRA — 3.6×10^9 — 61,000 km/s

(a)

(b)

Figure 27.2
Hubble's constant. (a) The red shifts of the spectra from galaxies show that their velocities are proportional to their distances, $v \propto d$. If velocity versus distance is plotted, Hubble's constant H can be determined ($v = Hd$). (b) The spectra of galaxies, like the examples shown here, provide the data necessary to determine Hubble's constant. The arrow below each horizontal streak of spectrum shows how far the spectral line being considered is red shifted. The spectrum of a known source located inside the telescope building appears as vertical lines above and below each galactic spectrum to provide a comparison with a red shift known to be zero.

generally to follow a simple linear relationship between speed and distance, or $v \propto d$ (Fig. 27.2). That is, if one galaxy is twice as far from us as another, it will be moving away twice as fast; if three times as far, the more distant galaxy will be moving away three times as fast, and so on. This relationship can be expressed mathematically by an equation that is called **Hubble's law**:

$$v = Hd$$

where H is called Hubble's constant.

■ ARE WE AT THE CENTER OF THE UNIVERSE?

If all the galaxies are generally moving away from us, as observations indicate, the question asked in this section heading quickly follows. Is the Earth at the center of the universe as the old geocentric theory predicts (Chapter 5)? The answer is that we can't say for certain, but probably not. If you were on a planet in another galaxy, you would see the distant

(a)

(b)

Figure 27.3
Expansion. When the balloon is blown up and expands, the dots get farther apart. An observer on any dot would see the other dots moving away, analogous to what we observe for galaxies from Earth.

Figure 27.4
Expansion. During baking, every raisin in a raisin cake would "see" all others moving away at a speed that depends on its distance. Each raisin would be at the center of expansion, measured from its own position. The universe is analogous to an infinitely large cake.

galaxies receding from you in the same way as we observe from Earth.

A common analogy used to demonstrate this is to picture a simplified model of the universe in which galaxies are represented by points on the surface of a balloon (Fig. 27.3). If the balloon is blown up, each point (galaxy) will move away from all other "galaxies." In addition, the speed at which any two galaxies separate will be proportional to the distance between them. In effect, an observer at any point on the balloon will see what we see from our Earth in the Milky Way galaxy "point."

Of course, we can't prove that this view represents the universe as a whole. As geocentric as we might be, it is highly improbable that we are at the center of the cosmos. Keep in mind that although the balloon model illustrates Hubble's law, the universe is not a surface like that of a balloon. A surface is "carved" out of three-dimensional space. The universe is three-dimensional, and a more realistic model must be carved out of four-dimensional space-time.

A three-dimensional analogy of the expanding universe is given by an expanding raisin cake during baking (Fig. 27.4). Every raisin would "see" every other raisin moving away from it at a speed that depends on the distance from it. Thus, each raisin would be at the center of the expansion, measured from its own position.

THE BIG BANG THEORY

Around 1927 Father Georges Lemaître, a Belgian astrophysicist educated as a Jesuit priest, suggested what has come to be known as the **"Big Bang"** theory of the universe. This theory was furthered and pro-

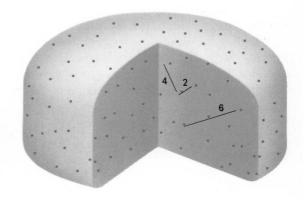

moted by George Gamow, a Russian-born physicist who emigrated to the United States in 1936. According to the Big Bang theory, the universe had a "beginning" and is approaching an "end." The universe began in this theory with all the matter of the cosmos in a primordial, high-density "cosmic egg," which exploded—in a "big bang." From that instant on, the universe expanded. As galaxies formed, they too shared in the expansion.

The concept of the formation of the universe is mind-boggling. Astronomers looking into space have noticed that on a very large scale the universe looks the same in all directions. This means that the universe had no special direction, and it is assumed that matter is distributed evenly throughout *all* space. As a result, we cannot consider that the big bang took place at the center of the universe, even though we commonly think of an "explosion" to have a central origin or location. We have to change our thinking a bit and think big and cosmological. As one astronomer explains it:

> Many students ask whether the fact that there was a big bang means that there was a center of the universe from which everything expanded. The answer is no; first of all, the big bang may have been the creation of space itself. Furthermore, the matter of this primordial cosmic egg was everywhere at once. There may be an infinite amount of matter in the universe, so it is possible that at the big bang an infinite amount of matter was compressed to an infinite density while taking up all space. We would have to imagine our expanding raisin cake extending indefinitely in all directions, with no edge.*

Gamow showed that the abundance of elements in the universe is what might be expected for the element formation in an ancient, expanding fireball. The density of the primordial substance of the cosmic egg, which Gamow called "ylem," † was very high, with all the mass of the universe being squeezed together like the particles in the atomic nucleus. The temperature also was very high— trillions of degrees. In the early stages of the expan-

sion after the Big Bang, the temperature dropped, and the protons or hydrogen nuclei of the "primordial plasma" began to fuse into helium nuclei. It is estimated that 20 to 30 percent of the hydrogen nuclei could have fused into helium during these early stages.

As the expansion continued and the temperature dropped further, very few heavier elements, such as lithium, were formed. The Big Bang theory does not explain the existence of the approximately 90 heavier elements. This is explained in the theory of stellar evolution, which will be discussed later. With further expansion and cooling, neutral atoms began to form, which were, for the most part, hydrogen atoms. Eventually, swirling masses of hydrogen condensed into galaxies and into stars within the galaxies (Fig. 27.5).

At present, as the evidence indicates, the universe is expanding—an expansion that, according

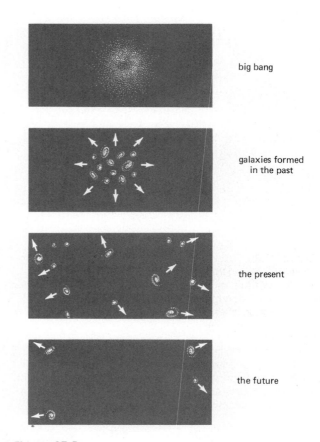

big bang

galaxies formed in the past

the present

the future

Figure 27.5
The Big Bang theory. According to this theory, the universe had a "beginning" and is approaching an "end." In the end, the cosmos will fade away into darkness.

---

* Pasachoff, Jay M., *Astronomy: From the Earth to the Universe,* Third Edition, Saunders College Publishing, Philadelphia, 1987.

† The name Aristotle gave to the basic substance from which all matter was believed to be derived.

to the Big Bang theory, will continue indefinitely. However, as this goes on, the density of matter in the universe decreases. In addition, as will be learned shortly, the hydrogen is being used up in the formation of stars that eventually "die." As fewer new stars are formed and the lights of the old stars go out, the galaxies will grow dim. In the "end" of the universe in the Big Bang theory, the cosmos will fade away into darkness.

Notice that the theory does not answer or address such questions as, Where did the primordial matter come from? What happened before the Big Bang? Why did the Big Bang occur?

THE AGE OF THE UNIVERSE

According to the Big Bang theory, the universe began at the time of the "big bang." If we knew when this occurred, we would know the theoretical age of the universe. An estimate can be obtained from Hubble's law ($v = Hd$). If the expansion had always proceeded at the same rate, then the time traveled by any galaxy would be $t = d/v$. Comparing this with Hubble's law, $d/v = 1/H$. The age of the universe with a Big Bang birth would be given by the value of $1/H$, where H is Hubble's constant.

Using the generally accepted value of H, this turns out to be about 15 to 20 billion years. However, the expansion rate would be expected to slow somewhat as a result of gravitational attraction, so that is the *maximum* age of the universe. Recently, newer techniques of estimating the distances to galaxies have given a value of H that places the maximum age of the universe between 10 and 12 billion years.

THE STEADY-STATE THEORY

In the late 1940's a group of British astrophysicists introduced another theory of cosmology, in which the universe has no beginning or end. In this theory, as the universe expands, new matter (hydrogen) is steadily created — out of nothing — throughout the cosmos. The newly created hydrogen in galaxies would provide the raw material for the formation of new stars to replace the old, as well as the formation of new galaxies in the vacant spaces left by expansion (Fig. 27.6). Hence, the universe remains steady or relatively unchanged in the **steady-state theory**.

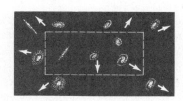

galaxies dispersing

new galaxies form to fill the gaps

Figure 27.6
The steady-state theory. In the steady-state theory, the universe had no beginning and will have no end. As the universe expands, new galaxies (matter) are created to fill the gaps, keeping the density of matter constant.

The idea of new matter just "appearing" at first seems unreasonable, but the necessary creation rate is very modest — only one hydrogen atom per cubic meter of space every billion years. Remember, the Big Bang theory has all the matter of the universe "appearing" in the cosmic egg all at once.

Many theories, when first proposed, were met with skepticism but are now accepted. Keep in mind that any idea or theory can be proposed, but the test comes in the scientific method.

THE TESTS OF THEORIES

The steady-state theory was quite popular during the 1950's and 1960's. It was considerably different from the Big Bang theory. Both predicted an expanding universe, so other predictions had to be considered to test the theories. The rate of expansion of the universe is one consideration.

Immediately after a Big Bang, the expansion would have been exceedingly rapid. With time, the expansion would slow down because of the attraction or backward-pulling force of gravity on the outermost receding parts of the expansion. This means that the Hubble constant would have a smaller value today than it had in the past. The steady-state theory, on the other hand, predicts the expansion rate or Hubble's constant to be "steady" or constant throughout time.

The possible change in H in a few years or a lifetime would be too small to detect. Perhaps if we could go back in time a billion years or so, there would be an observable change. But no recorded data are available from that far in the past. How about observing it directly, and even further in the past? Going back in time might at first seem impossible, but if you think about it, a telescope is a "time machine" that looks into the past.

Although light travels through space at a very fast speed, 3×10^8 m/s, it still takes time to cover astronomical distances. The Andromeda galaxy is about 2 million light years away, so the light we now receive from that galaxy is 2 million years "old," or we are receiving a "picture" of Andromeda as it was 2 million years ago (Fig. 27.7). Similarly, the picture of the Whirlpool Galaxy in Figure 27.7 is how it appeared 12 million years ago. In effect, you are looking back in time!

By looking closely at the receding speeds of galaxies at different distances and computing the Hubble constant, we can see if the Hubble constant has changed with time, in particular, whether it has become smaller. This was done for galaxies up to 6 billion light years away, and the results indicate that the Hubble constant was larger in the past than it is today. This evidence supports the Big Bang theory over the steady-state theory.

Other evidence, found somewhat accidentally, makes the Big Bang theory the winner of the contest. Around 1965 two Bell Laboratory physicists, Arno Penzias and Robert Wilson, who were using a radio antenna in experiments on a communications satellite noticed an annoying background microwave "noise" (Fig. 27.8). At first, they thought it was an instrument problem, but finally they established that this background radiation came from the sky and from *all* directions.

This information happened to be conveyed to some Princeton University researchers, in particular, Robert Dicke, who had predicted a feature of the Big Bang to be "fireball radiation." If the universe

(a)

(b)

Figure 27.7
Old pictures. (a) The Andromeda Galaxy is about 2 million light years away, so we receive a "picture" of Andromeda as it was 2 million years ago. (b) The Whirlpool Galaxy as it appeared 12 million years ago. In looking at the universe, we are in effect looking back in time.

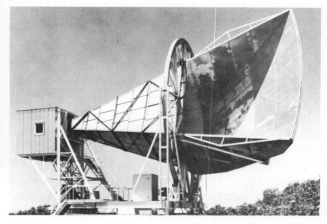

Figure 27.8
Detecting radiation. The large horn-shaped antenna used by Penzias and Wilson in the discovery of background radiation believed to have originated from the Big Bang.

had been formed in a Big Bang, it would have looked like a fireball, not unlike that seen when a hydrogen bomb explodes. As the fireball expanded and cooled, a great deal of radiation would have been emitted, and some of the original fireball radiation should be detectable today with a radio antenna. This radiation would be distinctive, inasmuch as it would fill the universe uniformly and hence should be detectable with the same intensity from *all* directions.

Penzias and Wilson had stumbled on evidence for the primordial fireball of the Big Bang (which won them a Nobel Prize). The **cosmic background radiation** is sometimes called **3-K radiation** because it appears to have the same spectrum as that coming from an ideal radiating body at a temperature of 3 K (see Chapter 24). This may seem strange, since the radiation from the cooling fireball came from a very hot environment. After about the first million years of expansion, the fireball had cooled to approximately 3000 K, and neutral atoms began to form (Fig. 27.9).

Prior to this, photons continually ionized any atoms that might form. A great deal of radiation was lost when neutral matter formed. But the radiation that is left doesn't look like 3000-K radiation because the observed radiation has been Doppler-shifted as a result of the expansion of the Big Bang material. The universe has expanded by a factor of about 1000 since this era, so the radiation appears to be from a body with a temperature about 1000 times lower, or 3-K radiation.

THE OSCILLATING UNIVERSE THEORY

Evidence for the Big Bang theory seems to dominate that for the steady-state theory. However, it is difficult not to think about or ignore what existed before the Big Bang. In the single "bang" of this theory, the mysterious explosion mixed matter and radiation in a primordial "soup," and any possible evidence of the conditions of an earlier universe was erased. Accordingly, there is no way of knowing the conditions that existed before the Big Bang. Moreover, the universe ends or fades away into darkness in the Big Bang theory.

If you prefer a more eternal outlook, as well as a prenatal universe, there is an oscillating theory of the universe, which modifies or extends the Big Bang theory such that the universe is self-renewing. When you throw a rock upward, it slows down and returns under the influence of gravity, or it is "pulled back" by gravity. Similarly, gravity acting through the universe pulls back on the expanding galaxies and slows their outward motions. If this retarding force were strong enough, it could bring the expansion to a halt at some time in the future! — like a rock thrown upward stopping at its highest point.

Then, like a rock returning under the pull of gravity, the universe would start to collapse or contract because of the ever-present force of gravity (Fig. 27.10). This would occur slowly at first, but as the galaxies gathered momentum on their "inward fall," they would rush together almost as violently as they flew apart from their explosive "birth." The contraction and compression of the "big crunch" would cause the universe to return to the form of another high-temperature, high-density cosmic egg.

The question then is, Will the egg "hatch" or explode again? No one knows for sure, but assuming it did, a new universe would be born in another Big Bang. The reborn universe would go through another expansion, followed by collapse, another creation, and so on. Thus, in this oscillating theory the universe "oscillates" through a never-ending cycle

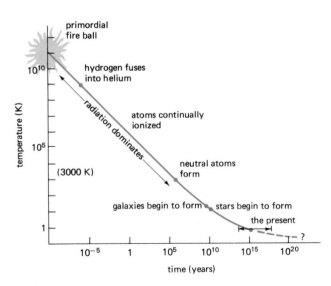

Figure 27.9
The cooling universe. A graph of temperature versus time. Using the general value of Hubble's constant, the big bang is computed to have occurred around 15 billion years ago. Newer measurements place the age of the universe to be nearer 10 to 12 billion years.

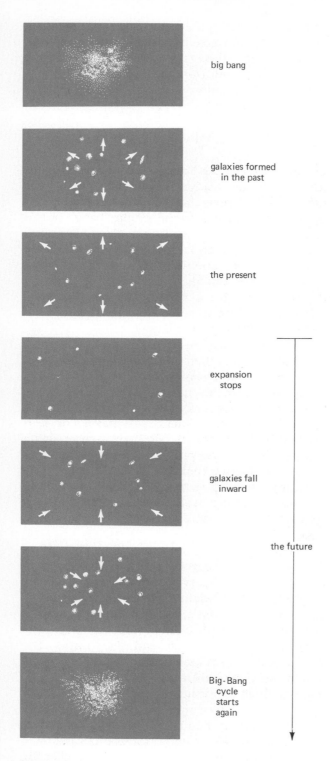

big bang

galaxies formed in the past

the present

expansion stops

galaxies fall inward

the future

Big-Bang cycle starts again

Figure 27.10
The oscillating theory of the universe. In this theory, the universe "oscillates" through never-ending cycles of birth, life, and death, with a new and different universe being formed with each new cycle.

of birth, life, and death, with a new and different universe being formed with each rebirth.

If you don't like explosions, a variation of the oscillating theory suggests that the approaching masses do not collide or implode but simply pass through each other and out the other side, thus giving rise to an oscillating universe without explosions. This is conceivable since the universe is mostly empty space.

All of this sounds nice, but how can such theories be tested? Certainly we can't wait around to see what happens, since any halt in the universe's expansion would occur many billions of years in the future. The critical point of the oscillating theory lies in whether there is enough mass in the universe so that the gravitational force is strong enough to eventually stop the current expansion. This means that a critical average density is required to reverse the expansion. Calculations show this to be about 5×10^{-27} kg/m³, which is equivalent to about three hydrogen atoms per cubic meter.

If the average density is computed from all the visible matter in the universe, it is found that this amount of matter is only about 10 percent of that needed. There is a great deal of radiation in the universe, which has a mass equivalence by Einstein's $E = mc^2$ equation. Taking this into account increases the amount of mass by only 1 or 2 percent.

Of course, we don't see everything in the universe. Not everything is luminous, for example, burned-out, dead stars, distant stars with small intensities that we cannot see, and black holes. Estimates of this "invisible" mass have been made through the study of motions in a galactic cluster and the required gravitational force involved, which gives an indication of the total mass—visible and invisible. Although the results indicate that there is much more mass in the universe than we can see, the total mass is still about 10 percent short of the critical mass needed for an oscillating universe.

Recent data indicate that an elementary particle called the neutrino, which had been thought to be massless, may have a tiny mass. Neutrinos result from nuclear reactions in stars, and they pervade the universe. If this presumed mass is added, the universe's mass is near the boundary between being opened or closed. More measurements are needed for a definite answer, but we have plenty of time for further investigations.

Cosmologists distinguish between the Big Bang theory and the oscillating theory by talking about open and closed universes. An open universe is composed of space with no boundaries, in which the expansion of the galaxies continues forever. The open-ended Big Bang theory predicts an open universe. A closed universe has limits or boundaries, in which galaxies can recede only a limited distance from each other before coming together. The oscillating theory predicts a closed universe.*

Recall from Chapter 5 that in Einstein's general relativity model, gravity is not treated as a "force." Instead, mass is pictured as distorting or warping four-dimensional space-time. If there is enough matter in the universe, then there is a closure of space-time. In effect, this means that a galaxy would travel around a "closed universe" and return to its starting point. This is analogous to a particle traveling on a spherical surface in three-dimensional space.

Another way of looking at this is to consider the effect of gravity on light beams (Chapter 23). In a closed universe, two parallel light beams would eventually meet (Fig. 27.11). In an open universe, the parallel light beams would eventually diverge. Should the universe have exactly the critical amount of mass so that it is neither open nor closed, the beams would remain parallel.

■ STELLAR EVOLUTION—THE BIRTH, LIFE, AND DEATH OF A STAR

As gaseous matter expanded from the universe's explosive birth, swirling "pockets" of gas were randomly formed, being held together by gravity. In time, these gas pockets gravitationally collapsed into isolated, dense clumps of gas, called protogalaxies (galaxies in formation). Inside the protogalaxies, smaller "sub-clumps" of gas came together, which were protostars (stars in formation). We now see the millions of galaxies that resulted from protogalaxies. One of these is our own Milky Way Galaxy, which is an aggregate of some 100 billion separate stars held together by mutual gravitational attraction (Fig. 27.12).

Protostars eventually become stars, one of which is our Sun. The Sun is about 4.6 billion years old, whereas the Milky Way sports an age of 12 to 14 billion years. This implies that not all the stars of a

* If the universe did not oscillate by never rebounding from the "big crunch," this would also be a closed universe.

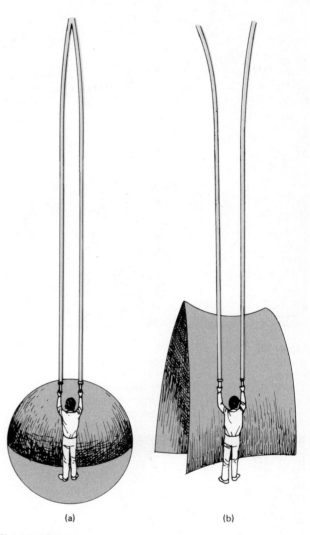

(a)   (b)

Figure 27.11
Open or closed universe? In a closed universe, two parallel light beams would eventually meet (a). In an open universe, the parallel light beams would eventually diverge (b).

galaxy are formed at the time of galactic birth. In fact, stars are being born now, as well as dying. This process of stellar evolution forms one of the most interesting topics of astrophysics.

Stellar Birth

A star is conceived in large clouds of interstellar gas (Fig. 27.13). By some random fluctuation, enough gas atoms or molecules come together in a clump of gas. This cluster of predominantly hydrogen atoms,

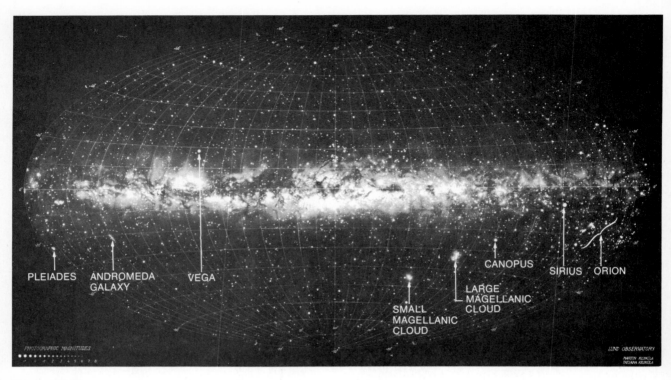

Figure 27.12
The Milky Way. A drawing showing an equatorial view of the Milky Way Galaxy. It is a vast, rotating, wheel-shaped system of approximately 100 billion stars, with a diameter on the order of 100,000 light years. Where are we located? The Sun is about two thirds of the way from the center out to the rim of the wheel in an orbital revolution about the galactic center that takes about 200 million years. Unlike a wheel, the galaxy does not rotate as a solid body.

held together by the grip of its own gravity, is a protostar.

The continual pull of gravity within the clump of gas causes the protostar to shrink. That is, the protostar collapses under its own gravitational attraction. When it does so, the density and temperature of the protostar increase. The protons or hydrogen nuclei falling inward gain speed, and hence the temperature of the gas increases. After about 10 million years of collapse, the interior temperature of the protostar reaches about 10 million kelvin.

At this critical temperature, protons have enough speed to overcome their mutual electrical repulsion, and nuclear fusion is initiated at the center of the contracting mass — a star is born! The release of nuclear energy from the fusion of hydrogen nuclei into helium nuclei generates an outward pressure that balances the inward pressure due to

Figure 27.13
Interstellar gas. The gaseous nebula in the constellation Orion.

the force of gravity, the gravitational collapse stops, and the shining, newborn star is stable.

Small pockets of contracting gas may not have enough mass and gravity for the collapse to result in the initiation of fusion and become a star. These objects are sometimes called **brown dwarfs**. They are faint, small objects radiating the heat of compression in infrared radiation.

Stellar Life

The hydrogen "burning" stage of a star, such as the Sun, accounts for about 90 percent of its lifetime. The Sun is "middle-aged." Being about 4.6 billion years old, it has 4 to 5 billion years to live.

As with humans, some stars have relatively short lifetimes, whereas others exist considerably longer. The lifetime of a star is related to its mass. Surprisingly, the largest stars have the shortest lifetimes. They have more fuel to burn, but they burn it faster than smaller stars. This is due to higher temperatures resulting from greater gravitational attraction, which increases the reaction rate. A star 15 times more massive than the Sun may live only 10 million years (a lifetime 1000 times shorter than that of the Sun). On the other hand, a star with only a fraction of the Sun's mass may live 100 million years or even a trillion years.

Stellar Old Age

The helium produced by the fusion or burning of a star's hydrogen accumulates at the center of the star, where most of the nuclear reactions take place. When a star's hydrogen fuel starts to run low and it is predominantly composed of a helium core, it begins to show signs of old age.

The temperature of the helium core is not high enough for its fusion, so with no release of energy the core begins to collapse under the influence of gravity. The shrinking helium core gets hotter, which causes the inner portion of the surrounding hydrogen shell to "burn" faster. It might be thought that the star would become brighter, but much of the energy is absorbed by the gaseous hydrogen shell, and it expands outward.

With a greater surface area to emit the energy that escapes, the surface temperature of the star drops, and it appears distinctly red. When the expanded hydrogen shell cannot absorb any more energy, the energy escapes as radiation, and the bright-ness or luminosity of the star soars. In this condition of being an expanded, brilliant red star, it is called a **red giant**.

Stellar Death

Any star exhausts its original hydrogen fuel eventually and then heads toward demise. How this happens depends on the mass of the star. Let's first consider the death of a relatively small star.

As the red-giant stage of any star progresses, the helium core continues to collapse. The critical temperature for stellar hydrogen fusion is about 10 million kelvins, but getting helium nuclei to fuse requires temperatures on the order of 100 million kelvins. When the temperature of the star's core grows large enough for helium fusion, there is a sudden burst of liberated energy, called the helium flash. The star then continues to fuse helium nuclei into carbon, and the burning of helium prevents further collapse (Fig. 27.14).

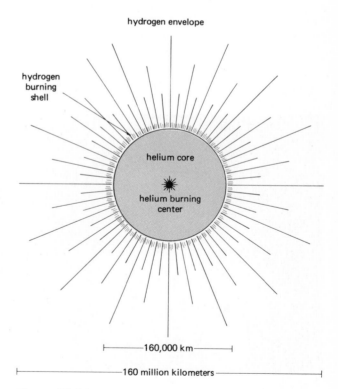

Figure 27.14
Red-giant star. With a huge hydrogen envelope at a relatively low temperature, the star appears red. Further gravitational collapse causes helium fusion to be initiated in the core.

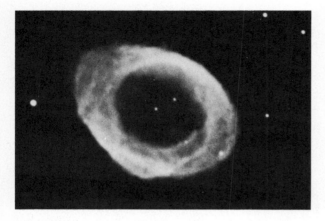

Figure 27.15
The Ring Nebula. A planetary nebula in the constellation Lyra.

end, its outer layers expand outwardly, forming a planetary nebula (Fig. 27.15). So called because they appeared similar to the outer planets when first discovered with small telescopes, these fuzzy (nebular) forms have nothing to do with planets but are signs of aging stars.

As a relatively small star approaches the end of its energy generation and life, its shrunken core becomes extremely dense, with a volume about as big as that of a good-sized planet. Although such a star is very faint, its surface is quite hot, in fact, white-hot, and we call it a **white dwarf**. Calculations show that if a star's mass is less than about 1.4 solar masses at this critical point in its life, it will become a white dwarf. The Sun will almost certainly become a white dwarf, which quietly and increasingly slowly radiates its energy into space and cools. Eventually ceasing to "shine" after billions of years, a white dwarf becomes a **black dwarf**, or a cold hulk in the "graveyard" of stars.

A different fate awaits a more massive star — an explosive death. For stars a few times more massive than the Sun, this occurs with a "carbon flash" as soon as the carbon burning begins. In still more massive stars, carbon fusion sets in and continues. Successive contractions and fusion reactions pro-

It might be thought that this cycle would continue with carbon fusion, and so on, but this does not happen for a small star that does not have sufficient mass for gravitational compression to raise the temperature of the core to about 600 million kelvins. Eventually, as the star's helium fuel nears its

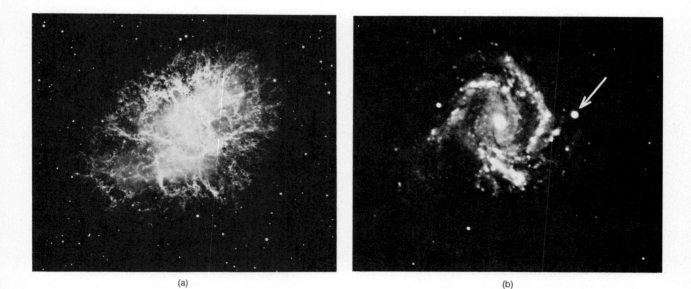

(a)

(b)

Figure 27.16
Supernovas. (a) The Crab Nebula is a great expanding cloud of gas, the remnant of a supernova. (b) The supernova that appeared in the galaxy NGC 4303 in 1961 shone for several weeks with an absolute brightness 300 million times that of the Sun.

duce heavier elements until iron is formed. Iron is very special, inasmuch as it marks the limit of exoergic fusion.

The formation of heavier elements would require energy input (endoergic). That is, iron is the demarcation point between fusion and fission (see Fig. 26.7). Elements heavier than iron would not release energy if their nuclei underwent or could undergo fission. With no more fusion, the massive star gravitationally collapses, then explosively rebounds like a giant compressed spring.

Such an exploding star is called a **supernova.*** It is the most energetic of all stellar explosions. They occur somewhat frequently on an astronomical time scale. The earliest reported supernova was observed by Chinese astronomers in A.D. 1054. This "guest star," as they called it, exceeded the brightness of any object in the night sky for several weeks. The remnant of this supernova is now observed as a great expanding cloud of gas called the Crab Nebula (Fig. 27.16). Its expansion rate is on the order of 1600 km/s (1000 mi/s).

Many supernovas have been observed in other galaxies (Fig. 27.16). The last observed supernova in our galaxy was seen in 1604, which was five years before the telescope was invented. Only five supernovas have been seen in the Milky Way in 1000 years. However, in 1987 astronomers got the treat of observing Supernova 1987A, just 170,000 light years away in the Greater Magellanic Cloud, which is just outside the edge of the Milky Way. (See color plate in this section.)

In the cataclysmic explosions of supernovas, heavier elements are made by nuclear reactions. In the final moments of these stars' lives, it is assumed that all of the elements found in nature are made, even those heavier than iron, for example, gold, lead, and uranium. The heavier elements above iron are relatively rare compared with the lighter elements, as might be expected.

All the elements are scattered into space and become mixed with interstellar hydrogen, which will eventually be condensed into protostars. With the evolution of stars and, presumably, of planet "spin-offs" such as our Earth, the elements are recycled. According to this theory, both the Earth and you are made up in large part of "stardust."

---

* A nova or "new" star is a brilliant exploding star that blows off a small amount of mass. This is believed to be associated with an exchange of mass in a binary star (a two-star system) that has white-dwarf and red-giant components.

■ PULSARS, NEUTRON STARS, AND BLACK HOLES

In late 1967 at a radio telescope installation in England, a graduate student, Jocelyn Bell, noticed that a celestial radio source emitted regular pulses or "beeps" of radio waves every 1.33733 s. At first it was thought that these might be signals from an intelligent life form in space. Several other beeping celestial "radio stations" were soon found. Not surprisingly, the sources were at first dubbed LGMs (for Little Green Men). However, subsequent investigations excluded this possibility, and the term **pulsar** replaced LGM.

An explanation of pulsars came from a previous prediction. In the explosion of a supernova, a tremendous inward pressure would compress the core of the star. This would be so great that individual electrons and protons would combine to form neutrons. Hence, it was predicted that an incredibly dense ball of pure neutrons, or **"neutron star,"** would be left at the center of a supernova explosion. A comparison of some stellar densities is shown in Figure 27.17. In 1968 a pulsar was discovered at the center of the Crab Nebula, right where a neutron star would be expected to be located.

But what causes the regular pulses or beeping? Scientists believe that violent, localized disturbances occur on the surfaces of neutron stars (which are only about 25 to 30 km in diameter), similar to those on the surface of the Sun. These localized storms emit radiation in well-defined directions — like beams of light. If a neutron star were spinning rapidly, the beam would be observed as a series of pulses or beeps, much the same as from a rotating flasher seen on police cars. Hence, pulsars are believed to be rotating neutron stars that resulted from supernova explosions.

Calculations show that a neutron star cannot exist with a mass three to five times greater than that of the Sun. When a very massive star becomes a supernova, it is speculated that the large core continues to collapse right past the density of a neutron star and forms a **black hole**. Recall that the gravitational attraction of a black hole is so great that it prevents light or anything else from leaving it. Hence, black holes must be invisible, and we can only search for them by indirect means, as described in Chapter 5. The fascinating subject of black holes is treated in more detail in Special Feature 27.1 at the end of the chapter.

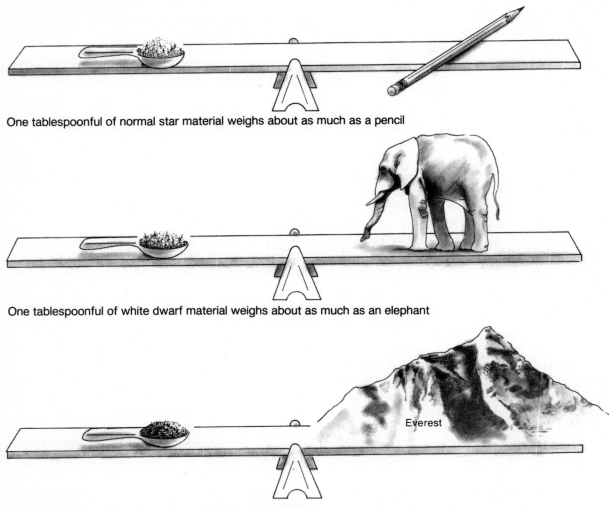

One tablespoonful of normal star material weighs about as much as a pencil

One tablespoonful of white dwarf material weighs about as much as an elephant

One tablespoonful of neutron star material weighs about as much as Mount Everest

Figure 27.17
A comparison of stellar densities.

QUASARS

Before leaving our study of astrophysics, let's take a brief look at something you may have heard and wondered about — quasars. When radio telescopes became a common astronomical tool (see Chapter 22), astronomers busied themselves viewing the "radio sky." In addition to many galaxies and nebulas being radio sources, some "stars" were found to emit strong radio signals — ordinary stars that were assumed to be in our galaxy. It was not understood how an ordinary star could emit such large amounts of radio waves, but for the moment, the strange objects were dubbed "radio stars."

Investigations of the visible spectra of these radio stars gave rise to even more intrigue. No spectra like these had been seen before coming from ordinary stars or galaxies. It was finally realized that the problem of nonrecognition occurred because the spectra were extremely shifted toward the red end of the spectrum. If this is a Doppler shift, it indicates that the "radio stars" are receding from us at very high speeds.

The first two "stars" studied showed recession speeds of about 15 and 30 percent of the speed of light. Others have red shifts that indicate they must be moving at speeds greater than 90 percent of the speed of light! Using Hubble's law, we find that this puts these "stars" at distances of 3, 6, and 16 billion

Black Holes

The idea of black holes has caught the public's fancy. Imagine — something so dense that neither light nor anything else can escape from it. Black holes are becoming increasingly common in science fiction, but in actuality they are still largely theoretical. Any object, even you, could become a black hole if it shrank enough, but in most cases this would be highly unlikely. The most favored candidates to become (or that may have become) black holes are collapsing stars that have too much mass to become neutron stars. Such stars have three to five times the mass of the Sun.

As we said, any object could theoretically become a black hole if it collapsed enough to a sufficiently small size or radius. This, of course, would depend on the mass of the object. As a star collapses under the influence of gravity, the important point to focus on is the gravitational force at the *surface* of the shrinking star.

The force gets larger because the distance from the surface to the center becomes less (recall that $F \propto 1/r^2$, Chapter 5). If a star's radius shrinks by a factor of ½, then the attractive gravitational force at the surface would *increase* by a factor of four. Similarly, if the star shrinks so that its radius is 10 times smaller, the gravitational force at the surface is 100 times stronger, and so on.

Recall from the chapter that a photon leaving a star loses some energy as a result of gravity, which gives rise to a gravitational red shift. For a photon leaving the Sun, this energy loss is about 0.0005 percent. If the Sun should collapse, the increased gravity at the surface would rob more energy from the photons.

Finally, if the Sun shrank to a radius of about 3 km, the escaping photons would lose 100 percent of their energy. In other words, they would not escape at all, and the Sun would become a black hole. Of course, the Sun is not expected to do this, as mentioned previously. If the Earth could shrink to become a black hole, its gravitational force at the surface would not let light escape when the Earth had a radius of 1 cm! (On a relative scale, what would be your size when your mass became a black hole?)

The idea of light rays escaping at various stages from a shrinking star is illustrated in Figure 1. At the size of a neutron star, the light rays (or photons) still escape, but they are bent considerably, owing to the strong gravitational field at the surface. As the star shrinks further, the rays are bent more, and finally some of the more horizontal rays are bent back into the star and do not escape. Only the rays emitted within a certain **event cone** can escape. As the star contracts still further past a certain critical point or radius, light and

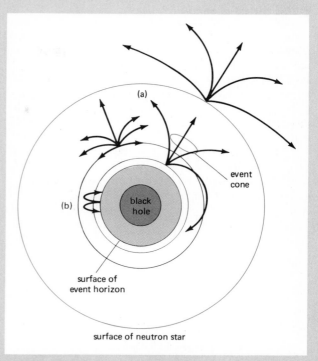

Figure 1
Light and a shrinking star. At (a) light is bent by dense stars. With further contraction, only light within an event cone escapes from the surface. At a critical (Schwarzschild) radius, which defines the surface of the event horizon (b), light can no longer escape, and a black hole is born. The mass of a black hole may continue to shrink.

everything else are trapped inside — a black hole is born!

The critical size or radius at which a shrinking star becomes a black hole is obtained by setting the escape velocity at the surface equal to the speed of light. The surface at this point is called the **event horizon,** and the critical radius is called the **Schwarzschild radius.*** As noted previously, this would be 3 km for a star the mass of the Sun. For a star with ten solar masses, the Schwarzschild radius would be about 30 km, making the density of a black hole incredibly large.

The boundary of a black hole is the surface of the event horizon at its Schwarzchild radius. The matter within could continue to contract, but the force of gravity at the event horizon would still be the same. Thus, the boundary of a black hole is not a sphere of

---

* Named for Karl Schwarzchild, a German astronomer and mathematician.

Black Holes (*Continued*)

matter, but the radius at which the gravitational force is sufficiently strong to keep light from escaping. The actual radius of the matter of a black hole may be much less. What form matter takes inside a black hole is unknown. If we ever determine the location of a black hole in space, any probe we might send to check this out could never report back when it arrived. Think about it.

The experimental proof of the existence of black holes is another thing. How do you observe a black hole? The most likely possibility comes from binary star systems, which are quite common. Cygnus X-1, the first X-ray source discovered in the constellation Cygnus,

provides the best evidence so far for a black hole in our Milky Way Galaxy. (Approximately 100 or more X-ray sources have been discovered by observations from rockets and satellites. X-rays do not penetrate the Earth's atmosphere.) If we look at the location of the Cygnus X-ray source visibly (Fig. 2), we can also observe a giant star whose spectra show Doppler shifts, indicating an orbital motion.

It is speculated that one member of a binary star system has become a black hole. Matter drawn from the other member would cause an **accretion disk** of spiraling matter around the black hole (Fig. 3). The matter falling into the disk would be accelerated and heated. Collisions would give rise to X-rays, and the black hole would be an X-ray source.

So, maybe black holes exist and maybe they don't. Time and investigations (the scientific method) will tell. In the meantime, there are a lot of speculations. Matter directed toward a black hole would be gone forever. Perhaps black holes are "vacuum cleaners" in space, gobbling up any matter that comes within its event horizon. As a black hole gains more matter, its horizon would expand. However, the consumption of matter does not sit well with our idea of conservation of mass.

Where does the matter go? This question has generated speculation that perhaps matter going into a black hole "pops out" somewhere in a parallel universe from what has been termed a **"white hole."** This may sound like a support of the steady-state theory of the universe, but the idea is questionable in terms of a black hole being the ultimate result of a collapsing star. Perhaps the situation can be summed up in the words of the scientist John Haldane: "My suspicion is that the universe is not only queerer than we suppose, but queerer than we can suppose."

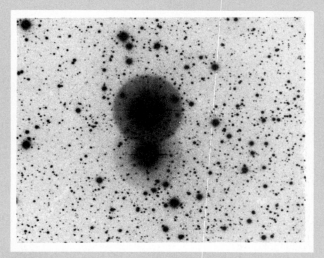

Figure 2
Cygnus X-1. The overexposed large dark spot at the center is a giant star believed to be a companion of the X-ray source and possible black hole, Cygnus X-1.

Figure 3
X-rays and black holes. Matter drawn from the other member of a binary star system forms a spiraling accretion disk around the hole. Matter falling into the disk is accelerated, and collisions give rise to X-rays.

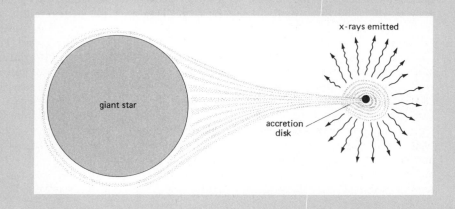

light years, respectively. The latter would be the most distant object known. Thus, it would seem that quasars are on the forefront of the expanding universe.

The fact that we can see these objects at such tremendous distances means that they would have to have enormous energy outputs. In fact, they can't be stars as we know them, since their energy outputs are comparable to those of galaxies. Hence, it appears that the "radio stars" are not stars at all. Since these strange objects looked somewhat like stars, they were given the name *quasi-stellar radio sources.* The *quasi-stellar* part of the name means "star-like." The name was quickly shortened to quasar.

Quasars at present are an astronomical puzzle or mystery. If they are at such great distances, then their energy outputs must be up to 100 times greater than those of large galaxies such as the Andromeda Galaxy and our own Milky Way Galaxy. Furthermore, data indicate that quasars are very small — only light days to light weeks in diameter. (The Milky Way Galaxy is about 100,000 light years across.) Such an energy source is difficult to imagine.

Perhaps the red shift is not really a Doppler shift. In Einstein's general theory of relativity, gravity affects (light) photons, and the photons lose a fraction of their energy in escaping from a star. If gravity robs the photons of energy, then by the energy-frequency relationship, $E = hf$, the frequency becomes smaller and the wavelength becomes longer — a red shift. This is called a gravitational red shift. However, to radiate as much energy and to produce the observed large red shift of light from quasars, an object would have to collapse quickly into a black hole. This makes the gravitational mimicking of the effects of a Doppler shift unlikely.

Other questions arise: Is Hubble's law valid for large distances? Could quasars really be nearby objects, which would then put them more in line with galactic energy outputs? Could the energy of a distant quasar result from the supernova of two colliding stars in a distant galaxy? All these questions have been asked, with no definite (proven) answers.

At present, the consensus is that quasars have giant black holes at their centers. These may be giant versions of the large black holes that may well exist at the centers of our own and other galaxies. As mass falls into the black hole, energy is given off (see Special Feature 27.1). Spectrum data have tended to support such a theoretical model. The riddle of the quasars is just one more example of the unknown that makes science so interesting and fascinating.

SUMMARY OF KEY TERMS

Astrophysics: physics applied to astronomical situations.

Cosmology: the study of the nature, structure, and origin of the universe.

Cosmos: the world or universe regarded as an orderly system.

Hubble's law: a relationship between the speed and distance of receding objects, $v = Hd$, where H is Hubble's constant.

Big Bang theory: a theory of cosmology that views the universe in the process of expanding from an explosive beginning. The expansion is indefinite, and the universe will eventually fade away into darkness.

Steady-state theory: a theory of cosmology that views the universe as having no beginning or end, but being replenished by the appearance or creation of matter out of nothing.

3-K background radiation: radiation permeating space that is believed to have come from the original fireball of the Big Bang beginning of the universe.

Oscillating universe theory: a theory of cosmology that predicts that the expanding universe will stop, contract, and be reborn with another explosion. The process continues indefinitely, and the universe oscillates between expansion and contraction.

Open universe: a universe with no boundaries, in which the expansion of the galaxies continues forever.

Closed universe: a universe with limits or boundaries in which galaxies can recede only a limited distance from each other before coming together.

Red giant: a star near the end of its hydrogen burning that has an expanded hydrogen shell that appears red.

White dwarf: a star at the end of its helium burning that is white-hot as a result of core contraction.

Black dwarf: the remnant of a white dwarf after it has radiated its energy into space and cooled.

Supernova: an exploding star.

Pulsars: pulsating radio sources believed to be rotating neutron stars that are formed and located at the center of supernovas.

Black hole: a gravitationally collapsed star that is so dense that light cannot escape from it.

Quasar (quasi-stellar radio source): a star-like object thought to be a galaxy, whose spectrum shows large red shifts. This indicates a large speed of recession, presumably because of the object's enormous distance from us.

QUESTIONS

The Expanding Universe

1. Considering the definition of the cosmos, explain the term used to describe particles called *cosmic* rays.
2. Describe what the known universe consisted of for early civilizations. What did the invention of the telescope provide? What is the present-day boundary of the known universe?
3. Is it possible that some of the stars we see no longer exist? Explain.
*4. If you plotted values of the Hubble constant determined from galaxies at various distances versus time, would you get a horizontal straight line? Explain.

Theories of the Universe

5. Are we at the center of the universe?
6. Can we determine exactly where the Big Bang took place? Explain.
7. Suppose three cars started from the same place and then moved along a straight road at different rates. Considering yourself to be in the middle car or at the center of things, what would the other cars appear to be doing relative to you? What would the drivers of the other cars see?
*8. Discuss the average density of the universe in terms of the (a) Big Bang theory, (b) steady-state theory, and (c) oscillating universe theory.
9. Is there a violation of the conservation of mass-energy in any of the theories of cosmology? Explain.
*10. Explain how Hubble's constant would be expected to be affected according to the oscillating universe theory.
11. How is the appearance of the railroad tracks in Figure 27.18 analogous to the behavior of parallel light beams in a closed universe?
12. Does the steady-state theory predict an open or closed universe? How would the continual creation of matter required by the theory affect this?

Stellar Evolution — The Birth, Life, and Death of a Star

13. When does a protogalaxy become a galaxy? A protostar become a star?
14. Explain why a star has such a long gestation period.
*15. How does stellar evolution support the Big Bang theory over the steady-state theory?
16. Will the Sun ever be able to have carbon fusion? Explain.
*17. Stars having up to four solar masses can become white dwarfs, but a white dwarf itself cannot have more than 2.4 solar masses. How is the mass lost?
18. When the Sun becomes a red giant, it will expand to a diameter of about 100 million miles. Considering this, what can you say about the future of Earth?

Figure 27.18
Do the tracks meet? See Question 11.

19. What determines whether a star becomes a white dwarf, a neutron star, or a black hole?

20. Give a brief account of what the matter in the universe has been doing for the last 15 billion years and what it will be doing in the future.

•21. Are there any similarities between the lifespans of very massive stars and those of overweight persons?

22. Assuming that a star collapsed directly into a black hole, the black hole would have the same amount of mass as the star. Why would the gravitational effects be any different?

23. Determining the distances to quasars using Hubble's constant is equivalent to linearly extrapolating Hubble's law in Figure 27.2 to many more billions of light years in distance. Is this really justified? (*Hint:* Think in terms of extrapolating Hooke's law for a spring, $F = kx$.)

24. Distinguish between a quasar, pulsar, neutron star, and black hole.

25. (a) Quasars are believed to be only light days to light weeks in diameter. What are these lengths in kilometers and miles ($c = 3 \times 10^5$ km/s $= 1.86 \times 10^5$ mi/s, or 1.1×10^9 km/h $= 6.7 \times 10^8$ mi/h)? (b) The diameter of the Sun is on the order of 10 km. What is this length in an appropriate light-distance unit such as light seconds?

26. How is it possible to detect a black hole when no light escapes from it?

APPENDIX

An Extended
View

The exercises in this Appendix are optional, provided here with extended discussion of topics from selected chapters. This additional material is for instructors who feel that it could benefit their students.

It is sometimes desirable or necessary to convert a measurement from a unit in one system to a unit in another, e.g., feet to meters. Some approximate conversions were involved in the chapter questions, and it is quite easy to make more accurate conversions. Just as you need to know equivalent words to translate from one language to another, e.g., *table* (English) and *mesa* (Spanish), you need to know a **conversion factor** to convert from one unit to another. If you didn't know the Spanish word for table, you could look it up in an English-Spanish dictionary. Similarly, we look up conversion factors for units in tables such as Table A.1. For example, note that 1 ft is equivalent to 0.305 m.

You have made unit conversions within our customary system, probably doing them in your head in many instances. Suppose you were asked how many yards there are in a length of 9 ft. You would no doubt give a prompt answer of 3 yd. How about how many inches are there in 2 ft? Easy—24 in.

What did you do in these cases? In the first conversion 9 ft was divided by 3 because 1 yd = 3 ft, or 1 yd/3 ft (read "1 yd per 3 ft"), which is a conversion factor. In the second case, 2 ft × 12 = 24 in., because 1 ft = 12 in., or 12 in./ft. So, to make a conversion you multiply (or divide) by a conversion factor.

So there is no confusion as to what to do, it is convenient to "let the units do it for you." For example, in the latter case previously, we may write

$$2 \, \cancel{ft} \quad \left(\frac{12 \text{ in.}}{1 \, \cancel{ft}} \right) = \quad 24 \text{ in.}$$

(quantity) (conversion factor) (converted quantity)

Table A.1
Basic Conversion Factors

Metric	British
1 cm	0.394 in.
1 m	1.094 yd
	3.28 ft
	39.37 in.
1 km	0.62 mi
1 g (weighs)	0.035 oz
1 kg (weighs)	2.2 lb
1 L	1.06 qt

British	Metric
1 in.	2.54 cm
1 ft	0.305 m
1 yd	0.914 m
1 mi	1.61 km
1 oz (has a mass of)	28.35 g
1 lb (has a mass of)	0.455 kg
1 qt	0.94 L

Note that the units cancel similar to numbers and the inch unit is left, which is what is wanted.

Try this one. What is the height of a 6-ft person in meters? (Could you do this in your head? Probably not.) From Table A.1, we see that 1 ft = 0.305 m; i.e., there is 0.305 m "per" ft, or (0.305 m/ft). [The "1" for the quantity in the denominator of a conversion factor, e.g., 1 ft, is often omitted.] Then,

$$6 \, \cancel{ft} \left(\frac{0.305 \text{ m}}{\cancel{ft}} \right) = 1.83 \text{ m}$$

where the mathematics may be quickly done on a calculator. Note that we could also have written (1 ft/0.305 m). But with the conversion factor in this form, the units would not have canceled properly. Hence, unit cancellation tells you whether to multiply or divide.

We can also see from Table A.1 that 1 m = 3.28 ft or (1 m/3.28 ft). This conversion factor could also be used:

$$6 \text{ ft} \left(\frac{1 \text{ m}}{3.28 \text{ ft}} \right) = 1.83 \text{ m}$$

The units will tell you how to write the conversion factor. Generally, the unit you want to convert to is on the top of the conversion factor ratio, and the unit to be converted is on the bottom. The rest is simple arithmetic.

Example: A student weighs 160 lb. What is his mass in "kilos"?

Solution: Note in Table A.1 that 1 lb has an equivalent mass of 0.455 kg, and

$$160 \text{ lb} \left(\frac{0.455 \text{ kg}}{\text{lb}} \right) = 73 \text{ kg}$$

or alternatively using the factor that 1 kg has a weight of 2.2 lb,

$$160 \text{ lb} \left(\frac{1 \text{ kg}}{2.2 \text{ lb}} \right) = 73 \text{ kg}$$

See how easy! Just make sure the units cancel properly.

Example: Given that 1 in. = 2.54 cm, what is the length in inches of a table that measures 1.27 m?

Solution: Here we do a double conversion using the well-known fact that there are 100 cm in a meter.

$$1.27 \text{ m} \left(\frac{100 \text{ cm}}{\text{m}} \right) \left(\frac{1 \text{ in.}}{2.54 \text{ cm}} \right) = 50 \text{ in.}$$

In some instances, a direct conversion factor may not be known. A series of conversion factors may be used to get to the unit wanted, as illustrated in this example for a double conversion. Any number of conversion factors may be used. As long as the units cancel properly, you're OK. Note that the conversion factor of 1 in. = 2.54 cm is a convenient factor to remember to convert between the metric and British lengths, a factor that "bridges the gap," so to speak. The decimal base of the metric system makes it easy to change a length to centimeters, and the conversion factors for inches, feet, and yards are well known.

■ EXERCISES

1. If a meter were divided into three equal parts as a yard is divided into 3 ft, what would be the length of such a metric "foot"?

   Answer: 0.333 m or 33.3 cm

2. At a particular signpost a town is listed to be 30 mi away. How would the distance to the town appear if posted in metric units?

   Answer: 48 km

3. How many pints are contained in a 2-L bottle of soda?

   Answer: 4.2 pt

4. Which has a greater weight and by how much, a metric tonne or a customary ton (2000 lb)?

   Answer: metric tonne, by 200 lb

5. At a metric weight-watchers club, how would each of the following weight losses be reported?
   (a) 22 lb, (b) 11 lb, (c) 2.2 lb, (d) 1.1 lb, (e) 0.55 lb

   Answer: (a) 10 kg, (b) 5.0 kg, (c) 1.0 kg, (d) 0.50 kg, (e) 0.25 kg

6. (a) A female student is 5 ft, 2 in. tall and weighs 120 lb. What are her height and mass in metric units? (b) Give your own height and mass in metric units.

   Answer: (a) 1.57 m and 54.5 kg

7. A homemaker ordinarily plans a meal with ¼ lb of meat per serving. Twelve people will be at a meal on Sunday. However, the butcher shop just went metric. About how many kilos of meat should be purchased?

   Answer: 1.4 kg

8. Convert each of the units in the following sayings to their metric equivalents.
   a. I wouldn't touch that with a 10-foot pole.
   b. The cowboy wore a 10-gallon hat.
   c. Give him an inch and he'll take a mile.
   d. This thing must weigh a ton.

   Answer: (a) 3.05 m, (b) 37.6 L, (c) 2.54 cm, 1.61 km, (d) 909 kg

9. Determine whether each of the following is likely or unlikely:
   a. One student is 20 cm taller than another student.
   b. A dog's tail is 300 mm long.
   c. The height of a room ceiling is 2 m.
   d. The normal speed on a rural interstate highway is 65 kph.
   e. A 10-L container of lemonade should serve 50 thirsty people nicely.

10. A room floor measures 10 ft by 12 ft. How many square meters of carpet would be needed to cover the floor?

    Answer: 11.16 m²

11. Noah, of Biblical fame, was instructed to build an ark 300 cubits long, 50 cubits wide, and 30 cubits high (see Fig. A.1). What are the approximate dimensions and (rectangular) volume of the ark in the metric system? (A cubit is 18 in. or half a yard, approximately the length of the forearm—elbow to fingertip.)

    Answer: 137 m × 23 m × 14 m ≈ 44,000 m³

Figure A.1
Biblical units. Noah was given the dimensions of the ark in cubits. See Exercise 11.

We have seen how motion is described in terms of speed, velocity, and acceleration. Speed $= d/t$ is usually an average and gives general information. However, velocity and acceleration can give more detail. To show how valuable these quantities are in the aspect of computing distances, let's take the special case of motion in a straight line. This allows only two directions, one way or the other, which are commonly designated as plus ($+$) and minus ($-$) and used to indicate vector directions.

First let's assume that we have motion with a constant velocity (v_o), and hence a constant speed. (The plus sign is understood and the motion is in the positive direction.) Then, the distance traveled in a time t is

$d = v_o t$ (constant velocity)

If this is plotted on a v versus t graph, as illustrated in Fig. A.2(a), a straight line is obtained. However, notice that the distance traveled is graphically equal to the area under the line (the length $\times$ width of the rectangle).

Suppose something is set into motion from rest with a constant acceleration. In this case, the velocity is changing. The distance traveled under these conditions involves the acceleration

$d = \tfrac{1}{2} at^2$ (constant acceleration, starting from rest)

On a v versus t graph, a constant acceleration gives a straight line as illustrated in Fig. A.2(b) — a nonhorizontal straight line. Why? The line starts at $v = 0$ for $t = 0$, which shows that the motion starts from rest.

The distance traveled in the time t is again equal to the area under the line, although it may not be so obvious in this case. The area of a triangle is equal to one half the height $\times$ the base. From the triangle on the graph, this is $\tfrac{1}{2} vt$ (height v, base t). But, recall that acceleration is velocity per time, or $a = v/t$. Then, $v = at$, and substituting this into the area equation, $\tfrac{1}{2} vt = \tfrac{1}{2}(at)t = \tfrac{1}{2} at^2 =$ distance d.

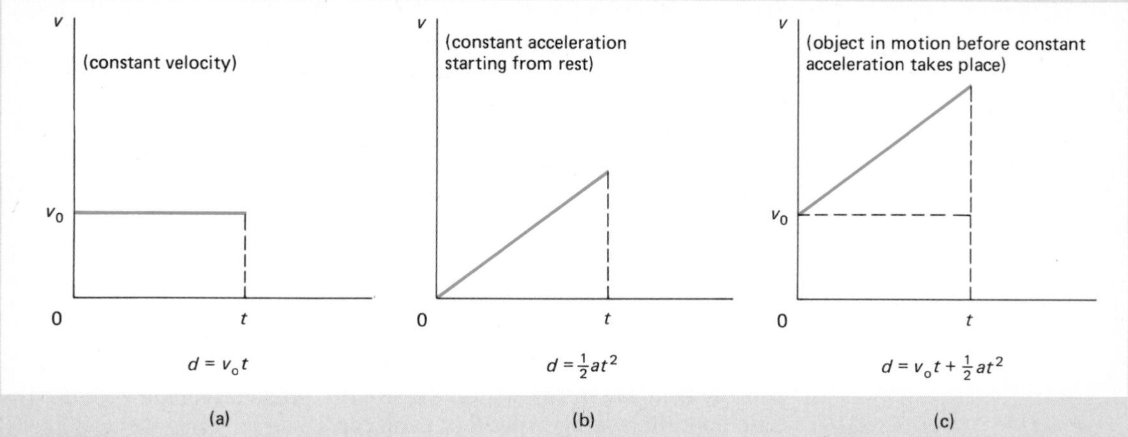

CHAPTER 1

Describing Motion

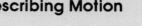

Figure A.2 graphical panels:

(a) (constant velocity) — v_0 horizontal line — $d = v_o t$

(b) (constant acceleration starting from rest) — $d = \frac{1}{2}at^2$

(c) (object in motion before constant acceleration takes place) — v_0 — $d = v_o t + \frac{1}{2}at^2$

Figure A.2
Graphical description of motion. See text for description.

Example: An object is dropped from a tall building. If it takes 3 s to reach the ground, from what height was it dropped?

Solution: This is another way of asking how far the object fell in $t = 3$ s. For free fall, we know that $a = g = 9.8$ m/s² (or 32 ft/s²), where downward is taken to be the positive direction. Then, using the previous equation since the object started from rest,

$$d = \tfrac{1}{2} a t^2 = \tfrac{1}{2} g t^2 = \tfrac{1}{2} (9.8 \text{ m/s}^2)(3 \text{ s})^2 = 44.1 \text{ m}$$

$$[\text{or} = \tfrac{1}{2} (32 \text{ ft/s}^2)(3 \text{ s})^2 = 144 \text{ ft}]$$

Finally, suppose an object is initially in motion with a velocity v_o and it receives a constant acceleration. [For example, you might be going down the road in your car with a constant velocity, and then (at $t = 0$) you step on the accelerator so that the velocity changes constantly.] The distance traveled in a time t is then

$$d = v_o t + \tfrac{1}{2} a t^2 \qquad \text{(object in motion before constant acceleration takes place)}$$

Notice that this equation is the same as the previous one if $v_o = 0$. Can you guess what the v versus t plot might be? As can be seen from Fig. A.2(c), it looks like a combination of the preceding cases. Also, the distance is again equal to the area under the line or curve. (Can you prove this?) If there is a slowing deceleration, or the acceleration is in the opposite direction of the motion or velocity, then the plus (+) sign in the equation becomes a minus (−).

Example: An automobile travels with an initial velocity of 5.0 m/s (about 18 km/h) and then accelerates at a constant rate of 4.0 m/s² for 3.0 s. How far does the car travel during this time?

Solution: We have $v_o = 5.0$ m/s, $a = 4.0$ m/s², and $t = 3.0$ s. This is an example of case (c) in the figure, so

$$\begin{aligned} d &= v_o t + \tfrac{1}{2} a t^2 \\ &= (5.0 \text{ m/s})(3.0 \text{ s}) + \tfrac{1}{2}(4.0 \text{ m/s}^2)(3.0 \text{ s})^2 \\ &= 15 \text{ m} + 18 \text{ m} = 33 \text{ m} \end{aligned}$$

Notice how the units cancel so as to leave a length. If this were not the case, something would be wrong. Using units in this manner is called unit dimensional analysis. Both sides of an equation must be equal not only in magnitude (numbers), but also in units.

EXERCISES

1. Two cars, A and B, travel the same 100-km distance when driven by students going home on a spring break. Car A makes the trip in 2 h and car B in 2½ h. (a) Which car has the greater average speed? (b) Calculate the average speed of each car.

    Answer: (a) A, (b) 50 km/h and 40 km/h

2. An automobile travels 50 km in 30 min, and a train travels 90 km in 1 h. (a) Which has the greater average speed? (b) What is the average speed of each?

   Answer: (a) Auto, (b) 100 km/h and 90 km/h

3. A person walks with a constant speed of 0.50 m/s. How far does he or she walk in (a) 1 s, (b) 20 s, (c) 1 min?

   Answer: (a) 0.50 m, (b) 10 m, (c) 30 m

4. An airplane flies eastward with a constant velocity of 300 km/h. Given the plane's position at a particular time, where is the plane 2 h later?

   Answer: 600 km, east of original position

5. An object moves with a constant velocity of 2.0 m/s in the x-direction. How far will it travel in 10 s?

   Answer: 20 m

6. A modern "bullet" train travels with a constant velocity of 160 km/h. How long would it take for the train to travel (a) 160 km, (b) 100 km, and (c) 1 km?

   Answer: (a) 1 h, (b) 0.625 h, (c) 0.00625 h

7. While maintaining a constant cruising speed on a straight, rural interstate highway in the Midwest, a car travels 35 km in 20 min. Was the speed limit exceeded?

   Answer: Not enough to get a ticket in a 65 mph zone.

8. A runner starting from rest can accelerate at a rate of 4.0 m/s² for 5.0 s before reaching a maximum constant velocity. How far does the runner travel before reaching this velocity, and what is it?

   Answer: 50 m, 20 m/s

9. When a stone is dropped, it falls with a constant acceleration due to gravity ($a = g = 9.8$ m/s²). Let a stone be dropped from the top of a building from a height of 122 m above the ground. (a) Make a table of time and distance, and compute how far the stone falls each second. (b) Estimate how long it takes for the stone to reach the ground.

   Answer: (b) 5.0 s

10. Two dragsters leave the starting line from rest at the same time. One accelerates at a rate of 8.0 m/s² and the other at 10 m/s². How far is the faster dragster ahead of the other at the end of 5.0 s?

    Answer: 125 m − 100 m = 25 m

11. A car accelerates from rest at a uniform rate to a constant velocity. This velocity is maintained for a short time and then the car is brought uniformly to a stop. Sketch the general v versus t graph for this situation. What is the distance traveled on the graph?

12. A horizontal line on a v versus t graph represents a constant velocity or zero acceleration. On the same v versus t plot, show three general constant accelerations labeled (1), (2), and (3), where (2) is greater than (1), and (3) is greater than (2). What would a vertical line on a v versus t plot imply?

13. (a) Describe the motion represented on the *v* versus *t* graph in Figure A.3. (b) What was the distance traveled? (Compute from the graph.)

Answer: (b) 52 m

14. If an object moving with an initial velocity of 2.5 m/s is accelerated at a rate of 0.50 m/s² for a period of 8.0 s, how far does it travel during this time?

Answer: 36 m

15. A skier traveling initially with a constant velocity of 8.0 m/s accelerates down a straight, inclined slope at 4.0 m/s². If it takes 10 s for the skier to reach the bottom, what is the length of the slope?

Answer: 280 m

16. A car traveling initially with a velocity of 12 m/s receives a constant acceleration of 8.0 m/s² in the opposite direction. In a table of *t* and *d*, compute the distances for the times *t* = 1 s, 2 s, 3 s, 4 s, and 5 s. On the basis of the distances obtained, explain what is happening physically. (*Hint:* Remember *d* = 0 at *t* = 0. That is, distance is measured from the location of the car at *t* = 0. Try *t* = 2.4 s, 2.5 s, and 2.6 s, and see what you get for *d*. Does this help?)

Answer: 1 s, *d* = 8 m; 5 s, *d* = −40 m

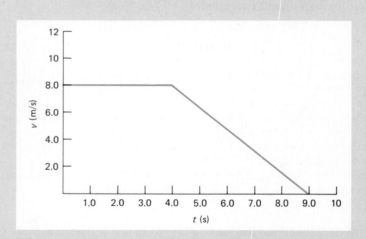

Figure A.3
What is the acceleration? See Exercise 13.

You may wonder why we focused on the conservation of mechanical energy in the chapter. After all, this applies only to *ideal* conservative systems. In the real world energy is always lost to friction or some other cause, and the sum of the potential and kinetic energies is not conserved. The example of an ideal pendulum was given to illustrate the conservation of mechanical energy in the chapter [See Fig. 2.10(a)]. However, we all know that in actual practice a pendulum is not ideal and it tends to "run down." The conservation of *total* energy always applies to a system, but it is often difficult to know how much mechanical energy was "lost."

However, in certain situations the energy no longer available for useful work or as mechanical energy can be found. This occurs when the total energy and the mechanical energy are initially equal, which is an instructive special case to consider. Keep in mind that we can always write by the conservation of total energy that

Total energy = mechanical energy + other energy

or $$E = (KE + PE) + Q$$

For our special case, if $Q = 0$ initially, then the total energy is equal to the mechanical energy initially. As time progresses in a real dynamic system, Q gets larger and represents the energy lost to friction or some other cause. The following examples illustrate this situation.

> *Example:* A 0.10-kg pendulum bob is released from rest from a height of 0.25 m relative to its lowest position. After several swings, the bob is 0.20 m above this reference when it swings back to its instantaneously stationary position (Fig. A.4). How much energy was lost?
>
> *Solution:* It is given that $h_o = 0.25$ m and later $h = 0.20$ m for the bob with $m = 0.10$ kg. Initially, the total energy is all potential energy, but some of this is lost, as indicated by the lower maximum height of swing. Looking at the time when the bob is momentarily at rest, we do not have to consider kinetic energy.

CHAPTER 2

The Conservation of Energy in Nonconservative Systems

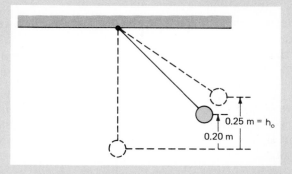

Figure A.4
What is the energy? See Example.

Then, writing the conservation of total energy for the two times as "before" and "after,"

$$PE_o \;=\; PE + Q$$

(total energy   (total energy
before)   after)

or $\qquad mgh_o = mgh + Q$

So,

$$Q = mg(h_o - h)$$
$$= (0.10 \text{ kg})(9.8 \text{ m/s}^2)(0.25 \text{ m} - 0.20 \text{ m}) = 0.049 \text{ J}$$

This energy is lost primarily to friction between the string and support.

We can also use the conservation of energy to find the speed or velocity of an object in some cases. For an ideal pendulum, the speed at the bottom of its swing can be found by equating the potential energy at its maximum height to the kinetic energy at the bottom of the swing ($h = 0$). That is,

$$PE_o = KE$$

$$mgh_o = \tfrac{1}{2} mv^2$$

and $\qquad v = \sqrt{2gh}$

If the pendulum in the previous example were ideal (or to an approximation for the first swing, why?), this would be

$$v = \sqrt{2gh} = \sqrt{2(9.8 \text{ m/s}^2)(0.25 \text{ m})} = 2.2 \text{ m/s}$$

We could also approximate the speed of the pendulum at its lowest point after it had lost the energy computed in the previous example. In this case, we would write

$$PE_o = \tfrac{1}{2} mv^2 + Q$$

Using the computed value of Q, you then solve for v. However, the real v would be a bit smaller. Why?

Example: A student relaxing on spring break starts down a water slide from rest from a height of 25 m above a level surface (Fig. A.5). If the student has a mass of 60 kg and a speed of 8.0 m/s (about 18 mi/h) at the bottom of the slide, how much energy is lost?

Solution: It is given that $h_o = 25$ m, $v = 8.0$ m/s, and $m = 60$ kg. At the top of the slide the total energy is all potential, and at the bottom it is all kinetic ($h = 0$). Then in terms of the conservation of total energy,

$$PE_o \;=\; KE + Q$$

(total energy =   (total energy
before)   after)

or $\qquad mgh_o = \tfrac{1}{2} mv^2 + Q$

Figure A.5
Down we go! See Example.

So,

$$Q = mgh_o - \tfrac{1}{2}mv^2$$
$$= (60 \text{ kg})(9.8 \text{ m/s}^2)(25 \text{ m}) - \tfrac{1}{2}(60 \text{ kg})(8.0 \text{ m/s})^2$$
$$= 14{,}700 \text{ J} - 1{,}920 \text{ J} = 12{,}780 \text{ J}$$

The idea of "lost" energy may give you a better insight into the idea of efficiency. The "lost" mechanical energy generally corresponds to the energy lost by a machine or motor. In both cases this energy is lost to friction or some other cause. So, in the preceding example, the "efficiency" of going down the slide in terms of energy is

$$(\text{Efficiency}) \ Eff = \frac{\text{energy out}}{\text{energy in}}$$

$$= \frac{1920 \text{ J}}{14{,}700 \text{ J}} = 0.13 \ (\times 100\%) = 13\%$$

So, going down the water slide isn't very efficient. This means that 87 percent of the initial energy was *lost*.

$$Q/E_o = 12{,}780 \text{ J}/14{,}700 \text{ J} = 0.87$$

As stated previously, efficiency literally means "what you get out (useful work output) for what you put in (work or energy input)." Do you have any suggestions as to how "efficiency" might be phrased in the academic sense for the learning process? What is your efficiency for this?

EXERCISES

1. A 0.15-kg ball is dropped from a height of 3.0 m. It strikes the ground with a speed of 7.6 m/s. Was the ball in free fall? If not, how much energy was lost?

   Answer: No, 0.08 J

2. A 0.10-kg pendulum bob is released from a height of 0.20 m relative to its lowest position. If it loses 0.050 J of energy after several swings, what would be its maximum height of swing at this time?

   Answer: 0.15 m

3. If the pendulum in Exercise 2 were ideal (a) what would be its speed at the bottom of the swing? (b) What would be its approximate speed at this position if it had lost the energy as stated?

   Answer: (a) 2.0 m/s, (b) 1.7 m/s

4. A child on a sled starts down a snowy hill from rest. After descending a vertical distance of 10 m, the sled and its passenger have a speed of 5.0 m/s. If the combined mass of the sled and the child is 30 kg, how much energy was lost?

   Answer: 2940 J − 375 J = 2565 J

5. A 0.50-kg rubber ball is dropped on a sidewalk from a height of 1.5 m. (a) If it falls freely, with what speed will it strike the sidewalk? (b) If it bounces back to a height of 1.2 m, how much energy was lost? (c) What was the "efficiency" of the bounce?

   Answer: (a) 5.4 m/s, (b) 1.5 J, (c) 80%

6. Referring to Fig. 2.10(b), let the initial height of the 10-kg block be 5.0 m. If it loses 25 percent of its energy sliding down the hill, what will its speed be at the bottom?

   Answer: 8.6 m/s

7. Motors are generally rated in terms of output power. Using the maximum electric motor efficiency in Table 2.1, what is the input power for each of the motors in Figure 2.3?

   Answer: 1 hp motor, 785 J/s

8. In starting from rest and reaching a speed of 90 km/h (or 25 m/s), a 850-kg sports car burns enough gasoline to release 1.80 MJ (megajoules) of energy. What is the efficiency of the car?

   Answer: 15%

The conservation conditions associated with the different types of collisions, elastic and inelastic, provide a means to find things out or tell what happens. Such processes are major tools in atomic and subatomic physics where particles are studied and identified through collisions. However, let's take a look at some more macroscopic examples. Although you might not think it to be a collision, we'll start with an example of the conservation of momentum.*

Example: Suppose after two skaters "push off," as illustrated in Figure 3.5 (reproduced here for convenience), the 50-kg boy has a velocity of 2.0 m/s to the left. What is the velocity of the 40-kg girl?

Solution: Since there are no external forces (friction of ice considered negligible), the total momentum is conserved, and

Total momentum before = total momentum after

or
$$0 = p_2 - p_1$$

where the total momentum before is zero ($P = 0$) since there is no motion and the minus sign indicates that p_1 is in the opposite direction of p_2 (vector addition).

Then,
$$p_2 = p_1$$

or
$$m_2 v_2 = m_1 v_1$$

and
$$(50 \text{ kg})(2.0 \text{ m/s}) = (40 \text{ kg}) v_1$$

Solving for v_1

$$v_1 = 100 \text{ m/s}/40 = 2.5 \text{ m/s}$$

So, the velocity of the girl is 2.5 m/s to the right in the figure.

CHAPTER 3

More on Collisions:

Conservation of Momentum—Every Time

Conservation of Kinetic Energy—Sometimes

$P = 0$ $P = p_2 - p_1 = 0$

(a) Before   (b) After

Figure 3.5
Conservation of total momentum. (a) The total momentum P of the skaters' system is initially zero. (b) With no external forces ("pushing off" involves internal forces), the total momentum is conserved and remains zero.

* Technically, a collision is an interaction in which there is an exchange of momentum (and energy).

Figure A.6
A rear-end collision. How fast does the "Bug" take off after collision? See Example.

The following example describes a more familiar, if unfortunate, kind of collision. See Figure A.6.

Example: A van with a mass of 2 *m* going 90 km/h hits a car of mass *m* stopped at an intersection (Fig. A.6). After the collision, the van has a velocity of 50 km/h. What would be the velocity of the car immediately after collision?

Solution: We assume that the conservation of momentum holds during the collision, so the total momentum before is equal to the total momentum *immediately* after collision. Then,

Total momentum before = total momentum after

or $$2\,m \times 90 \text{ km/h} = (2\,m \times 50 \text{ km/h}) + mv_{car}$$

Canceling the *m*'s on each side of the equation (or dividing both sides of the equation by *m*), we have

$$2 \times 90 \text{ km/h} = (2 \times 50 \text{ km/h}) + v_{car}$$

$$180 \text{ km/h} = 100 \text{ km/h} + v_{car}$$

and it can be seen that the velocity of the car after collision must be 80 km/h in the original direction of the van. (Maybe there will be a case of "whiplash." Do you remember why from Chapter 1?)

It is easy to show that kinetic energy was not conserved or that some was lost in the collision. Where did it go? Correct! Into the work of bending the fenders, the bumper, and maybe a couple of other things in the rear engine of the VW.

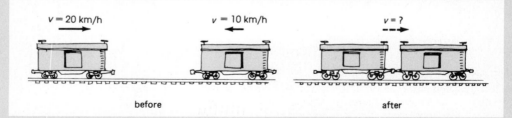

Figure A.7
Kaboom! A completely inelastic collision. In such a collision, the objects stick together. Here, two railroad cars are coupled during collision. What happens after collision? See Example.

Finally, let's try a completely inelastic case.

Example: Two railroad cars of equal mass roll toward each other as illustrated in Figure A.7. If the cars become coupled during collision, what happens after the collision?

Solution: Applying the conservation of momentum,

$$\text{Total momentum before} = \text{total momentum after}$$

or $\quad (m \times 20\ \text{km/h}) - (m \times 10\ \text{km/h}) = (m + m) \times v$

Here, the negative sign indicates that the second car is moving in the opposite direction to the first (positive direction to the right in the figure). Then, canceling the m's,

$$20\ \text{km/h} - 10\ \text{km/h} = 2 \times v$$

or $\quad\quad\quad\quad 10\ \text{km/h} = 2 \times v$

and v can be seen to be 5 km/h. Since v is positive, the coupled cars move to the right. (If the velocity had come out to be negative, the motion would be in the designated negative direction or to the left.)

Can you tell what would happen if the railroad cars approached each other with equal speed?

EXERCISES

1. In applying a reverse thrust in a jet engine, suppose a gas molecule of mass m and velocity v is deflected directly opposite its original direction. The change in momentum (impulse) of the molecule is then 2 mv. Show that this is correct. (*Hint:* The change in momentum is a vector addition [or subtraction], and direction is a consideration.)

Answer: 2 mv

2. Two masses on a horizontal frictionless surface with a spring between them are pushed together, and a string is attached to hold them together (Fig. A.8). If the string is then burned and the 3M mass moves to the right with a speed of 2 m/s, what is the velocity of the other mass? What kind of force does the spring apply to the system? (Disregard the mass of the spring.)

Answer: 6 m/s

3. A person who weighs 80 kilos skates with a constant velocity of 6 m/s and catches a child who has fallen from a window. If the child weighs 20 kilos, what is their velocity after the catch?

Answer: 4.8 m/s

4. A car traveling at 50 km/h hits another car of the same mass from behind while the other car is stopped at an intersection. If the collision is completely inelastic, what is the velocity of the cars immediately after collision?

Answer: 25 km/h

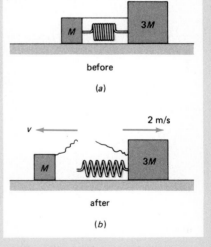

Figure A.8
Together and then apart. See Exercise 2.

(Questions 5 and 6 work as a pair.)

5. Referring to Figure A.6, if the mass of the van were three times that of the car, what would be the velocity of the car immediately after collision?

    Answer: 120 km/h

6. Compute the percentage of the kinetic energy lost in this situation. (*Hint:* 120 km/h is 33.3 m/s, 90 km/h is 25.0 m/s, and 50 km/h is 13.9 m/s.)

    Answer: 10%

7. A single railroad car with a velocity v rolls toward two coupled railroad cars moving with a velocity of $v/2$ in the opposite direction. If the track is level and all the cars have equal mass, what happens after collision?

    Answer: $v = 0$

8. Referring to Figure A.7, if the car traveling at 20 km/h carried a load so that its mass were twice that of the other car, what would happen after collision?

    Answer: $v = 10$ km/h

9. Let the railroad car in Figure A.7 traveling at 10 km/h be three times as massive as the other car. What happens after collision?

    Answer: $v = -2.5$ km/h

10. A railroad car with a speed of 20 km/h rolls on a level track in the same direction as another car of the same mass that has a speed of 10 km/h. The faster car overtakes the slower one, and they couple on collision. What is the velocity of the cars immediately after collision?

    Answer: $v = 15$ km/h

Inverse-square laws and relationships are common in nature. As we have seen, the expression for the gravitational force is an inverse-square law. The electric force is also described by such a relationship, and the intensity of light and sound "fall off" as $1/r^2$.

We say a quantity "falls off" to emphasize how quickly it becomes smaller with increasing distance (greater r). This is illustrated in Figure A.9. As can be seen, if the distance is doubled, the magnitude of the force is one quarter of its original value; if the distance is tripled, the force is then only one ninth as great; and so on.

Of course, looking at it the other way, a quantity "grows" as $1/r^2$ as the distance decreases (smaller r). If you halved the distance, the magnitude of, say, an inverse-square force would increase by a factor of four; if the distance were cut to a third, the force would be nine times as great.

The factor of change either way can be easily found by using a ratio:

$$\frac{F_2}{F_1} = \frac{\dfrac{1}{r_2^2}}{\dfrac{1}{r_1^2}} = \frac{r_1^2}{r_2^2} = \left(\frac{r_1}{r_2}\right)^2$$

That is, $$\frac{F_2}{F_1} = \left(\frac{r_1}{r_2}\right)^2$$

The following examples show how to use this formula.

Example: By what factor is the gravitational force between two objects changed if the separation distance is (a) decreased to one fifth the original distance, and (b) increased from $3r$ to $4r$?

CHAPTER 5

Inverse-Square Laws

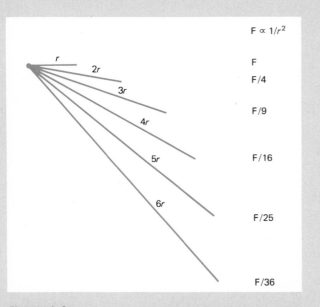

$F \propto 1/r^2$

F

$F/4$

$F/9$

$F/16$

$F/25$

$F/36$

Figure A.9
An illustration of the inverse-square relationship. Notice that when r doubles, F decreases by a factor of 4; when r triples, F decreases by a factor of 9; and so on.

Solution: One needs only to identify r_2 and r_1 and use the formula given previously. For (a) we have $r_2 = (1/5)r_1 = r_1/5$ and for (b), $r_1 = 3r$ and $r_2 = 4r$. Then, (a) with $r_2 = r_1/5$, or $5r_2 = r_1$, or $r_1/r_2 = 5$ we have

$$\frac{F_2}{F_1} = \left(\frac{r_1}{r_2}\right)^2 = (5)^2 = 25$$

and $F_2 = 25\ F_1$

or the force is increased by a factor of 25.
(b) Very simply,

$$\frac{F_2}{F_1} = \left(\frac{r_1}{r_2}\right)^2 = \left(\frac{3r}{4r}\right)^2 = \left(\frac{3}{4}\right)^2 = (0.75)^2 = 0.56$$

and $F_2 = (0.56)F_1$

or the force is reduced by a factor of 0.56.

Try this one — going from force to distance.

Example: An inverse-square force is increased by a factor of five. By what factor is the separation distance *decreased* (right?)?

Solution: It is given that $F_2 = 5\ F_1$ (force increased by a factor of five). Then, taking the square root of both sides of our equation and turning it around,

$$\frac{r_1}{r_2} = \sqrt{\frac{F_2}{F_1}}$$

With $F_2 = 5\ F_1$ or $F_2/F_1 = 5$,

$$\frac{r_1}{r_2} = \sqrt{\frac{F_2}{F_1}} = \sqrt{5} = 2.2$$

and $r_2 = \dfrac{r_1}{2.2}$

or the separation distance was decreased by a factor of 1/2.2. Suppose r_1 were 3.0 m, then $r_2 = r_1/2.2 = 3.0$ m$/2.2 = 1.4$ m.

EXERCISES

1. The distance between two masses is increased by a factor of 1.5. How is the gravitational force affected?

   Answer: $F_2 = (0.44)F_1$

2. If the distance between two masses is changed from 10 m to 4.0 m, how is the gravitational force affected?

   Answer: $F_2 = (6.25)F_1$

3. Sketch a graph of F versus r for the gravitational force. Let $F = F_o$ at $r = r_o$ and label the axes in terms of these values, e.g., r_o, $2r_o$, $3r_o$.

4. If the Earth shrank to one half of its present radius but kept its mass constant, what effect would this have on the acceleration due to gravity at the Earth's surface, and what would be its value?

   Answer: $g = 39.2$ m/s²

5. By what factor would the distance between two particles have to be changed (a) to increase the gravitational force by a factor of 10 and (b) to decrease the gravitational force by one tenth?

   Answer: (a) $r_2 = r_1/3.2$, (b) $r_2 = (1.05)r_1$

6. If a planet had the same mass as the Earth but had an acceleration due to gravity on its surface of (a) 4.9 m/s² and (b) 29.4 m/s², what would be the radius of the planet?

   Answer: (a) $r = (1.4)R_e$, (b) $r = (0.58)R_e$

7. Would it be possible to increase the gravitational force between two masses by a factor of ∞? If so, how?

8. The gravitational force between two objects is reduced from 500 N to 200 N by moving them. How is their separation distance affected?

   Answer: $r_2 = (1.6)r_1$

9. Set up a ratio for the acceleration due to gravity that incorporates both changes in mass and the radius of a planet.

   Answer: $\dfrac{g_2}{g_1} = \left(\dfrac{m_2}{m_1}\right)\left(\dfrac{r_1}{r_2}\right)^2$

10. If the mass of the Earth were reduced by one half and its radius tripled, (a) how would the acceleration due to gravity be affected on its surface? (b) What would be the value of *g*?

    Answer: (a) $g_2 = g_1/18$

11. What is the gravitational force per unit mass due to the Earth at a distance of 60 Earth radii? This is the distance of the moon from the Earth. Interpret physically what this force per unit mass is or does.

    Answer: 0.0027 m/s²

Density is a measure of the compactness of matter, i.e., how much matter there is in a given volume. As noted in the chapter, it is defined

$$\text{Density} = \frac{\text{mass}}{\text{volume}}$$

or in symbol form

$$\rho = \frac{m}{V}$$

Of course density is usually an average value, unless the substance is pure and homogenous.

To find the density of a substance, we take an amount of it, determine the mass, find the volume, and take the ratio. Mass is usually determined on a scale that is calibrated in mass units, or from weight (w) measurements. Recall $w = mg$, so $m = w/g$.

Determining the volume of a liquid (or gas) is easy, since it assumes the shape of the container. The volume of a solid can be computed from length measurements if it has a nice shape, for example, a cube. But, what if you have something that is irregularly shaped, say a rock? Easy. Put some water in a graduated container, note the volume, then put in the rock and find the change in the volume. This will be the volume of the rock, since it displaces its own volume. (If you have something that floats, then you simply sink it taking into account the volume of the sinker.)

> **CHAPTER 8**
>
> **Density and Volume**

Example: A metal cube 10 cm on a side has weight of 113 N. What is the density of the metal?

Solution: The length of a side of the cube is $l = 10$ cm $= 0.10$ m. Then its volume is $V = l \times l \times l = 0.10$ m $\times 0.10$ m $\times 0.10$ m $= 0.001$ m³. The mass is determined from the weight,

$$m = \frac{w}{g} = \frac{113 \text{ N}}{9.80 \text{ m/s}^2} = 11.5 \text{ kg}$$

so

$$\rho = m/V = 11.5 \text{ kg}/0.001 \text{ m}^3 = 11,500 \text{ kg/m}^3$$

From Table 8.1, we see that the metal was probably lead.

In general, when we say density we mean *mass* density. In the British system, a *weight* density (D) is sometimes used, since this is convenient in a gravitational (force or weight) system. The weight density is simply the weight per unit volume

$$D = \frac{w}{V}$$

Example: Suppose you are asked to get 10 lb of water. How large a container would you need?

Solution: From Table 8.1, the weight density of water is $D = 62.4 \text{ lb/ft}^3$. Then, with $w = 10$ lb, you can find the volume from the weight density equation,

$$V = \frac{w}{D} = \frac{10 \text{ lb}}{62.4 \text{ lb/ft}^3} = 0.16 \text{ ft}^3$$

So, you'd have to have a container that holds at least 0.16 ft³ or about 1.2 gal.

EXERCISES (See Table 8.1 for densities.)

1. Using a scale, a student determines the mass of rock to be 0.50 kg and then using the water-displacement method finds it has a volume of 0.96 m³. What is the density of the rock?

   Answer: 0.52 kg/m³

2. Estimate your average density. (*Hint:* Find your mass from your weight, 2.2 lb = 1 kg.)

3. Show that the weight density D and the mass density ρ are related by $D = \rho g$.

4. An "empty" rectangular box has the dimensions of 1.0 m × 0.80 m × 0.75 m. (a) What mass of air does it contain? (b) What is the weight of the air in pounds?

   Answer: (a) 0.77 kg, (b) 1.7 lb

5. How much larger would a 1-kg cube of aluminum be than a 1-kg cube of iron? (*Hint:* Use a ratio.)

   Answer: $V_{Al} = (2.9) V_{Fe}$

6. When 1 kg of water freezes to ice, how much larger is its volume? (*Hint:* Use a ratio.)

   Answer: $V_{ice} = (1.09) V_w$

7. What would be the volume of 100 lb of water? Could you take a soaking bath in this amount of water?

   Answer: $V = 1.6 \text{ ft}^3$ (12 gal)

8. A cubic centimeter of gold sits on one side of a double pan balance. What would be the size (volume) of brass weights placed on the other pan that would just balance the gold?

   Answer: 2.27 cm³

9. Given a 2-gal bucket, how much lift force would you have to apply to carry it if it were full of water? (*Hint:* 1 gal = 0.13 ft³.)

   Answer: 16 lb

10. An aquarium measures 36 in. × 18 in. × 12 in. If it were filled with seawater instead of freshwater so as to make a saltwater aquarium, how much more would it weigh?

    Answer: 8.6 lb

On occasion you might wonder and would like to know what a Celsius temperature would be in Fahrenheit or vice versa. No need to wonder. It's very easy to convert from one temperature scale to the other. The Celsius and Fahrenheit scales are related by the equation

$$T_F = \tfrac{9}{5}T_C + 32$$

where T_F and T_C are the Fahrenheit and Celsius temperatures, respectively.

So, given a Celsius temperature T_C, you multiply it by $\tfrac{9}{5}$ (or 1.8) and then add 32 to the result to get the equivalent Fahrenheit temperature. Let's illustrate this with an example — one for which you already know the answer.

CHAPTER 11

Fahrenheit to Celsius — and Back

Example: What is the Fahrenheit temperature for 100°C?

Solution: $T_C = 100°C$, and using the formula,

$$T_F = \tfrac{9}{5}T_C + 32 = \tfrac{9}{5}(100) + 32$$
$$= 9(20) + 32 = 180 + 32 = 212°F$$

(Notice that this is the boiling point of water if you haven't already.)

How about going the other way, i.e., from Fahrenheit to Celsius? The same relationship is used. It is convenient to rewrite the equation as

$$T_C = \tfrac{5}{9}(T_F - 32)$$

Example: What is the normal human body temperature of 98.6°F in degrees Celsius?

Solution: $T_F = 98.6°F$, and this is simply plugged into the formula. Note that $\tfrac{5}{9}(T_F - 32)$ means that 32 is subtracted from T_F, and then the result is multiplied by 5/9 (or 0.56).

$$T_C = \tfrac{5}{9}(T_F - 32) = \tfrac{5}{9}(98.6 - 32)$$
$$= \tfrac{5}{9}(66.6) = 37°C$$

On the Celsius scale, normal body temperature comes out to be an even number, exactly 37°C.

In the chapter it was stated that there was one temperature at which the readings on the Fahrenheit and Celsius scales were equal, namely −40°. That is, −40°F = −40°C. This may seem difficult to believe since the temperatures on the scales are so different elsewhere, e.g., 212°F = 100°C, so let's prove it.

Example: Can the temperature be the same on the Celsius and Fahrenheit scales, and if so, what is the temperature?

Solution: In physics when such a question arises and you have an equation that relates the quantities, you "ask" the equation if this is possible. ("Hey, equation, is it possible that $T_C = T_F$?")

The equation might respond by saying, "Why don't you try it and see!" What this means is that we will assume that this is the case ($T_c = T_F$) and put this into our equation relationship. If we obtain a sensible answer, then our assumption is correct. However, if our assumption is not correct, the equation will tell us so by giving a ridiculous answer that makes no sense — maybe something like a temperature below $-273°C$ (absolute zero).

So, let's put $T_c = T_F$ into the equation and see what happens. Putting T_F in for T_c (since we're assuming $T_c = T_F$),

$$T_F = \tfrac{9}{5}T_c + 32 = \tfrac{9}{5}T_F + 32$$

and rearranging,

$$\tfrac{9}{5}T_F - T_F = -32$$

or

$$\tfrac{4}{5}T_F = -32$$

so

$$T_F = \tfrac{5}{4}(-32) = 5(-8) = -40°F$$

Hence, $T_F = T_c = -40°$.

Finally, if you want to express a temperature in terms of the absolute Kelvin scale, it's very simple to convert from the Celsius reading. Recall from the chapter

$$T_K = T_c + 273$$

and the absolute temperature is in kelvins (K).

EXERCISES

1. What are the following temperatures on the Fahrenheit scale? (a) 0°C, (b) 30°C, and (c) −10°C

   Answer: (a) 32°F, (b) 86°F, (c) 14°F

2. What are the following temperatures on the Celsius scale? (a) 0°F, (b) 114°F, and (c) −13°F

   Answer: (a) −18°C, (b) 46°C, (c) −25°C

3. In a weather forecast, it is reported that the high temperature for the next day is expected to be 15°C. How would you plan to dress when you go to class tomorrow? (What is the Fahrenheit temperature?)

   Answer: Warmly (59°F)

4. A European visitor reads that the average temperatures of two different places in the United States are (a) 77°F and (b) 59°F. He asks you what these temperatures are in Celsius. Could you help him? If so, what would you tell him?

   Answer: (a) 25°C, (b) 15°C

5. What is the present outside temperature in degrees Celsius?

6. A child has a temperature of 40°C. Is this serious? Explain in terms of Fahrenheit temperature.

   Answer: 104°F

7. What is the Kelvin temperature when the Fahrenheit and Celsius temperatures are equal?

   Answer: 233 K

8. A strict user of the SI system tells you the temperature that day is 288 K. What is the temperature on the Fahrenheit scale?

   Answer: 59°F

9. On a very hot day the temperature gets up to 104°F. A strict SI buff has his Kelvin thermometer along. What does it read?

   Answer: 313 K

■ EXERCISES

1. For a particular auto with a four-cylinder engine, the tachometer indicates that the engine is turning at a rate of 3000 rpm at a normal speed. (a) How many power strokes or outputs does the engine have each minute? (b) Would the tachometer rate be the same for a six- or eight-cylinder engine for the same power output? Explain.

   Answer: (a) 6000/min, (b) no

2. It is possible to use the heat of the ocean to supply the energy to turn a generator and produce electricity. A diagram of such a system is shown in Figure A.10. In certain tropical areas, the surface layer of the ocean is about 25°C and the lower layers are about 5°C. Answer the following questions about the system:
   a. Is the system a heat engine?
   b. Why is ammonia used as a working fluid, and how does it produce mechanical work?
   c. What would be the ideal efficiency of the system?
   d. Do you think ocean thermal energy conversion will ever be economically feasible?

   Answer: (c) 6.7 percent

CHAPTER 13

Thermodynamics, Heat Engines, and Heat Pumps

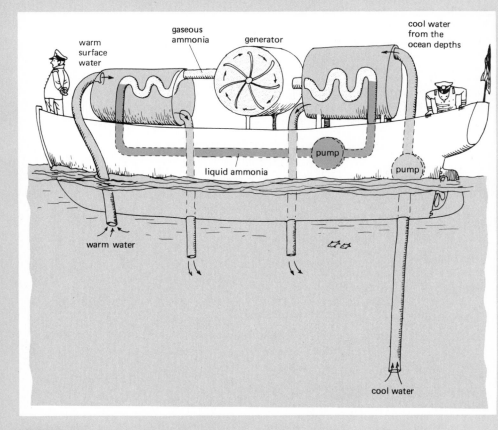

Figure A.10
A floating heat engine? See Exercise 2.

3. In electrical generation, the maximum practical temperature limit of superheated steam used in the turbine is about 540°C, owing to material limitations. (a) If the steam condenser or low-temperature reservoir operated at room temperature, what would be the ideal efficiency of the system? (b) The actual efficiency is about 35 to 40 percent. What does this tell you?

   Answer: (a) 64 percent

4. An engineer plans a heat engine that is to operate ideally with high and low temperature reservoirs of 250°C and 60°C, respectively. It is then decided to redesign the engine so that its ideal efficiency doubles when using the same low-temperature reservoir. What would the temperature of the high-temperature reservoir be in this case?

   Answer: 916°C

5. Heat pumps are rated in terms of coefficient of performance (cop) rather than in terms of efficiency. Thermal cop = heat removed (from low-temperature reservoir)/work. Ideal cop = $T_{cold}/(T_{hot} - T_{cold})$, where the T's are absolute temperatures.
   a. Show that the ideal cop for normal operating temperatures of refrigerators is greater than one.
   b. The actual or thermal cop of a refrigerator is around 2.5. What does this tell you about the amount of heat pumped per work input?

      Answer: $Q_{out} = 2.5\ W_{in}$

   c. On moderately cold days, a heat pump used to heat a house has a thermal cop of about 3.0. If the work input was 4000 J, how much heat would be delivered to the home? In terms of energy, how much of this is cost-free?

      Answer: $Q_{out} = 12,000$ J, $Q_{in} = 8,000$ J

   d. How does the cop of a heat pump used for home heating vary with the outside temperature?

▪ EXERCISES

1. If an object oscillates with a frequency of 10 Hz, how long does it take to complete one cycle?

   Answer: 0.10 s

2. Suppose it takes you one minute to walk around a house twice. (a) What is your period? (b) What would be your frequency in Hz?

   Answer: (a) 30 s, (b) 1/30 Hz

3. A mass oscillates on a spring with a frequency of 5 Hz. (a) How many cycles does it go through in 2 s and 0.6 s? (b) What is the period of oscillation for each of these times?

   Answer: (a) 10 and 3, (b) 0.2 s

4. Radio station frequencies (radio waves) are in the kHz and MHz (kilohertz and megahertz) ranges. What are the corresponding ranges of the periods of these radio waves?

   Answer: 10^{-3} s and 10^{-6} s

5. After a motorboat passes by on a lake, an observer on the shore notices that the waves hit the shore every 2 s and that the distance between the crests of the waves is about 2 m. What is the speed of the water waves? Does this depend on the speed of the boat? Explain.

   Answer: 1 m/s, no

6. If the sine wave curves in Figure 14.17 represent the 3 s of oscillation, what are the period and frequency for each curve?

   Answer: (a) ¾ s, ⁴⁄₃ Hz

7. Light waves in vacuum travel at a speed of 300,000 km/s. The frequency of visible light (an electromagnetic wave) is on the order of 10^{14} Hz (1 followed by 14 zeros). What is the general magnitude of the wavelength of visible light?

   Answer: 10^{-6} m

8. How would the intensity you receive from a point sound source vary if you (a) moved four times as far from the source? (b) moved two thirds of the distance closer to the source?

   Answer: (a) $I_o/16$, (b) $9\,I_o$

9. When a fellow student gets on a swing that you have agreed to push, she informs you that the resonant frequency of the swing is 0.33 Hz. How often should you push the swing to give a smooth, high ride?

   Answer: every 3 s

CHAPTER 14

Vibrations and Waves

■ EXERCISES

1. About how many times faster in air is the speed of light than the speed of sound? (*Hint:* The speed of light is about 300 million meters per second.)

   Answer: 1 million times

2. How much faster does sound travel on a warm day when the temperature is 25°C than on a cold day when the temperature is 0°C?

   Answer: 15 m/s

3. By how much would the temperature have to decrease to have a reduction of (a) 6 m/s and (b) 10 m/s in the speed of sound?

   Answer: (a) 10°C, (b) 16.7°C

4. The distance between two condensations of a sound wave is 2 m. Could this sound be heard by a human ear if it were intense enough? (Take $v_s = 340$ m/s and recall that $\lambda f = v$.)

   Answer: Yes, $f = 170$ Hz

5. What are the wavelength limits of the audible hearing range for humans? (Use $v_s = 340$ m/s.)

   Answer: $\lambda = 17$ m to 0.017 m

6. A rocket travels at Mach 1.5 in the lower atmosphere, where the speed of sound is 330 m/s. What is the speed of the rocket in meters per second?

   Answer: 495 m/s

CHAPTER 15

Sound and Music

In general, anything placed in a circuit, other than a voltage source, acts as a resistance. This may be a commercial resistor or perhaps an appliance. It is important to be able to calculate the current in a circuit with a given voltage. The current might be too large and if the actual circuit were set up, a component might burn out.

To compute the current in a circuit (supplied by a voltage source), we need to know the total effective resistance of the circuit. This depends on how the resistances are connected in the circuit — in series, parallel, or a combination thereof.

As noted in the chapter, when two or more resistances are connected in series, the total effective resistance (R_s) is simply the sum of the resistances,

$$R_s = R_1 + R_2 + R_3 + \ldots$$
(resistances in series)

This is quite simple. You might expect the same would be true for resistances in parallel, but not so. It turns out (your instructor can show you why) that for two or more resistances in parallel, the total effective resistance (R_p) is given by

$$\frac{1}{R_p} = \frac{1}{R_1} + \frac{1}{R_2} + \frac{1}{R_3} + \ldots$$
(resistances in parallel)

This formula gives a somewhat unexpected result. The effective total resistance R_p turns out to always be *less* than the *smallest* component resistance.

Before illustrating this with an example, let's make sure you understand what is meant by total effective resistance. Suppose three resistances connected in parallel gave a particular R_p. Then, the three resistances could be replaced with a single resistance of R_p and the same current would be drawn from the voltage source. The same is true for R_s of a series arrangement (and the total effective resistance of a series-parallel combination arrangement).

CHAPTER 17

Series and Parallel Resistances

Example: Three resistances of 10 Ω, 15 Ω, and 30 Ω in parallel are connected to a 12-V battery. How much current is drawn from the battery or flows into the circuit?

Solution: With $R_1 = 10\ \Omega$, $R_2 = 20\ \Omega$, and $R_3 = 30\ \Omega$, we have

$$\frac{1}{R_p} = \frac{1}{R_1} + \frac{1}{R_2} + \frac{1}{R_3} = \frac{1}{10\ \Omega} + \frac{1}{20\ \Omega} + \frac{1}{30\ \Omega}$$

$$= \frac{3}{30\ \Omega} + \frac{2}{30\ \Omega} + \frac{1}{30\ \Omega}$$

$$= \frac{6}{30\ \Omega} = \frac{1}{5\ \Omega}$$

So, $\frac{1}{R_p} = \frac{1}{5\ \Omega}$, and inverting,

$$R_p = 5\ \Omega$$

Then, using Ohm's law ($V = IR$), the current is

$$I = \frac{V}{R_p} = \frac{12 \text{ V}}{5 \text{ }\Omega} = 2.4 \text{ A}$$

Note: When working with parallel resistances don't forget that it is $1/R_p$ in the formula, so you have to invert or turn the fraction answer "upside-down" to get R_p. If there are just two resistances in parallel, then R_p may be written

$$R_p = \frac{R_1 R_2}{R_1 + R_2}$$

and the inversion is unnecessary.

The total effective resistance of a circuit can also be used to find the total power delivered to or expended in the circuit. Recall $P = IV = I^2 R = V^2/R$ as written in various forms using Ohm's law (see footnote in electric power section of the chapter). Of course, one can go the other way. For example, what is the resistance of a 60-W light bulb? Assuming this is the power rating for common 120-V household voltage, we have $P = 60$ W and $V = 120$ V. Then using $P = V^2/R$, we have

$$R = V^2/P = (120^2 \text{ V})/60 \text{ W} = 240 \text{ }\Omega$$

Could you find how much current flows through the bulb?

Example: A circuit with three 60-W light bulbs, two connected in parallel and the other in series with the parallel arrangement, is plugged into a 120-V source. How much power is expended in the circuit?

Solution: First sketch a circuit diagram for yourself. You might think that the power expended is just 3×60 W or 180 W. But be careful! The voltage across each bulb is not 120 V. Why?

Let's first find the total equivalent resistance of the circuit. From the foregoing light bulb example, a 60-W bulb has a resistance of 240 Ω. (This is fixed by the resistance of the filament.) For two bulbs in parallel ($R_1 = R_2 = 240$ Ω),

$$R_p = \frac{R_1 R_2}{R_1 + R_2} = \frac{(240 \text{ }\Omega)(240 \text{ }\Omega)}{240 \text{ }\Omega + 240 \text{ }\Omega} = 120 \text{ }\Omega$$

This effective resistance R_p is in series with $R_3 = 240$ Ω, and

$$R_s = R_p + R_3 = 120 \text{ }\Omega + 240 \text{ }\Omega = 360 \text{ }\Omega$$

which is the *total* effective resistance of the circuit.

Then, with $V = 120$ V, we have

$$P = \frac{V^2}{R_s} = \frac{(120 \text{ V})^2}{360 \text{ }\Omega} = 40 \text{ W}$$

Note that we could have found the current drawn from the source (Ohm's law, $V = IR$) and then computed the I^2R loss:

$$I = \frac{V}{R_s} = \frac{120 \text{ V}}{360 \text{ }\Omega} = \frac{1}{3}\text{ A}$$

and $\quad R = I^2R = \left(\frac{1}{3}\text{ A}\right)^2 (360 \text{ }\Omega) = 40 \text{ W}$

and the power comes out the same, which is a good thing.

EXERCISES

1. Given a 4-Ω resistor and a 6-Ω resistor, what is the total effective resistance when they are hooked in (a) series and (b) parallel?

   Answer: (a) 10 Ω, (b) 2.4 Ω

2. Show that for two resistances connected in parallel that the total effective resistance is given by $R_p = R_1R_2/(R_1 + R_2)$.

3. What is the total effective resistance of three resistances, 1 Ω, 2 Ω, and 3 Ω, when connected in (a) series and (b) parallel?

   Answer: (a) 6 Ω, (b) 0.55 Ω

4. Given three different resistances, R_1, R_2, and R_3, how many different values of resistance could you get by using them individually and connecting them in different arrangements for effective resistance? (*Hint:* Don't forget series-parallel arrangements.)

   Answer: 14

5. A 2-Ω resistor and a 4-Ω resistor are connected in parallel, and a 1-Ω resistor and a 3-Ω resistor are also connected in parallel. The two parallel arrangements are then connected in series. (a) What is the total effective resistance of the resistors arranged in this manner? (b) If this resistor arrangement was connected to a 12-V battery, how much current would the battery deliver to the circuit?

   Answer: (a) 2.1 Ω, (b) 5.7 A

6. (a) How much power is expended by a common 100-W light bulb? (b) What current does it draw, and (c) what is its resistance? (Use 120 V for household voltage.)

   Answer: (a) 100 W, (b) 0.83 A, (c) 144 Ω

7. Two 100-Ω resistances connected in series are hooked to a 120-V source. (a) How much current is drawn from the source? (b) How much power is expended in the circuit? (c) Can you tell how much current is in each resistance and how much power each expends? Suppose the resistors were not equal?

   Answer: (a) 0.60 A, (b) 72 W

8. A 5-Ω resistance and a 20-Ω resistance are connected in parallel and hooked to a 12-V source. How much (a) current is drawn from the

source and (b) power is expended in the circuit? (c) How much current flows through each resistance? (*Hint:* Remember the current divides proportionately.) (d) How much power is expended by each resistance?

Answer: (a) 3 A, (b) 36 W, (c) 2.4 A and 0.60 A, (d) 28.8 W and 7.2 W

9. (a) How much power does a can opener use with a rating as shown in Figure 17.11? (b) What is its resistance?

Answer: (a) 168 W, (b) 86 Ω

10. A 100-W color television is used on the average of 2 h each day. If the cost of electricity is 8 cents per kilowatt-hour, how much of your monthly (30 days) electric bill is for television use? (*Hint:* Recall that $P = W/t$ or $W = Pt$.)

Answer: 48 cents

ILLUSTRATION CREDITS

Various photographs and line drawings are courtesy of:

Figure (b) in Special Feature 9.1, Figures 12.1, 16.18(b), and 16.19

Kuhn, K., and J. Faughn, *Physics in Your World,* Second Edition, Saunders College Publishing, Philadelphia, 1980.
Figures 2.21, 4.16, 6.5(b), 6.24, 7.12, 8.3, 8.19(b), 10.4, 10.17, 11.6, 14.12, 15.7, 15.9(b), 16.12 (photo), 18.1, 19.8, 19.14, 21.14(a), and 22.6

Ladd, Gary
Chapter 16 introductory photo

Lick Observatory
Figures 27.13 and 27.16(b)

Lund Observatory, Sweden
Figure 27.12

Luray Caverns Corporation
Chapter 8 introductory photo

Masterton, W. L., et al., *Chemical Principles,* Fifth Edition, Saunders College Publishing, Philadelphia, 1981.
Figures 7.3(a), 7.7, 7.9(b), 8.8 and 8.9(a)

Merken, M., *Physical Science,* Third Edition, Saunders College Publishing, Philadelphia, 1984.
Figure in Special Feature I.1, Figures 11.1, 16.17(a), and Chapter 18 introductory photo

Metrologic Instruments, Inc., 143 Harding Ave., Bellmawr, NJ 08031. Used by permission.
Figure 24.15(c)

National Aeronautics and Space Administration (NASA)
Figures 3.7 *(bottom),* 3.16, Chapter 5 introductory photo, Figure 5.11, figure in Special Feature 5.1, Figures 5.17, 5.18, and space color plates

National Oceanic and Atmospheric Administration (NOAA)
Figures 12.10, 18.18
(Vic Hessler, University of Alaska)

Nave, C., and B. Nave, *Physics for the Health Sciences,* W. B. Saunders Co., Philadelphia, 1975.
Figure 10.12

New York Times Pictures
Figure in Special Feature 16.2

Palomar Observatory Photographs
Figures 22.16, 27.1, 27.2, 27.7, 27.15, 27.16(a), and Figure 1 in

Special Feature 27.1

Pasachoff, J., *Contemporary Astronomy,* Third Edition, Saunders College Publishing, Philadelphia, 1981.
Figures 27.4 and 27.8

Pierce, Bill, Time Magazine
Color plate of swinging superconductor

Philadelphia Orchestra
Part IV introductory photo

Photo Researchers, Inc.
Chapter 7 introductory photo, Chapter 10 introductory photo, Figure 14.8(b), Chapter 17 introductory photo, and Chapter 20 introductory photo

PSCC Physics, Fourth Edition, 1976; D.C. Heath and Company, with Education Development Center, Inc., Newton, MA.
Chapter 21 introductory photo

Saunders, Jocelyn, Lander College
Figure 9.2

Serway, R., *Physics for Scientists and Engineers,* Saunders College Publishing, Philadephia, 1981.
Figure 1.16, Chapter 4 introductory photo, Figures 14.3 and 14.15(b), and Chapter 15 introductory photo

Stampf, Edward, Lander College
Figure 10.10

Stevenson, R., and R. B. Moore, *Theory of Physics,* W. B. Saunders Co., Philadelphia, 1967.
Figures 1.13(b), 4.3(a), 21.2 (photo), and 21.3 (photo)

Tass/Sovfoto
Figure 2(b) in Special Feature 26.1

Turk, J., and A. Turk, *Environmental Science,* Third Edition, Saunders College Publishing, Philadelphia, 1984.
Figure 2.19(b)

Turk, J., and A. Turk, *Physical Science,* Second Edition, Saunders College Publishing, Philadelphia, 1981.
Figure in Special Feature 1.3, Figures 1.23, 1.24, 3.13(a), 4.1(b), 6.9, 6.23, 7.11, 7.13, 8.15, 8.17, 13.10, 13.11, 13.12, 15.18, 19.19, 21.9(d), 21.10(b), 21.13(a), 21.22, Figure (b) in

Special Feature 21.1, Figures 22.12, 24.14, 25.20, Figure 1 in Special Feature 26.1, Figures 27.17 and A.8

United Press International
Chapter 26 introductory photo

University of Chicago
Figures 26.2 and 26.12

U.S. Coast Guard
Figure in Special Feature 9.2

U.S. Department of Energy
Figure 19.11, figure in Special Feature 25.2, Figures 26.11(c), 26.14 (with NRC data), 26.15, 26.17(b), 26.18, 26.19, 26.20, and energy research color plates

Welch Scientific Co.
Spectrum chart adaption, color plate

Wicks Organ Company
Figure 15.15 (photo)

Williams, J., F. Trinklein, and H. Metcalfe, *Modern Physics,* Holt, Rinehart and Winston, New York, 1980.
Figure 1.22(b), Figure 1(b) in Special Feature 6.2, Figures 7.3(b), 8.2, 9.11, 9.12, 11.15, 15.12(b), 15.14, 20.10, and 25.17(a)

Wilson, J., *Technical College Physics,* Second Edition, Saunders College Publishing, Philadelphia, 1987.
Chapter 2 introductory photo, Figures 4.4(a), 4.9, 5.1, 6.9(b), 6.19, 8.13, 9.7, 9.8(b), 9.9, 9.14, figure in Special Feature 9.3, Figures 9.19, 9.20, 10.6, 10.7, 10.21, 11.2(b), 11.5, figure in Special Feature 11.1, Figure 11.11, Figures 1–4 in Special Feature 11.3, Figures 11.13, 12.16, 13.4, 13.5, 13.6, 13.8, 13.9, figure in Special Feature 13.2, Figure 14.14, figure in Special Feature 14.2, Figures 1–3 in Special Feature 15.1, Figures 15.8, 15.9(c), figure in Special Feature 15.3, Figure 16.5, Figures 1 and 2 in Special Feature 16.3, Figures 16.17, 17.5, 17.17, 17.18, 17.21, 18.3, 18.9, 18.10, 18.11(a) and (b), figure in Special Feature 18.2, Figures 18.13, 18.14, 18.15,

19.8(a), 19.10 (photos), 19.13, 20.2, 20.3, 20.5(b), 20.7, 20.8, 20.14, 21.6, 21.7(a), 21.13(a), 21.16, 21.19, 22.1, 22.10, 22.15 Special Feature 22.3, Figures 24.6, (photos), Figures 1 and 2 in 24.7(b), figure in Special Feature 24.1, Figure 24.17, Figure 2 in Special Feature 24.2, and Figure 26.10

Winters, Charles D.
Part III introductory photo and color plate of combustion
Yerkes Observatory
Figure 22.14(b)

CHAPTER 1

ANSWERS TO
SELECTED
QUESTIONS

4. Determining speed by radar involves the reflection of radio waves from the moving vehicle. The reflection takes place over a very short time, so the computer-calculated speed is the approximate instantaneous speed. (More on radar detection in Chapter 15 when the Doppler effect is studied.)

7. Yes. This is a deceleration that slows the car. Taking the direction of the car's motion to be positive, the acceleration would be negative, $-a$, indicating that its direction is opposite to that of the velocity.

12. This involves the same principle as the air-bubble level accelerometer discussed in the chapter, with the helium balloon analogous to the bubble and the air analogous to the liquid in the level. The balloon goes forward because of the greater inertia of the air in the car.

15. Yes. An object can have a constant velocity when there is no force or the net force is zero.

19. Because of the inertia of the roll, it doesn't respond quickly to a quick jerk and the towel sheet tears along the perforation. A larger roll works better because the more mass, the more inertia.

22. Because of the person's inertia or opposition to changes in motion, (a) goes backward, (b) goes backward, (c) stationary, (d) goes forward, (e) continues in a straight line but feels a sense of being thrown outward.

23. (a) $\$ \propto n$. (b) No, only the more bottles returned, the more money one gets. (c) $\$ = (0.05)n$. (d) $\$ = (0.05)(10) = 0.50$, or 50 cents.

25. (a) $a = F/m = 2 \text{ N}/1 \text{ kg} = 2 \text{ m/s}^2$
    (b) $a = F/m = 1 \text{ N}/0.5 \text{ kg} = 2 \text{ m/s}^2$
    (c) $a = F/m = 2 \text{ N}/2 \text{ kg} = 1 \text{ m/s}^2$

27. Give each a slight push. The empty container with less mass will accelerate or move faster.

30. No. $g = 9.8 \text{ m/s}^2$ means that the velocity changes 9.8 m/s each second, so a falling object travels different distances each second, $d \propto t^2$.

31. The diver in the fetal position will have to open his or her chute first, since with less air resistance this diver would fall faster.

34. Yes. Air resistance depends on shape and size, as well as speed. With the proper shape and size, a more massive object could have the same terminal velocity as a lighter object.

39. The block exerts equal and opposite forces on the wall and the person's hand. The forces acting *on* the block, the reaction force of the wall and the force of the hand, are equal and opposite, so the net force on the block is zero.

40. (a) The tires exert a force on the road, and the road exerts equal and opposite forces on the tires. (b) Vertical force pairs as in (a). Also, because of internal power transmitted to the wheels, there are horizontal reaction forces on the road surface and equal and opposite forces of static friction on the tires in the direction of the motion that accelerates the car. This is the only external force acting on the car parallel to the motion. (A complicated force of rolling friction also acts to oppose the motion.)

43. (a) $F = mg = (1 \text{ kg})(9.8 \text{ m/s}^2) = 9.8 \text{ N}$. The scale would read this force as can be seen by imagining yourself holding one of the ropes or strings. (b) If several parts of the rope had different tension forces, the parts would accelerate differently and the rope would not remain straight; i.e., one part of the rope would overlap another. When a connecting rope or string is moving, as in a pulley system, there is a net tension difference between the ends of the rope that accelerate it as part of the system. In this case, the force is transmitted

undiminished only in the approximation that the rope has zero or negligible mass.

45. (a) The downward reaction force to the upward arm force causes the scale reading to increase (while the arms are in motion or accelerating). (b) The forces are reversed, and the scale reading decreases.

■ CHAPTER 2

5. Yes. Negative work means that the force and the motion are in opposite directions. That is, a force acts or does work to slow down a moving object. (See Question 8.)

8. (Negative) work is done by friction. Dusting lubricates the board so that the puck slides more easily and farther, with less applied force.

10. The pay for "piece" work depends on power, i.e., the amount of work or number of pieces done per time. Being paid a fixed hourly rate implies, but does not demand, that work should be done.

12. (a) The power outputs are not the same, since with $2F$ the acceleration is greater and the work is done faster. (b) For P_2 to be twice as large as P_1 requires $t_2 = t_1$. ($2P_1 = 2Fd/t_1$ and $P_2 = (2F)d/t_2$.) Since t_2 is less than t_1, P_2 is more than twice as large as P_1.

16. All are correct from their own zero points of reference, which each takes to be at his or her floor.

21. Internal energy is used to start and increase the jump height with successive bounces. (Think about jumping off the ground.) On a trampoline, the elasticity of the cover stores energy when the jumper is in contact. The degree of elasticity and/or energy exchange determines the maximum height.

24. Since $K \propto v^2$, with $3v$, we have $K \propto (3v)^2 = 9v^2$, so the kinetic energy increases by a factor of 9.

27. Potential energy stored in the chemical bonds of compounds is released, and the products of the reaction have less energy. The released energy manifests itself as heat or in the kinetic energy of the molecules.

31. (a) Chemical energy into electrical energy. (b) Wind kinetic energy into boat kinetic energy. (c) Radiant light energy into electrical energy. (d) Gravitational potential energy into electrical energy.

36. With a smaller acceleration due to gravity on the moon, less work would be required to lift the bob to its release point, and it would have less potential energy. The speed of the bob would be less, but the swing heights would be the same, assuming conservation of energy. (There are no air resistance considerations on the moon, which has no atmosphere.)

39. (a) Yes, relative to the ground or bottom of the loop. (b) The initial height must be great enough so that the potential difference between this height and the top of the loop is sufficient to give the coaster enough kinetic energy or speed to get it around the loop.

41. With no friction, the work output equals the work input, $F_oL_o = F_iL_i$. Then the mechanical advantage = $F_o/F_i = L_i/L_o$.

44. When a load is moved up a ramp, work is done against a component of gravity, so less input force is required in moving the load a greater distance against gravity up the plane than in lifting it directly to the top. The mechanical advantage increases with decreasing incline angle.

47. $E_A = W_o/W_i = 500$ J/700 J = 5/7, and $E_B = 900$ J/1400 J = 9/14 = 4.5/7, so machine A is more efficient.

■ CHAPTER 3

5. It would be assumed that this would be some sort of advanced rocket-type engine that uses impulse ($F\Delta t$) to change momentum and accelerate the *Enterprise,* perhaps a nuclear-propelled rocket. What would be more interesting is how "warp drive" works. Warp-2 is two times the speed of light; Warp-3, three times; and so on, like Mach numbers and the speed of sound (Chapter 15). However, the theory of relativity says an object cannot go faster than the speed of light (Chapter 23).

10. This is an example of jet propulsion. Air is forced from the mouth of the balloon, and the balloon experiences an equal and opposite force. The mouth extension is nonuniform and flexible, so the escaping air varies in direction as does the balloon.

13. A rocket launcher is essentially an open tube, so the exhaust gas goes out the back, causing little reaction or recoil on the launcher. The exhaust blast is quite powerful (necessary to project the shell) and can be dangerous if a person is standing too close to the rear.

14. The rifle has a system by which some of the exhaust gas is ejected with a component in the forward direction, giving a "reverse thrust" to reduce the recoil. (See Question 15.)

18. The exhaust gases are directed toward the moon. The force on the rocket is in the opposite direction and hence decelerates or slows the rocket. (Think about the rocket acceleration and exhaust gas directions on blast-off.)

19. No, the momentum is not conserved. Before collision it is in one direction, and afterward in the opposite direction. Hence, the momentum is not the same before and after, although the magnitude is the same. The change in momentum is $\Delta p = p_2 - p_1 = mv - (-mv) = 2mv$. For momentum to be conserved, $\Delta p = 0$.

21. When the fan blows air into the cloth sail, most of the air is trapped and slowed almost to a stop, so an impulse is imparted to the sail. However, when the fan blows air, an opposite impulse is imparted to the fan. Both being parts of the boat system, the effects cancel. With fan directed backward, there is jet propulsion—the air goes one direction and the boat the other. (As an interesting sidelight, if a rigid sail were used with the fan blowing on it and the air molecules bounced backward, there would be a net change in momentum in the forward direction and the boat would move forward. See Question 19.)

24. With equal masses, if the momentum is conserved, so is the velocity, since it is the vector quantity (the masses cancel in the conservation equation). The velocity is not conserved if the masses are different, since the magnitudes of the velocities will be different so as to conserve momentum.

28. (a) Yes. In an elastic collision, kinetic energy may be converted to elastic potential energy during collision. But after collision the potential energy is again zero and the kinetic energy is the same, so mechanical energy is conserved. (b) No, it is not possible to lose all the kinetic energy. If all kinetic energy were lost, there would be no motion or speed after collision (K is a scalar quantity). With no motion or (zero) momentum, the momentum could not be conserved since there was momentum before collision.

■ CHAPTER 4

2. The observer would see the coin follow a parabolic path as a projectile. In moving down the road with the car, the coin has a constant horizontal velocity, and a vertical velocity is given when the coin is flipped upward.

7. (a) Air resistance on a vertical projection would result in a lower maximum height and a longer time of fall. (b) The range would be less. (c) A combination of (a) and (b).

12. $a_c = v^2/r = (\text{m/s})^2/\text{m} = \text{m}^2/\text{s}^2/\text{m} = \text{m/s}^2$.

13. In a washing machine, the spin rate is fast enough for the clothes to be "held" against the drum. The reaction force of the drum on the clothes provides the centripetal force, and the friction force is great enough so the clothes don't fall—similar to a Rotor ride. In a dryer, the rotation speed is not as great. Gravity provides more force than is needed for the centripetal force, and the clothes accelerate downward when free to do so. (A washer rotates with a vertical spin axis, and a dryer rotates with a horizontal spin axis.)

15. On a banked curve a component of the reaction force of the road on the car is directed inward toward the center of curvature and provides a contribution to the centripetal force that is greater than the reduced component of friction.

20. While the projectile is traveling directly southward from the North Pole, the Earth rotates beneath it and the projectile would land to the west of the original meridian. To an observer on the rotating Earth, in terms of Newton's laws, some sort of force is necessary to deflect the projectile to the right. The situation in the Southern Hemisphere is reversed; i.e., the deflection is to the left as viewed along an object's path of motion.

24. The orbital speed would not vary. The radius of orbit would be constant, and so would the gravitational attraction of the Sun and the orbital speed.

26. $T^2/R^3 = (1 \text{ year})^2/(1 \text{ AU})^3 = 1 \text{ y}^2/\text{AU}^3$, so k would have a magnitude of 1, and $T^2 = R^3$ in these units.

■ CHAPTER 5

2. Because the moon also has tangential velocity, it moves sideways while falling inward, so to speak. The combined motions produce a nearly circular orbit.

7. The force of gravity is an inverse-square law, $F \propto 1/r^2$. We are much closer to the center of the Earth than the center of the Sun, 4000 miles compared to 93 million miles. So with a small r^2, the force of gravity of the Earth is much larger than that of the Sun, even though the latter is much more massive.

9. The gravitational attraction within a body depends on the mass interior to an object. For an object falling down the hole, the effective mass and the gravitational attraction become smaller, going to zero at the center. As the object "falls up" the other side, the attraction increases. The stone would rise to the top of the other side and fall back again and would oscillate through the center of the Earth. If the hole is through the center of the Earth, it would come out not in China, but rather in the Indian Ocean. Check it out.

11. Because of frictional drag and nonuniform surface features.

12. Spring tides occur when the Earth, moon, and Sun are lined up (as viewed from above), and this condition occurs at new and full moons. Neap tides occur when the moon and Sun form a right angle with the Earth. This occurs at first and third-quarter moons.

17. The relatively small planet would be very far away. Even if you knew about where to look, the reflected sunlight might be so dim that the planet could not be seen or distinguished with our current telescopes.

19. $g \propto 1/R_e^2$ and $g' \propto 1/(3R_e)^2 = 1/9(1/R_e^2)$, or 1/9 of g at the Earth's surface.

24. Since $g = GM/r^2$, this means that r^2 for Saturn is 94 times greater than for

Earth, i.e., $g_s = G(94 M_e)/94(R_e)^2 = GM_s/R_s^2 = g$. Hence, Saturn is much larger, with a radius of $R_s = \sqrt{94} R_e \approx 10 R_e$ (a radius about ten times larger).

27. At the surface region of the original star. The gravitational force there is as if all the mass were concentrated at the center, which is the condition a black hole approaches.

31. No, the fact that it is falling indicates there is acceleration due to gravity and weight. The cup is falling at the same rate as the water, so no water comes out the hole. This is similar to an astronaut in an orbiting spacecraft.

36. No, less tangential velocity would be required. For an orbit with a greater radius, less centripetal force is required ($F = mv^2/r$) and less is supplied by gravity, so the tangential velocity or orbital speed is less.

39. The gravitational attraction of the moon was not strong enough to hold the energetic molecules of its atmosphere, and they escaped.

■ CHAPTER 6

1. Yes. The cm and the cg correspond in a *uniform* gravitational field. The Earth's gravitational field varies with altitude, so technically the cm of a sky-scraper might be above its cg, but this is usually negligible.

6. Directly below its point of suspension.

7. You shouldn't. A particle can't travel in a straight line and a circle at once.

11. (a) 2π, (b) π, (c) $\pi/2$.

12. (a) Angular speed $= 2\pi$ rad/60 s $= \pi/30$ rad/s. (b) Angular speed $= 2\pi$ rad/60 min $= \pi/30$ rad/min, or since 1 min $= 60$ s, $\pi/1800$ rad/s. (c) Angular speed $= 2\pi$ rad/12 h $= \pi/6$ rad/h, or since 1 h $= 3600$ s, $\pi/21,600$ rad/s.

15. No. Even though all parts of the record travel with the same angular speed, the outer portions have greater tangential speeds and the outer bands would be smaller. Think of length or music per time.

16. Yes. With larger tires, a greater distance would be traveled in one revolution and the speed would be greater than normal.

19. Consider the distance of the mass from the axis. The comparative rotational inertia for (a) is smallest, for (b) is largest, and for (c) is intermediate.

20. The braking causes a torque, and the center of gravity tends to rotate.

22. Maximum on the downward push when pedal bar is horizontal. (Occurs twice each revolution.) Minimum and zero at top and bottom.

23. Try the one with the larger handle. This would give a larger radius and increased torque when you twist or apply a force on it.

27. The torques balance each other, so $r_1 F_1 = r_2 F_2$, where r is measured from the center of gravity or the point of suspension. So, $r_1 = (F_2/F_1) r_2 = (4\,\text{N}/1\,\text{N}) r_2$, and $r_1 = 4r_2$ and the cg is 80 cm from the (center of the) small ball. Without math you could say 4 N is to r_2 as 1 N is to r_1 to see that r_1 is four times r_2.

29. The component of the force of gravity or weight force of the cylinder acting down the plane produces a torque and motion. This acts on the cg and is analogous to the component of the weight force of a block sliding down a frictionless inclined plane.

32. The center of gravity of the longer pole is higher, but it also has a greater moment of inertia that resists the toppling rotational motion and gives you time to recover or move your finger under the cg.

35. Eleven bricks counting the bottom one would be stable. The twelfth brick would put the center of gravity at the edge of the base of support, and the stack would be in unstable equilibrium and probably topple.

39. It increases the effective area of the base of your support in the lateral or sideways directions, but not forward or backward.

42. Stable, 6 sides; and unstable, 8 end points and 12 edges, for a total of 26.

46. Turntable would rotate in the opposite direction by conservation of angular momentum, with the rate depending on the moment of inertia. If the moments were the same, there would be no change in position (same angular speeds) to an outside observer. With a large turntable inertia, a person would progress around the table. Walking inward in a spiral would decrease the radius and increase the angular speed.

47. An opposite net torque due to forces on the air foils (wings and tail). A small plane actually flies slightly tilted.

■ CHAPTER 7

3. There are about 20,000 billion billion molecules in a liter of air. Comparatively, the U.S. population of 250 million, or a quarter of a billion, and the world population of 5 billion are rather small numbers.

4. (a) With 10^{22}, or 10,000 billion billion, molecules per half liter of air, 6 liters is 12×10^{22}, or 120,000 billion billion molecules. (b) In general yes, but many of the oxygen molecules have combined with carbon to form carbon dioxide (CO_2). The nitrogen, which makes up more than 70 percent of the air, goes along for the ride.

7. Yes, if it is a neutral atom and no electrons have been added or removed to form an ion.

11. (a) Copper, 29; (b) thorium, 90; (c) nobelium, 102.

13. The protons determine the element, so U-238 with 92 protons less 2 protons gives thorium, Th-234 (2 neutrons also removed, so $238 - 4 = 234$).

15. Sucrose is $C_{12}H_{22}O_{11}$, so with the decomposition of 12 molecules there are $12 \times 12 = 144$ carbon atoms and $12 \times 11(H_2O) = 132$ water molecules.

17. Temperature (thermal energy) and pressure.

18. Yes. Solid-liquid, melting and freezing, or solidification; liquid-gas, vaporization and condensation; and solid-gas, sublimation, and although not given in the text, going from the gas to solid phase is sometimes called deposition.

21. They are quickly annihilated.

■ CHAPTER 8

2. Basically a cubic lattice, but with (a) an atom at the center of each face of the cube and (b) an atom at the center of the cube.

6. Yes. There is more mass in a smaller volume, so the average density increases.

9. $5 \text{ cm} \times 10 \text{ cm} \times 20 \text{ cm} = 1000 \text{ cm}^3$. From the chapter table of densities, those of gold and lead are 19.3 g/cm³ and 11.5 g/cm³, respectively. Then, density = mass/volume or mass = density × volume, and the gold brick would weigh 19,300 g, or 19.3 kg, and the lead brick 11.5 kg.

10. Because there are $100 \text{ cm} \times 100 \text{ cm} \times 100 \text{ cm} = 1,000,000$, or 10^6, cm³ in 1 m³. This would be 10^6 g of water, or 1000 kg (per m³).

12. Total area (of square) = 4 cm × 4 cm = 16 cm². Total area of circles = $4 \times \pi(1 \text{ cm})^2 = 4\pi = 12.6 \text{ cm}^2$. Empty area = 16 cm² − 12.6 cm² = 3.4 cm², and 3.4 cm²/16 cm² (×100%) = 21%.

16. They "evaporate" or sublime.

22. Not really since it is not a mixture.

23. Because the density of the other metal in the alloy, tin, has a lower density (6.5 g/cm³).

25. Still 2 cm. The stretched length will be the same for the same mass, so each spring will stretch 1 cm.

27. Because stinging vibrations could be transmitted to the hand. Rubber mallets are used when forceful blows are wanted without deformation to the pounded object.

29. Clockwise from upper left box, Ir, Pb, C, and Al.

■ CHAPTER 9

3. $\rho = m/V$, and $\rho g = mg/V = w/V = D$.

4. The flow from the lower hole would be greater. $\frac{1}{2}\rho v^2 = \rho gh$, and $v = \sqrt{2gh}$, where h is measured from the surface down. The hole size does make a difference. The greater the hole or area, the lower the flow speed.

5. Both have the same pressure on the bottom, since the depths are the same. The container on the right has pressure on its sides. Think of a column in the center of each container with an area equal to the base of the smaller container.

7. $p = p_o + \rho gh = 2000$ kPA $+ (9800$ N/m³$)(9.8$ m/s²$)(1.0$ m$) = 2000$ kPa $+ 96,000$ Pa $= 2096$ kPa.

10. Since the area of the larger piston is 100 times that of the smaller piston, it would move 0.50 cm when the small piston moved 50 cm, or 1/100 as far. Think of the small piston displacing 50 cm³ of fluid. Spread over the larger piston, this would have a depth of 0.50 cm.

11. The mechanical advantage would be 1/100, or 0.01. This means that there is a distance multiplication or a force reduction; i.e., the small piston moves farther than the larger piston. When the larger piston is moved 10 cm, the smaller one would move 1000 cm. (See Question 10.)

13. Higher in the ocean, since if the same volume displacement of water is made, salt water has a greater density and the buoyant force would be greater.

16. The kinetic energy gained from the acceleration due to the buoyant force.

21. (a) The numbers decrease vertically, since the hydrometer would float lower in a less dense liquid with a specific gravity less than 1.000. (b) The specific gravity gives a relative indication of concentration, and the hydrometer would float higher if the battery were charged and had a lot of sulfuric acid, and vice versa for the antifreeze.

25. When the boat fills, this is like replacing the water displaced, so the buoyant force decreases. Bailing helps because it reduces the amount of water in the boat or increases the displacement.

28. Soapy water films have more surface tension than plain water.

32. With the detergent, the water wet the feathers and filled the internal air space that makes a duck overall larger and buoyant—sort of like a life preserver.

34. The height of a column depends on the weight or density of the fluid, so the lighter liquid would rise higher.

38. To have a fluid pressure greater than the patient's blood pressure. If not, blood would flow into the needle.

40. Closing the valve partway wouldn't reduce the pressure, only the amount of water. The pressure is transmitted undiminished (Pascal's principle).

44. No, the surface tension of water also is reduced with increasing temperature. This is why we generally use hot water for cleaning so that the water can wet better.

46. You'd probably have trouble starting your car on a cold morning because the oil would be thick and viscous.

CHAPTER 10

2. (a) The density is doubled for a volume decrease of one half. (b) The pressure would double, assuming constant temperature. To compress the gas to a smaller volume would require that work be done.

5. The water flow from the hole would stop. After the hand was applied and some water ran out, the pressure above the liquid would be less than atmospheric pressure, and the pressure difference between the external pressure on the hole area and the space gives an upward force.

7. To the point that the pressure due to the weight of the column of water in the glass equals atmospheric pressure. Think of a barometer.

10. The pressure on the tires due to the weight of a rider is relatively large because of the small areas of the tires. An equal and opposite pressure force is needed if the tires are not to flatten. (Typical automobile tire pressure is about twice atmospheric pressure.)

11. (a) In filling, the bulb is first squeezed and then released. Releasing increases the volume inside the bulb and decreases the air pressure therein. The greater force of the atmospheric pressure acting on the liquid then "pushes" the liquid up the dropper. The bulb is squeezed to force the liquid out by reversing the pressure difference. (b) Similarly, when a syringe is filled, the plunger is pulled outward, which increases the volume and gives a lower pressure in the syringe. Pressure is exerted on the plunger in dispensing to reverse the pressure difference. Air is usually first injected into a medicine vial to increase the pressure, because fluid removal decreases the air pressure in the vial and a reverse pressure difference would retard filling the syringe.

13. Atmospheric pressure acting on the card holds the water in. (Gravity could easily overcome any surface tension effects.) This is similar to the barometer principle. Atmospheric pressure would still act on the card if the glass were turned horizontal (Pascal's principle).

16. (a) Approximately 114 cm or 45 in. (b) 38 cm or 15 in.

19. Yes, atmospheric pressure will only support a column of liquid so tall. For mercury, this is 76 cm. Water, with a density of 1.00 g/cm³ (or 1.00×10^3 kg/m³), is 13.6 times lighter or less dense than mercury, so the maximum column height for water is 76 cm $\times$ 13.6 = 1033 cm = 10.33 m—a pretty impractical straw. (This assumes that you could aspirate the straw to near vacuum. In practice, the pressure difference and the column height would be less.)

20. The atmospheric pressure decreases faster than the temperature with increasing altitude. Although the volume may increase, the air becomes less dense with increasing altitude, and the buoyant force decreases.

23. So as to displace a large enough volume of air to give the proper buoyant force. A blimp has motors and fins that allow it to navigate directionally. Hot-air balloon control is only up and down. There is no control over which direction the wind blows a balloon.

25. Things are never sucked in. They are "pushed" in as a result of the force due to a pressure difference.

26. The pulse rate is the number of times the heart beats (about 60 times per minute). The figure 120/80 expresses the pressure in the arteries due to the beating or pumping of the heart.

32. The blades of the rotor force air downward (like a fan), and there is an upward or opposite force.

35. When the vehicles are in close proximity, drag causes the relative air flow speed between them to be greater and the pressure is reduced (Bernoulli's principle). The pressure difference between the inside and outside of the car gives an inward force. Think of the car as an air foil in the vertical plane. Ships do not pass each other too closely because of a similar effect in water.

36. When the bulb is squeezed, air is forced over the top of a tube extending into the liquid. This reduces the pressure, and the liquid rises in the tube (similar to the draft of a chimney).

■ CHAPTER 11

2. The flow is from an object with a higher thermal (temperature) energy, but not necessarily from one of greater total internal energy as the *Hint* implies.

5. No. Helium (He) is a monatomic (single atom) gas and oxygen (O_2) is a diatomic (double atom molecule) gas, so oxygen can have internal vibrations and greater internal energy at the same temperature.

6. The ice would slide off as the strip bends downward. See Fig. 11.2(a) and compare heating and cooling.

10. Stress due to nonuniform thermal expansion causes the plate or glass to crack.

13. When the ball is heated, it expands and no longer fits through the ring. When the ring is also heated, it expands with the internal hole getting larger and the ball then fits through.

18. No. Below $4°C$ the column would rise with decreasing temperature, and below $0°C$ it would be impractical, probably breaking the glass because of the expansion of ice.

19. The glass bulb in contact with the hot liquid initially expands faster than the liquid (heat must be conducted inward), and the liquid column falls slightly before greater liquid expansion causes it to rise.

21. (a) Degree Celsius ($1°C = 9/5°F$), (b) a kelvin, since it is equivalent to $1°C$, (c) the same.

23. 100 kelvins (273 K to 373 K). For $20°C$, $T_K = 20 + 273 = 293$ K.

25. To double the internal energy, the *absolute* temperature is doubled. $T_1 = 20°C + 273 = 293$ K, and $T_2 = 2 \times 293$ K $= 586$ K. On the Celsius scale this is $T_C = 586$ K $- 273 = 313°C$.

27. A joule is the amount of energy that would raise the temperature of 1 g of water $4.2°C$ (or 4.2 g of water $1°C$).

30. Less because the heat of combustion of alcohol is less than that of gasoline.

33. The container with twice the temperature change has one half as much water (mass).

34. Since copper has a specific heat (0.093 cal/g–$°C$) that is three times that of lead (0.031 cal/g–$°C$), it takes three times as much heat to raise the temperature of the copper an equal amount (equal masses).

CHAPTER 12

3. The ground with air spaces is a relatively good insulator.

5. An aluminum pan because of its higher thermal conductivity.

7. Probably to some degree for looks, but copper does have a relatively high thermal conductivity and would promote the overall heat transfer capability of an iron pot.

9. It is a transfer of heat, but not in the sense that it arises from a temperature difference.

12. The foam helps break up convection cycles that could transfer heat between the walls.

14. Things would get much too hot since there would be no heat loss from the atmosphere into space. Eventually, the atmosphere would be lost as the air molecules gained enough energy to overcome gravity and escape.

16. Primarily to provide more area for conduction to the air. Radiation losses are also increased.

19. The evacuated space insulates by having no convection cycles to transfer heat conducted through the walls, and the silvering prevents radiation losses by reflection.

21. The greater the temperature of the coffee, or the greater the temperature difference with the surroundings, the faster heat is lost. Adding cream first, giving it more internal energy and reducing the temperature of the mixture, slows the heat loss and would help the coffee mixture to stay warm longer than adding cool cream to cooled coffee.

24. The internal molecular bonds are different, which would also be expected to give rise to different latent heats.

27. Evaporation, promoted by the relative air motion due to the motion of the car, cools the water.

30. $-6°F$.

31. $30°F$.

33. The ground in the valley loses heat by radiation at night, cooling the air in the valley to below its dew point.

35. The frozen coffee is placed in a vacuum chamber and the pressure reduced. The water sublimes, leaving the coffee granules.

38. The vapor pressure over the water is increased, which lowers the boiling point. (Think of the pressure cooker principle.)

40. (a) To raise the boiling point of the coolant so it can operate at a higher temperature without boiling. (b) The sudden reduction in pressure could cause the coolant to "flash" boil, with hot liquid erupting from the radiator. *Don't do this.* Severe burns can result.

41. A water-antifreeze mixture is adequate to prevent freezing and is cheaper.

42. Alcohol has a lower boiling point and it eventually could be lost.

CHAPTER 13

2. Heat input is from fluid (reheated by atmosphere) and heat output is through the beak.

3. Watts, or probably kilowatts for practicality. 8 hp (746 W/hp) = 5969 W ≈ 6 kW.

6. With two revolutions for all cyclinders to fire, a six-cylinder engine fires every one third of a revolution and an eight-cyclinder engine every one quarter of a revolution.

9. For initial heating. A cold cylinder could conduct heat from the compressed air and lower its temperature so the cylinder would not fire properly.

11. Thermal efficiency = work/heat in = $(\Delta Q - \Delta U)$/heat in, since by the first law, $\Delta Q = \Delta U + W$. ΔQ is the heat out and for a cycle, $\Delta U = 0$.

15. Greater than 100 percent.

17. No. There is work input.

19. No. Work is done by the expanding air (and gravity).

23. No. Heat is expelled from the refrigerator into the room—not only that removed from the interior of the refrigerator, but also that from the work input.

26. There is a smaller temperature difference through which to pump heat for the water than for the outside air, but water systems are more costly and require extra maintenance. Also, water has a higher specific heat (capacity) than air. The water is the high-temperature reservoir in the winter.

28. Ice is a good thermal insulator and impairs the loss of heat. With less heat removed, the compressor works continually and inefficiently and may burn out.

31. It decreases (has more available energy owing to work of compression).

■ CHAPTER 14

1. Place a felt-tipped pen in a swivel-mounted sleeve on the bob (so pen can adjust length to surface below bob) and move a paper horizontally as the pendulum oscillates perpendicularly to the direction of the paper motion. (A pen light on the bob and light-sensitive paper would also work nicely.)

4. (a) Both oscillate together—same direction and displacement at the same time. (b) Out of phase by 90°. For example, as one mass goes upward through the equilibrium position, the other would be at the top amplitude position starting downward. (c) Out of phase by 180°. The masses would travel in opposite directions and be at opposite amplitude positions at the same time and pass each other at the equilibrium position.

7. The springs absorb the energy and reduce the impulse to the chassis, and the hydraulic shock absorbers damp the spring oscillations.

9. We say that the energy is stored in the electric and magnetic fields of the wave as the disturbance from an accelerated electrically charged particle propagates through space (see Chapter 19).

12. Greater amplitude, (a); greater frequency, (b); greater wavelength, (a); greater period, (a).

15. How the ocean floor slopes up to the beach influences how waves break or a surf forms. Tides have great influences on the movement of ocean water and surfs. (Ocean storms also generate disturbances and water motion.)

17. In the intermolecular bonds of the medium. The waveform is "destroyed."

19. The combination of mutually perpendicular longitudinal and transverse motions gives rise to circular or elliptical motions.

21. Only odd multiples of quarter wavelengths (half of a loop) fit the boundary condition, i.e., $\lambda/4$, $3\lambda/4$, $5\lambda/4$, etc., so the characteristic frequencies are f_o, $3f_o$, $5f_o$, and so on.

24. Multiples of the single characteristic frequency just mean the swing is not

driven every time it returns to its driving position; i.e., $2f_o$, every other time; $3f_o$, every third time; and so on.

CHAPTER 15

1. A disturbance and a medium are definitely needed. It might be said a detector is needed to know that the sound is there.

4. The energy of the disturbance dissipates and becomes part of the internal energy of the medium.

8. The normal pitch or sound of one's voice depends on resonances set up in the vocal and nasal cavities. These have characteristic dimensions and accommodate certain wavelengths. Sound travels faster in helium, and with $f = v/\lambda$, when v increases so does f.

10. The soft snow absorbs sound.

12. Because of reflections and resonance.

15. Although introduced with sound, the Doppler effect applies to any type of periodic wave, e.g., the Doppler red shift of light.

17. No. There has to be relative motion.

20. No. The sonic boom is due to the bow wave that is continually formed by a supersonic aircraft.

23. No. The beat of the music refers to tempo or timing, not interference effects.

25. Yes. A negative dB value would correspond to an intensity below the threshold intensity.

28. Energy is transmitted to the tabletop, and it acts as a sounding board. The bodies of stringed instruments have a similar effect, e.g., the top and bottom of a guitar or violin.

30. The lighter string should have less than normal tension, so it has a lower frequency.

34. A pure tone has a single frequency. A musical note as played by an instrument usually has some overtones.

36. Two speakers give spatial variation that is characteristic of the source, such as a band. Adding more speakers enhances the effect.

CHAPTER 16

2. (a) No, there would always be a net force for either a positive or negative charge. (b) No, there would always be a net force because the test charge is not equidistant from the charges at the end of the meterstick.

3. $q = ne$, so $n = q/e = 1$ C$/1.6 \times 10^{-19}$ C$/e \approx 10^{19}$ electrons.

7. Reduced by one fourth, $1/(2r)^2 = 1/4r^2$.

9. Bring the object near the electroscope (or vice versa). Leaves would diverge if the object is charged. If the object is charged, contact with the electroscope would charge it, and then the known charge, e.g., on a rod rubbed with fur, could be used to identify the charge on the electroscope.

12. (a) Leaves would collapse as the charge is conducted away through the finger. (b) Leaves would collapse as the positively charged rod would attract electrons to the electroscope bulb.

18. If the electric force is stronger than the gravitational force.

20. The sphere would be neutralized by charge transfer.

23. Bringing a charged rod near the spheres, say a negatively charged rod, electrons would be repelled to the opposite sphere. Then, using the insulating supports to separate the spheres, one would be negatively charged (sphere farther from the rod) and the other positively charged.

28. In general, lightning rods provide contact and a path for the downward ionized leader to be conducted to ground.

32. Because by convention a positive test charge is used to define the field.

34. Because of the equilibrium distribution of charges on the outside of the object. Imagine a sphere completely covered with negative charge on the outside. All external fields would end on the charge, and there would be no field or force inside.

37. When insulated from the ground and touching the charged Van de Graaff sphere, the boy is also charged, and his hair acts like electroscope leaves.

40. The object would have twice as much electric potential energy *per unit charge*.

■ CHAPTER 17

5. The electric field alternates back and forth and does work.

9. $I = q/t = 10 \text{ C}/2.0 \text{ s} = 5.0 \text{ A}$.

12. Yes, there could be a complete circuit with a potential difference. The direct current could run either way. However, it usually is not possible to get good contact with the batteries backward.

16. (a) Single-bulb circuit, (b) three-bulb circuit. All have the same voltage, so the more bulbs, the greater the resistance and the less the current (less joule heating and less bright).

20. The 75-W bulb has greater resistance because less current flows (less bright, less joule heating) for the same voltage. If the filaments were the same length, then the 100-W bulb filament would be thicker or have a greater cross-sectional area so more current could flow.

21. When connected in parallel, the voltage drop across each bulb is the same (as the voltage rise of the battery), and more current flows in the 100-W bulb. (See Question 20.)

26. Moisture in the hot dog makes it conductive, and joule heating cooks it. A multiple-dog cooker would be wired in parallel so that the same voltage is across each dog and they all cook generally at the same rate.

27. (a) The oxygen in the air would cause the filament to oxidize and burn out at high temperatures. (b) Metal is "boiled" away from the hot filament, and the vapor condenses on the inside of the glass bulb near the filament. This is called vapor deposition, and the process is used in some instances to coat objects with a metal film.

31. Series. One light still burns when the other burns out or is removed.

33. (a) Series batteries, $I = V/R = (V_1 + V_2)/R = (6.0 + 6.0)/4.0 = 3.0$ A. (b) Parallel batteries, $I = V/R = 6.0/4.0 = 1.5$ A. (c) Series batteries, $I = V/R = (V_1 + V_2)/(R_1 + R_2) = (6.0 + 6.0)/(4.0 + 2.0) = 2.0$ A; parallel batteries, $I = V/R = V/(R_1 + R_2) = 6.0/(4.0 + 2.0) = 1.0$ A. (d)—(a) 12 V, (b) 6.0 V, (c) series, $V_1 = IR_1 = (2.0)(4.0) = 8.0$ V, $V_2 = 4.0$ V; parallel, 6.0 V.

34. So there is not a voltage or potential difference in part of the circuit when the switch is off.

37. They don't make contact with the other wire to complete a circuit. On one wire, the bird has the same potential as the wire.

43. You wouldn't be able to make contact with both hands and get a potential difference across or have a current flow through your chest, which could be fatal. Contact made across the fingers of one hand might cause burns but would probably not kill.

■ CHAPTER 18

2. Yes, because opposite polarities are induced in the nail at each end of the magnet.

7. The magnetic field is in concentric circles with a clockwise sense of direction looking northward along the current-carrying wire (see Fig. 18.6). (a) East, (b) west, (c) north (because of the Earth's magnetic field, wire field is up and down).

10. Since this would allow a separation of magnetic poles or magnetic "charges," ask him to show you evidence of a magnetic monopole.

14. (1) Needle attracted to the magnet. (2) Magnet suspended on string points north-south (compass). (3) Magnet would attract unmagnetized bar at center (magnet placed perpendicularly to unmagnetized bar at mid-length point). No attraction if situation is reversed. (Magnetic field is perpendicular to the bar and does not align domains along the axis.)

17. Both ac and dc would work. The alternating current would produce an alternating magnetic field (oppositely induced in scrap metal), but with sufficient frequency, metal would not have time to fall free when current goes to zero value.

18. Force would still be upward (right-hand rule).

21. By the right-hand rule (reversed for negative charge), (a) force on proton to right (on electron to left). (b) Zero force for both. (c) Upward force on proton (downward force on electron).

25. When the button switch is pushed, the circuit is closed, the electromagnet is activated, and the metal armature is attracted to the magnet (causing the clapper to strike and ring the bell). However, the movement breaks the electrical contact at the contact screw and spring leaf, so current goes to zero, the armature is released, and it springs back to the original position. The cycle repeats as long as the button is depressed.

27. An ammeter measures the (series) in-line current flow. A low resistance is needed so as to affect the current flow as little as possible with the measurement. A voltmeter measures the voltage across (parallel to) a circuit element. A high resistance is needed so as to prevent any appreciable current flow from the circuit through the voltmeter.

29. The absorbing D layer has dissipated, and transmission is better.

■ CHAPTER 19

2. Because the Earth's magnetic field is relatively weak, it wouldn't be economically practical—more input than output.

6. (a) Yes, the flux (number of field lines through the loop) is increasing because of the area increase. The induced magnetic field is opposite to *B*, as shown in the figure. (b) Induced current and resulting magnetic field reversed from (a).

7. Toward far end of bar by right-hand rule ($v \times B$, for *positive charge*). This is the direction of the induced electric current.

8. Energy-wise, from work done in turning the loop.

11. (a) A deflection first one way and then the other. (b) Less in both cases since there are an induced repulsion and attraction that oppose the magnet (Lenz's law).

16. (a) Alternating. (b) Save your money. The oscillation would be damped and stopped by opposing induced repulsions and attractions (Lenz's law).

23. $12,000/120 = 100$ times more.

25. It could be used as either, depending on how the windings were connected.

28. Because of losses due to corona discharge at very high voltages.

33. All are composed of oscillating electric and magnetic fields, but they differ in frequency and wavelength.

36. Clouds absorb infrared radiation (heat rays) and transmit ultraviolet (tanning rays).

41. Because the electrons in an X-ray tube are decelerated (negative acceleration).

■ CHAPTER 20

2. A resonance vibration in the crystal has a definite frequency and period to which time is referenced.

5. The mercury spectrum has strong emission lines in the blue region of the visible spectrum, and the sodium spectrum has some strong emission lines in the orange-yellow region.

11. It depends on what the media are on each side of the film and what phase shifts take place. A soap bubble with a quarter-wavelength thickness and air on both sides would have constructive interference because of different phase shifts of reflected waves at each interface.

13. Thinner, since blue light is at the shorter wavelength end of the visible spectrum.

17. Consider air-film-glass interfaces with the light incident from the air side. There would be a 180° phase shift on reflection at the air-film interface. If the glass is more optically dense than the film, there would be a 180° phase shift at the film-glass interface, so a half-wavelength film thickness would be needed for constructive interference. For the case of 180° and 0° phase shifts, respectively, on reflection, a quarter-wavelength film thickness would give constructive interference.

22. By looking through one lens of each pair and rotating, you would observe alternate darkening and brightening if the glasses were polarized. If no darkening and brightening were observed, one or both of the pairs of glasses could be unpolarized. To determine if an individual pair is polarized, you'd have to take the lenses out of the glasses, look through both, and rotate to see if there was any darkening. (This assumes using only the glasses. One might try looking at the partially polarized skylight on a sunny day.)

23. (a) Twice. (Draw arrows for the polarizing directions for the rotation to see this. With the arrow directions initially either parallel or antiparallel, they cross twice when one is rotated through 360°.) (b) Four times. (Draw arrows again.) (c) When rotated at the same speed in the same direction, there would be no change. At different speeds, the frequency of the darkening would de-

pend on the relative speed. Darkening would occur each time the relative speed produced a 90° rotation difference.

CHAPTER 21

1. Wave optics considers the wave aspects of light, and geometrical optics represents light as rays.

5. Light cannot be reflected "sideways," i.e., be incident in one plane and bounce off to the right or left.

6. (a) Normal or direct incidence. (b) Grazing incidence or along the surface.

12. After one reflection 95 percent, after two reflections 0.95×0.95 (95 percent of 95 percent), and so on, and for four mirrors, $0.95 \times 0.95 \times 0.95 \times 0.95 = 0.8145$ or 81.45 percent.

17. (a) No. Zero degrees is direct incidence. (b) No. A grazing incidence of 90° is along the surface.

21. Considering that the Sun can be seen when slightly below the horizon because of atmospheric refraction, without an atmosphere the daylight hours would be slightly shorter.

23. No. The incident medium must be more optically dense than the second one.

27. (a) No, because of refraction you would have only a "cone of view." Draw some rays from the water to the air considering internal reflection and then look at reverse ray tracing. (The cone comes out to have an angle of about 49° as your instructor can show you.) (b) A fish would see a panoramic view of the "above-water" world in all directions (0° to 90°), but it would be all from within a cone at 49°. Consider reverse ray tracing again.

29. Another single prism would not recombine the light components, but would also disperse them. (Three more prisms are actually needed to converge the dispersed light from the original prism. Essentially, prism combinations act as a converging lens.)

31. Because of the angle conditions for the refraction and internal reflection, light satisfying these conditions comes to an observer from spherical raindrops on a circular arc. There are two internal reflections for a secondary rainbow, and more energy is lost. (Internal reflection is not *total* inasmuch as some light is transmitted.)

36. An image of the Sun. The lens converges or concentrates the light falling on its surface on a small area, increasing the intensity to the point that the temperature of an absorbing object in that region can rise above its combustion temperature.

38. (a) Spherical aberration. (b) Spherical and chromatic aberrations.

CHAPTER 22

2. By analogy, the eye "camera" has both black and white "film" (the rods) and color "film" (the cones). Also, the eye has a very "fast" film, i.e., very sensitive, and takes pictures quickly.

8. In general, for nearsightedness a diverging lens, and for astigmatism an appropriate curvature correction.

10. Moonlight is generally not intense enough to activate the color-sensitive cones. Also, the surface of the moon has little distinguishing color.

15. A cyan object appears dark since cyan pigment is "minus red." A yellow object appears slightly red (pink), since red is close enough to yellow to be reflected by the yellow "oscillators."

17. (a) Green with yellow sunglasses since the yellow sunglasses transmit green (close enough to yellow). (b) The green board would appear dark with rose-colored sunglasses. (c) The colors of the glasses.

22. Because of subtraction. For example, blue is the complement of yellow, but subtractively, yellow is "minus blue."

28. Light is absorbed by rather than reflected from the larger droplets of a dense nimbus cloud, or the Sun is at an angle where there is no incident light.

31. High-pressure air masses are generally associated with fair weather since air does not rise and form clouds. When there is a "high" to the west, this can give a "red sky at night" as a result of Rayleigh scattering (see Question 30). "Red sky at morning" indicates a high-pressure air mass to the east, and a poor-weather, low-pressure air mass follows from the west. (In the conterminous United States, we are in the westerly wind zone. The general air circulation pattern is from west to east, so air masses and weather conditions generally move eastward.)

35. The distance an object can be seen or detected depends on the amount of light gathered by the telescope. A distant bright object may be seen, but a dim object at the same distance may not be. To increase the overall limit, the collecting mirror or lens of a telescope is made larger. Mirrors of reflecting telescopes have a weight advantage because a large lens for a reflecting telescope is heavy and difficult to grind and fabricate without flaws. (New technology is using several small mirrors that focus the collected light into one image.)

38. $d = 200$ in. (2.54 cm/in.) $= 508$ cm $= 5.08$ m. Also, $d = 200$ in. (1 ft/12 in.) $= 16.7$ ft. Then, $A = \pi d^2/4 = \pi(16.7 \text{ ft})^2/4 = 218 \text{ ft}^2$. This would correspond roughly to a room 22 ft $\times$ 10 ft, which is quite a large dorm room — probably a double.

■ CHAPTER 23

1. (a) $t = d/c = 93{,}000{,}000$ mi/186,000 mi/s $= 500$ s $= 8.3$ min. (b) 4.2 years.

2. $d = ct = (3 \times 10^8 \text{ m/s})(0.2 \text{ s}) = 0.6 \times 10^8 \text{ m} = 6 \times 10^4$ km.

9. It is unitless. The units in $(v/c)^2$ cancel.

11. Twice as great. $L = L_o/\gamma = L_o/2$, so $\gamma = 2$ and $m = \gamma m_o = 2m_o$.

15. In the proper frame, no change in any of the properties. For an observer on Earth, (a) slower pulse, (b) increased mass, and (c) decrease in volume (length contraction).

19. Because the horizontal speed of light is so great and would travel about 186,000 mi in this direction in 1 second.

21. In the rotating system, clock C is accelerating toward A (centripetal acceleration) and hence runs slower than A or B, which run together.

■ CHAPTER 24

3. (a) Blue light, since it has a greater frequency. (b) Visible light quanta, since they have greater frequencies. (Red end of the spectrum has longer wavelengths and lower frequencies, $f = c/\lambda$.)

7. The amount of water in the incident bucket (photon with energy hf) is less than that required to fill the other bucket (work function of the photoelectron, hf_o).

12. Discrete spectra come generally from gases. In solids, the atoms are closely packed, and interactions influence the electron states to the point that an electron from one atom can jump to a level of another atom. Excitation occurs randomly and the atoms emit at random frequencies, giving a continuous spectrum.

18. Six (if all transitions are allowed). For $(n_i - n_f)$, these are $(4 - 3)$, $(4 - 2)$, $(4 - 1)$, $(3 - 2)$, $(3 - 1)$, and $(2 - 1)$.

21. So there can be a net emission or amplification of photons. Without an inverted population, and with more atoms in the ground state than in excited states, there would be a net absorption of photons.

26. Athletes can run the 100-m dash in about 11 to 12 s. Let's make it 20 s for the average person, and $v = 100$ m/20 s $= 5.0$ m/s. Say you have a mass of 70 kg (154 lb). Then, $\lambda = h/mv \approx 10^{-34}/(70)(5.0) = 2.9 \times 10^{-37}$ m $= 2.9 \times 10^{-35}$ cm, which is very small.

CHAPTER 25

2. (a) $^1_1H_0$, $^2_1D_1$, and $^3_1T_2$. (b) HOH (or commonly H_2O), HOD, HOT, DOD, DOT, and TOT.

9. An electric field could be used. The positively charged alpha particles would be deflected in the direction of the electric field, and the negatively charged beta particles in the opposite direction. The uncharged gamma rays would not be deflected.

14. (a) $^{222}_{86}Rn \longrightarrow {}^{218}_{84}Po + {}^4_2He$

   (b) $^{218}_{84}Po \longrightarrow {}^{214}_{82}Pb + {}^4_2He$, and

      $^{218}_{84}Po \longrightarrow {}^{218}_{85}At + {}^0_{-1}e$

20. A time of 25,000 to 30,000 years is five to six half-lives for C-14. After this many half-lives, the activity is very small and difficult to measure accurately.

21. There would be more C-14 than had been thought and a greater activity when measured. This would indicate fewer half-lives as the ages of objects would be computed to be less.

24. Six flavors; up (u), down (d), strange (s), charmed (c), beauty or bottom (b), and truth or top (t). For each flavor there are three colors or editions, so $6 \times 3 = 18$, and with each quark having an antiquark, $2 \times 18 = 36$.

CHAPTER 26

3. No. Manhattan was a code name. Major components of the project were carried out at the University of Chicago and Los Alamos, New Mexico.

6. Yes. With less positive charge, protons would be repelled less and penetrate the nucleus more easily.

9. $^1_0n + {}^{235}_{92}U \longrightarrow {}^{94}_{37}Rb + {}^{139}_{55}Cs + 3({}^1_0n)$

17. Yes, to some extent. By controlling the moderation and the number of "slow" neutrons, fissioning and the rate of the chain reaction are affected.

23. $^1_0n + {}^6_3Li \longrightarrow {}^3_1H + {}^4_2He$ (see Fusion and Fusion Reactors section in text).

28. Gravity collapses the stellar material (initially hydrogen gas), which in-

creases the temperature. When this becomes sufficiently high, fission is initiated.

■ CHAPTER 27

4. No, a horizontal straight line would indicate no change or a constant value of H with time. Data indicate that Hubble's "constant" is smaller today than it was in the past owing to the expansion of the universe.

8. (a) The average density decreases with the Big Bang theory since the universe is expanding. (b) It remains steady in the steady-state theory since new matter is continually being created to fill the voids left by the expansion. (c) The average density would "oscillate" (get smaller on expansion and larger on contraction) in the oscillating universe theory.

10. In an oscillating universe, Hubble's constant would also "oscillate," i.e., get smaller on expansion and larger on contraction.

15. It really doesn't. Stellar evolution is allowed in both theories.

17. Primarily through the radiation of energy resulting from the fusion process in which mass is converted into energy.

21. In the sense that overweight persons tend to have shorter lifetimes, as do very massive stars.

25. (a) For a light day, $d = ct = (1.1 \times 10^9 \text{ km/h})(24 \text{ h}) = 2.6 \times 10^{10}$ km. With 1 km = 0.62 mi, $d = 2.6 \times 10^{10}$ km (0.62 mi/km) = 1.6×10^{10} mi. A light week is seven times a light day, or 18×10^{10} km = 11×10^{10} mi. (b) Light travels 3×10^5 km in 1 second, so this length is equal to a light second. Then, $d \approx 10^6 \text{ km}/(3 \times 10^5 \text{ km/ls}) = 3.3$ light seconds.

Index

(Numbers followed by *n* or *t* refer to pages with terms in footnotes or tables, respectively.)